Plant Enzyme Systems and Their Biochemical and Environmental Interactions

Plant Enzyme Systems and Their Biochemical and Environmental Interactions offers a comprehensive exploration of the pivotal role plant enzymes play in sustaining life, driving metabolic pathways, and enabling plants to thrive under both favorable and adverse conditions. As climate change, soil degradation, drought, and salinity continue to challenge global agriculture, this book provides an in-depth understanding of how plant enzymes sense, adapt, and respond to environmental cues. It bridges fundamental plant biochemistry with practical environmental applications, revealing how enzymatic networks govern photosynthesis, respiration, nutrient assimilation, defense responses, and stress resilience. Drawing on the latest research, the book highlights innovative strategies that harness enzyme systems for sustainable agriculture, crop improvement, and ecological restoration. It is an essential resource for MSc and PhD scholars, researchers, plant biologists, and environmental scientists seeking to enhance plant productivity and develop solutions for global food security and environmental sustainability.

Salient features of the book include the following:

- Offers a unique blend of fundamental plant biochemistry and applied environmental science.
- Explores how plant enzymes regulate vital biochemical processes and interact with environmental factors.
- Addresses global challenges such as climate change, drought, and soil degradation, offering innovative enzyme-based strategies.
- Provides cutting-edge research and practical solutions, making it essential for researchers, biochemists, agricultural scientists, and environmentalists.

CRC Plant Chemistry Series

Overview

The *Plant Chemistry* book series provides a comprehensive and interdisciplinary exploration of the chemical dimensions of plant biology, positioned at the intersection of plant science, chemistry, biotechnology, and nanotechnology. It emphasizes the intricate molecular and biochemical frameworks that support plant growth, development, and adaptation. Core topics include the structure, function, and biosynthetic pathways of primary and secondary metabolites, which play vital roles in defense, signaling, pigmentation, and developmental regulation. The series examines various plant hormones, enzymology, and molecular mechanisms that govern key physiological processes in photosynthesis, respiration, and defense. It gives significant attention to the biosynthesis and ecological functions of volatile organic compounds, allelopathic interactions, and pigment biochemistry, especially regarding light absorption, stress signaling, and biotic communication. The series also explores emerging fields like sustainable and green chemistry, emphasizing environmentally responsible innovations in plant-based technologies. Research in nanotechnology is featured, with a focus on the application of cutting-edge nanomaterials. Additional topics in terms of chemistry, include abiotic and biotic stress tolerance, plant-microbe interactions, mineral nutrient dynamics, phytoremediation, and the role of plants in ecological restoration. Further, with the applications in drug discovery, functional foods, and sustainable agriculture, the series highlights overall plant chemistry as a critical field for advancing environmental and human well-being. Through this broad and interdisciplinary lens, the series fosters innovation in plant-based product development and the sustainable use of various botanical resources. It serves as a valuable reference for researchers, academics, and professionals working to advance the frontiers of plant sciences, chemistry, and related fields.

Target Audience

The book series targets a wide, interdisciplinary audience across academia, research, and industry. It serves as a key resource for researchers, faculty, and upper-level of students in plant science, chemistry, biotechnology, nanotechnology, and environmental sciences. Biotechnologists, agricultural scientists, pharmacologists, and chemists find valuable insights into plant-derived innovations, natural product formulation, and applications. Environmental scientists and ecologists benefit from content on phytoremediation, pollution control, and sustainable plant production. Additionally, professionals in the food, fragrance, and functional food industries gain from its applied research and practical strategies for plant-based product development.

Unique Selling Points

- Bridges plant science, chemistry, biotechnology, and nanotechnology to provide an integrated perspective on the chemical foundations of plant biology.
- Focuses on the structure, function, and biosynthesis of primary and secondary metabolites central to plant defense, signaling, pigmentation, and growth regulation.
- Highlights emerging and applied themes such as green chemistry, plant-based nanotechnology, and phytoremediation to advance sustainable environmental and agricultural solutions.
- Investigates intricate biochemical and enzymatic processes that underpin key physiological functions, including photosynthesis, respiration, stress adaptation, and hormonal signaling.
- Facilitates innovation across fields like natural product research, drug discovery, functional foods, ecological restoration, sustainable agriculture, and so on.

Brief Biography of Series Editor

Professor Azamal Husen is a globally recognized academic and researcher, celebrated for over two decades of pioneering scientific leadership and academic distinction. His unparalleled scholarly influence is validated by his consistent recognition as one of the World's Top 2% Scientists by Stanford University since 2020. Dr. Husen's career showcases significant international engagement, highlighted by his tenure as a Full Professor and Head of the Department of Biology at the University of Gondar, Ethiopia, and as a strategic Foreign Delegate, for which he received official honors. In India, he currently serves as an Adjunct Professor at Graphic Era University, building on previous academic roles at Sankalchand Patel University and the prestigious Forest Research Institute. A commanding influence in R&D, he has successfully led major projects secured through elite global and national funding bodies, including the World Bank, ICAR, and JBIC. Furthermore, he is instrumental as a curriculum architect, modernizing BSc, MSc, and PhD programs. He actively shapes the scientific discourse in his field as a distinguished Book Series Editor and a frequent Guest Editor for leading international publishers and journals.

For more information on books in this series please visit https://www.routledge.com/Plant-Chemistry/book-series/PE

Plant Enzyme Systems and Their Biochemical and Environmental Interactions

Edited by
Azamal Husen

CRC Press
Taylor & Francis Group
Boca Raton London New York

CRC Press is an imprint of the
Taylor & Francis Group, an **informa** business

Designed cover image: Shutterstock.

First edition published 2027
by CRC Press
2385 NW Executive Center Drive, Suite 320, Boca Raton FL 33431

and by CRC Press
4 Park Square, Milton Park, Abingdon, Oxon, OX14 4RN

CRC Press is an imprint of Taylor & Francis Group, LLC

ISBN: 978-1-041-13461-9 (hbk)
ISBN: 978-1-041-13462-6 (pbk)
ISBN: 978-1-003-66992-0 (ebk)

DOI: 10.1201/9781003669920

Typeset in Times
by KnowledgeWorks Global Ltd.

Contents

Preface

Enzymes are the silent architects of life, directing an intricate network of biochemical reactions that underpin every aspect of plant growth, development, and survival. They are nature's most skilled catalysts, orchestrating the transformation of energy, regulating metabolic pathways, and enabling plants to adapt to a constantly changing environment. Every movement of a leaf toward sunlight, every nutrient absorbed from the soil, and every defensive response to stress is driven by enzymatic activity. Understanding these molecular engines is therefore fundamental to unlocking the mechanisms that sustain plant life and, ultimately, the ecosystems and food systems that support humanity. The book *Plant Enzyme Systems and Their Biochemical and Environmental Interactions* under the *Plant Chemistry* series explores this fascinating molecular world in which enzymes act as both regulators and responders to internal and external cues. It delves into how enzymes maintain physiological balance, mediate stress responses, and enable plants to function efficiently under diverse environmental conditions. As the global community faces unprecedented challenges from climate change, land degradation, and diminishing soil fertility, there has never been a more critical time to deepen our understanding of plant enzyme systems. These systems represent the biochemical bridge between plant metabolism and environmental adaptation, forming the foundation for sustainable agricultural and ecological practices.

This volume opens with an introduction to the nature, diversity, and physiological roles of plant enzymes. It explains how enzymes function within the cellular framework, controlling metabolic fluxes that influence energy generation, nutrient assimilation, and growth regulation. Through these molecular details, readers can understand how precisely enzymes work. Each reaction happens under carefully controlled conditions that maintain the right balance of speed, efficiency, and regulation. The initial chapters describe the biosynthesis and modulation of enzyme activity, revealing the molecular mechanisms that determine how enzymes are produced, activated, and degraded. Such knowledge is essential for appreciating the dynamic equilibrium that allows plants to respond to subtle shifts in their environment. Subsequent sections of the book explore key enzymatic processes involved in photosynthesis, respiration, and energy metabolism. These processes form the very basis of plant life, converting sunlight into chemical energy and channeling it toward the synthesis of essential biomolecules. The discussion extends to the role of enzymes in carbon fixation, nitrogen metabolism, and secondary metabolite biosynthesis, emphasizing how these pathways contribute to plant performance and resilience. Special attention is given to the influence of environmental factors such as light, temperature, water availability, and soil composition on enzymatic activity. By linking these external variables to internal biochemical responses, the book offers a comprehensive view of the plant–environment dialogue that defines ecosystem balance. One of the most compelling aspects of this volume is its focus on enzyme-mediated stress tolerance. Plants are continually exposed to a wide range of stresses (both abiotic and biotic) that threaten their survival. Enzymes serve as the first line of biochemical defense, mitigating the effects of drought, salinity, heat, and oxidative stress while also protecting plants from pathogens and herbivores. The chapters illustrate how antioxidant enzymes such as superoxide dismutase, catalase, and peroxidases help neutralize reactive oxygen species, thereby preventing cellular damage. Likewise, enzymes involved in osmolyte biosynthesis and nutrient adaptation help plants withstand water and nutrient stress. The volume further explores how enzyme systems participate in plant immune responses, producing defensive metabolites and signaling molecules that coordinate resistance. Equally significant is the discussion on enzymes that mediate beneficial plant–microbe interactions. Through these enzymatic pathways, plants form symbiotic relationships that enhance nutrient acquisition, improve soil structure, and promote growth. Such interactions demonstrate that plant enzymes are not isolated biochemical tools but are integrated components of a complex ecological network. The book also addresses enzyme responses to environmental pollutants and toxins, explaining how plants use enzymatic detoxification systems to cleanse soils and maintain ecological integrity. This biochemical versatility highlights the potential of plant enzyme systems in bioremediation and environmental sustainability. Moving from fundamental understanding to applied perspectives, the later chapters of the book present advances in biotechnology and genetic engineering that exploit enzyme systems for crop improvement. By combining enzymology with molecular biology, scientists are now able to manipulate enzyme expression and activity to enhance photosynthetic efficiency, nutrient uptake, and stress resilience. The book discusses how metabolic engineering, gene editing, and synthetic biology can be used to design plants that are better adapted to future environmental challenges. It also examines the role of enzyme-based technologies in bioenergy production, sustainable agriculture, and environmental restoration. These applications illustrate the profound societal and economic importance of enzyme research in ensuring food security and ecological balance.

Beyond its scientific content, this book stands out for its interdisciplinary vision. It integrates principles from plant biochemistry, molecular genetics, physiology, agronomy, and environmental science to create a cohesive narrative on enzyme function. Each chapter, authored by leading scholars in their respective fields, combines rigorous experimental data with insightful analysis and forward-looking perspectives. Together, these contributions form a rich resource for researchers, educators, and students interested in plant biology and environmental biochemistry. The text also identifies key knowledge gaps and future directions, underscoring the importance of emerging analytical tools such as genomics, proteomics, metabolomics, and computational modeling in advancing enzyme studies. By weaving together fundamental principles, experimental discoveries, and applied innovations, this volume contributes to a broader scientific mission, understanding and utilizing the molecular machinery that drives plant life. Enzymes not only sustain individual plants but also regulate ecological processes that determine the health of our planet. Their study bridges the microscopic world of molecules with the macroscopic realities of agriculture and environmental management. The knowledge contained within these pages offers practical insights for improving crop performance, conserving biodiversity, and developing strategies for climate resilience.

Ultimately, this work celebrates the elegance and complexity of plant enzyme systems as central forces in biology and ecology. Through understanding these dynamic catalysts, we gain the power to design better agricultural systems, restore degraded landscapes, and ensure food and environmental security for future generations. The collaboration of distinguished contributors and the support of the scientific publishing community have made this endeavor possible. I extend my heartfelt gratitude to all authors whose expertise and dedication have shaped this comprehensive and timely volume. Their collective effort reflects the spirit of scientific curiosity and commitment to sustainable progress that this book seeks to inspire. I am also deeply thankful to Taylor & Francis Group for their continued support in bringing forth this important contribution to the field of plant science.

About the Editor

Professor Azamal Husen is an eminent academic and researcher whose career spans over two decades of outstanding contributions to teaching, research, and scientific leadership. His academic journey has been marked by distinction and international engagement, having served as Full Professor and Head of the Department of Biology at the University of Gondar, Ethiopia, and Foreign Delegate at Wolaita Sodo University, Ethiopia. In India, he is currently an Adjunct Professor at Graphic Era University, and has previously served as Professor at Sankalchand Patel University and Visiting Faculty at several reputed Indian institutions, including the Forest Research Institute, Dehradun. His research has been supported by several prestigious national and international organizations, such as the World Bank, National Agricultural Technology Project, Indian Council of Agricultural Research, Indian Council of Forestry Research and Education, and the Japan Bank for International Cooperation. A prolific scholar, Prof. Husen has authored a vast number of research papers, books, and book chapters published by leading international publishers. His commitment to advancing global science is reflected in his editorial and peer review work for numerous high-impact international journals. His expertise has been widely acknowledged through invitations to evaluate book proposals for leading global publishers such as Springer, Taylor & Francis Group, and Elsevier, as well as to review research projects for prestigious international funding agencies, including the Czech Science Foundation (Czechia) and the National Research, Development and Innovation Office (Hungary). He is a Fellow of the Plantae Group under the American Society of Plant Biologists, and a member of several prestigious scientific organizations, including the International Society of Root Research, the Asian Council of Science Editors, and the International Natural Product Sciences Taskforce. Throughout his distinguished career, Prof. Husen has been a keynote and invited speaker at numerous international conferences and symposia. He has played a significant role in academic development, contributing to curriculum design, faculty training, and educational module development across undergraduate, postgraduate, and doctoral programs. His outstanding achievements in teaching, research, and community service earned him special recognition from the University of Gondar. In the realm of scientific publishing, he serves as Editor-in-Chief of the *American Journal of Plant Physiology* and as Series Editor for several globally recognized academic collections, including *Exploring Medicinal Plants; Plant Chemistry; Plant Biology, Sustainability, and Climate Change; Smart Nanomaterials Technology;* and the *Handbook Series on Nanotechnology and Its Applications.* Consistently recognized among the World's Top 2% Scientists (2020 onwards) by Stanford University, Prof. Husen has also served as Guest Editor for several influential special issues, including *Advances in Plant Metabolomics* (*Plant Physiology and Biochemistry*), *Exploring Nanotechnology in Medicinal Plants* (*Journal of Herbal Medicine*), *Integrating Nanotechnology with Plant Compounds* (*Frontiers in Plant Science*), *Nanobiotechnology for Sustainable Agriculture and Crop Improvement* (*Discover Applied Sciences*) among others. Through his scholarly excellence, editorial leadership, and global collaborations, Prof. Husen continues to advance cutting-edge research at the intersection of plant science, nanotechnology, and sustainability, establishing himself as a leading figure in contemporary biological and environmental research.

Contributors

Zainul Abideen
Dr. Muhammad Ajmal Khan
Institute of Sustainable Halophyte Utilization
University of Karachi
Karachi, Pakistan
Sharjah, Seed Bank and Herbarium
University of Al Dhaid
Sharjah, United Arab Emirates

Parisa Aghaei-Gharachorlou
Department of Plant Eco-physiology
Faculty of Agriculture
University of Tabriz
Tabriz, Iran

Khulood Fahad Alabbosh
Department of Biology
College of Science
University of Hail
Saudi Arabia

Faridul Alam
Soils Unit
Natural Resources Management Division
Bangladesh Agricultural Research Council
Farmgate, Dhaka, Bangladesh

Md. Ashraful Alam
Plant Breeding Division
Spices Research Centre
Bangladesh Agricultural Research Institute
Bogura, Bangladesh

Morteza Alami-Milani
Department of Plant Eco-physiology
Faculty of Agriculture
University of Tabriz
Tabriz, Iran

Ahmad Ali
Department of Life Sciences
University of Mumbai
Mumbai, Maharashtra

Hina Alim
Department of Life Sciences
University of Mumbai
Mumbai, Maharashtra

Abdel Rahman Mohammad Al-Tawaha
Department of Biological Sciences
Al-Hussein Bin Talal University
Maan, Jordan

Hamzeh Amiri
Department of Biology
Faculty of Sciences
Lorestan University
Khorram Abad, Iran

Farwa Basit
Department of Biology
College of Science and Technology
Wenzhou-Kean University
Wenzhou, China

Nargiz Bayramova
LANDAU School
Baku, Azerbaijan

Javaid Akhter Bhat
Key Laboratory of Soybean Molecular Design Breeding
Northeast Institute of Geography and Agroecology
Chinese Academy of Sciences
Changchun, China
Research Center for Life Science Computing
Zhejiang Lab
Hangzhou, China

Alexander Machado Cardoso
Department of Biology
Rio de Janeiro State University (UERJ)
Rio de Janeiro, Brazil

Salar Farhangi-Abriz
Cotton Research Institute of Iran
Agricultural Research Education and Extension Organization (AREEO)
Gorgan, Iran

Glaucio Dire Feliciano
Laboratory of Chemical and Biological Analysis (LAQB)
State University of Rio de Janeiro (UERJ)
Rio de Janeiro, Brazil

Xianzhong Feng
Key Laboratory of Soybean Molecular Design Breeding
Northeast Institute of Geography and Agroecology
Chinese Academy of Sciences
Changchun, China

K. Gayathri
Department of Microbiology
Sacred Heart College (Autonomous) Tirupattur
Tamil Nadu, India

Kavisa Ghosh
Department of Biotechnology
The Oxford College of Science
Bengaluru
Karnataka, India

Wang Hao
Department of Biology
College of Science and Technology
Wenzhou-Kean University
Wenzhou, China

Ahmad Numery Ashfaqul Haque
Soils Unit
Natural Resources Management Division
Bangladesh Agricultural Research Council
Farmgate, Dhaka, Bangladesh

Akbar Hossain
Soil Science Division
Bangladesh Wheat and Maize Research Institute
Dinajpur, Bangladesh

Muhammad Ijaz
Institute of Agronomy
Bahauddin Zakariya University
Multan, Pakistan

Muhammad Imran
College of Agriculture and Life Sciences
Department of Applied Biosciences
Kyungpook National University
Daegu, South Korea

Md. Injamum-Ul-Hoque
Department of Agriculture
Gopalganj Science and Technology University
Gopalganj, Bangladesh

Babar Iqbal
School of Environment and Safety Engineering
Jiangsu University
Zhenjiang, China

Hina Jabeen
Department of Biotechnology
Women University Mardan
Pakistan

Qaiser Javed
Institute of Agriculture and Tourism
Poreč, Croatia

Md. Parvez Kabir
Soil Science Division
Bangladesh Wheat and Maize Research Institute
Dinajpur, Bangladesh

Fatima Zahar Kamal
Health Care and Biology-Health Team
Laboratory
Higher Institute of Nursing and Health Techniques of Casablanca (ISPITS)
Ministry of Health and Social Protection
Morocco

Sang-Mo Kang
College of Agriculture and Life Sciences
Department of Applied Biosciences Kyungpook National University
Daegu, South Korea

Kashif
Department of Botany
Center of Plant Sciences and Biodiversity
University of Swat
Charbagh, Pakistan

Said Kewedar
Department of Entomology
Penn State University
University Park, Pennsylvania, USA

Abdul Latif Khan
Department of Engineering Technology
Cullen College of Engineering
University of Houston
Texas, USA

Amir Abdullah Khan
School of Environment and Safety Engineering
Jiangsu University
Zhenjiang, China

Ibrahim Khan
Department of Applied Biosciences
Kyungpook National University
Daegu, Republic of Korea

Muhammad Aaqil Khan
International Joint Research Laboratory for Global Change Ecology
School of Life Sciences
Henan University
Kaifeng, Henan, China

In-Jung Lee
College of Agriculture and Life Sciences
Department of Applied Biosciences
Kyungpook National University
Daegu, South Korea

Yuvalakshmi Logesan
Department of Microbiology
Sacred Heart College (Autonomous) Tirupattur
Tamil Nadu, India

B. Lokeshwari
Department of Microbiology
Sacred Heart College (Autonomous), Tirupattur
Tamil Nadu, India

M. Manigandan
Department of Microbiology
Sacred Heart College (Autonomous) Tirupattur
Tamil Nadu, India

S. Monisha
Department of Microbiology
Sacred Heart College (Autonomous) Tirupattur
Tamil Nadu, India

Seyyed Amirreza Mousavi
Department of Plant Eco-physiology
Faculty of Agriculture
University of Tabriz
Tabriz, Iran

Elshan Musazade
Key Laboratory of Soybean Molecular Design Breeding
Northeast Institute of Geography and Agroecology
Chinese Academy of Sciences
Changchun, China

Komal Pandey
Department of Microbiology
College of Basic Sciences & Humanities
Chaudhary Charan Singh Haryana Agricultural University
Hisar, India

Nimisha Patel
Department of Life Sciences
J C Bose University of Science and Technology
YMCA Faridabad
Haryana, India

Odongkara Peter
Department of Applied Biosciences
Kyungpook National University
Daegu, Republic of Korea

Md. Marufur Rahman
Regional Station
Bangladesh Institute of Research and Training on Applied Nutrition
Rangpur, Bangladesh

Umair Rasheed
Department of Agronomy
University of Agriculture
Faisalabad, Pakistan

Akhtar Rasool
Center for Biotechnology and Microbiology
University of Swat
Pakistan

L Jahnavi Reddy
Credora Life Sciences
Bangalore, Karnataka, India

Md. Selim Reza
Soil Science Division
Bangladesh Institute of Nuclear Agriculture
Mymensingh, Bangladesh

R. Ridhi
Department of Physics
Mehr Chand Mahajan DAV College for Women
Chandigarh, India

Baljeet Singh Saharan
Department of Microbiology
College of Basic Sciences & Humanities
Chaudhary Charan Singh Haryana Agricultural University
Hisar, India

P. Saranraj
Department of Microbiology
Sacred Heart College (Autonomous) Tirupattur
Tamil Nadu, India

Abdul Sattar
Institute of Agronomy
Bahauddin Zakariya University
Multan, Pakistan

Riyaz Sayyed
Department of Biological Science and Chemistry
University of Nizwa
Oman

Rabiha Seboussi
College of Veterinary Medicine
University of Al Dhaid
Sharjah, United Arab Emirates

Shifa Shaffique
College of Agriculture and Life Sciences
Department of Applied Biosciences
Kyungpook National University
Daegu, South Korea

Anis Ali Shah
Department of Botany
University of Education
Lahore, Pakistan

Nadia Sharif
Department of Biotechnology
Women University Mardan
Pakistan

Ahmad Sher
Institute of Agronomy
Bahauddin Zakariya University
Multan, Pakistan

Falguni Akter Shraboni
Department of Genetics and Plant Breeding
Gazipur Agricultural University
Gazipur, Bangladesh

P. Sivasakthivelan
Tamil Nadu Agricultural University
Agricultural Research Station Bhavanisagar
Tamil Nadu, India

Nargiz Sultanova
Department of Technology of Organic Substances and High-Molecular Compounds
Faculty of Chemical Technology
Azerbaijan State Oil and Industry University
Baku, Azerbaijan

Javier A. Trujillo-Rugamas
College of Agriculture and Life Sciences
Department of Applied Biosciences
Kyungpook National University
Daegu, South Korea

Amit B. Verma
Department of Life Sciences
University of Mumbai
Mumbai, Maharashtra, India

Weria Weisany
Department of Agriculture Science and Engineering
SR.C.
Islamic Azad University
Tehran, Iran

Dan-Dan Zhao
Crop Environment Research Division
Department of Crop Sciences
National Institute of Crop and Food Science
Wanju-gun, Republic of Korea

Kebaichun Zhu
Department of Biology
College of Science and Technology
Wenzhou-Kean University
Wenzhou, China

1 Introduction to Plant Enzyme Systems and Their Role in Plant Physiology

Amir Abdullah Khan, Khulood Fahad Alabbosh, Kashif, Babar Iqbal, and Qaiser Javed

1.1 INTRODUCTION

1.1.1 Definition and General Importance of Enzymes

Plant enzymes are biological catalysts essential for numerous biochemical and physiological activities. They significantly influence plant growth, development, and tolerance to environmental changes (Bano et al., 2023). They enable and govern metabolic pathways, serving a vital function in fundamental processes such as photosynthesis, respiration, and nutrient assimilation, in addition to more specialized roles such as hormone regulation, stress responses, and the biosynthesis of diverse metabolites (Karimi & Jacobs, 2021; Montecchiarini et al., 2021). Enzymes are crucial to nearly every aspect of a plant's life cycle, encompassing seed germination to fruit ripening (Kumar et al., 2024). They facilitate reactions by lowering the activation energy of specific processes, enabling them to occur at rates essential for life (Eberhart et al., 2025; Fedoreyeva, 2024). Enzymes exhibit high specificity, typically catalysing a singular reaction or a closely related collection of processes (Wang et al., 2025a, 2025b, 2025c). Plant enzymes are essential for numerous physiological activities. They are critical for both primary and secondary metabolic processes. Photosynthesis (e.g., RuBisCO), respiration, and nutrient absorption are essential components of primary metabolism necessary for plant survival and growth (Khan et al., 2025b; Su et al., 2024). Secondary metabolism encompasses the synthesis of specialized chemicals such as alkaloids, terpenoids, and flavonoids, which assist plants in defence, signalling, and environmental adaptation (Akbar et al., 2024). Phytohormones regulate plant growth, development, and stress responses. Enzymes are crucial for the synthesis, degradation, and modification of these hormones, thereby controlling their levels and action (Khan et al., 2025a; Montecchiarini et al., 2021). Enzymes like SABATH methyltransferases regulate hormonal function (Wang et al., 2024a). For instance, abscisic acid (ABA), a vital phytohormone, relies on enzyme synthesis to enhance protein and enzyme production and optimize the photosynthetic system (Islam et al., 2024). Plants encounter various forms of environmental stress, including drought, salt, and heavy metal toxicity. Enzymes facilitate the elimination of toxic chemicals, the synthesis of protective metabolites, and the regulation of stress-responsive genes (Khan et al., 2025b, 2025c). For instance, oxalate oxidase (OXO) enzymes facilitate the decomposition of oxalate into hydrogen peroxide, a key process during periods of elevated oxidative stress (Gómez-Espinoza et al., 2025). Similarly, plants possess antioxidant systems reliant on enzymes such as peroxidase and dismutase to safeguard against oxidative stress induced by heavy metal exposure (Nguyen et al., 2025).

1.1.2 Brief History of Enzyme Discovery in Plants

Significant advancements in biochemistry, molecular biology, and synthetic biology have defined the dynamic and growing field of plant enzyme discovery history (Heckmann & Paradisi, 2020; Seshadri et al., 2025). Early research focused on identifying and characterizing plant-derived enzymes that are vital for defence, signalling, and metabolism, such as amylases and proteases (Meshram et al., 2019). Understanding plant enzymatic processes has been greatly improved by developments in structural biology and enzyme kinetics (Heckmann & Paradisi, 2020). Enzymes have been recognized by humanity for centuries. For instance, they were employed in baking and brewing (Verma et al., 2022). Long before people realized what enzymes were, they used plant parts and their extracts to perform activities like tenderizing meat with papaya leaves (Choudhary et al., 2025; Mohd Azmi et al., 2023). In the 19th century, enzymes were first formally studied (Lanfranchi et al., 2025). Amylase, invertase, papain, bromelain, and lipoxygenase are among the plant-derived enzymes that are crucial for the production of foods such as syrups, baked goods, alcoholic beverages, and dairy products (Zhang et al., 2024). A re-evaluation of the mechanism by which plants produce glutamate from ammonia through the cooperative action of glutamine synthetase (GS) was driven by the identification of glutamate synthase, an enzyme necessary for ammonium assimilation in plants (Ibarra & Reynoso, 2025). Several significant techniques have enabled advancements in plant molecular biology. In the 1980s, researchers were able to fully utilize molecular biology methods and techniques by using Arabidopsis thaliana as a model plant, which allowed them to integrate plant sciences with other fields of study (Yaschenko et al., 2025).

1.1.3 Relevance of Enzymes in Plant Life and Agriculture

Enzymes have a crucial role in agriculture by facilitating crop growth, maintaining soil health, and safeguarding plants against pests and diseases (Xing et al., 2025). In

DOI: 10.1201/9781003669920-1

modern crop biotechnology, enzymes are widely employed in recombinant DNA technology to genetically modify plants for beneficial traits, such as pest resistance, herbicide tolerance, and enhanced nutritional value (Yadav et al., 2024). Research indicates that the overexpression of transcription factors such as GhDof1 enhances cotton's resistance to salinity and cold temperatures while simultaneously increasing the oil content of the seeds (Li et al., 2024; Wang et al., 2024a). Microbial enzymes, including cellulases, xylanases, and proteases, decompose organic matter in the soil ecosystem, thereby rendering nutrients accessible for plant absorption (Datta, 2024; Singh et al., 2024). These enzymes facilitate the recycling of essential elements such as nitrogen, phosphorus, and zinc, which are crucial for plant growth and health (Khan et al., 2025a, 2025b). Zinc-solubilizing bacteria (ZSB) enhance the bioavailability of zinc in the organism. Zinc is a micronutrient essential for enzyme activation, protein synthesis, and stress resilience (Khan et al., 2025c). Enzymes play a crucial role in biofertilizers, facilitating sustainable crop growth by redistributing nutrients and enhancing total soil fertility (Xing et al., 2025).

1.2 MAJOR PLANT ENZYME CLASSIFICATION AND THEIR PHYSIOLOGICAL FUNCTIONS

Several biochemical reactions essential to plant growth, development, and environmental responsiveness are made possible by plant enzymes acting as biocatalysts (Singh et al., 2024). In addition to producing essential chemicals through primary metabolism, they also participate in specialized metabolism, which produces a variety of secondary metabolites for defence and adaptability as described in Table 1.1 (Goura et al., 2025). According to Angeli et al. (2024), these enzymes are classified according to the kinds of reactions they catalyse. Plant enzymes are divided into six major classes based on the Enzyme Commission's (EC) classification scheme (McDonald & Tipton, 2023).

1.2.1 Oxidoreductases

In plants, oxidoreductases are a fundamental class of enzymes that regulate critical physiological functions related to growth, metabolism, and stress tolerance as described in Table 1.1 (Cárdenas-Moreno et al., 2023). Their primary function is to accelerate oxidation–reduction reactions by facilitating the transfer of electrons between molecules. This is a crucial aspect of maintaining cellular viability and facilitating plant adaptation to fluctuating conditions (Wang et al., 2024c). Plants frequently experience oxidative stress when subjected to abiotic conditions such as salt, drought, heavy metal toxicity, and severe temperatures. This occurs when reactive oxygen species (ROS), such as superoxide radicals, hydroxyl radicals, hydrogen peroxide, and singlet oxygen, accumulate excessively (Wang et al., 2024b). Elevated levels of ROS can inflict significant harm on cellular macromolecules such as lipids, proteins, and nucleic acids, thus impeding growth and development (Adhikari, 2024). Oxidoreductases are crucial for plant defence against oxidative stress since they maintain stable levels of ROS. Superoxide dismutase (SOD), catalase (CAT), and peroxidases are crucial antioxidant enzymes that collaboratively eliminate ROS (Sood, 2025). SOD rapidly converts superoxide radicals into hydrogen peroxide. Subsequently, CAT and peroxidases decompose hydrogen peroxide into water and molecular oxygen. This

TABLE 1.1
Different Classes of Plant Enzymes and Their Key Plant Roles

EC Class	Representative Enzymes	Main Reaction	Key Plant Roles	Representative References
Oxidoreductases	SOD, CAT, peroxidases	Electron transfer; ROS detoxification	Redox homeostasis; stress tolerance (salinity, drought, metals)	Cárdenas-Moreno et al. (2023) and Wang et al. (2025b)
Transferases	Glycosyltransferases, aminotransferases, methyltransferases, COMT	Transfer of functional groups (sugar, amino, methyl)	Hormone/metabolite modification; N-metabolism; detoxification (GSTs)	Huang et al. (2024) and Ikram et al. (2025)
Hydrolases	GH1, endo-β-mannanase, chitinases, lipases, proteases	Hydrolysis of bonds (glycosidic, ester, peptide)	Cell-wall remodelling, reserve mobilization, defence, hormone activation	Panicker and Sayyed (2022); Shobade et al. (2025)
Lyases	ACC synthase, AOS, HPL, pyruvate decarboxylase, ATP-citrate lyase	Non-hydrolytic bond cleavage (e.g., decarboxylation)	Biosynthesis of ethylene, oxylipins, VOCs; carbon flux	Kumar et al. (2024), Bajguz and Piotrowska-Niczyporuk (2023)
Isomerases	Glucose-6-P isomerase, xylose isomerase, IPI, (Z)/(E)-hexenal isomerase	Intramolecular rearrangements (stereochemical)	Glycolysis/gluconeogenesis flux, sugar use, isoprenoid precursor balance, volatiles	Li et al. (2025)
Ligases	DNA ligases; E3 ubiquitin ligases (CRL, PUB, ATL); SUMO E3 (SIZ1)	ATP-dependent bond formation (e.g., DNA sealing; ubiquitin/SUMO conjugation)	DNA repair, regulated proteolysis, PTM (SUMOylation)—growth and stress control	Yaschenko et al. (2025)

coordinated enzymatic system maintains the equilibrium between the production and elimination of ROS, hence preserving cellular redox balance and preventing oxidative damage (Wang et al., 2025a). Plants utilize non-enzymatic antioxidants and signalling compounds alongside enzymatic defences to enhance the specificity of their stress responses. Melatonin has been demonstrated to enhance drought tolerance by increasing the activity of antioxidant enzymes, reducing oxidative damage, and improving redox stability (Kathirvel et al., 2024). Oxidoreductases and other defence mechanisms collaborate to enable plants to endure challenging situations.

1.2.2 Transferases

Transferases play a crucial role in primary and secondary plant metabolism by facilitating the movement of functional groups between molecules (Ikram et al., 2025). Among the most significant families are methyltransferases, which alter DNA, RNA, and proteins and thereby impact gene expression and epigenetic states; aminotransferases, which regulate the interconversion of amino acids and nitrogen metabolism (Table 1.1) (Yu & Li, 2024); and glycosyltransferases, which alter the solubility, stability, and activity of hormones, metabolites, and secondary compounds (Zhang et al., 2025b). Growth and structural integrity are affected by lignin and melatonin, which are produced by enzymes such Caffeic Acid O-methyltransferase (COMT) (Huang et al., 2024). Transferases that modify cell walls, however, affect the structure and growth of the wall (Delmer et al., 2024). Additionally, transferases are involved in growth and stress adaptation. Through hormone and transcriptional circuits, they aid in root and flower patterning (Mas-ud et al., 2025). They also aid in detoxification through glutathione S-transferases (GSTs), which connect glutathione to oxidative products and xenobiotics, increasing plants' resistance to heavy metals, herbicides, and other stresses (Bose Mazumdar & Chattopadhyay, 2025).

1.2.3 Hydrolases

Hydrolases are a significant and beneficial category of enzymes that cleave chemical bonds by hydrolysis (Table 1.1). Among their various functions are the degradation of macromolecules, the alteration of cell walls, the transfer of energy, and cell defence (Panicker & Sayyed, 2022). GH1 family members and endo-β-mannanases are examples of glycoside hydrolases that break glycosidic bonds to restructure polysaccharide networks during development, germination, and growth (Bhattacharya et al., 2022). In the apoplast, many hydrolases, including chitinases, also function as first-line defences by rupturing pathogen cell walls or producing antimicrobial fragments (Zhang et al., 2022). By activating or deactivating conjugated hormones, hydrolase activities alter hormone reservoirs and signalling, which impacts stress reactions and developmental processes (Široká et al., 2025). To aid in seedling growth or the movement of nutrients, lipases, proteases, and phospholipases release stored reserves during germination and senescence (Zhan et al., 2024). Plants adjust the expression of hydrolases in response to abiotic stress in order to remove harmful substances (for example, by means of the glyoxalase system) and to draw nutrients out of complex substrates (Chen et al., 2024). According to Lu et al. (2024), the alpha/beta hydrolase fold serves as the foundation for a number of enzymatic and signalling processes in phytohormone pathways.

1.2.4 Lyases

Lyases enable non-hydrolytic (often oxidative) bond cleavages and are essential to both primary and specialized metabolism in plants (Kumar et al., 2024). Lyases, including pyruvate decarboxylase and ATP-citrate lyase, facilitate carbon flux, energy metabolism, and the synthesis of essential precursors for development by catalysing activities such as decarboxylation and the creation or cleavage of C–C and C–O bonds (Spormann, 2023). Many lyase-dependent pathways demonstrate sensitivity to the cellular ionic state (K^+ availability), underscoring the relationship between ion homeostasis and catalytic activity (Munteanu et al., 2025). Lyases additionally facilitate the synthesis of signalling and defensive chemicals. ACC synthase, a lyase, is crucial for ethylene production, while enzymes such as allene-oxide synthase and hydroperoxide lyase participate in the oxylipin and jasmonate pathways that generate defence signals and volatile organic compounds in response to herbivore attacks or infections (Bajguz & Piotrowska-Niczyporuk, 2023). Lyases have facilitated the evolution of several secondary metabolite pathways by recruiting and modifying genes associated with primary metabolism (Weng et al., 2021). This has enabled plants to react to both biotic and abiotic stresses. Consequently, lyases are crucial for metabolic flexibility, signalling, and stress resilience in plants (Table 1.1).

1.2.5 Isomerases

Isomerases are enzymes that alter the stereochemistry or connectivity of metabolites by rearranging them inside a single molecule. They are crucial for both primary and secondary metabolic processes in plants (Table 1.1) (Li et al., 2025). Enzymes such as glucose-6-phosphate isomerase facilitate the interconversion of glucose-6-phosphate and fructose-6-phosphate within central carbon metabolism. This is a crucial step for glycolysis, gluconeogenesis, and optimal carbon use (Wang et al., 2025b). Isomerases influence the synthesis of cell walls and the transport of substances. For instance, xylose isomerases facilitate the conversion of xylose into xylulose, enabling its utilization in metabolic networks ((Miyamoto et al., 2022). Isomerases regulate critical junctions in specialized metabolism. Isopentenyl diphosphate isomerase (IPI) interconverts IPP and DMAPP, supplying the C_5 precursors essential for all isoprenoids synthesized via the MVA and MEP pathways

(Zografos & Anagnostaki, 2016). Post-cyclization isomerases modify the structures of triterpenoids, resulting in structural variants with distinct ecological and physiological functions (Chuang et al., 2023). Isomerases play a crucial role in small-molecule signalling and the synthesis of fragrances. (Z)-3:(E)-2-hexenal isomerases produce green leaf volatiles that facilitate plant interactions and insect communication (Qian et al., 2024).

1.2.6 Ligases

Ligases are essential for maintaining genomic stability in plants, regulating protein levels, and delivering adaptive signals Table 1.1 (Su et al., 2024). DNA ligases repair breaks in both single and double strands to maintain genomic integrity. DNA LIGASE VI is essential for repairing damage incurred during seed imbibition, hence enhancing seed longevity and germination efficacy (Tan et al., 2025). Ubiquitin ligases (E3s) confer substrate specificity to the ubiquitin–proteasome system, hence regulating hormone signalling, developmental patterning, and stress responses (Liu et al., 2021). Cullin-RING ligases (CRLs), U-box/PUB proteins, and extended plant families such as ATL represent significant classes of E3 ligases that influence organ growth, hormone receptor degradation, and adaptive proteome reconfiguration(Wu et al., 2023). For instance, CRLs regulate kernel growth, while PUB25/26 influences organ size (Ban & Estelle, 2021; Karthik et al., 2025; Li et al., 2020). E3s localized to the endoplasmic reticulum (ER) and specialized ATL members assist in maintaining cellular stability during stress from biotic or abiotic factors (Liu et al., 2021). SUMO ligases, such as SIZ1, promote SUMOylation, a reversible post-translational modification that alters protein stability, localization, and interactions, hence governing development and stress tolerance (Singh et al., 2022). Ligases collaborate to regulate growth, signalling, and resilience in plants by mending DNA, selectively degrading proteins, and implementing regulatory modifications post-translation.

1.3 ENZYMES DRIVING PLANT METABOLISM

1.3.1 A Brief Overview of Plant Enzymes Involved in Primary Metabolism

Plant enzymes provide the primary components of metabolic networks that supply energy, diminish power, and generate the essential building blocks required for growth, development, and survival (Zheng et al., 2025). Photosynthesis, glycolysis, the pentose-phosphate pathway (PPP), the tricarboxylic acid (TCA) cycle, amino acid biosynthesis, and lipid metabolism constitute components of primary metabolism. Collectively, these processes generate carbon skeletons, ATP, NAD(P)H, and precursors for synthesizing macromolecules (Sharkey, 2024).

In photosynthesis, chloroplast enzymes such as Rubisco facilitate the fixation of CO_2 in the Calvin cycle. Simultaneously, the protein complexes of Photosystem II and I facilitate the light-driven processes that generate ATP and reducing equivalents (Khan et al., 2024). Glycolysis degrades glucose into pyruvate within the cytosol. Hexokinase, phosphofructokinase, and pyruvate kinase are the principal regulatory sites (Zheng et al., 2025). The PPP produces NADPH and pentose precursors via the enzymatic activities of glucose-6-phosphate dehydrogenase and 6-phosphogluconate dehydrogenase (Trejo-Solís et al., 2025). Mitochondrial TCA-cycle enzymes, including citrate synthase, isocitrate dehydrogenase, α-ketoglutarate dehydrogenase, and malate dehydrogenase, oxidize acetyl-CoA to generate NADH and $FADH_2$ for respiration and biosynthesis (Zhang & Fernie, 2023). GS and glutamate synthase (GOGAT) collaborate with transaminases to interconvert amino acids (Leng et al., 2024). Acetyl-CoA carboxylase (ACCase) initiates lipid metabolism, whereas the fatty-acid synthase complex and lipases facilitate the synthesis and transport of stored lipids (Schmid, 2021). Allosteric effectors, post-translational modifications, and compartmentation meticulously regulate activity in these pathways to ensure that metabolic fluxes align with developmental and environmental requirements (Bhattacharya, 2022).

1.3.2 A Brief Overview of Plant Enzymes Involved in Secondary Metabolism

Specialized enzyme networks synthesize secondary metabolites such as terpenoids, phenolics, and alkaloids. These facilitate self-defence, environmental adaptation, and interactions with other living organisms (Somkuwar Subhash et al., 2024). The synthesis and function of these enzymes are meticulously controlled by environmental stimuli (e.g., UV radiation, pathogens, nutrient availability), hormonal signals (particularly jasmonates), developmental phases, and post-transcriptional processes such as noncoding RNAs, collectively enabling plants to dynamically adjust their metabolite profiles (Altaf et al., 2024; Shobade et al., 2025). Environmental signals (such as UV radiation, pathogens, and nutrition availability), hormonal signals (notably jasmonates), developmental stages, and post-transcriptional processes (including noncoding RNAs) together regulate the synthesis and function of these enzymes. This enables plants to modify their metabolite profiles instantaneously (Martín-Cardoso & San Segundo, 2025). Secondary metabolites produced by plants serve to defend against herbivores and diseases, attract pollinators and seed dispersers, and assist in coping with abiotic challenges such as ultraviolet radiation and drought. This links enzyme biochemistry directly to ecological function and agricultural significance (Akbar et al., 2024).

1.4 ENZYMES IN GROWTH, DEVELOPMENT, AND SIGNALLING

1.4.1 Plant Enzymes in Growth and Development

Enzymes predominantly govern plant growth, development, and signalling (Figure 1.1). They are molecular switches that

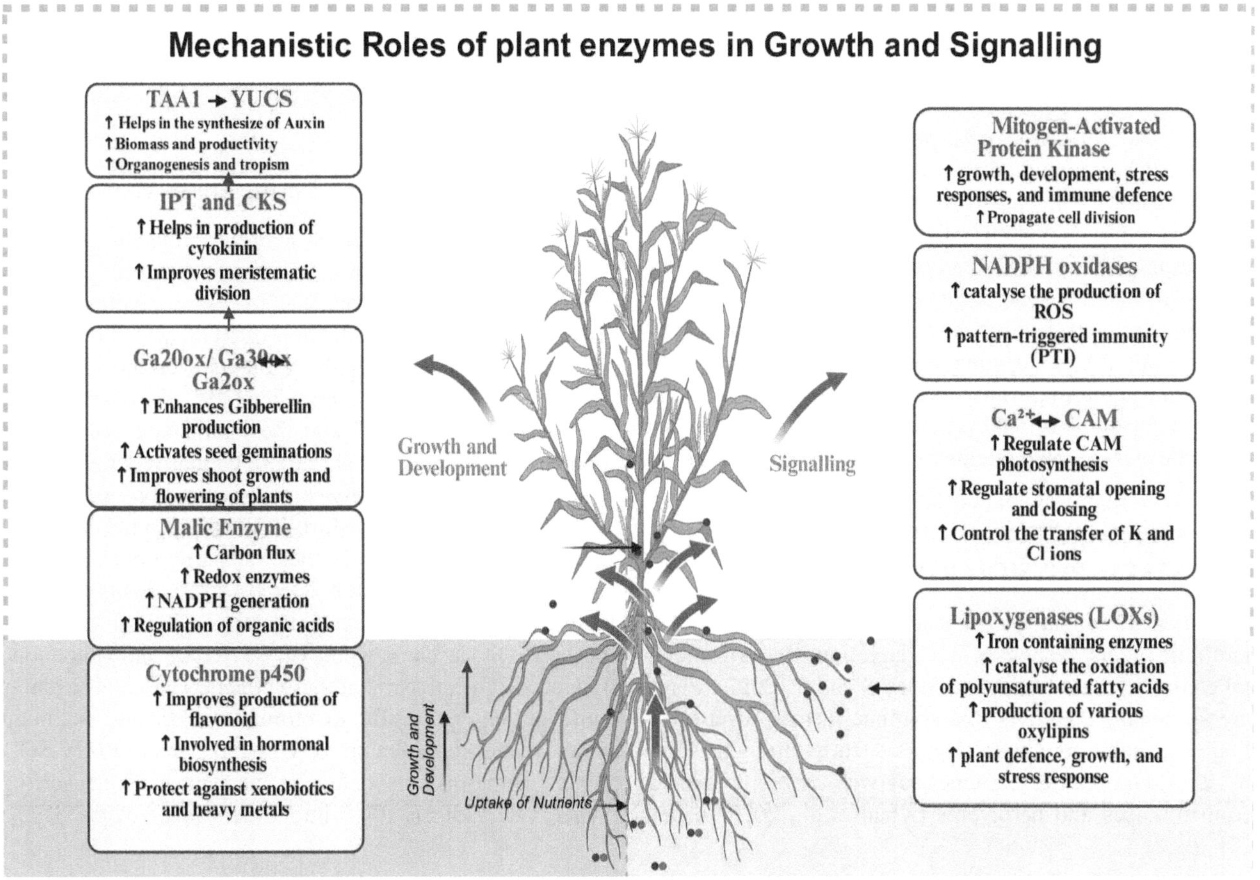

FIGURE 1.1 Mechanistic role of plant enzymes in improving growth, development and the mechanisms driving different signalling pathways in plants.

link hormonal balance, metabolic state, and environmental stimuli to developmental outcomes (Kurepa & Smalle, 2025). Enzymes govern the timing and location of cellular division, development, differentiation, or apoptosis by catalysing biosynthesis, activation, and inactivation activities within hormonal pathways and by regulating metabolite pools. Auxin biosynthesis enzymes (TAA1, YUCs) regulate localized auxin production, influencing organ patterning and tropic responses (Gao et al., 2024). Cytokinin-metabolizing enzymes govern meristem function and organogenesis by orchestrating synthesis, conjugation, and hydrolysis. Gibberellin metabolic enzymes govern stem elongation, seed germination, and flowering via precisely regulated activation and deactivation cycles (Mandal et al., 2022). Alongside hormones, metabolic enzymes like malic enzymes enable the transfer of carbon and redox fluxes across plastids, mitochondria, and the cytosol. Cytochrome P450 monooxygenases and lipoxygenases produce bioactive chemicals and oxylipin signals that regulate development, defence, and aging (Igamberdiev & Bykova, 2023). Aldehyde dehydrogenases and other detoxification enzymes protect cellular machinery from reactive compounds generated by stress. Regulated proteases promote reserve mobilization and tissue remodelling during germination and senescence. Numerous enzymes operate solely in the presence of cofactors and post-translational modifications (Dorion et al., 2021). These two parameters provide spatial-temporal specificity and adaptability to enzymes. Enzymes serve as ideal targets for efforts to improve agricultural production, resilience, and nutritional value due to their synergistic roles (Figure 1.1).

1.4.2 Role of Enzymes in Plant Signalling

Enzymes are essential elements in the signalling mechanisms of plants (Figure 1.1). They operate as both sensors and effectors, transforming environmental stimuli and developmental signals into precise metabolic responses (Thilakarathne et al., 2025). Mitogen-activated protein kinase (MAPK) cascades transmit various signals via a sequence of phosphorylation events that amplify and refine ensuing transcriptional and physiological responses. Redox-active enzymes modulate the signalling of ROS (Mooli et al., 2022). Regulated ROS formation acts as a second messenger, while antioxidant enzymes, including SODs and peroxidases, modulate the signal's intensity and duration to safeguard cells from damage. Calcium-binding proteins, including calmodulin, transmute transitory Ca^{2+} spikes into precise protein phosphorylation (Zandi & Schnug, 2022). They also engage with other substances, such as melatonin, to augment CaM-dependent functions. Enzymes linked to nitric oxide (NO) facilitate S-nitrosylation, a reversible modification that alters protein activity throughout the phases from germination to senescence (Singh et al., 2025). Lipid-derived signals, peptide phytocytokines, and traditional hormone-receptor systems illustrate concurrent pathways that connect growth and defence with metabolic status to improve regulation (Dodueva et al., 2021). Bacteria inhabiting plants produce and break down signalling molecules in the rhizosphere. Such activity affects the host's communication transmission and its ability to handle stress (Khan et al., 2023). Together, these enzyme networks form a dynamic signalling framework that enables temporal and spatial control. This platform aids plants in adaptation and offers farmers many methods to improve their yields.

1.5 ENZYME'S ROLE IN PLANT STRESS PHYSIOLOGY

Enzymes are crucial for plant stress physiology since they facilitate plants' responses and adaptations to environmental changes (Figure 1.2) (Chaudhry & Sidhu, 2022). Plants are stationary creatures that continually face several difficulties, including abiotic factors such as drought, salinity, and temperature extremes, as well as biotic stresses from diseases and herbivores (Khan et al., 2022). Plants employ a complex assortment of enzymatic and non-enzymatic antioxidant defence systems to mitigate the harmful effects of ROS and maintain cellular functions (Chaudhry & Sidhu, 2022).

1.5.1 Enzyme's Role in Abiotic Stress Management

Plants employ an intricate system of antioxidant enzymes to mitigate the adverse effects of drought, salinity, and heat stress (Khan et al., 2025a, 2025b). These abiotic circumstances promote the production of ROS, which can cause oxidative damage to lipids, proteins, and nucleic acids (Wang et al., 2024b). Antioxidant enzymes are crucial for the removal of ROS and the maintenance of cellular homeostasis (Jomova et al., 2023). Oxidative stress occurs when the equilibrium between the generation of ROS and antioxidant defences is disturbed, resulting in the deterioration of cellular macromolecules, including lipids, proteins, and nucleic acids (Dumanović et al., 2021). Plants employ a network of enzyme antioxidants, such as SOD, CAT, peroxidases (POX), ascorbate peroxidase (APX), and glutathione reductase (GR), to combat these effects. These antioxidants function synergistically to eliminate ROS and maintain redox homeostasis (Fujita & Hasanuzzaman, 2022). SOD serves as the primary defensive mechanism by converting superoxide radicals (O_2^-) into hydrogen peroxide (H_2O_2)

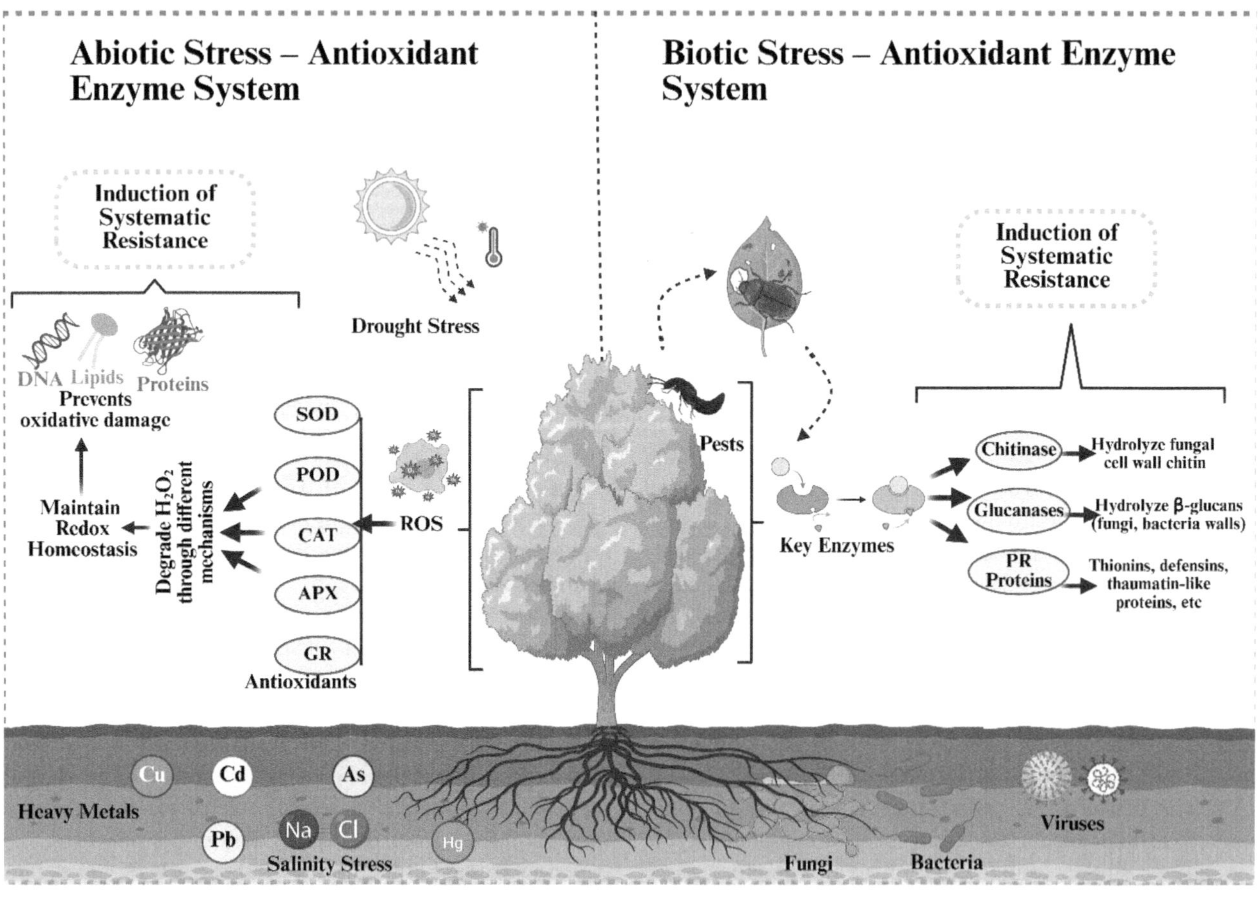

FIGURE 1.2 Enzyme's role against abiotic and biotic stress alleviation in crops.

and oxygen (O_2). This process prevents the accumulation of superoxide radicals and initiates the detoxification process for ROS (Liu et al., 2022). Multiple SOD gene families in *Brassica rapa* are crucial for growth, development, and environmental adaptability (Sarcheshmeh et al., 2023; Su et al., 2021). CAT mitigates H_2O_2 by transforming it into water and oxygen, so averting oxidative damage caused by excessive H_2O_2 accumulation (Anwar et al., 2024). Peroxidases (POX), such as guaiacol peroxidases, reduce H_2O_2 levels by utilizing phenolic and other electron-donating substrates. Class III peroxidases are crucial as they generate and decompose ROS, hence regulating ROS levels within cells (da Silva et al., 2023). APX is crucial, as it catalyses the conversion of H_2O_2 into water through the oxidation of ascorbate. This enzyme is a component of the glutathione–ascorbate cycle, which is crucial for chloroplasts and other organelles to maintain stable ROS levels (Foyer & Kunert, 2024). GR maintains the reservoir of reduced glutathione (GSH), a crucial non-enzymatic antioxidant, by converting oxidized glutathione (GSSG) back to its reduced form utilizing NADPH (Gřešková & Petřivalský, 2024). The activity of these enzymes is meticulously regulated by environmental factors. For instance, when wheat seedlings experience drought stress, their antioxidant defences increase throughout the cellular compartments, enhancing their drought resistance (Figure 1.2) (Haghpanah et al., 2024). Excessive salt intake elevates the levels of ROS. Enzymes such as SOD and APX operate to maintain redox balance. The overexpression of stress-related genes, including BchFSD2, improves salt tolerance by increasing antioxidant activity (Zhang et al., 2025a; Zhou et al., 2024).

1.5.2 Enzyme-mediated Pathogen Defence in Plants

Plants utilize enzyme-mediated defence mechanisms, with pathogenesis-related (PR) proteins, especially chitinases and glucanases, being pivotal in pathogen resistance (Figure 1.2) (Bahmani et al., 2023). Plant defence is regulated by both permanent and transient systems, with PR proteins serving as a crucial component of the transient response (Roychowdhury et al., 2024). Subsequent to pathogen infection, these proteins amass in both intracellular and extracellular compartments, obstructing pathogen proliferation and promoting the establishment of systemic acquired resistance (SAR) (Su et al., 2024). PR proteins were initially discovered in the 1970s, and currently, there are 17 distinct families (PR1–PR17). Many of these reach the apoplast, where they function as molecular identifiers and serve as the primary line of defence for the plant immune system (Sharma et al., 2021). Overexpression of PR genes, including chitinases, glucanases, thionins, thaumatin-related proteins, and defensins, has been demonstrated to markedly improve the resistance of many crop species to infections (Yu et al., 2022). Chitinases catalyse the hydrolysis of chitin, a primary structural element of fungal cell walls, by severing β-1,4-glycosidic bonds, leading to cell wall destabilization, lysis, and diminished pathogen proliferation (Thakur et al., 2023). Plants recognize chitin fragments via membrane-localized receptors, triggering defensive signalling pathways that augment the synthesis of chitinases and other PR proteins (Meresa et al., 2024). Transgenic methods reveal that chitinase overexpression provides substantial resistance to fungal infections (Chouhan et al., 2023), whereas empirical research underscores their efficacy in integrated pest control (Kebe et al., 2023). Functional instances encompass the accumulation of chitinases in tomato apoplasts in response to *Cladosporium fulvum* (Pinto et al., 2025), and insect-interactive functions, including the inhibition of maize defence by fall armyworm frass-derived chitinases (Vassão et al., 2018). Furthermore, chitinases derived from Trichoderma viride can disrupt the peritrophic matrix of Lepidoptera, adversely affecting insect physiology (Berini et al., 2024), whereas certain fungal infections utilize β-N-acetylhexosaminidases in conjunction with chitinases to infiltrate insect cuticles (Lu et al., 2024). Glucanases represent an additional category of PR proteins. They decompose β-glucans, which are constituents of the cell walls of bacteria and fungi. This diminishes the virulence of viruses, reducing their likelihood of causing illness in individuals (Mészáros et al., 2024). Glucanases directly eliminate bacteria and produce oligosaccharide fragments during hydrolysis that function as elicitors, initiating defensive signalling pathways and enhancing plant immune responses (Nayak et al., 2021).

1.6 APPLICATIONS OF PLANT ENZYME RESEARCH

Research on plant enzymes has produced multiple applications in diverse domains, such as food biotechnology, agriculture, and environmental research (Streimikyte et al., 2022). Plant-derived enzymes are increasingly preferred as sustainable and biocompatible substitutes for animal and microbial enzymes due to their low immunogenicity, environmental efficiency, substrate specificity, and stability under moderate circumstances (Barbinta-Patrascu et al., 2024).

1.6.1 Application in Food Biotechnology

Amylase, invertase, papain, bromelain, and lipoxygenase are notable plant-derived enzymes essential in food biotechnology (Zhang et al., 2024). These enzymes are essential for the manufacture of syrups, baked items, alcoholic drinks, and dairy products (Radadiya et al., 2025). Enzyme technology is progressively preferred in the food business owing to its effectiveness, accuracy, and safety (Pang et al., 2021). Recent breakthroughs have enabled the creation of new gadgets that allow individuals to alter the texture and nutritional makeup of food. These instruments are intended to tackle prevalent concerns regarding texture, nutrition, and flavour in upcoming "future foods" (Pang et al., 2020). Enzymes are employed to extract flavourings and colorants from botanical sources (Bashir et al., 2024).

Enzyme-assisted extraction methods enhance extraction efficiency while safeguarding target molecules by degrading cellulose and other cell wall constituents (Tizón Alba et al., 2023).

1.6.2 Application in Agriculture

Plant biotechnology, which includes enzyme research, is a crucial tool for improving agriculture by increasing food supply through various methods such as tissue culture, molecular biology, and crop enhancement (Akram et al., 2025). Enzymes are essential for enhancing the functionality of plant proteins, which is crucial for producing plant-based foods with appropriate forms and textures (Gasparre et al., 2025). Enzymes can rectify the inadequate and inflexible functionalities prevalent in seed storage proteins, which are the primary plant proteins consumed by humans (Kumar et al., 2025). Proteins involved in plant defence, including PR proteins, chitinases, glucanases, protease inhibitors, lectins, and antimicrobial peptides, are crucial for protecting plants from biotic and abiotic stresses (Han & Schneiter, 2024). Advancing plant biology and developing novel biotechnological applications necessitates a more profound understanding of the functions and mechanisms of these proteins (Akinsemolu et al., 2024). Precision plant genotype fingerprinting (PPGF) integrates Phyto-enzyme identification and quantification with electro-sensor technologies. This method analyses the genetic composition of plant species by examining their unique Phyto-enzyme signatures (Rachappanavar, 2025). Enzymes enhance the capacity of plants and microorganisms to remediate contaminated soils (Jia et al., 2024). Modifying enzymes that assist plants in the absorption and degradation of substances, in conjunction with molecular docking and dynamics, can facilitate the remediation of soils contaminated with polychlorinated biphenyls (PCBs) (Boro et al., 2025).

1.6.3 Enzymes Application in Environmental Research

Lignin-degrading enzymes (LDEs) obtained from fungi like *Phanerochaete chrysosporium* exhibit significant potential for many environmental applications, including bioremediation and industrial wastewater treatment (Akbar Aly et al., 2025). Lignin peroxidase (LiP), manganese peroxidase (MnP), and laccase are enzymes that decompose lignin, a complex polymer present in plant cell walls (Gałązka et al., 2025). Bacterial enzymes facilitate the degradation of cellulose and hemicellulose, which are essential components of plant cell walls (da Silva Pereira et al., 2025). These enzymes offer eco-friendly alternatives to chemical methods in several biotechnological applications (Seth & Meena, 2024). Zymography techniques enable the visualization of enzyme activities in soil and facilitate the identification and mapping of hotspots. This is beneficial for examining and enhancing the enzymatic remediation of polluted soils (Loginova & Bilyera, 2025).

1.7 FUTURE PERSPECTIVES AND CHALLENGES

Plant enzyme systems are essential for growth, development, and stress response; nevertheless, their complexity makes them difficult to apply in agriculture. Future expansion will depend on applying advanced methods to identify new functionalities and use them to augment agricultural yields.

1.7.1 Omics Approaches

High-throughput genomics, transcriptomics, proteomics, and metabolomics have accelerated the identification and characterization of enzyme activities. Multi-omics integration provides a comprehensive overview of enzyme regulation, metabolic flux dynamics, and their association with phenotypic characteristics. This data can enhance agricultural resilience and productivity.

1.7.2 Synthetic Biology and Enzyme Design

Synthetic biology and protein engineering enable the modification of enzymes to enhance their stability, improve their functionality, or impart novel catalytic properties. CRISPR-mediated genome editing enables precise modifications of endogenous enzymes, whereas metabolic engineering in plants or microbes facilitates the environmentally sustainable production of beneficial chemicals. These novel concepts may enhance yield, stress resilience, and nutritional quality.

1.8 CHALLENGES

Despite improvements, certain issues remain to be addressed. Regulating enzymes is complex due to the involvement of feedback mechanisms, post-translational modifications, and environmental interactions. Translating knowledge from model systems to crops can be challenging due to the distinct requirements of various species. Prior to widespread implementation, one must meticulously consider biosafety, environmental concerns, and the applicable regulations.

1.9 CONCLUSION

Plant enzymes are essential for survival and productivity, as they are involved in primary and secondary metabolism, growth regulation, signalling, and stress resilience. Recent breakthroughs in molecular biology, omics technologies, and synthetic biology have significantly enhanced our comprehension of enzyme functions and their prospective uses in agriculture. Nonetheless, the complex systems that regulate enzymes and their environmental interactions continue to provide substantial hurdles. Effective utilization of plant enzyme systems necessitates the integration of fundamental research with practical techniques in a cohesive manner. Plant enzymes have considerable potential to improve crop resilience, productivity, and adaptability in the face of global issues like climate change and food security. This is attributable to their relevance in precision breeding, metabolic engineering, and sustainable agricultural methodologies.

REFERENCES

Adhikari, D. (2024). Biotic and abiotic stresses and reactive oxygen species (ROS)-mediated signalling in higher plants: Regulatory mediators and their mechanism of actions. *International Journal of Trend in Scientific Research and Development* (ijtsrd)@www. ijtsrd. com eISSN, 2456–6470.

Akbar, R., Sun, J., Bo, Y., Khattak, W. A., Khan, A. A., Jin, C., …, & Liu, W. (2024). Understanding the influence of secondary metabolites in plant invasion strategies: A comprehensive review. *Plants*, *13*(22), 3162.

Akbar Aly, A. B., Shanmugaraj, B., & Ramalingam, S. (2025). Industrial applications of *Phanerochaete chrysosporium* lignin-degrading enzymes: Current status, production challenges, and future directions. *World Journal of Microbiology and Biotechnology*, *41*(5), 171.

Akinsemolu, A. A., Onyeaka, H., Odion, S., & Adebanjo, I. (2024). Exploring *Bacillus subtilis*: Ecology, biotechnological applications, and future prospects. *Journal of Basic Microbiology*, *64*(6), 2300614.

Akram, J., Naeem, N., Manzoor, M., Ali, Q., Marwat, U., & Zia, T. (2025). Role of modern biotechnology in improving crop protection. *Bulletin of Biological and Allied Sciences Research*, *2025*(1), 100.

Altaf, F., Parveen, S., Farooq, S., Lone, M. L., Haq, A. U., & Tahir, I. (2024). Enigmas of senescence: A reappraisal on the hormonal crosstalk and the molecular mechanisms. *Theoretical and Experimental Plant Physiology*, *36*(1), 51–81.

Angeli, C., Atienza-Sanz, S., Schroder, S., Hein, A., Li, Y., Argyrou, A., …, & Schmidt, S. (2024). Recent developments and challenges in the enzymatic formation of nitrogen–nitrogen bonds. *ACS Catalysis*, *15*(1), 310–342.

Anwar, S., Alrumaihi, F., Sarwar, T., Babiker, A. Y., Khan, A. A., Prabhu, S. V., & Rahmani, A. H. (2024). Exploring therapeutic potential of catalase: Strategies in disease prevention and management. *Biomolecules*, *14*(6), 697.

Bahmani, R., More, P., Babarinde, S., Zhou, M., & Prithiviraj, B. (2023). Seaweeds for plant disease management: Current research advances and future perspectives. *Phytoparasitica*, *51*(4), 783–802.

Bajguz, A., & Piotrowska-Niczyporuk, A. (2023). Biosynthetic pathways of hormones in plants. *Metabolites*, *13*(8), 884.

Ban, Z., & Estelle, M. (2021). CUL3 E3 ligases in plant development and environmental response. *Nature plants*, *7*(1), 6–16.

Bano, A., Singh, K., Singh, S. P., & Sharma, P. (2023). Abscisic acid: Metabolism, signaling, and crosstalk with other phytohormones under heavy metal stress. *Stresses*, *3*(4), 665–686.

Barbinta-Patrascu, M.-E., Bita, B., & Negut, I. (2024). From nature to technology: Exploring the potential of plant-based materials and modified plants in biomimetics, bionics, and green innovations. *Biomimetics*, *9*(7), 390.

Bashir, I., Pandey, V. K., Dar, A. H., Dash, K. K., Shams, R., Mir, S. A., …, & Zahoor, I. (2024). Exploring sources, extraction techniques and food applications: A review on biocolors as next-generation colorants. *Phytochemistry Reviews*, *23*(4), 1–26.

Berini, F., Montali, A., Liguori, R., Venturini, G., Bonelli, M., Shaltiel-Harpaz, L., …, & Casartelli, M. (2024). Production and characterization of *Trichoderma asperellum* chitinases and their use in synergy with *Bacillus thuringiensis* for lepidopteran control. *Pest Management Science*, *80*(7), 3401–3411.

Bhattacharya, A. (2022). Lipid metabolism in plants under low-temperature stress: A review. In: *Physiological processes in plants under low temperature stress* (pp. 409–516). https://doi.org/10.1007/978-981-16-9037-2_5

Bhattacharya, A., Majtorp, L., Birgersson, S., Wiemann, M., Sreenivas, K., Verbrugghe, P., …, & Stålbrand, H.. (2022). Cross-feeding and enzymatic catabolism for mannan-oligosaccharide utilization by the butyrate-producing gut bacterium *Roseburia hominis* A2-183. *Microorganisms*, *10*(12), 2496.

Boro, M., Chettri, D., Bhadra, S., & Verma, A. K. (2025). An insight into pollutants and their degradation by microbial enzymes. *Biocatalysis and Biotransformation*, 1–25.

Bose Mazumdar, A., & Chattopadhyay, S. (2025). Unfolding the role of glutathione to combat environmental stresses through "Omics"-based approaches/studies. *Journal of Plant Growth Regulation*, *44*(2), 419–435.

Cárdenas-Moreno, Y., González-Bacerio, J., García Arellano, H., & del Monte-Martínez, A. (2023). Oxidoreductase enzymes: Characteristics, applications, and challenges as a biocatalyst. *Biotechnology and Applied Biochemistry*, *70*(6), 2108–2135.

Chaudhry, S., & Sidhu, G. P. S. (2022). Climate change regulated abiotic stress mechanisms in plants: A comprehensive review. *Plant Cell Reports*, *41*(1), 1–31.

Chen, J., Wan, H., Zhao, H., Dai, X., Wu, W., Liu, J., & Zeng, C. (2024). Identification and expression analysis of the *Xyloglucan transglycosylase*/hydrolase (XTH) gene family under abiotic stress in oilseed (*Brassica napus* L.). *BMC Plant Biology*, *24*(1), 400.

Choudhary, R., Kaushik, R., Chawla, P., & Manna, S. (2025). Exploring the extraction, functional properties, and industrial applications of papain from *Carica papaya*. *Journal of the Science of Food and Agriculture*, *105*(3), 1533–1545.

Chouhan, R., Ahmed, S., & Gandhi, S. G. (2023). Over-expression of PR proteins with chitinase activity in transgenic plants for alleviation of fungal pathogenesis. *Journal of Plant Pathology*, *105*(1), 69–81.

Chuang, L., Liu, S., & Franke, J. (2023). Post-cyclization skeletal rearrangements in plant triterpenoid biosynthesis by a pair of branchpoint isomerases. *Journal of the American Chemical Society*, *145*(9), 5083–5091.

da Silva, R. M., Canellas, N. A., Olivares, F. L., Piccolo, A., & Canellas, L. P. (2023). Humic substances trigger plant immune responses. *Chemical and Biological Technologies in Agriculture*, *10*(1), 123.

da Silva Pereira, M., Alcantara, L. M., de Freitas, L., de Oliveira Ferreira, M., & Leal, A. L. (2025). Microbial rumen proteome analysis suggests Firmicutes and Bacteroidetes as key producers of lignocellulolytic enzymes and carbohydrate-binding modules. *Brazilian Journal of Microbiology*, *56*(2), 817–833.

Datta, R. (2024). Enzymatic degradation of cellulose in soil: A review. *Heliyon*, *10*(1).

Delmer, D., Dixon, R. A., Keegstra, K., & Mohnen, D. (2024). The plant cell wall—Dynamic, strong, and adaptable—Is a natural shapeshifter. *The Plant Cell*, *36*(5), 1257–1311.

Dodueva, I., Lebedeva, M., & Lutova, L. (2021). Dialog between kingdoms: Enemies, allies and peptide phytohormones. *Plants*, *10*(11), 2243.

Dorion, S., Ouellet, J. C., & Rivoal, J. (2021). Glutathione metabolism in plants under stress: Beyond reactive oxygen species detoxification. *Metabolites*, *11*(9), 641.

Dumanović, J., Nepovimova, E., Natić, M., Kuča, K., & Jaćević, V. (2021). The significance of reactive oxygen species and antioxidant defense system in plants: A concise overview. *Frontiers in Plant Science*, *11*, 552969.

Eberhart, M. E., Alexandrova, A. N., Ajmera, P., Bim, D., Chaturvedi, S. S., Vargas, S., & Wilson, T. R. (2025). Methods for theoretical treatment of local fields in proteins and enzymes. *Chemical Reviews*, *125*(7), 3772–3813.

Fedoreyeva, L. I. (2024). ROS as signaling molecules to initiate the process of plant acclimatization to abiotic stress. *International Journal of Molecular Sciences*, *25*(21), 11820.

Foyer, C. H., & Kunert, K. (2024). The ascorbate–glutathione cycle coming of age. *Journal of Experimental Botany*, *75*(9), 2682–2699.

Fujita, M., & Hasanuzzaman, M. (2022). *Approaches to enhancing antioxidant defense in plants* (Vol. 11, pp. 925): MDPI.

Gałązka, A., Jankiewicz, U., & Orzechowski, S. (2025). The role of ligninolytic enzymes in sustainable agriculture: Applications and challenges. *Agronomy*, *15*(2), 451.

Gao, J., Zhuang, S., & Zhang, W. (2024). Advances in plant auxin biology: Synthesis, metabolism, signaling, interaction with other hormones, and roles under abiotic stress. *Plants*, *13*(17), 2523.

Gasparre, N., Rosell, C. M., & Boukid, F. (2025). Enzymatic hydrolysis of plant proteins: Tailoring characteristics, enhancing functionality, and expanding applications in the food industry. *Food and Bioprocess Technology*, *18*(4), 3272–3287.

Gómez-Espinoza, O., Rojas-Villalta, D., Zúñiga-Pereira, A. M., Chacón-Díaz, C., Bravo, L. A., & Reyes-Diaz, M. (2025). Plant oxalate oxidases: Key enzymes in redox and stress regulation. *Journal of Experimental Botany*, *76*(17), 4896–4909.

Goura, K., Legrifi, I., Kallali, N. S., Taoussi, M., Kenfaoui, J., Meddich, A., …, & Lahlali, R. (2025). Beyond survival: The role of secondary metabolites in plant defense mechanisms. *Journal of Crop Health*, *77*(4), 121.

Gřešková, A., & Petřivalský, M. (2024). Thioredoxin system in insects: Uncovering the roles of thioredoxins and thioredoxin reductase beyond the antioxidant defences. *Insects*, *15*(10), 797.

Haghpanah, M., Hashemipetroudi, S., Arzani, A., & Araniti, F. (2024). Drought tolerance in plants: Physiological and molecular responses. *Plants*, *13*(21), 2962.

Han, Z., & Schneiter, R. (2024). Dual functionality of pathogenesis-related proteins: Defensive role in plants versus immunosuppressive role in pathogens. *Frontiers in Plant Science*, *15*, 1368467.

Heckmann, C. M., & Paradisi, F. (2020). Looking back: A short history of the discovery of enzymes and how they became powerful chemical tools. *ChemCatChem*, *12*(24), 6082–6102.

Huang, E., Tang, J., Song, S., Yan, H., Yu, X., Luo, C., …, & Zhou, J. (2024). Caffeic acid O-methyltransferase from *Ligusticum chuanxiong* alleviates drought stress, and improves lignin and melatonin biosynthesis. *Frontiers in Plant Science*, *15*, 1458296.

Ibarra, E., & Reynoso, M. A. (2025). Plant trihelix transcription factors and their functions in development and stress. *Journal of Experimental Botany*, *76*(17), 4849–4868.

Igamberdiev, A. U., & Bykova, N. V. (2023). Mitochondria in photosynthetic cells: Coordinating redox control and energy balance. *Plant Physiology*, *191*(4), 2104–2119.

Ikram, M., Batool, M., Ullah, M., Khalid, B., El-Badri, A. M., Mohamed, I. A., …, & Zhao, J. (2025). Molecular alchemy: Converting stress into resilience via secondary metabolites and calcium signaling in rice. *Rice*, *18*(1), 32.

Islam, S. N., Kouser, S., Hassan, P., Asgher, M., Shah, A. A., & Khan, N. A. (2024). Gamma-aminobutyric acid interactions with phytohormones and its role in modulating abiotic and biotic stress in plants. *Stress Biology*, *4*(1), 36.

Jia, L., Sun, N., Fang, L., Cheng, X., & Huang, X. (2024). Enhancing the phytoextraction capacity of chromium-contaminated soil by co-addition of garbage enzymes and microelectrolytic iron-carbon fillers. *Environmental Pollution*, *363*, 125224.

Jomova, K., Raptova, R., Alomar, S. Y., Alwasel, S. H., Nepovimova, E., Kuca, K., & Valko, M. (2023). Reactive oxygen species, toxicity, oxidative stress, and antioxidants: Chronic diseases and aging. *Archives of Toxicology*, *97*(10), 2499–2574.

Karimi, M., & Jacobs, T. B. (2021). GoldenGateway: A DNA assembly method for plant biotechnology. *Trends in Plant Science*, *26*(1), 95–96.

Karthik, H. N., Parmar, S., Gawande, N. D., & Sankaranarayanan, S. (2025). Multifaceted roles of U-box E3 ligases in plant development. *Plant and Cell Physiology*, *66*(8), 1123–1136.

Kathirvel, A. K., Karuppasami, K. M., Dhashnamurthi, V., Vellingiri, G., Muthurajan, R., Venugopal, A., …, & Alagarsamy, S. (2024). Enhancement of grain yield in rice under combined drought and high-temperature stress conditions by maintaining photosynthesis through antioxidant enzyme activities by melatonin. *Plant Physiology Reports*, *29*(2), 262–277.

Kebe, A. A., Hameed, S., Farooq, M. S., Sufyan, A., Malook, M. B., Awais, S., …, & Abbas, N. (2023). Enhancing crop protection and yield through precision agriculture and integrated pest management: A comprehensive review. *Asian Journal of Research in Crop Science*, *8*(4), 443–453.

Khan, A. A., Wang, T., Nisa, Z. U., Alnusairi, G. S., & Shi, F. (2022). Insights into cadmium-induced morphophysiological disorders in *Althea rosea* Cavan and its phytoremediation through the exogeneous citric acid. *Agronomy*, *12*(11), 2776.

Khan, A. A., Wang, Y.-F., Zeb, A., Hayat, K., Alhoqail, W. A., & Soliman, M. H. (2025a). Beneficial elements and their roles against soil pollution. In: *Beneficial elements for remediation of heavy metals in polluted soil* (pp. 1–32). Elsevier.

Khan, A. A., Wang, Y.-F., Akbar, R., & Alhoqail, W. A. (2025b). Mechanistic insights and future perspectives of drought stress management in staple crops. *Frontiers in Plant Science*, *16*, 1547452.

Khan, A. A., Yongfeng, W., Soliman, M. H., Alhoqail, W. A., ALHaithloul, H. S., Alghanem, S. M., & Latef, A. A. H. A. (2025c). PGPR in agriculture: A sustainable approach to increasing climate change resilience. In: *Plant-microbe interactions for environmental and agricultural sustainability* (pp. 129–154). Springer.

Khan, M., Ali, S., Al Azzawi, T. N. I., Saqib, S., Ullah, F., Ayaz, A., & Zaman, W. (2023). The key roles of ROS and RNS as a signaling molecule in plant–microbe interactions. *Antioxidants*, *12*(2), 268.

Khan, N., Choi, S.-H., Lee, C.-H., Qu, M., & Jeon, J.-S. (2024). Photosynthesis: Genetic strategies adopted to gain higher efficiency. *International Journal of Molecular Sciences*, *25*(16), 8933.

Kumar, N., Hong, S., Zhu, Y., Garay, A., Yang, J., Henderson, D., …, & Li, Y. (2025). Comprehensive review of chickpea (*Cicer arietinum*): Nutritional significance, health benefits, technofunctionalities, and food applications. *Comprehensive Reviews in Food Science and Food Safety*, *24*(2), e70152.

Kumar, P., Sharma, R., & Kumar, K. (2024). A perspective on varied fungal virulence factors causing infection in host plants. *Molecular Biology Reports*, *51*(1), 392.

Kumar, P., Sharma, S., Sharma, S., Verma, P., & Irfan, M. (2024). Hormonal, biochemical, and genetic regulations of walnut fruit development and ripening: An integrated perspective. *Acta Physiologiae Plantarum*, *46*(11), 100.

Kurepa, J., & Smalle, J. (2025). The evolution of plant hormones: From metabolic byproducts to regulatory hubs. *International Journal of Molecular Sciences*, *26*(15), 7190.

Lanfranchi, E., Ferrario, V., Gandomkar, S., Payer, S. E., Zukic, E., Rudalija, H., …, & Schachtschabel, D. (2025). Transforming a historical chemical synthetic route for vanillin starting from renewable eugenol to a cell-free bi-enzymatic Cascade. *ChemSusChem*, *18*(11), e202500387.

Leng, X., Wang, H., Cao, L., Chang, R., Zhang, S., Xu, C., …, & Xu, Z. (2024). Overexpressing GLUTAMINE SYNTHETASE 1; 2 maintains carbon and nitrogen balance under high-ammonium conditions and results in increased tolerance to ammonium toxicity in hybrid poplar. *Journal of Experimental Botany*, *75*(13), 4052–4073.

Li, H., Liu, Y., Wu, J., Chang, K., Zhang, G., Zhao, H., …, & Bao, Y. (2024). Overexpression of GhGSTF9 enhances salt stress tolerance in transgenic Arabidopsis. *Genes*, *15*(6), 695.

Li, L., Rodríguez-Concepción, M., & Al-Babili, S. (2025). Update on carotenoid and apocarotenoid metabolisms and functions in plants. *Plant Physiology*, *198*(4), kiaf304.

Li, Y., Zhao, L., Guo, C., Tang, M., Lian, W., Chen, S., ... & Chen, L. (2024). OsNAC103, an NAC transcription factor negatively regulates plant height in rice. *Planta*, *259*(2), 35.

Liu, M., Sun, X., Chen, B., Dai, R., Xi, Z., & Xu, H. (2022). Insights into manganese superoxide dismutase and human diseases. *International Journal of Molecular Sciences*, *23*(24), 15893.

Liu, R., Xia, R., Xie, Q., & Wu, Y. (2021). Endoplasmic reticulum-related E3 ubiquitin ligases: Key regulators of plant growth and stress responses. *Plant Communications*, *2*(3).

Loginova, I., & Bilyera, N. (2025). Novel approach in soil zymography to assess enzyme activity in the rhizosphere. *Rhizosphere*, *33*, 101006.

Lu, L., Dong, Z., Yin, X., Chen, S., & Mehvish, A. (2024). Integration of phenotypes, phytohormones, and transcriptomes to elucidate the mechanism governing early physiological abscission in coconut fruits (*Cocos nucifera* L.). *Forests*, *15*(8), 1475.

Lu, Z., Zhu, Q., Bai, Y., Zhao, X., Wang, H., Peng, X., …, & Zhang, Y. (2024). A fungal pathogen secretes a cell wall-associated β-N-acetylhexosaminidase that is co-expressed with chitinases to contribute to infection of insects. *Pest Management Science*, *80*(9), 4699–4713.

Mandal, S., Ghorai, M., Anand, U., Roy, D., Kant, N., Mishra, T., …, & Tiwari, R. K. (2022). Cytokinins: A genetic target for increasing yield potential in the CRISPR era. *Frontiers in Genetics*, *13*, 883930.

Martín-Cardoso, H., & San Segundo, B. (2025). Impact of nutrient stress on plant disease resistance. *International Journal of Molecular Sciences*, *26*(4), 1780.

Mas-ud, M. A., Juthee, S. A., Chowdhury, M. R., Hosenuzzaman, M., Hossain, M. F., Zaki, M., …, & Matin, M. N. (2025). Biosynthesis, signaling, homeostasis, transport, and regulatory mechanisms of cytokinin in rice and Arabidopsis response to heat stress. *Journal of Plant Growth Regulation*, 1–19.

McDonald, A. G., & Tipton, K. F. (2023). Enzyme nomenclature and classification: The state of The art. *The FEBS Journal*, *290*(9), 2214–2231.

Meresa, B. K., Ayimut, K.-M., Weldemichael, M. Y., Geberemedhin, K. H., Kassegn, H. H., Geberemikael, B. A., & Egigu, E. M. (2024). Carbohydrate elicitor-induced plant immunity: Advances and prospects. *Heliyon*, *10*(15).

Meshram, A., Singhal, G., Bhagyawant, S. S., & Srivastava, N. (2019). Plant-derived enzymes: A treasure for food biotechnology. In: *Enzymes in food biotechnology* (pp. 483–502). Elsevier.

Mészáros, Z., Kulik, N., Petrásková, L., Bojarová, P., Texidó, M., Planas, A., …, & Slamova, K. (2024). Three-step enzymatic remodeling of chitin into bioactive chitooligomers. *Journal of Agricultural and Food Chemistry*, *72*(28), 15613–15623.

Miyamoto, R. Y., de Melo, R. R., de Mesquita Sampaio, I. L., de Sousa, A. S., Morais, E. R., Sargo, C. R., & Zanphorlin, L. M. (2022). Paradigm shift in xylose isomerase usage: A novel scenario with distinct applications. *Critical Reviews in Biotechnology*, *42*(5), 693–712.

Mohd Azmi, S. I., Kumar, P., Sharma, N., Sazili, A. Q., Lee, S.-J., & Ismail-Fitry, M. R. (2023). Application of plant proteases in meat tenderization: Recent trends and future prospects. *Foods*, *12*(6), 1336.

Montecchiarini, M. L., Silva-Sanzana, C., Valderramo, L., Alemano, S., Gollán, A., Rivadeneira, M. F., …, & Podesta, F. E. (2021). Biochemical differences in the skin of two blueberries (*Vaccinium corymbosum*) varieties with contrasting firmness: Implication of ions, metabolites and cell wall related proteins in two developmental stages. *Plant Physiology and Biochemistry*, *162*, 483–495.

Mooli, R. G. R., Mukhi, D., & Ramakrishnan, S. K. (2022). Oxidative stress and redox signaling in the pathophysiology of liver diseases. *Comprehensive Physiology*, *12*(2), 3167–3192.

Munteanu, C., Galaction, A. I., Onose, G., Turnea, M., & Rotariu, M. (2025). The Janus face of oxidative stress and hydrogen sulfide: Insights into neurodegenerative disease pathogenesis. *Antioxidants*, *14*(3), 360.

Nayak, S. K., Nayak, S., Mohanty, S., Sundaray, J. K., & Mishra, B. B. (2021). Microbial chitinases and their applications: An overview. In: B. B. Mishra, S. K. Nayak, S. Mohapatra, & D. Samantaray (Eds.), *Environmental and agricultural microbiology: Applications for sustainability* (pp. 313–340). https://doi.org/10.1002/9781119525899.ch15

Nguyen, D. H., Tran, Q. M., Tran, Q. T., & Nguyen, H. V. (2025). Changes in the oxalate oxidase activity during germination of three cultivars of paddy rice (*Oryza sativa* L.). *Journal of Food Measurement and Characterization*, 1–9.

Pang, C., Yin, X., Zhang, G., Liu, S., Zhou, J., Li, J., & Du, G. (2021). Current progress and prospects of enzyme technologies in future foods. *Systems Microbiology and Biomanufacturing*, *1*(1), 24–32.

Pang, Q., Chen, X., Lv, J., Li, T., Fang, J., & Jia, H. (2020). Triacontanol promotes the fruit development and retards fruit senescence in strawberry: A transcriptome analysis. *Plants*, *9*(4), 488.

Panicker, S., & Sayyed, R. (2022). Hydrolytic enzymes from PGPR against plant fungal pathogens. In: *Antifungal metabolites of rhizobacteria for sustainable agriculture* (pp. 211–238). Springer.

Pinto, L., Soler-López, L., Serrano, A., & Sánchez-Rodríguez, C. (2025). Between host and invaders: The subcellular cell wall dynamics at the Plant–Pathogen interface. *Annual Review of Plant Biology*, *76*(1), 255–284.

Qian, Q., Cui, J., Miao, Y., Xu, X., Gao, H., Xu, H., …, & Zhu, P. (2024). The plant volatile-sensing mechanism of insects and its utilization. *Plants*, *13*(2), 185.

Rachappanavar, V. (2025). Exploring plant diversity through enzyme-mediated analysis using electro-carbon sensors. *Plant Breeding*, *144*(1), 70–85.

Radadiya, H., Patel, R., & Kothari, R. (2025). The use of microbial enzymes in the food industries: A global perspective. *Microbial Enzymes: Production, Purification and Industrial Applications*, *1*, 207–224.

Roychowdhury, R., Mishra, S., Anand, G., Dalal, D., Gupta, R., Kumar, A., & Gupta, R. (2024). Decoding the molecular mechanism underlying salicylic acid (SA)-mediated plant immunity: An integrated overview from its biosynthesis to the mode of action. *Physiologia Plantarum*, *176*(3), e14399.

Sarcheshmeh, M. K., Abedi, A., & Aalami, A. (2023). Genome-wide survey of catalase genes in Brassica rapa, Brassica oleracea, and Brassica napus: Identification, characterization, molecular evolution, and expression profiling of BnCATs in response to salt and cadmium stress. *Protoplasma*, *260*(3), 899–917.

Schmid, K. M. (2021). Lipid metabolism in plants. In: *Biochemistry of lipids, lipoproteins and membranes* (pp. 121–159). Elsevier.

Seshadri, K., Abad, A. N., Nagasawa, K. K., Yost, K. M., Johnson, C. W., Dror, M. J., & Tang, Y. (2025). Synthetic biology in natural product biosynthesis. *Chemical Reviews*, *125*(7), 3814–3931.

Seth, R., & Meena, A. (2024). Enzymes-based nanomaterial synthesis: An eco-friendly and green synthesis approach. *Clean Technologies and Environmental Policy*, 1–24.

Sharkey, T. D. (2024). The end game (s) of photosynthetic carbon metabolism. *Plant Physiology*, *195*(1), 67–78.

Sharma, A., Sharma, A., Kumar, R., Sharma, I., & Vats, A. K. (2021). PR proteins: Key genes for engineering disease resistance in plants. In: *Crop improvement* (pp. 81–98). CRC Press.

Shobade, S. O., Nilsen-Hamilton, M., & Zabotina, O. A. (2025). Plant defense proteins: Recent discoveries and applications. *Plants*, *14*(13), 2069.

Singh, M., Singh, A., Yadav, N., & Yadav, D. K. (2022). Current perspectives of ubiquitination and SUMOylation in abiotic stress tolerance in plants. *Frontiers in Plant Science*, *13*, 993194.

Singh, N., Kalra, G., & Bhatla, S. C. (2025). NO-mediated protein tyrosine nitration crosstalks with melatonin homeostasis and HO-1 activity in mitigating salt stress in sunflower seedlings. *Journal of Plant Biochemistry and Biotechnology*, *34*(2), 526–546.

Singh, S., Chhatwal, H., & Pandey, A. (2024). Deciphering the complexity of terpenoid biosynthesis and its multi-level regulatory mechanism in plants. *Journal of Plant Growth Regulation*, 1–17.

Singh, S., Singh, A. K., Pradhan, B., Tripathi, S., Kumar, K. S., Chand, S., …, & Shahid, M. K. (2024). Harnessing *Trichoderma* mycoparasitism as a tool in the management of soil dwelling plant pathogens. *Microbial Ecology*, *87*(1), 158.

Široká, J., Ament, A., Mik, V., Pospíšil, T., Kralová, M., Zhang, C., …, & Nishizato, Y. (2025). Amide conjugates of the jasmonate precursor cis-(+)-12-oxo-phytodienoic acid regulate its homeostasis during plant stress responses. *Plant Physiology*, *197*(1), kiae636.

Somkuwar Subhash, R., Meshram, J. H., Gogle, D., Chaudhary, R., Mahakhode, R., Patil, V. N., …, & Chaudhary, R. R. (2024). Nanoelicitors mediated abiotic stresses in plant defense response mechanisms: Current review and future perspectives. In : G.C. Nikalje, M. Shahnawaz, J. Parihar, H. A. Qazi, V. N. Patil, & D. Zhu (Eds.), *Plant secondary metabolites and abiotic stress* (pp. 591–621). https://doi.org/10.1002/9781394186457

Sood, M. (2025). Reactive oxygen species (ROS): Plant perspectives on oxidative signalling and biotic stress response. *Discover Plants*, *2*(1), 187.

Spormann, A. M. (2023). Prototypic reactions of prokaryotic carbon catabolism. In: *Principles of microbial metabolism and metabolic ecology* (pp. 137–173). Springer.

Streimikyte, P., Viskelis, P., & Viskelis, J. (2022). Enzymes-assisted extraction of plants for sustainable and functional applications. *International Journal of Molecular Sciences*, *23*(4), 2359.

Su, F., Zhao, B., Dhondt-Cordelier, S., & Vaillant-Gaveau, N. (2024). Plant-growth-promoting rhizobacteria modulate carbohydrate metabolism in connection with host plant defense mechanism. *International Journal of Molecular Sciences*, *25*(3), 1465.

Su, J., Song, Y., Zhu, Z., Huang, X., Fan, J., Qiao, J., & Mao, F. (2024). Cell–cell communication: New insights and clinical implications. *Signal Transduction and Targeted Therapy*, *9*(1), 196.

Su, W., Raza, A., Gao, A., Jia, Z., Zhang, Y., Hussain, M. A., …, & Zou, X. (2021). Genome-wide analysis and expression profile of superoxide dismutase (SOD) gene family in rapeseed (*Brassica napus* L.) under different hormones and abiotic stress conditions. *Antioxidants*, *10*(8), 1182.

Su, Y., Ngea, G. L. N., Wang, K., Lu, Y., Godana, E. A., Ackah, M., …, & Zhang, H. (2024). Deciphering the mechanism of E3 ubiquitin ligases in plant responses to abiotic and biotic stresses and perspectives on PROTACs for crop resistance. *Plant Biotechnology Journal*, *22*(10), 2811–2843.

Tan, S., Cao, J., Li, S., & Li, Z. (2025). Unraveling the mechanistic basis for control of seed longevity. *Plants*, *14*(5), 805.

Thakur, D., Bairwa, A., Dipta, B., Jhilta, P., & Chauhan, A. (2023). An overview of fungal chitinases and their potential applications. *Protoplasma*, *260*(4), 1031–1046.

Thilakarathne, A. S., Liu, F., & Zou, Z. (2025). Plant signaling hormones and transcription factors: Key regulators of plant responses to growth, development, and stress. *Plants*, *14*(7), 1070.

Tizón Alba, A., Aliaño-González, M. J., Palma, M., Fernández Barbero, G., & Carrera, C. (2023). Enhancing efficiency of enzymatic-assisted extraction method for evaluating bioactive compound analysis in mulberry: An optimization approach. *Agronomy*, *13*(10), 2548.

Trejo-Solís, C., Escamilla-Ramírez, Á, Gómez-Manzo, S., Castillo-Rodriguez, R. A., Palomares-Alonso, F., Castillo-Pérez, C., …, & Gallardo-Pérez, J. C. (2025). The pentose phosphate pathway (PPP) in the glioma metabolism: A potent enhancer of malignancy. *Biochimie*, *232*, 117–126.

Vassão, D. G., Wielsch, N., Gomes, A. M. M. M., Gebauer-Jung, S., Hupfer, Y., Svatoš, A., & Gershenzon, J. (2018). Plant defensive β-glucosidases resist digestion and sustain activity in the gut of a lepidopteran herbivore. *Frontiers in Plant Science*, *9*, 1389.

Verma, V., Singhal, G., Joshi, S., Choudhary, M., & Srivastava, N. (2022). Plant extracts as enzymes. In: *Plant extracts: Applications in the food industry* (pp. 209–223). Elsevier.

Wang, K., Li, X., Guo, S., Chen, J., Lv, Y., Guo, Z., & Liu, H. (2025b). Metabolic reprogramming of glucose: The metabolic basis for The occurrence and development of hepatocellular carcinoma. *Frontiers in Oncology*, *15*, 1545086.

Wang, P., Liu, W. C., Han, C., Wang, S., Bai, M. Y., & Song, C. P. (2024b). Reactive oxygen species: Multidimensional regulators of plant adaptation to abiotic stress and development. *Journal of Integrative Plant Biology*, *66*(3), 330–367.

Wang, P., Wang, D., Li, Y., Li, J., Liu, B., Wang, Y., & Gao, C. (2024c). The transcription factor ThDOF8 binds to a novel cis-element and mediates molecular responses to salt stress in Tamarix hispida. *Journal of Experimental Botany*, *75*(10), 3171–3187.

Wang, W., Guo, H., Bowman, J. L., & Chen, F. (2024a). Plant SABATH methyltransferases: Diverse functions, unusual reaction mechanisms and complex evolution. *Critical Reviews in Plant Sciences*, *43*(4), 291–312.

Wang, X., Wang, T., Su, N., Liu, W., Gao, M., Li, R., …, & Bi, Y. (2025a). L-phenylalanine preharvest spraying effectively enhances the reactive oxygen metabolism and antioxidant capacity in postharvest broccoli. *Plant Physiology and Biochemistry*, 110185.

Wang, Z., Xie, D., Wu, D., Luo, X., Wang, S., Li, Y., …, & Zheng, L. (2025c). Robust enzyme discovery and engineering with deep learning using CataPro. *Nature Communications*, *16*(1), 2736.

Weng, J.-K., Lynch, J. H., Matos, J. O., & Dudareva, N. (2021). Adaptive mechanisms of plant specialized metabolism connecting chemistry to function. *Nature Chemical Biology*, *17*(10), 1037–1045.

Wu, M., Musazade, E., Yang, X., Yin, L., Zhao, Z., Zhang, Y., …, & Guo, L. (2023). ATL protein family: Novel regulators in plant response to environmental stresses. *Journal of Agricultural and Food Chemistry*, *71*(51), 20419–20440.

Xing, Y., Wang, X., & Mustafa, A. (2025). Exploring the link between soil health and crop productivity. *Ecotoxicology and Environmental Safety*, *289*, 117703.

Xing, Y., Xie, Y., & Wang, X. (2025). Enhancing soil health through balanced fertilization: A pathway to sustainable agriculture and food security. *Frontiers in Microbiology*, *16*, 1536524.

Yadav, V., Pal, D., & Poonia, A. K. (2024). A study on genetically engineered foods: Need, benefits, risk, and current knowledge. *Cell Biochemistry and Biophysics*, *82*(3), 1931–1946.

Yaschenko, A. E., Alonso, J. M., & Stepanova, A. N. (2025). Arabidopsis as a model for translational research. *The Plant Cell*, *37*(5), koae065.

Yu, H., Wang, Y., Fu, F., & Li, W. (2022). Transgenic improvement for biotic resistance of crops. *International Journal of Molecular Sciences*, *23*(22), 14370.

Yu, X., & Li, S. (2024). Specific regulation of epigenome landscape by metabolic enzymes and metabolites. *Biological Reviews*, *99*(3), 878–900.

Zandi, P., & Schnug, E. (2022). Reactive oxygen species, antioxidant responses and implications from a microbial modulation perspective. *Biology*, *11*(2), 155.

Zhan, Y., Wang, J., Zhao, X., Zheng, Z., & Gan, Y. (2024). Arachis hypogaea monoacylglycerol lipase AhMAGL3b participates in lipid metabolism. *BMC Plant Biology*, *24*(1), 1278.

Zhang, S., Li, C., Si, J., Han, Z., & Chen, D. (2022). Action mechanisms of effectors in plant-pathogen interaction. *International Journal of Molecular Sciences*, *23*(12), 6758.

Zhang, W., Cheng, Y., Shao, L., Li, Z., Sarwar, R., Wei, Q., …, & Tan, X. (2025b). The glycosylation status of small molecules impacts different aspects of plant immunity. *Physiologia Plantarum*, *177*(3), e70292.

Zhang, Y., & Fernie, A. R. (2023). The role of TCA cycle enzymes in plants. *Advanced Biology*, *7*(8), 2200238.

Zhang, Y., Li, J., Nardin, C., Villalonga, R., & Simpson, B. (2024). Enzymes in food processing: Present uses and future prospects. In: *Improving and tailoring enzymes for food quality and functionality* (pp. 111–134). Elsevier.

Zhang, Y., Wang, R., Wang, H., & Wang, H. (2025a). LuCSD3 enhances salt stress tolerance in flax: Genome-wide profiling and functional validation of the SOD gene family. *Frontiers in Plant Science*, *16*, 1609085.

Zheng, K., Martinez, M. P., Bouzid, M., Balparda, M., Schwarzländer, M., & Maurino, V. G. (2025). Regulation of plant glycolysis and the tricarboxylic acid cycle by posttranslational modifications. *The Plant Journal*, *122*(1), e70142.

Zhou, Y., Li, S., Ran, S., Xu, Y., Hou, M., Han, M., & Zhong, F. (2024). Genome-wide identification and characterization of the superoxide dismutase (SOD) gene family in Pakchoi and the role of the BchFSD2 gene in the salt stress toleran. *Agronomy*, *14*(2), 384.

Zografos, A. L., & Anagnostaki, E. E. (2016). From acetate to mevalonate and deoxyxylulose phosphate biosynthetic pathways: An introduction to terpenoids. *From Biosynthesis to Total Synthesis: Strategies and Tactics for Natural Products*, 187–195.

2 Biosynthesis of Plant Enzymes Molecular Pathways and Regulatory Mechanisms

R. Ridhi

2.1 INTRODUCTION

Current digital world has incentivized the role of natural products as a significant part of human population. Organic plants and their products are an irreplaceable source for mankind that is gaining attention nowadays due to increase in food adulterations and other impeding sources (Li et al. 2025). However, current depletion of natural renewable resources due to over population and growth of urban sector has made it an indispensable need to avert natural resources via suitable paradigm programs (Singh et al. 2025). Secondary metabolites play significant role for ecological interactions, amending the amenability of adaptability of organisms and hence human health (Dinglasan et al. 2025). For instance, they act as defense mechanisms for plants and other living organisms against pathogens, herbivores and other competing plants. In addition, they act as a viable source for sustainable agriculture in the form of natural herbicides and pesticides (Erb & Kliebenstein 2020; Selwal et al. 2023; Ridhi et al. 2024a, 2024b). Therapeutic properties of secondary metabolites like penicillin antibiotic, pain-relieving alkaloids for cancer and acting as antioxidants make them amenable source for human health and medicine (Sreenivasulu et al. 2022; Kumari et al. 2024). An outpacing of reproductive growth due to the contribution of secondary metabolites in seed dispersal and enhanced nutrient absorption due to their tendency to undergo mutualistic interactions integrates another asset of secondary metabolites (Tripathi et al. 2024). Secondary metabolites are classified in three major groups, phenolics like flavonoids, phenolic acids, lignins and tannins; nitrogen-containing compounds, including glycosylates and alkaloids and terpenes such as sterols, plant volatiles and carotenoids (Reddy et al. 2021). Mao et al. (2025) have recently reviewed the paradigm of plant flavonoids with their regulatory mechanisms coupled with biosynthesis techniques. Flavonoids are derived from the pathway of phenylpropane metabolism uniquely characterized by C6-C3-C6 carbon skeleton. They are assets for medical applications in addition to plant pigment production and other biological applications (Zhang et al. 2023a, 2023b; Mao et al. 2025). The varied parameters consorting regulation of plant flavonoids include plant hormones, non-coding endogenous small RNAs (miRNAs) and related transcription factors(TF) (Mao et al. 2025). These factors play pivotal role in their metabolism integrating in specialized metabolites (SM) to identify TF and other structural genes involved in their respective domain (Zhang et al. 2023; Zhao et al. 2023).

This chapter deals with various molecular pathways, regulatory mechanisms and related cumulative factors responsible for biosynthesis of plant enzymes. For instance, Gámez et al. (2020) have emphasized the role of recombinant enzymes for understanding kinetic, regulatory and molecular properties of *Escherichia coli* plants. Enzymes act as a sustainable source for plants while serving as successive biological catalysts (Zheng et al. 2024). For instance, enzymes such as RuBisCO (ribulose-1,5-bisphosphate carboxylase/oxygenase) facilitate in Calvin cycle via fixation of atmospheric CO_2 into consorting organic molecules (Sharma et al. 2024; Zheng et al. 2024). Similarly pyruvate kinase, phosphofructokinase and hexokinase enzymes aid in respiration process that facilitates efficient energy production (Moorhead et al. 2013; Wang et al. 2024). For successful nutrient uptake that is imperative for root cells, phosphatases and nitrate reductase aid in converting phosphorous and nitrogen into respective bioavailable forms requisite for plant reproductive growth and development (Waratadar et al. 2022).

2.2 BIOSYNTHESIS MECHANISM OF PLANT ENZYMES

An integrated and enzyme-catalytic process assorts in the biosynthesis of plant enzymes where living organisms form consortium of complex molecules those are imperative for their reproduction, growth and maintenance (Li et al. 2023). These biological pathways convert the precursors into lipids, proteins, nucleic acids and carbohydrates that are significant requisite for the working of plant cells (Zabotina et al. 2021). For instance, glycosyltransferase (GT) enzymes are utilized as catalyzing agents for biosynthesis of polysaccharide via forming multiprotein complexes (Zabotina et al. 2021). Bioengineering can serve as a sustainable source for the development of sustainable energy sources (Mirsalami & Mirsalami 2025). For instance, fermentation of microorganisms can be an asset for renewable alternative to fossil fuels and averting emission of harmful greenhouse gases (Carruthers & Lee 2022; Joshi et al. 2025; Mirsalami & Mirsalami 2025). Birchfield and McIntosh (2020) have elucidated the significance of metabolic engineering of natural organic products to integrate co-culture and polyculture to form successful biosynthesis paradigm approaches. Glucosinolates, alkaloids, betalains and flavonoids are four major categories of secondary metabolites of plants imperative for significant contributions in metabolic engineering

DOI: 10.1201/9781003669920-2

(Birchfield & McIntosh 2020; Courdavault et al. 2021). Microbial host organism is an imperative source to consider as a precursor material for the development of proliferating paradigm for the biosynthesis of plant compounds. For instance, some microbes like *Saccharomyces cerevisiae*, *Corynebacteriumglutamicum*, *Yarrowialipolytica* or *E.coli* are relatively easy to grow with ample research details availability on metabolic engineering protocols (Birchfield et al. 2020; Ndochinwa et al. 2024). Similarly for enhancing biochemical properties of varied biocatalysts, protein engineering mechanism may serve as a revitalizing approach (Chatterjee et al. 2023; Ndochinwa et al. 2024). Efficiencies and selectivity of plant cells have a strong dependence on the enzymes and their catalytic mechanism (Barinje 2025). Therefore, selection of suitable enzyme with regulatory mechanisms acts as promulgating source to enhance efficacy and sensitivity of a plant cell toward a target analyte (Stephens et al. 2023; Ridhi et al. 2024). In view of above contemplating innovations need to be shackled and tapped to design effective regulatory mechanisms and substantive molecular pathways for biosynthesis of plant enzymes. In addition, the sustainability concerns, overexploitation and low natural abundance of plants' natural products offer another impeding challenge in this domain (Szwarc et al. 2024). A recent review by Cheng et al. (2025a, 2025b) has emphasized the role of utilizing complete genome assemblies to produce diverse flora varieties from existing current plant varieties while outpacing research and development in the field of medicinal plant genomics. The role of medicinal plants has gained significance for the development of new plant varieties due to their non-toxicity and high selectivity (Yin & Yang 2024).

2.3 SIGNIFICANCE OF LIGHT FOR GROWTH, DEVELOPMENT AND METABOLISM IN PLANTS

Light is an asset for plants as it acts as a modulating parameter for the physiology of plants and is an integrating primary paradigm for the photosynthesis phenomena. The various photoreceptors utilized by the plants to articulate different regions of wavelength of light include Ultraviolet (UV) Resistance Locus 8 (UVR8), cryptochromes (CRY1 and CRY2) and phytochromes (PHYA-PHYE) (Liu et al. 2023). For instance, CRY1 and CRY2 detect blue/UV-A light; UVR8 detects UV-B light and PHYA-PHYE sense red and far-red light (Liu et al. 2023). These photoactivators with ensued activation via light sensitivity act as transducers and transmit signals to the interrelated components downstream for regulating varied light-induced biological processes (Liu et al. 2023; Dong & Qi 2025). Dong & Qi (2025) have recently reviewed the proliferation in research of triterpene synthesis while elucidating its regulatory mechanisms and biosynthetic pathways. The sustainable utilization of plant triterpenoids on account of their unique structure and varied biological activities integrates another significant asset to it (Dong & Qi 2025). For instance,cucurbitacins produced in cucurbit plants act as a defensive triterpenoid and exhibit antitumor pharmacological, anti-inflammatory and hepatoprotective effects (Dong & Qi 2025). Similarly antimicrobial activity of avenacin A-1 found in oat roots offers other promising asset of plant triterpenoid (Cheng et al. 2025; Dong & Qi 2025). The significance of terpenoids in biofuels, nutraceuticals and pharmaceuticals integrates another factor to pipeline research and development in this domain (Cheng et al. 2025). For instance, Cheng et al. (2025) have highlighted the key enzymes, novel terpenoid precursors, biosynthetic networks and progress in fundamental pathways for the biosynthesis of terpenoid plants. They have elucidated the role of regulation of metabolic network for biosynthesis of terpenoid plants (Cheng et al. 2025). The transduction paradigm of photoreceptor is promulgated by the set of evolutionary proteins that are central repressors of photomorphogenesis (Guan et al. 2025). Suitable maintenance of carbon and nitrogen nutrition is imperative for substantive growth and development of a plant (Guan et al. 2025). C/N balance aids in homeostasis of C and N metabolism, nitrogen assimilation, carbon fixation, growth and resistance, maintaining material and energy allotment to the plants (Drosou & Pantazis 2024; Guan et al. 2025). Water and carbondioxide are converted into hexoses via chloroplasts utilizing light energy, that is light regulates the uptake of nitrogen, its utilization and assimilation (Guan et al. 2025). The intensity of light has a significant impact on photosynthesis mechanism. For instance, optimum light intensity is required to sufficiently energize chlorophyll molecules to make proportionality between the rate of photosynthesis and the number of photons available till the light saturation point. Similarly, intensity of light must not be excessive enough to avert photoinhibition that may damage photosynthetic machinery and diminish the efficacy of photosynthesis (Wu et al. 2025). The unique light signaling components include Elongated Hypocotyl 5 (HY5), constitutive photomorphogenic 1 (COP1), Phytochrome Interacting Factors (PIFs), the B-box proteins (BBX), De-Etiolated 1(DET1) and suppressor of PHYA-105 (SPA) (Liu et al. 2023). More than 2 lakhs SM are synthesized by plants that serve as chemical defenses to avert plants from varied abiotic and biotic stresses. Inherited therapeutic properties of SM also play pivotal role for the benefits of living population (Liu et al. 2023). Hashim et al. (2021) have emphasized the role of artificial light stimulators to enhance SM production in cultured medicinal plants.

2.4 NATURAL PRODUCTS AS SUSTAINABLE SOLUTION FOR METABOLITE BIOSYNTHESIS OF HERBAL AND MEDICINAL PLANTS

Optimization of synthetic parameters,including temperature, pH values, incubation period and many more, aids in stable formation of metal nanoparticles (NPs) utilizing plant biosynthesis mechanism (Zhao et al. 2023). Nanoscale metal NPs play a vital role in therapeutics of varied diseases like cancer and also employed for prosthetic implants (Chopra & Dhingra 2021). Among the marketed medicines, 60% of extracted by natural resources (Chopra & Dhingra 2021) and 22% are accounted by higher plant-derived natural products.

Conrado et al. (2024) have highlighted the role of plant secondary metabolites and bioactive crude extracts as an imperative assorts for emerging research in drug delivery paradigm mechanism for successful synthesis of therapeutic agents. However, there are several impeding challenges like scaling up the supply of bioactive molecules, acquiring selective and specific plant materials, identifying the availability of respective high-throughput screening bioassays and many more that should be obviated to extract natural products from medicinal plants (Chaachouay & Zidane 2024). For instance, Chen (2024) has demonstrated the role of synthetic biology, biosynthetic pathways and metabolic engineering for obtaining abundance of natural products in synergistic manner. There is an imperative need to have research thrust in the domain of plant natural products (PNP) in the pre-genomic era with emphasis on identifying bioactive molecules having substantive pharmaceutical activities and individual genes trickling the track toward biosynthesis (Guo et al. 2022). A recent review by Ahmad et al. (2025) has explored the marine and plant resources as sustainable support for drug research and development on account of having consortium of complex metabolites. Table 2.1 lists the various natural product types with their molecular pathways and their role in secondary metabolites with suitable examples.

NPs act as a proliferating asset for the modulating the synthesis of anti-aging secondary metabolites with better stability, bioavailability and efficacy as compared to traditional synthesis methods (Holghoomi & Colagar 2025). For instance, silver nanoparticles (AgNPs) are utilized for outpaced synthesis of secondary metabolites within plants (Ho et al. 2022). Synthesis of *Artemisia* utilizing AgNPs is a promising asset for mitigating cardiac function and correlating with senescence in aging stem cells is a contemplated example of significance of NPs for synthesizing secondary metabolites (Ho et al. 2022). Marine NPs include a high degree of structural novelty with higher incidence of bioactivity that act as an outpacing source for successful synthesis of marine drugs (Ahmad et al. 2025). Food and Drug Administration (FDA) has approved various anticancer marine drugs like vidarabine, trabectedin and eribulin mesylate, different antiviral drugs,including iota carrageenan and Cytrabine, ziconotide drug for neuropathic pain and hypertriglyceridemia marine drug omega-3-acid ethyl esters. Atanasov et al. (2021) have discussed various impeding challenges of natural products for drug delivery and implication in pharmaceutical industry,including technical bottlenecks to screening, optimization and articulate characterization. However, Alum et al. (2025)have addressed these issues via narrating analytic approach metabolomics for quantification and identification of complex herbal mixtures. Similarly, Theodoridis et al. (2023)have proposed certain specific spatial indicators for promulgating the exposure of natural products and resources toward global environmental change. It requires the need for establishing potential societal benefits, economic values of native medicinal plant and establishing capacity.

2.5 PARADIGM APPROACHES TO BIOSYNTHESIS TYPES

Collective biosynthesis via protein engineering of imperative natural product spirooxindole alkaloids offers an incentivizing approach for sustainable research, development and production of complex natural products (Chu et al. 2025). Figure 2.1 depicts the various biosynthesis techniques suitable for the synthesis of plant enzymes and natural products. Nitrogen-rich alkaloids having significant pharmaceutical asset trickle the track toward diversity of metabolic chemistry in plants that aids in metabolic engineering strategies (Villanueva & Nett 2025). The secondary metabolites produced by the plants act as defense molecules and the TF initiate their biosynthesis (Godbole et al. 2022).

TABLE 2.1
Different Types of Natural Products with Their Molecular Pathways and Role in Secondary Metabolism

Natural Product Type	Molecular Pathway for Synthesis	Role in Secondary Metabolites	Examples	Key Reference
Phenolics	Phenylpropanoid	Avert from harmful ultraviolet radiation	Curcumin (turmeric), resveratrol	Sun and Shahrajabian (2023)
Flavonoids	Polyketide and phenylpropanoid	Pigmentation	Quercetin, kaempferol	Hasnat et al. (2024)
Alkaloids	Amino acids	Signal transducers and defense mechanism	Morphine (papaver), vincristine	Heinrich et al. (2021)
Glycosides	Glycosylation of secondary metabolites	Storage and transport of bioactive forms	Stevioside	Ceunen and Geuns (2013)
Terpenoids	Mevalonate	Defense, regulation and growth	Artemisinin (*Artemisia*), Taxol	Hu et al. (2025)

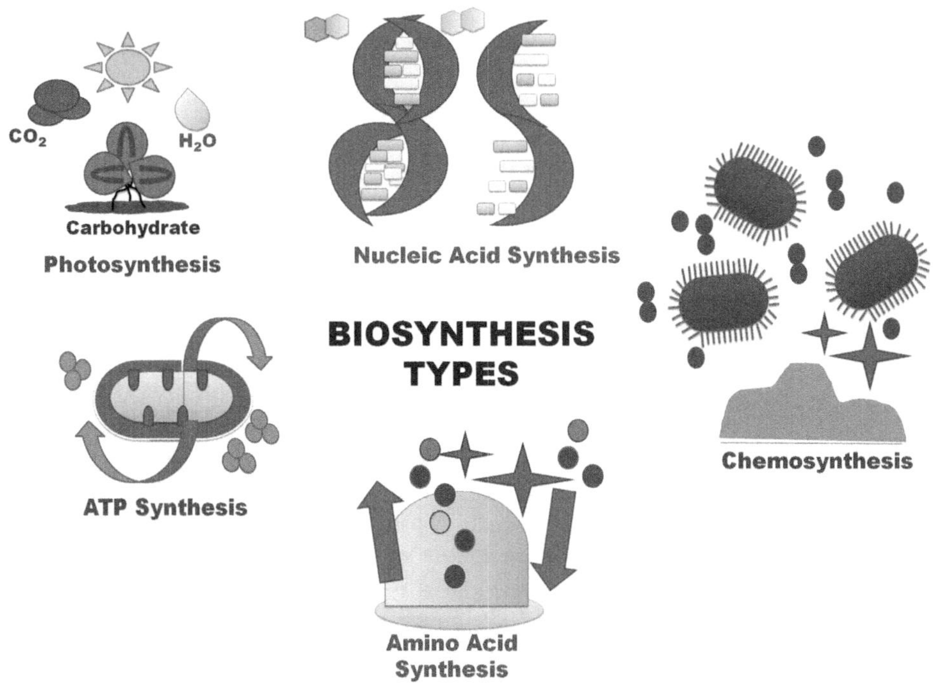

FIGURE 2.1 Various biosynthesis types for the synthesis of plant enzymes and natural products.

Phenylalanine, arginine and ornithine are some of the substrates utilized for the biosynthesis of plants (Bhakshu & Pullaiah 2025). Bernonville et al. (2020) have elucidated omics-based strategies for the identification of genes among complex metabolic pathways enhancing the development of molecular engineered natural products. Lichman (2021) has reviewed amino acid-derived SM constituting nitrogen heterocycle. Amino acid metabolism is the origin of nitrogen in complex alkaloids.

2.5.1 Photosynthesis

Photosynthetic machinery is a consortium of protein network assemblies functioning in chloroplasts that aid in energy conversion process under severe climatic conditions and abiotic stresses (Kallio et al. 2025). MicroRNAs and plant growth regulators are significant signal transducers for understanding the paradigm mechanism of photosynthesis (Chauhan et al. 2023). Amendment of photosynthetic proteins levels via dwindling pigmentation with enhancement in light utilization amenability per leaf acts as a substantive source to increase efficacy of plant synthesis (Leister 2023). A balanced ratio of NADPH/adenosine triphosphate (ATP) is regulated and optimized via linear electron flows and photosynthetic cyclic as per the adaption of plant needs (González et al. 2023; Ridhi 2025). For instance, redox-based dithiol–disulfide interchange plays pivotal role for optimizing photoprotective mechanisms (González et al. 2023). Appropriate need to optimize photosynthetic electron transport chain (PETC) is imperative for averting plant enzymes and natural products under adverse climatic conditions (Nishiyama et al. 2006).

2.5.2 Nucleic Acid Synthesis

NPs-mediated DNA and RNA act as substantive asset for enhancing the delivery of nucleic acids for plant genetic engineering (Liu et al. 2025). They play a crucial role in trickling the track toward protein synthesis and plant resistance to abiotic and biotic stresses (Zhang et al. 2023). Mesoporous silica nanoparticles (MSNs) are useful asset for better penetration of essential nutrients on account of their porous structure that allows better interaction of particles (Ridhi et al. 2023; Mathieu et al. 2025). Gold-plated MSNs were the first NPs to be utilized for the delivery of plasmid DNA (pDNA) into intact tobacco (Yong et al. 2024). Similarly, carbon-based NPs, peptide-based NPs and magnetic NPs have been applied for pDNA delivery vehicles (Liu et al. 2025). The tendency of short nucleotide termed as aptamers is to have high binding affinity and selectivity toward a different range of target molecules (Sanjay et al. 2025). Hataya (2021) has reported polymerase chain reaction (PCR) for the extraction of nucleic acids from plant tissues for the detection of plant diseases and viruses.

2.5.3 Chemosynthesis

Photosynthesis technique is common in plant synthesis, but chemosynthesis is promulgating under research and development domain for plant synthesis (Smith 2012). Especially for sea plants where the range of solar radiation and photolight is restricted (Smith 2012). Aerobic chemosynthetic microorganisms need CO_2 to synthesize organic carbon and catalysis is significant aspect for artificial photosynthesis (Machín et al. 2023). Chemosynthesis is need of an hour for sustainable solution for energy security in unpredictable climate risks in current scenario (Yang & Lin 2025). Diverse marine organisms impart additional energy from inorganic precursors like ammonia, nitrite, hydrogen, iron and sulfide via specific and selective enzymes (Ricci et al. 2024). Hydrogen-oxidizing bacteria act as an asset for the plants to derive energy for their growth, reproduction and fertilization effects (Islam et al. 2023). Some of the ocean regions utilize chemosynthesis for energy production and security, ranging from oligotrophic below-shelf waters to highly productive geologically fueled hydrothermal vents (Ricci et al. 2024).

2.5.4 ATP Synthesis

Plants synthesize ATP via majorly two paradigm approaches namely, oxidative phosphorylation in mitochondria during the process of cellular respiration and photophosphorylation in chloroplasts via trickling the track through photosynthesis. Crop quality is significantly dependent on cellular ATP to activate global metabolic responses (Xiao et al. 2024). Modulation of enzyme activity is also optimized via unique redox switch present on chloroplast ATP synthase in some higher plants that enables ATP hydrolysis at night (Yang et al. 2020). ATP synthases are consortium of supramolecular structures homo-oligomeric species and dimers that facilitate the energy conversion (Seelert &Denche 2011). ATP synthesis has rich research data available for animals; however, for the case of plants, research data is limited. However, a recent study by Guo et al. (2025)has identified 40 ATP synthase family members in rubber tree in which cis-elements, phylogenetic relationship and expression patterns are analyzed and evaluated. Ample cis parameters, including stress resistance, phytohormones and light, were present in the promulgators of ATP synthase genes in rubber trees with maximum percentage of light signal (Guo et al. 2025).

2.5.5 Amino Acid Synthesis

Amino acids are fundamental building blocks of enzymes with assorts of one or more polypeptide chains forming protein molecules. A recent review by Heidarzadeh (2025) has elucidated the role of individual amino acids in affecting biochemical and physiological processes in plants. The molecular pathways,including aspartate and shikimate and interaction mechanism of amino acids with hormones like indole-3-acetic acid, salicylic acid and jasmonic acid, strongly affect the plants activation (Heidarzadeh 2025). Nutrient assimilation, plant defense mechanism activation, stress perception, response and developmental regulation are some of the parameters affected under biotic and abiotic stresses (Sowmya et al. 2023; Heidarzadeh 2025). Amino acids serve as precursors for primary and secondary metabolites that play pivotal roles in nutraceutical compounds (Trovato et al. 2021). Nitrogen is an indispensable need for the biosynthesis of amino acids; therefore, extraction, storage and transport parameters of nitrogen are essential research assorts of amino acid biology (Neill & Lee 2020).

2.6 CONCLUSIONS

This chapter emphasizes on various biosynthesis techniques and their consorting parameters that assorts to form amenable natural products imperative for regulation, growth and development for plants. Different biosynthesis techniques, including natural and artificial photosynthesis, nucleic and amino acids and ATP are analyzed with specific domain and application. The significance of light and NPs for increasing the amenability and efficacy of natural products and for understanding articulate metabolic engineering mechanisms are highlighted in detail in the present work.

REFERENCES

Ahmad, M., Tahir, M., Hong, Z., Zia, M.A., Rafeeq, H., Ahmad, M.S., Rehma, S., Sun, J. (2025) Plant and marine-derived natural products: Sustainable pathways for future drug discovery and therapeutic development, Frontiers in Pharmacology 6(15), 1497668.

Alum, E.U., Manjula, V.S., Uti, D.E., Echegu, D.A., Ugwu, O.P.C., Egba, S.I., Agu, P.C. (2025) Metabolomics-driven standardization of herbal medicine: Advances, applications, and sustainability considerations, Natural Product Communications.

Atanasov, A.G., Zotchev, S.B., Dirsch, V.M., Supuran, T.C. (2021) Natural products in drug discovery: Advances and opportunities, Nature Reviews Drug Discovery 20, 200–216.

Bernonville, T.D., Papon, N., Clastre, M., Connor, S.E.O., Courdavault, V. (2020) Identifying missing biosynthesis enzymes of plant natural products, Trends in Pharmacological Sciences 41(3), 142–146.

Bhakshu, L.M., Pullaiah, T. (2025) Biosynthesis of Tropane Alkaloids, Tropane Alkaloids, 89–101. https://link.springer.com/chapter/10.1007/978-981-96-5805-3_4?fromPaywallRec=true

Birchfield, A.S., McIntosh, C.A. (2020) Metabolic engineering and synthetic biology of plant natural products – A minireview, Current Plant Biology 24, 100163.

Carruthers, D.N., Lee, T.S. (2022) Translating advances in microbial bioproduction to sustainable biotechnology, Frontiers in Bioengineering and Biotechnology 10. https://doi.org/10.3389/fbioe.2022.968437

Ceunen, S., Geuns, J.M.C. (2013) Steviol glycosides: Chemical diversity, metabolism, and function, Journal of Natural Products 76(6), 1201–1228.

Chaachouay, N., Zidane, L. (2024) Plant-derived natural products: A source for drug discovery and development, Drugs Drug Candidates 3(1),184–207.

Chatterjee,A., Puri, S., Sharma, P.K., Deepa, P.R., Chowdhury, S. (2023) Nature-inspired Enzyme engineering and sustainable catalysis: Biochemical clues from the world of plants and extremophiles, Frontiers in Bioengineering and Biotechnology 20 (11), 1229300. https://doi.org/10.3389/fbioe.2023.1229300

Chauhan, J., Prathibha, M.D., Singh, P., Choyal, P., Mishra, U.N., Saha, D., et al. (2023) Plant photosynthesis under abiotic stresses: Damages, adaptive, and signaling mechanisms, Plant Stress 10, 100296.

Chen, S. (2024) Biosynthesis of natural products from medicinal plants: Challenges, progress and prospects, Chinese Herbal Medicine 16(1), 1–2.

Cheng, L.T., Wang, Z.L., Zhu, Q.H., Ye, M., Ye, C.Y. (2025a) A long road ahead to reliable and complete medicinal plant genomes, Nature Communications 16, 2150.

Cheng, R. ,Yang, S., Wang, D., Qin, 3, Wang, F., Meng, S. (2025b) Advances in the biosynthesis of plant terpenoids: Models, mechanisms, and applications, Plants (Basel) 14(10), 1428.

Chopra, B., Dhingra, A.K. (2021) Natural products: A lead for drug discovery and development, Phytotherapy Research 35(9), 4660–4702.

Chu, D., Wang, H., Nie, Z., Li, K.L., Cao, J., Yang, M., Yin, Q., Gu, Y., Jiang, Y. (2025) Collective biosynthesis of plant spirooxindole alkaloids through enzyme discovery and engineering, Journal of the American Chemical Society 147(25), 21600–21609.

Conrado, G.G., Rosa, R., Reis, R.D., Pessa, L.R. (2024) Building natural product–based libraries for drug discovery: Challenges and opportunities from a Brazilian pharmaceutical industry perspective, Revista Brasileira de Farmacognosia, 34, 706–721.

Courdavault, V., O'Connor, S.E., Jensen, M.K., Papon, N. (2021) Metabolic engineering for plant natural products biosynthesis: New procedures, concrete achievements and remaining limits, Natural Product Reports 38, 2145–2153.

Dinglasan, J.L.N., Otani, H., Doering, D.T., Udwary, D., Mouncey, N.J. (2025) Microbial secondary metabolites: Advancements to accelerate discovery towards application, Nature Reviews Microbiology 23, 338–354.

Dong, H., Qi, H. (2025) Biosynthesis of triterpenoids in plants: Pathways, regulation, and biological functions, Current Opinion in Plant Biology 85, 102701.

Drosou, M., Pantazis, D.A. (2024) Comprehensive evaluation of models for ammonia binding to the oxygen evolving complex of photosystem II, The Journal of Physical Chemistry B128(6), 1333–1349.

Erb, M., Kliebenstein, D.J. (2020) Plant secondary metabolites as defenses, regulators, and primary metabolites: The blurred functional trichotomy, Plant Physiologists 84, 39–52.

Gámez, J.A.U., Azaz, J.E., Ávila, T., Cánovas, F.N. (2020) Enzymes involved in the biosynthesis of arginine from ornithine in maritime pine(*Pinus pinaster* Ait.), Plants 20209(10), 1271. https://doi.org/10.3390/plants9101271

Godbole, R.C., Pable, A.A., Singh, S., Barvkar, V.T. (2022) Interplay of transcription factors orchestrating the biosynthesis of plant alkaloids, 3 Biotech12 (250).https://link.springer.com/article/10.1007/s13205-022-03316-x?fromPaywallRec=true

González, M.C., Cejudo, F.J., Sahrawy, M., Serrato, A.J. (2023) Current knowledge on mechanisms preventing photosynthesis redox imbalance in plants, Antioxidants 10(11), 1789.https://doi.org/10.3390/antiox10111789

Guan, C., Zhang, D., Chu, C. (2025) Interplay of light and nitrogen for plant growth and development, The Crop Journal 13(3), 641–655.

Guo, B., Fan, S., Liu, M., Yang, H., Dai, L.,Wang, L. (2025) ATP synthase members of chloroplasts and mitochondria in rubber trees (*Hevea brasiliensis*) response to plant hormones, Plants 202514(4), 604.

Guo, L., Yao, H., Chen, W., Wang, X., Ye, P., Xu, Z., Zhang, S., Wu, H. (2022) Natural products of medicinal plants: Biosynthesis and bioengineering in post-genomic era, Horticulture Research 9, uhac223. https://doi.org/10.1093/hr/uhac223

Hashim, M., Ahmad, B., Drouet, S., Hano, C., Abbasi, B.H., Anjum, S. (2021) Comparative effects of different light sources on the production of key secondary metabolites in plants in vitro cultures, Plants 10(8), 1521. https://doi.org/10.3390/plants10081521

Hasnat, H., Shompa, S.A., Islam, M.M., Alam, S., Richi, F.T., Emon, N.U., et al. (2024) Flavonoids: A treasure house of prospective pharmacological potentials, Heliyon 10(6) e27533.

Hataya, T. (2021) An improved method for the extraction of nucleic acids from plant tissue without grinding to detect plant viruses and viroids, Plants 10(12), 2683.

Heidarzadeh, A. (2025) Role of amino acids in plant growth, development, and stress responses: A comprehensive review, Discover Plants 2, 237.

Heinrich, M., Mah, J., Amirkia, V. (2021) Alkaloids used as medicines: Structural phytochemistry meets biodiversity—An update and forward look, Molecules 26(7), 1836.

Ho, T.J., Goswami, D., Kuo, W.W., Kuo, C.H., Yen, S.C., Lin, P.Y., Lin, S.Z., Hsieh, D.J.Y., Shibu, M.A., Huang, C.Y. (2022) Artemisia argyi exhibits anti-aging effects through decreasing the senescence in aging stem cells, 14 (15), 6187–6201.

Holghoomi, R., Colagar, A.H. (2025) Nanoparticle-assisted synthesis of anti-aging secondary metabolites in medicinal plants, Plant Nano Biology 13, 100184.

Hu, Y., Zheng, T., Dong, J., Li, W., Ma, X., Li, J., et al. (2025) Regulation of the main terpenoids biosynthesis and accumulation in fruit trees, Horticultural Plant Journal 11(5), 1761–1776.

Islam, Z.F., Greening, C., Hu, H.W. (2023) Microbial hydrogen cycling in agricultural systems – Plant beneficial or detrimental? Microbial Biotechnology 16(8), 1623–1628.

Joshi, A., Verma, K.K., Rajput, V.D., Minkina, T., Arora, J. (2025) Recent advances in metabolic engineering of microorganisms for advancing lignocellulose-derived biofuels, Bioengineered 13(4), 8135–8163.

Kallio, P., Konert, G., Pyytövaara, S., Tikkanen, M. (2025) Molecular photosynthesis research facilitating technology development towards enhanced indoor Farming, Physiologia Plantarum 177(4), 70407. https://doi.org/10.1111/ppl.70407

Kumari, A., Sangal, R. (2024) Secondary metabolites in drug development: Tracing their historical and therapeutic impact, Emerging Trends in Metabolites 1(2), 51–55.

Leister, D. (2023) Enhancing the light reactions of photosynthesis: Strategies, controversies, and perspectives, Molecular Plant 16(1), 4–22.

Li, M.S., Pan, Y., Liu, B., Tan, C. (2025) Regulation of important natural products biosynthesis by WRKY transcription factors in plants, Journal of Advanced Research. https://doi.org/10.1016/j.jare.2025.01.009

Li, Z., Wang, Y., Xu, M., Liu, H.,Li, L., Xu, D. (2023) Molecular mechanism overview of metabolite biosynthesis in medicinal plants, Plant Physiology and Biochemistry 204, 108125.

Lichman, B.R. (2021) The scaffold-forming steps of plant alkaloid biosynthesis, Natural Product Reports 38, 103–129.

Liu, S., Zheng, Y., Pan, L., Wang, W., Li, Y., Liu, Z., Zhang, X. (2025) Nanodelivery of nucleic acids for plant genetic engineering, 20, 31.

Liu, Y., Singh, S.K., Pattanaik, S., Wang, H., Yuan, L. (2023) Light regulation of the biosynthesis of phenolics, terpenoids, and alkaloids in plants, Communications Biology 6, 1055.

Machín, A., Cotto, M., Ducongé, J., Márquez, F. (2023) Artificial photosynthesis: Current advancements and future prospects, Biomimetics 8(3), 298. https://doi.org/10.3390/biomimetics8030298

Mao, Y., Luo, J., Ca, Z. (2025) Biosynthesis and regulatory mechanisms of plant flavonoids: A review, Plants (Basel) 14(12), 1847.

Mathieu, S., Lesch, E., Garcia, S., Graindorge, S., Rüdinger, M.S., Hamman, K. (2025) De novo RNA base editing in plant organelles with engineered synthetic P-type PPR editing factors, Nucleic Acids Research 53(7), 279. https://doi.org/10.1093/nar/gkaf279

Mirsalami, S.M., Mirsalami, M. (2025) Advances in genetically engineered microorganisms: Transforming food production through precision fermentation and synthetic biology, Future Foods 11, 100601.

Moorhead, D.L., Rinkes, Z.L., Sinsabaugh, R.L., Weintraub, M.N., Weintraub (2013) Dynamic relationships between microbial biomass, respiration, inorganic nutrients and enzyme activities: Informing enzyme-based decomposition models, 4.https://doi.org/10.3389/fmicb.2013.00223

Najmi, A., Javed, S.A., Bratty, M.A., Alhazmi, H.A. (2022) Modern approaches in the discovery and development of plant-based natural products and their analogues as potential therapeutic agents, Molecules 27(2), 349. https://doi.org/10.3390/molecules27020349

Ndochinwa, O.G, Wang, Q.Y., Amadi, O.C., Nwagu, T.N., Nnamchi, C.I., Okeke, E.S., Moneke, A.N. (2024) Current status and emerging frontiers in enzyme engineering: An industrial perspective, Heliyon 10(11), e32673. https://doi.org/10.1016/j.heliyon.2024.e32673

Neill, K.C., Lee, Y.J. (2020) Visualizing genotypic and developmental differences of free amino acids in maize roots with mass spectrometry imaging, Frontiers in Plant Science Plant Metabolism and Chemodiversity 11. https://doi.org/10.3389/fpls.2020.00639

Nishiyama, Y.,Allakhverdiev, S.I.,Murata, N. (2006) A new paradigm for the action of reactive oxygen species in the photoinhibition of photosystem II. Biochimica et Biophysica Acta1757, 742–749.

Reddy, S., Sinha, A., Osborne, W.J. (2021) Chapter 1 – Microbial secondary metabolites: Recent developments and technological challenges, Volatiles and Metabolites of Microbes, 1–22. https://doi.org/10.1016/B978-0-12-824523-1.00007-9

Ricci, F., FrancescoGreening, C. (2024) Chemosynthesis: A neglected foundation of marine ecology and biogeochemistry, Trends Microbiology 32 (7), 631–639.

Ridhi, R. (2025) Nanosensors for wastewater treatment and disinfection, In Nanobiosensors for Agricultural and Other Related Sectors (pp. 307–330), https://doi.org/10.1007/978-981-96-6499-3_14.

Ridhi, R., Saini, G.S.S., Tripathi, S.K. (2023) Study of sensing mechanism of heterocyclic hazardous vapors with metal phthalocyanines, Dyes and Pigments 216, 111328.

Ridhi, R., Saini, G.S.S., Tripathi, S.K. (2024a) Nanotechnology as a sustainable solution for proliferating agriculture sector, Materials Science and Engineering: B304, 117383.

Ridhi, R., Saini, G.S.S., Tripathi, S.K. (2024b) Effect of central metal ions on interaction of metal Phthalocyanines composites with chemical analytes, Applied Physics A130, 359.

Sanjay, K.U., Vinay, C.M., Prabhu, N.B., Rai, P.S. (2025) Emerging trends in nucleic acid and peptide aptamers in plant science research, Planta 261(3), 63. https://doi.org/10.1007/s00425-025-04637-w

Seelert, H., Denche, N.A. (2011) ATP synthase superassemblies in animals and plants: Two or more are better, Biochimica et Biophysica Acta (BBA) – Bioenergetics 1807(9),1185–1197.

Selwal, N., Rahayu, F., Herwati, A., Latifah, E., SupriyonSuhara, C., Suastika, I.B.K., Mahayu, W.M., Wani, A.K. (2023) Enhancing secondary metabolite production in plants: Exploring traditional and modern strategies, Journal of Agriculture and Food Research 14, 100702.

Sharma, S., Bhatt, U., Shah, G., Soni, V. (2024) Assessment of waterlogging-induced changes in enzymatic antioxidants and carbohydrate metabolism in peanuts genotypes, Biochemistry and Biophysics Reports 39, 101794.

Singh, D., Singh, N., Dwivedi, S., Trivedi, P.K. (2025) Transcriptional regulation of secondary plant product biosynthesis: Insights into flavonoid, alkaloid, and terpenoid pathways, Plant Cell, Tissue and Organ Culture (PCTOC)160. https://doi.org/10.1007/s11240-024-02925-z

Smith, C. (2012) Chemosynthesis in the deep-sea: Life without the sun, Biogeosciences Discuss 9, 17037–17052. https://doi.org/10.5194/bgd-9-17037-2012

Sowmya,R.S., Warke, V.G., Mahajan, G.B., Annapure, U.S. (2023) Effect of amino acids on growth, elemental content, functional groups, and essential oils composition on hydroponically cultivated coriander under different conditions, Industrial Crops and Products 197, 116577.

Sreenivasulu, N., Fernie, A.R. (2022) Diversity: Current and prospective secondary metabolites for nutrition and medicine, Current Opinion in Biotechnology 74, 164–170.

Stephens, Z., Wilson, L.F.L., Zimmer, J. (2023) Diverse mechanisms of polysaccharide biosynthesis, assembly and secretion across kingdoms, Current Opinion in Structural Biology 79, 102564.

Sun, W., Shahrajabian, M.H. (2023) Therapeutic potential of phenolic compounds in medicinal plants—Natural health products for human health, Molecules 28(4), 1845.

Szwarc, S., Pogam, P.L., Beniddir, M.A. (2024) Emerging trends in plant natural products biosynthesis: A chemical perspective, Current Opinion in Plant Biology 82, 102649.

Theodoridis, S., Drakou, E.G., Hickler, T., Thines, M., Bravo, D.N. (2023) Evaluating natural medicinal resources and their exposure to global change, The Lancet Planetary Health 7(2),155–163.

Tripathi, A., Chauhan, N., Mukhopadhyay, P. (2024) Recent advances in understanding the regulation of plant secondary metabolite biosynthesis by ethylene-mediated pathways, Physiology and Molecular Biology of Plants 30, 543–557.

Trovato, M., Funck, D., Forlani, G., Okumoto, S., Amir, R. (2021) Editorial: Amino acids in plants: Regulation and functions in development and stress defense, Frontiers in Plant Science 12, 772810.

Villanueva, F.M., Nett, R.S. (2025) Emerging functions within the enzyme families of plant alkaloid biosynthesis, Phytochemistry Reviews 24, 2463–2484.

Wang, Q., Qi, Z., Fu, W., Pan, M., Ren, X., Zhang, X., Rao, Z. (2024) Research and prospects of enzymatic hydrolysis and microbial fermentation technologies in protein raw materials for aquatic feed, Fermentation 10(12), 648. https://doi.org/10.3390/fermentation10120648

Waratadar, A., Nirmalnath, P.J., Matiwade, P.S., Navi, V. (2022) Soil enzyme activity and nutrient uptake as influenced by arbuscular mycorrhizal fungi in tobacco rhizosphere under parasitic weed stress condition, International Journal of Plant & Soil Science 34(22), 984–995.

Wu, W., Chen, L., Liang, R., Huang, S., Li, X., Huang, B., Luo, H., Zhang, M., Wang, X., Zhu, H. (2025) The role of light in regulating plant growth, development and sugar metabolism: A review, Frontiers in Plant Science. https://doi.org/10.3389/fpls.2024.1507628. Corpus ID: 275364710.

Xiao, J., Zhou, Y., Xie, Y., Li, T., Su, X., He, J., et al. (2024) ATP homeostasis and signaling in plants, Plant Communications 5(4), 100834.

Yang, J.H., Williams, D., Kandiah, E., Fromme, P., Chiu, P.L. (2020) Structural basis of redox modulation on chloroplast ATP synthase, Communications Biology 3, 482.

Yang, W., Lin, R. (2025) Advances in photosynthesis research: Unlocking the potential for food security, renewable energy, and environmental sustainability, Journal of Integrative Plant Biology67(4), 879–881.

Yin, Q., Yang, M. (2024) Chemoproteomics approach to elucidating biosynthetic pathway of plant natural products, Frontiers in Plant Science Plant Metabolism and Chemodiversity 15.

Yong, J., Wu, M., Carroll, B.J., Xu, Z.P., Zhang, R. (2024) Enhancing plant biotechnology by nanoparticle delivery of nucleic acids, Trends in Genetics 40(4),352–363.

Zabotina, O.A., Zhang, N., Weerts, R. (2021) Polysaccharide biosynthesis: Glycosyltransferases and their complexes, Frontiers in Plant Science 12, 625307. https://doi.org/10.3389/fpls.2021.625307

Zhang, W., Xu, F., Kwon, M. (2023a) Editorial: Biosynthesis and regulation of plant specialized metabolisms, Frontiers in Plant Science 14, 1319666. https://doi.org/10.3389/fpls.2023.1319666

Zhang, Y., Chen, Y., Guo, Y., Ma, Y., Yang, M., Fu, R.,et al. (2023b) Proteomics revealed the effects of nucleic acid metabolism, transcription and translation on postharvest broccoli senescence under elevated O_2 stress, Postharvest Biology and Technology 195, 112110.

Zhao, C., Rehman, F.U., Shaikh, S., Qazi, R.M., Sajid, Z., Mian, A.A., He, N. (2023) Metallic nanoscale-knife application in cancer theranostics, Smart Materials in Medicine 4, 313–336.

Zheng, C.C., Gao, L., Sun, H., Zhao, X.Y., Gao, Z., Liu, J., Guo, W. (2024) Advancements in enzymatic reaction-mediated microbial transformation, Heliyon10 (19), 38187.

3 Enzyme Function in Photosynthesis
Key Enzymes of Carbon Fixation and Energy Production

Komal Pandey, Baljeet Singh Saharan, and Riyaz Sayyed

3.1 INTRODUCTION

Photosynthesis is a crucial biochemical process in autotrophic organisms, such as plants, algae, and certain bacteria, enabling the conversion of solar energy into chemical energy and thereby sustaining life on Earth. This complex process consists of two sequential stages: the light-dependent reactions and the light-independent reactions, also known as the Calvin–Benson cycle. The light-dependent reactions occur in the thylakoid membranes of the chloroplast, where pigment–protein complexes, primarily chlorophyll, absorb photons to drive electron transport. Through the splitting of water molecules, molecular oxygen is released, and NADPH and ATP, the crucial energy carriers required for carbon fixation, are produced.

In most plants, including major crops such as rice, wheat, and potatoes, carbon fixation occurs primarily through the Calvin–Benson cycle, a central component of the C3 photosynthetic pathway (Kubis & Bar-Even, 2019). Within this cycle, the enzyme RuBisCO catalyzes the incorporation of atmospheric CO_2 into ribulose-1,5-bisphosphate (RuBP), a five-carbon sugar phosphate. Two molecules of 3-phosphoglycerate (3-PGA) are then further processed with ATP and NADPH to produce glyceraldehyde-3-phosphate (G3P), which serves as a precursor for glucose biosynthesis. Although the C3 pathway is metabolically efficient, RuBisCO's ability to act as both a carboxylase and oxygenase can lead to photorespiration. This issue is pronounced in hot or dry environments, where high oxygen levels and limited CO_2 reduce photosynthetic efficiency by consuming energy and previously fixed carbon; Jaramillo et al., 2025; Yang & Lin, 2025).

To compensate for photorespiratory losses, certain plants have developed the C4 photosynthetic pathway, which is highly efficient under hot, intense light and water-limiting conditions. C4 photosynthesis features a unique leaf structure known as Kranz anatomy, which spatially separates initial CO_2 fixation from the Calvin cycle. Here, phosphoenolpyruvate carboxylase (PEPCase) first fixes CO_2 in mesophyll cells into oxaloacetate, a four-carbon compound, which is then transferred to bundle sheath cells. In these cells, concentrated CO_2 is released, allowing RuBisCO to function optimally with minimal photorespiration. Consequently, C4 plants, such as maize and sugarcane, exhibit enhanced water and nitrogen use efficiency, which contributes to their prevalence in tropical and disturbed habitats (Berry et al., 2016; Wang et al., 2017; Swift et al., 2024).

Crassulacean acid metabolism (CAM) represents a temporal adaptation to water scarcity by separating carbon fixation and the Calvin cycle over 24 hours. CAM plants open their stomata at night, fixing CO_2 into organic acids such as malate, which are stored in vacuoles. During the daytime, when stomata are closed to conserve water, stored carbon dioxide is utilized for photosynthesis. This nocturnal pathway enables CAM species, including many succulents and epiphytes, to thrive in extreme arid and semi-arid environments, thereby maximizing water use efficiency. Some species can switch between C3 and CAM pathways depending on environmental stress, a strategy known as facultative CAM. Modeling studies indicate that CAM plants can substantially reduce water use, emphasizing their potential in improving crop resilience under climate change (Gonçalves et al., 2020; Tan & Chen, 2023).

Photosynthetic strategies demonstrate a clear evolutionary trajectory shaped by environmental challenges. The earliest photosynthetic organisms likely relied on simple pigment-based systems, which evolved into the complex electron transport mechanisms observed today. The interaction of metabolic pathways, anatomical innovations, and regulatory systems, including retrograde signaling and epigenetic control, enables plants to adapt and optimize photosynthesis(Kodama et al., 2023). This thorough understanding drives current synthetic biology efforts to transfer C4 and CAM traits into C3 crops, aiming to improve productivity and resource efficiency in future agricultural systems (Kallio et al., 2025; Yang & Lin, 2025).

3.2 OVERVIEW OF PHOTOSYNTHETIC PATHWAYS

Photosynthetic pathways are the primary metabolic strategies by which autotrophic organisms harness solar energy and convert it into chemical energy, which is essential for life on Earth. These pathways rely on a diverse array of biochemical reactions and structural adaptations. The classic C3 pathway, common in temperate regions, and the specialized C4 and CAM pathways found in plants adapted to heat and water stress, illustrate how photosynthesis has evolved to optimize productivity under different environmental conditions. The overall process of photosynthesis is

DOI: 10.1201/9781003669920-3

divided into two distinct phases: light-dependent and light-independent reactions. In the thylakoid membranes of chloroplasts, pigment–protein complexes such as chlorophyll absorb photons, initiating a series of events in Photosystem II that lead to water splitting and the release of protons, high-energy electrons, and molecular oxygen. The flow of electrons along the electron transport chain establishes a proton gradient, driving ATP synthesis, while electrons ultimately reduce NADP+ to form NADPH. These compounds, ATP and NADPH, are then utilized to drive the assimilation of atmospheric carbon dioxide into organic molecules through the Calvin–Benson cycle, as illustrated in Figure 3.1 (Hartzell et al., 2018; Ermakova et al., 2021).

The C3 pathway, also known as the Calvin–Benson cycle, is the most widely distributed and evolutionarily ancient pathway for carbon assimilation. In this cycle, RuBisCO initially captures CO_2 and incorporates it into the five-carbon compound ribulose-1,5-bisphosphate (RuBP), generating two molecules of the three-carbon compound 3-phosphoglycerate (3-PGA). However, under hot or dry conditions, RuBisCO's oxygenase activity triggers photorespiration, which can waste fixed carbon and energy, thus reducing photosynthetic efficiency. C4 photosynthesis represents an evolutionary solution to minimize photorespiratory losses, especially in environments with intense light, high temperatures, and limited water. Here, the fixation of CO_2 and the subsequent reactions of the Calvin cycle are spatially separated. In C4 plants such as maize and sugarcane, CO_2 is initially fixed in mesophyll cells by phosphoenolpyruvate carboxylase (PEPCase) to form oxaloacetate, a four-carbon compound. This is then transported to bundle sheath cells, where CO_2 is released in high concentration for fixation by RuBisCO, minimizing photorespiration. This architecture significantly enhances water and nitrogen efficiency, thereby supporting the ecological dominance of C4 species in tropical grasslands. CAM photosynthesis is an adaptation to arid environments, separating carbon fixation and assimilation temporally rather than spatially. CAM plants, including many succulents and epiphytes, open their stomata at night to fix CO_2 into organic acids, mainly malate, which are stored in vacuoles. During the day, stomata close to conserve water, and stored CO_2 is released for use in the Calvin cycle. Some CAM plants can even switch between C3 and CAM modes, reflecting substantial metabolic flexibility in response to environmental changes (Hartzell et al., 2018; Gonçalves et al., 2020).

On a mechanistic level, each photosynthetic pathway is distinguished by its unique anatomical or temporal organization as well as its biochemical steps. In C3 plants, all reactions occur within mesophyll cells. C4 plants utilize

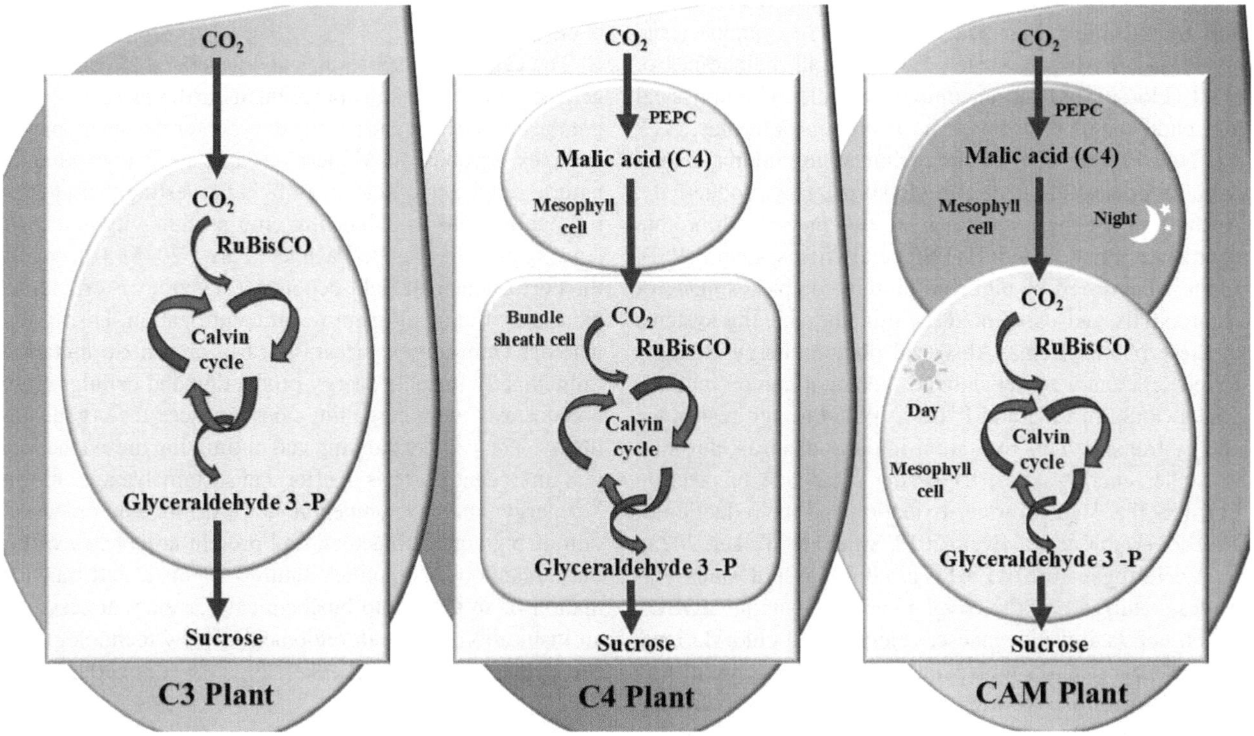

FIGURE 3.1 This figure compares the core strategies of carbon assimilation in C3, C4, and CAM plants. In C3 plants, CO_2 is directly fixed by RuBisCO into the Calvin cycle within mesophyll cells, producing glyceraldehyde-3-phosphate, which is then converted into sucrose. C4 plants first fix CO_2 into malic acid via phosphoenolpyruvate carboxylase (PEPC) in mesophyll cells; This is then shuttled to bundle sheath cells, where high CO_2 concentrations enhance RuBisCO efficiency and Calvin cycle activity. CAM plants temporally separate carbon fixation: At night, CO_2 is fixed into malic acid, stored in vacuoles, and decarboxylated during the day for entry into the Calvin cycle. These mechanisms enable C4 and CAM plants to efficiently minimize photorespiration and conserve water, thereby optimizing photosynthesis in diverse and challenging environments.

"Kranz anatomy," partitioning reactions between mesophyll and bundle sheath cells, while CAM plants separate key responses by the time of day, all within the same cell. These distinctions translate into varied ecological strategies and physiological capacities. Ongoing research continues to uncover the genetic and regulatory networks responsible for this diversity. Newly discovered metabolic pathways, such as C1 metabolism and the links between serine biosynthesis and methionine formation, highlight the broader role of photosynthesis as a metabolic integrator beyond conventional carbohydrate production. The ecological impact of these pathways is substantial. C3 plants dominate cooler, moist climates and are responsible for the majority of food crops. C4 plants contribute significantly to food, forage, and fuel production in warm and disturbed habitats, whereas CAM species maintain productivity and stabilize carbon cycling in the most extreme environments. Shifting global conditions are reshaping the balance among these pathways, influencing plant distribution and ecosystem function (von Caemmerer & Furbank, 2016; Ahmad et al., 2020; Kapoor, 2023).

3.2.1 Light-Dependent Reactions

Light-dependent reactions are a fundamental component of oxygenic photosynthesis, representing a remarkable evolutionary innovation that enables the conversion of solar energy into chemical energy, fuelling nearly all terrestrial and aquatic life. These reactions occur in cyanobacteria, algae, and the highly structured thylakoid membranes of plant chloroplasts, and encompass complex photophysical and photochemical events. The resulting molecules, ATP and NADPH, are vital for the carbon assimilation and anabolic processes that underpin global primary productivity. At the heart of these reactions are two major multiprotein complexes: Photosystem II (PSII) and Photosystem I (PSI). Light is harvested by pigment–protein complexes made of chlorophylls and carotenoids within the antenna systems of these photosystems. Absorbed photon energy is transferred efficiently to specialized "reaction centre" chlorophylls (P680 in PSII and P700 in PSI) through resonance energy transfer. This excitation initiates electron elevation to higher energy states, triggering a cascade of electron transfers that leads to water oxidation and the reduction of NADP+ (Kubis & Bar-Even, 2019; Strand & Walker, 2023).

A defining feature of PSII is its ability to split water molecules, catalyzed by the oxygen-evolving complex (OEC), which consists of manganese, calcium, and chloride ions. This process releases molecular oxygen as a byproduct and channels electrons through a series of carriers, including plastoquinone, the cytochrome b6f complex, and plastocyanin, ultimately passing them to PSI. As electrons traverse this pathway, concurrent proton translocation across the thylakoid membrane creates a proton motive force, which powers ATP synthase to generate ATP from ADP and inorganic phosphate, a process known as photophosphorylation. Upon reaching PSI, electrons receive a second energy boost from absorbed light and are eventually transferred by ferredoxin-NADP+ reductase (FNR) to reduce NADP+ into NADPH. This linear electron flow produces ATP and NADPH in ratios that meet the demands of the Calvin–Benson cycle. Additionally, cyclic electron flow around PSI diverts electrons back to the plastoquinone pool, increasing ATP output without forming NADPH or releasing oxygen. This flexibility allows the cell to fine-tune its ATP-to-NADPH production ratio, meeting variable metabolic needs. (Gonçalves et al., 2020; Oliveira-Filho et al., 2023).

Regulation of the light-dependent reactions occurs at multiple levels, enabling plants to adapt rapidly to fluctuating light environments. Adjustments in antenna complex configurations, state transitions, and post-translational modifications optimize energy distribution and minimize the generation of reactive oxygen species (ROS). In C4 plants, photosystem stoichiometry and light-harvesting modifications adapt to support distinct photosynthetic processes in mesophyll and bundle sheath chloroplasts (Cejudo et al., 2019).

Advances in molecular and structural biology, such as high-resolution cryo-electron microscopy, have revealed the intricacies of pigment–protein architecture and electron transfer pathways. The roles of auxiliary proteins, including psbS and PGR5, in protective energy dissipation and redox balancing are increasingly understood, as is the importance of alternative electron sinks, like the Mehler process and plastid terminal oxidase, in safeguarding against over-reduction and oxidative stress (Ma et al., 2021; Mlinarić et al., 2021).

The capacity to split water and generate molecular oxygen marked a pivotal turning point in Earth's history, reshaping the atmosphere and paving the way for the emergence of complex, aerobic life. Modern research seeks to expand the natural range and efficiency of light harvesting, with potential applications in enhancing crop productivity and resilience to climate change(Page and Parry, 2025). Ultimately, the performance of light-dependent reactions is crucial for plant growth and efficient resource utilization. Disruption, whether from abiotic stress, disease, or genetic mutation, immediately impacts energy production and cellular redox homeostasis, with cascading consequences for organismal fitness. Thus, understanding and optimizing these reactions remains a central focus of efforts in sustainable agriculture, bioenergy, and environmental stewardship. The orchestration of pigments, cofactors, and protein complexes within these reactions exemplifies nature's sophisticated transformation of sunlight into biochemical currency. It serves as an inspiration for the development of new technologies in energy and agriculture (Leister, 2023; Li et al., 2025).

3.2.2 Calvin Cycle (Light-Independent Reactions)

The Calvin cycle, also known as the reductive pentose phosphate cycle or light-independent reactions, is the primary process by which photoautotrophic organisms fix atmospheric carbon dioxide and drive primary productivity in both land and aquatic environments. Although traditionally called the "dark reactions," it depends on the ATP and

NADPH generated by light-dependent reactions to power the conversion of inorganic carbon into organic compounds. This complex process takes place in the chloroplast stroma and plays a crucial role in global carbon cycling, energy flow, and biosynthesis. At the molecular level, RuBisCO, considered the most abundant enzyme on Earth, catalyzes the carboxylation of ribulose-1,5-bisphosphate (RuBP), a five-carbon sugar, using CO_2 to produce two molecules of 3-phosphoglycerate (3-PGA), a three-carbon compound. This carbon fixation step repeats continuously, enabling massive global CO_2 uptake. Despite RuBisCO's catalytic efficiency, its oxygenase activity leads to photorespiration, which consumes energy and fixed carbon, thereby reducing overall efficiency. However, this trade-off is offset by its broad evolutionary conservation (Cejudo et al., 2019).

Following carboxylation, 3-PGA is phosphorylated by ATP and reduced by NADPH to form glyceraldehyde-3-phosphate (G3P), a key metabolic intermediate. G3P acts as a hub for biosynthesis, providing precursors for carbohydrates such as starch and sucrose, as well as for amino acids, lipids, and nucleic acids. The cycle also regenerates RuBP through a series of enzymatic rearrangements involving aldolase, transketolase, ribose-5-phosphate isomerase, and phosphoribulokinase, ensuring the cycle's continuous function. The Calvin cycle is finely tuned by redox signals, ATP and NADPH availability, and environmental factors including light intensity and CO_2 levels. Enzymes are regulated via thioredoxin-mediated reduction and other mechanisms to synchronize the cycle with photosynthetic activity. In C4 and CAM plants, biochemical and anatomical strategies enhance local CO_2 concentration at the site of RuBisCO, thereby reducing photorespiration and improving carbon fixation efficiency. In contrast, light availability, photorespiratory rate, and ambient CO_2 levels have a more direct influence on Calvin cycle efficiency in C3 plants, as shown in Figure 3.2 (Chollet et al., 1996).

Recent advances in metabolic engineering and genomics have revealed layers of regulation, such as post-translational enzyme modifications, inter-organelle metabolic exchanges, and retrograde chloroplast-to-nucleus signaling, that stabilize cycle performance under stress. Efforts to optimize Calvin cycle function through synthetic biology, such as modifying RuBisCO's catalytic traits or overexpressing key enzymes like phosphoribulokinase, aim to enhance carbon assimilation and crop yield, particularly under adverse or changing climatic conditions. Globally, the Calvin cycle mediates a substantial flux of inorganic carbon into the biosphere, underpinning food webs, biomass production, and the regulation of atmospheric CO_2. Its function bridges solar energy and the organic molecules essential for life,

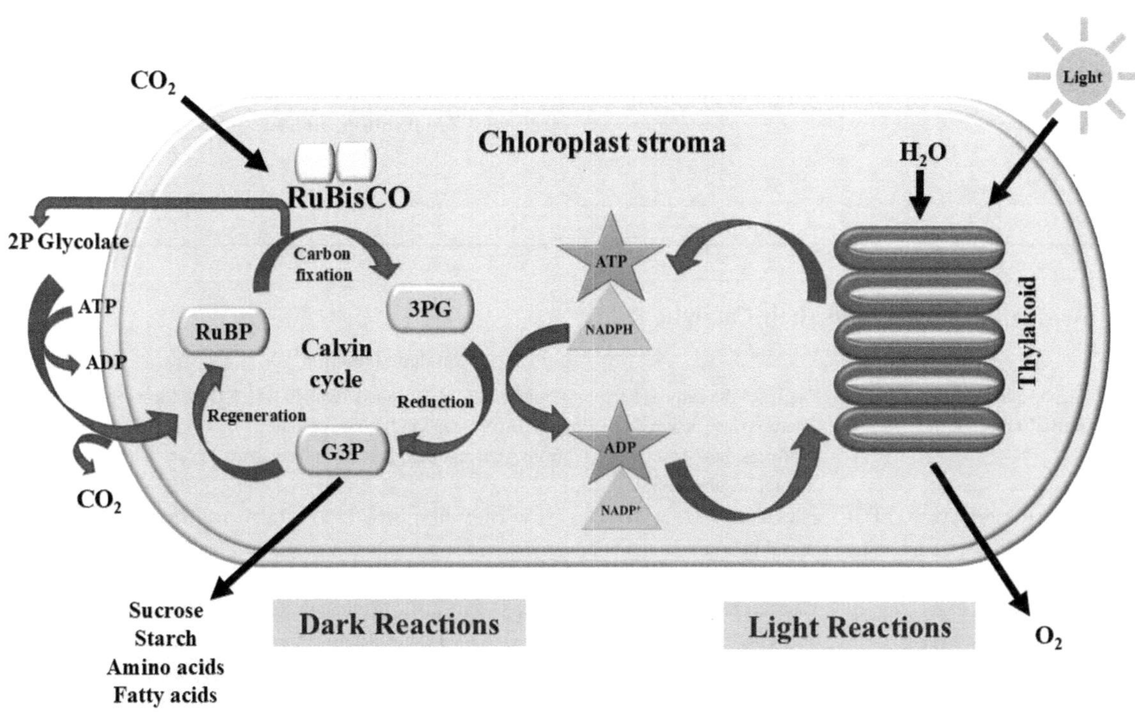

FIGURE 3.2 This figure illustrates the dual processes of photosynthesis within chloroplasts, emphasizing the interplay between light-dependent and light-independent (Calvin cycle) reactions. On the right, light energy is captured by thylakoid membranes, driving the photolysis of water and generating ATP and NADPH. These energy carriers are then utilized in the stroma, where the Calvin cycle is active. Here, the enzyme RuBisCO catalyzes the fixation of atmospheric CO_2 into RuBP, which subsequently forms 3-phosphoglycerate (3PG) and then glyceraldehyde-3-phosphate (G3P). This process culminates in the synthesis of key organic compounds, such as sucrose, starch, amino acids, and fatty acids. The figure also highlights photorespiration, which can redirect resources and impact overall efficiency under stress. Through these linked pathways, plants transform solar energy into the molecular currency that fuels growth and sustains ecosystems.

and alterations in its efficiency or environmental context have profound implications for ecosystem health, agriculture, and climate stability (Weissmann & Brutnell, 2012).

3.3 MAJOR ENZYMES INVOLVED IN PHOTOSYNTHESIS

Photosynthesis, a vital biochemical process that sustains life on Earth, depends on a suite of specialized enzymes that drive key reactions during light absorption, electron transport, carbon fixation, and metabolic integration. These enzymes play a central role in maintaining the global carbon cycle, supporting agriculture, and ensuring ecosystem stability by converting solar energy into chemical energy within chloroplasts and related organelles. Some enzymes play especially crucial roles due to their specific functions and contributions. Ribulose-1,5-bisphosphate carboxylase/oxygenase (RuBisCO) is the most abundant and well-studied enzyme in photosynthesis, as shown in Table 3.1. It catalyzes the crucial first step of the Calvin cycle by binding atmospheric CO_2 to ribulose-1,5-bisphosphate (RuBP), producing two molecules of 3-phosphoglycerate (3-PGA). Despite being essential, RuBisCO has a slow catalytic rate and can also react with oxygen, leading to photorespiration, which lowers photosynthetic efficiency. Thus, considerable research seeks to improve its selectivity and turnover through genetic and biochemical means. RuBisCO activity is dynamically regulated in response to environmental stimuli, adjusting photosynthetic capacity via changes in its activation state (Uchida et al., 2018; Yu et al., 2020).

ATP synthase is another key enzyme embedded in the thylakoid membrane, responsible for synthesizing ATP using the proton motive force generated during electron transport. It drives the production of ATP from ADP and inorganic phosphate through a rotary mechanism powered by the flow of protons. ATP produced fuels the Calvin cycle and other biosynthetic pathways, maintaining energy balance and maximizing carbon fixation (Oliveira-Filho et al., 2023; Mao et al., 2024).

Phosphoenolpyruvate carboxylase (PEPC) is critical in C4 and CAM plants, where it fixes bicarbonate (HCO_3^-) to phosphoenolpyruvate (PEP), forming oxaloacetate. This reaction efficiently concentrates CO_2 near RuBisCO, suppressing its oxygenase activity and reducing photorespiration. PEPC's activity is finely regulated by gene expression, post-translational modifications, and feedback from metabolites, such as malate. Although Photosystem II (PSII) and Photosystem I (PSI) are primarily pigment–protein complexes, they contain enzymatic components vital for photochemical reactions. PSII houses the OEC, a manganese-calcium cluster that catalyzes water splitting, generating oxygen, protons, and electrons. FNR catalyzes the final electron transfer to NADP+, thereby producing NADPH during the light reactions. Electron carriers, such as plastocyanin and cytochrome b6f, shuttle electrons efficiently, ensuring balanced electron flow and maintaining photosynthetic stability under variable environmental conditions (Weissmann & Brutnell, 2012).

Enzymes involved in RuBP regeneration, vital for continuous CO_2 fixation, include phosphoribulokinase (PRK),

TABLE 3.1
Key Photosynthetic Enzymes and Their Catalytic Roles

Enzyme	Catalytic Function
Ribulose-1,5-bisphosphate carboxylase/oxygenase (RuBisCO)	Catalyzes the carboxylation of ribulose-1,5-bisphosphate (RuBP) with CO_2 to form 3-phosphoglycerate (carboxylase activity); also catalyzes oxygenation (photorespiration)
ATP synthase	Catalyzes the synthesis of ATP from ADP and inorganic phosphate using the proton motive force during photophosphorylation
Phosphoenolpyruvate carboxylase (PEPC)	Fixes bicarbonate ion (HCO_a^-) to phosphoenolpyruvate (PEP) to form oxaloacetate, mainly in C4 and CAM plants
Photosystem II (PSII)	Catalyzes water splitting (photolysis) to release oxygen, protons, and electrons, initiating electron transport
Photosystem I (PSI)	Transfers electrons from plastocyanin to ferredoxin, facilitating NADP+ reduction to NADPH
Ferredoxin-NADP+ reductase (FNR)	Catalyzes the reduction of NADP+ to NADPH using electrons from ferredoxin
Phosphoribulokinase (PRK)	Catalyzes ATP-dependent phosphorylation of ribulose-5-phosphate to regenerate RuBP in the Calvin cycle
Sedoheptulose-1,7-bisphosphatase (SBPase)	Catalyzes dephosphorylation of sedoheptulose-1,7-bisphosphate to sedoheptulose-7-phosphate in the Calvin cycle regeneration phase
Fructose-1,6-bisphosphatase (FBPase)	Catalyzes dephosphorylation of fructose-1,6-bisphosphate to fructose-6-phosphate in the Calvin cycle
Transketolase	Catalyzes the transfer of two-carbon units between sugar phosphates during the regeneration of RuBP
Aldolase	Catalyzes aldol condensation and cleavage reactions in sugars during the Calvin cycle regeneration
NADP-malic enzyme (NADP-ME)	Catalyzes the decarboxylation of malate to pyruvate and CO_2 in C4 photosynthesis
Phosphoenolpyruvate carboxykinase (PCK)	Decarboxylates oxaloacetate to PEP and CO_2, part of the CO_2 concentration mechanism in some C4 plants

Sources: Chollet et al., 1996; Yin et al., 2008; Bar-Even, 2013; Locato et al., 2013; Weis & Kobilka, 2018; Chen et al., 2022; Taylor et al., 2022

fructose-1,6-bisphosphatase (FBPase), sedoheptulose-1,7-bisphosphatase (SBPase), transketolase, and aldolase. Their activities are tightly regulated through redox mechanisms, such as thioredoxin, to coordinate metabolic flux between the Calvin cycle and light-driven electron transport. Recent studies have highlighted additional enzymes, such as phosphoenolpyruvate carboxykinase (PCK) and NADP-malic enzyme (NADP-ME), that play crucial roles in C4 photosynthesis by catalyzing the release of CO_2 from organic acids, thereby facilitating efficient carbon concentration. Structural analyses reveal how subunit composition governs their enzymatic activity, reflecting a complex regulatory network that underpins carbon fixation across various plant types. PSII, a multiprotein complex in thylakoid membranes, initiates photosynthesis by catalyzing water oxidation and oxygen evolution. Its core proteins primarily include the D1 and D2 reaction center subunits, along with numerous antenna proteins that optimize light capture. PSII's dynamic assembly and repair enable plants to adapt to varying light intensities and minimize photodamage, ensuring sustained photosynthetic efficiency (Huang et al., 2016; Jiang et al., 2017; Itoga et al., 2021).

ATP synthase, a conserved enzyme complex found in mitochondria and chloroplasts, synthesizes ATP by harnessing proton gradients created during electron transport. It operates via a rotary mechanism linking proton translocation to catalytic conformational changes. Structural insights emphasize its role in cellular energy production and potential vulnerability to oxidative stress. FNR, a flavoenzyme, catalyzes the transfer of electrons from ferredoxin to NADP+, producing NADPH used in carbon fixation. FNR's structure supports reversible electron transfer involving flavin adenine dinucleotide (FAD), and emerging evidence links it to redox signaling and stress responses, offering targets for enhancing crop resilience. Glyceraldehyde-3-phosphate dehydrogenase (GAPDH) catalyzes the NAD+-dependent oxidation and phosphorylation of glyceraldehyde-3-phosphate, a crucial step in both glycolysis and the Calvin cycle. It exhibits its regulation via post-translational modifications and plays diverse roles beyond metabolism, including cellular signaling and stress response. Phosphoglycerate kinase (PGK) transfers phosphate groups during glucose metabolism, producing ATP and 3-phosphoglycerate. Its structural plasticity enables efficient catalysis, and it participates in various biological processes beyond classic energy metabolism. Together, these enzymes constitute a sophisticated biochemical network that drives photosynthesis, balancing energy conversion, carbon assimilation, and adaptive responses essential for plant survival and productivity (Ebanks & Chakrabarti, 2022; Miranda-Astudillo et al., 2022).

3.4 REGULATION OF PHOTOSYNTHETIC ENZYMES

The regulation of photosynthetic enzymes is a crucial aspect of photosynthesis, ensuring efficient energy conversion in plants while allowing them to adapt to environmental changes. This process involves multilayered control mechanisms at transcriptional, post-transcriptional, and post-translational levels to optimize photosynthetic efficiency, protect the photosynthetic machinery under stress, and balance metabolic demands. Enzyme complexes in the chloroplast thylakoid membranes, such as those in charge of light harvesting, electron transport, and carbon fixation through the Calvin–Benson cycle, are the primary source of energy for photosynthesis. To adapt photosynthetic rates to changing light intensities, nutrient availability, and other environmental conditions, these enzymes must be controlled. For example, nitrogen deficiency reduces photosynthetic activity as an adaptive response by rapidly downregulating genes encoding photosynthetic proteins and enzymes involved in light harvesting and the Calvin cycle. This transcriptional control lays the foundation for longer-term changes in photosynthetic ability. However, dynamic structural alterations and post-translational modifications in photosynthetic complexes are primarily responsible for short-term control. Reversible phosphorylation of light-harvesting complex proteins and enzymes is a key regulatory mechanism that prevents photodamage by modulating the flow of energy between photosystems. Kinases sensitive to the redox status of the plastoquinone pool, which serves as a sensor reflecting the equilibrium between photosystems I and II activity, orchestrate this type of phosphorylation (Hu et al., 2021). This regulation maintains an adequate distribution of light energy. It shields the photosynthetic machinery from excess energy that may otherwise produce ROS and induce oxidative stress, as described in Table 3.2 (Berry et al., 2016; Huang et al., 2016; Foyer, 2018; Simkin et al., 2019; Flexas & Perera-Castro, 2020; Storti et al., 2020; Cui, 2021; Liu et al., 2021).

Changes in the chloroplast microenvironment, such as pH and redox status, which fluctuate with light intensity and metabolic demands, tightly regulate the functioning of the Calvin cycle enzyme, especially Rubisco. An essential post-translational regulatory mechanism that connects electron transport to carbon fixation enzyme activity is thiol-based redox control, achieved through the reversible formation of disulfide bonds. To balance ATP and NADPH consumption with light-dependent electron flow, this redox regulation enables the quick modification of enzyme kinetics, thereby reducing wasteful reactions and ROS accumulation. Additionally, non-photochemical quenching (NPQ) methods are used by photosynthetic organisms. These mechanisms include photoinhibition repair, state transitions that move antenna proteins across photosystems, and energy dissipation as heat. Together, these systems allow plants to adapt to changing light levels by dynamically regulating enzyme and photosystem activity, preserving photosynthetic efficiency and reducing photodamage (Miranda-Astudillo et al., 2022; Emrich-Mills et al., 2025).

To balance the synthesis of protein subunits required for photosystem assembly, chloroplast sensor kinases and phosphorylation cascades control the expression and activity of photosynthetic enzymes at the molecular level. By ensuring that enzyme synthesis is in sync with complex formation

TABLE 3.2
Key Regulatory Factors Affecting Photosynthetic Enzyme Activities: Mechanisms and Impacts

Regulatory Factor	Mode of Regulation	Enzyme(s) Affected	Functional Outcome	Regulatory Mechanism(s)	Key References
Redox state	Controls enzyme activation/inactivation via the thioredoxin system	RuBisCO, FBPase, SBPase, PRK	Regulates carbon fixation and enzyme activity based on cellular redox balance	General chloroplast redox regulation	Cejudo et al. (2019)
Phosphorylation/dephosphorylation	Modifies enzyme activity and protein interactions	Light-harvesting complex, PSII proteins	Adjusts light energy distribution and photosystem efficiency	State transitions in photosystems	Tikkanen et al. (2008)
Transcription factors	Regulate gene expression of photosynthetic enzymes	Multiple photosynthetic genes	Coordinates enzyme levels in response to environmental stress	Stress-responsive gene regulation	Hu et al. (2023)
Electron transport regulation	Balances electron flow between PSII and PSI	Cyt b6f complex, plastocyanin	Maintains energy balance between ATP and NADPH production	Electron transport chain control	Oliveira-Filho et al. (2023)
Metabolite feedback	Allosteric inhibition or activation via metabolic intermediates	PEPC, RuBisCO	Adjusts activity based on metabolic demand and substrate availability	Feedback regulation by malate, ATP	Oliveira-Filho et al. (2023)
pH and ion concentrations	Influences enzyme structure and catalytic capacity	RuBisCO, PRK	Modulates enzyme efficiency, especially in the Calvin cycle	Stromal pH and Mg^{2+} effects	Zhang et al. (2022)
Light intensity and quality	Modulates energy absorption and protective responses	PSI, PSII, antenna proteins	Regulates photoprotection, photoinhibition, and energy flow	Light acclimation and stress response	Simkin et al. (2019)
Reactive oxygen species (ROS)	Induces protective enzyme systems and repair pathways	Antioxidant enzymes	Prevents oxidative damage to the photosynthetic apparatus	ROS management under high-light stress	Foyer (2018)

requirements, this assembly-dependent feedback keeps unassembled proteins from building up and endangering the photosynthetic apparatus. To prevent oxidative damage to photosynthetic enzymes and complexes, the ROS produced during photosynthesis function as signaling molecules that control antioxidant enzyme activity and repair processes, in addition to light and nutritional signals. Under extreme light stress, certain electron carriers, such as plastoquinone, have two functions: they transport electrons and protect Photosystem II against photoinhibition (Berry et al., 2016; Huang et al., 2016; Foyer, 2018; Storti et al., 2020).

3.5 GENETIC AND BIOTECHNOLOGICAL ADVANCES

Genetic and biotechnological advances are driving transformative changes in science, health, agriculture, and various industries, as evidenced by recent research and reviews. These breakthroughs build on our growing understanding of molecular biology and have led to the development of advanced techniques, including genetic engineering, genome editing, synthetic biology, and omics-based approaches (Yang et al., 2023).

3.5.1 Foundations of Genetic and Biotechnological Advances

Biotechnology broadly involves utilizing living systems or organisms to create or enhance products across various sectors, including environmental management, food production, and medicine. Precision genome modification techniques, including CRISPR/Cas9, TALENs, and zinc finger nucleases (ZFNs), have revolutionized genetic manipulation by enabling targeted gene editing with greater efficiency and fewer off-target effects. Omics technologies, including transcriptomics, proteomics, metabolomics, and genomics, have accelerated comprehensive biological investigations, transforming diagnosis, forensics, and trait development through tools such as whole-genome sequencing and gene expression profiling (Yang et al., 2023).

3.5.2 Applications in Medicine and Diagnostics

Biotechnological innovations have ushered in personalized medicine, revolutionizing disease treatment and prevention through gene therapy, recombinant vaccines, stem cell therapy, and monoclonal antibodies. Monoclonal antibodies play a crucial role in targeted cancer therapies, while polymerase chain reaction (PCR) and nucleic acid amplification tests (NAATs) facilitate the rapid and accurate detection of diseases. The Human Genome Project has laid the foundation for identifying disease-associated mutations and tailoring medical interventions. Biotechnology also advances pharmacogenomics and regenerative medicine, shaping the future of healthcare (Cui, 2021).

3.5.3 Impact on Agriculture and the Environment

Agriculture benefits significantly from genetically modified (GM) crops that exhibit enhanced yield, improved nutrition, increased pest resistance, and greater tolerance to environmental stress. Technologies such as transgenics, tissue culture, and marker-assisted selection facilitate the development of climate-resilient varieties, contributing to global food security. Genome editing and cloning also support biodiversity conservation by restoring genetic diversity that has been lost through habitat degradation. Synthetic biology enables the design of plants with reduced undesirable traits, such as bitterness or allergenicity. Environmental biotechnology utilizes microbial engineering and bioremediation to treat waste, control pollution, and restore ecosystems, including the development of sustainable microbial biosynthesis platforms (Kapoor, 2023; Yang et al., 2023).

3.5.4 Role of Enzyme Engineering in Crop Improvement

Enzyme engineering is emerging as a powerful strategy to enhance crop traits by modifying enzyme structure and function to improve catalytic efficiency, substrate specificity, stability, and plant compatibility. Techniques include rational design, which leverages detailed structural knowledge, and directed evolution, which iteratively selects improved enzyme variants with less prior information. Continuous directed evolution using systems like yeast has effectively optimized both native and foreign enzymes for use in plant systems. Enzyme engineering integrates into breeding via traditional transgenics and advanced genome editing, combined with guided evolution. Genome editing technologies, such as CRISPR, can introduce beneficial mutations into native enzymes without introducing foreign DNA, thereby enhancing traits like disease resistance and herbicide resistance, and improving regulatory acceptance. Screening innovations like FIND-IT enable rapid identification of valuable natural or induced mutations (Simkin et al., 2019; Ahmad et al., 2020; Liu et al., 2021).

Industrial and Societal Implications Genome editing tools, especially CRISPR-based systems, have revolutionized industrial biosynthesis by enabling efficient production of pharmaceuticals, biofuels, and specialty chemicals. Multi-gene platforms and biosensor-assisted CRISPR interference (CRISPRi) streamline strain engineering and manufacturing processes. Regulatory bodies, such as the FDA, EPA, and USDA, ensure the safe and ethical applications of these technologies. Rapid innovation foreshadows expanded use across agriculture, pharmaceuticals, and environmental sectors, fundamentally reshaping perspectives on sustainable development, food security, and health. Engineered enzymes have improved crop quality and resistance traits. For example, enzyme variants resistant to herbicide inhibition confer herbicide tolerance, while modifications that increase essential amino acids enhance nutritional quality. Other applications include enhancing grain malting quality, increasing dietary fiber, and modifying cell wall enzymes to extend fruit shelf life, improve texture, and facilitate harvesting. Enzyme engineering enables the rapid adaptation of crops to modern agricultural challenges, such as higher yield demands and climate change resilience (Mao et al., 2024; Pesaresi et al., 2025).

3.6 OUTLOOK AND FUTURE DIRECTION

Ongoing advances in enzyme engineering promise broader applications, faster crop improvement, and more substantial alignment with sustainable agriculture. Combining genome editing technologies with enzyme engineering can accelerate the development of crops with enhanced traits, reduced reliance on agrochemicals, and improved adaptability to environmental stresses. Future efforts will focus on identifying key enzyme targets that significantly impact yield, nutritional quality, and stress resilience. Integrating these cutting-edge tools into conventional breeding programs while addressing regulatory and societal concerns will be essential. As a result, enzyme engineering is playing an increasingly vital role in modern crop development, linking molecular innovations directly to practical agricultural benefits (Nowicka et al., 2018; Wang et al., 2025).

3.7 CONCLUSION

Nature's ability to adapt energy conversion mechanisms to environmental challenges is evident in the diversity of photosynthetic pathways. Through the C3, C4, and CAM pathways, plants employ unique biochemical and structural strategies to optimize carbon fixation, water use, and productivity across various climates. This metabolic flexibility not only sustains global primary productivity but also influences ecological distribution and evolutionary trends. Recent advances in environmental physiology, synthetic genetics, and molecular biology have significantly deepened our understanding of these complex systems, revealing opportunities to enhance agricultural efficiency and improve stress tolerance. Ultimately, gaining mastery over photosynthetic processes holds significant promise for advancing ecological restoration, renewable bioenergy, and sustainable agriculture, key goals for a changing world.

REFERENCES

Ahmad, N., Zaidi, S. S. A. and Mansoor, S. (2020). Alternative routes to improving photosynthesis in field crops. *Trends Plant Sci*, *25*, 958–960. https://doi.org/10.1016/j.tplants.2020.07.003

Bar-Even, A. (2013). Does acetogenesis really require especially low reduction potential? *Biochim Biophys Acta – Bioenerg*, *1827*, 395–400. https://doi.org/10.1016/j.bbabio.2012.10.007

Berry, J. O., Mure, C. M. and Yerramsetty, P. (2016). Regulation of rubiscogene expression in C4 plants. *Curr Opin Plant Biol*, *31*, 23–28. https://doi.org/10.1016/j.pbi.2016.03.004

Cejudo, F. J., Ojeda, V., Delgado-Requerey, V., González, M. and Pérez-Ruiz, J. M. (2019). Chloroplast redox regulatory mechanisms in plant adaptation to light and darkness. *Front Plant Sci*, *10*, 1–11. https://doi.org/10.3389/fpls.2019.00380

Chen, S., Peng, W., Ansah, E. O., Xiong, F. and Wu, Y. (2022). Encoded C4 homologue enzymes genes function under abiotic stresses in C3 plant. *Plant Signal. Behav*,*17*.

Chollet, R., Vidal, J. and O'Leary, M. H. (1996). Phosphoenolpyruvate carboxylase: A ubiquitous, highly regulated enzyme in plants. *Annu Rev Plant Physiol Plant Mol Biol*, *47*, 273–298. https://doi.org/10.1146/annurev.arplant.47.1.273

Cui, H. (2021). Challenges and approaches to crop improvement through C3-to-C4 engineering. *Front Plant Sci*, *12*. https://doi.org/10.3389/fpls.2021.715391

Ebanks, B. and Chakrabarti, L. (2022). Mitochondrial ATP synthase is a target of oxidative stress in neurodegenerative diseases. *Front Mol Biosci*, *9*, 1–7. https://doi.org/10.3389/fmolb.2022.854321

Emrich-Mills, T. Z., Proctor, M. S., Degen, G. E., Jackson, P. J., Richardson, K. H., Hawkings, F. R., Buchert, F. et al. (2025). Tethering ferredoxin-NADP+ reductase to photosystem I promotes photosynthetic cyclic electron transfer. *Plant Cell*, *37*, 1–18. https://doi.org/10.1093/plcell/koaf042

Ermakova, M., Arrivault, S., Giuliani, R., Danila, F., Alonso-Cantabrana, H., Vlad, D.et al. (2021). Installation of C4 photosynthetic pathway enzymes in rice using a single construct. *Plant Biotechnol J*, *19*, 575–588. https://doi.org/10.1111/pbi.13487

Flexas, J. and Perera-Castro, A. V. (2020). Faculty opinions recent advances in understanding and improving photosynthesis. *Fac Rev*, *9*, 5. https://doi.org/10.12703/b/9-5

Foyer, C. H. (2018). Reactive oxygen species, oxidative signaling and the regulation of photosynthesis. *Environ Exp Bot*, *154*, 134–142. https://doi.org/10.1016/j.envexpbot.2018.05.003

Gonçalves, A. Z., Latansio, S., Detmann, K. C., Marabesi, M. A., Neto, A. A. C., Aidar, M. P. M.et al. (2020). What does the RuBisCO activity tell us about a C3-CAM plant? *Plant Physiol Biochem*, *147*, 172–180. https://doi.org/10.1016/j.plaphy.2019.12.020

Hartzell, S., Bartlett, M. S. and Porporato, A. (2018). Unified representation of the C3, C4, and CAM photosynthetic pathways with the Photo3 model. *Ecol Modell*, *384*, 173–187. https://doi.org/10.1016/j.ecolmodel.2018.06.012

Hu, J., Wang, D., Chen, H. and Wang, Q. (2023). Advances in genetic engineering in improving photosynthesis and microalgal productivity. *Int J Mol Sci*, *24*. https://doi.org/10.3390/ijms24031898

Hu, Z., Li, J., Ding, S., Cheng, F., Li, X., Jiang, Y., Yu, J., et al.2021. The protein kinase CPK28 phosphorylates ascorbate peroxidase and enhances thermotolerance in tomato. *Plant Physiol*, *186*, 1302–1317. https://doi.org/10.1093/PLPHYS/KIAB120

Huang, C. F., Chang, Y. M., Lin, J. J., Yu, C. P., Lin, H. H., Liu, W. Y.et al. (2016). Insights into the regulation of C4 leaf development from comparative transcriptomic analysis. *Curr Opin Plant Biol*, *30*, 1–10. https://doi.org/10.1016/j.pbi.2015.12.011

Itoga, N. K., Martinez-Singh, K., Lee, J. T., John Harris, E., Baker, L. C. and Garcia-Toca, M. (2021). Analysis of medicare payments and patient outcomes with pre-operative imaging for carotidendarterectomy. *Ann Vasc Surg*, *76*, 179–184. https://doi.org/10.1016/j.avsg.2021.06.001

Jaramillo, A., Satta, A., Pinto, F., Faraloni, C., Zittelli, G. C., Silva Benavides, A. M., Torzillo, G.et al. (2025). Outlook on synthetic biology-driven hydrogen production: Lessons from algal photosynthesis applied to cyanobacteria. *Ener Fuels*, *39*(11), 4987–5006. https://doi.org/10.1021/acs.energyfuels.4c04772

Jiang, H. J., Underwood, T. C., Bell, J. G., Lei, J., Gonzales, J. C., Emge, L., et al. 2017. Mimicking lighting-induced electrochemistry on the early earth. *Proc Natl Acad Sci*, *120*, 2017. https://doi.org/10.1073/pnas

Kallio, P., Konert, G., Pyytövaara, S. and Tikkanen, M. (2025). Molecular photosynthesis research facilitating technology development towards enhanced indoor farming. *Physiol Planta*, *177*(4), e70407. https://doi.org/10.1111/ppl.70407

Kapoor, N. (2023). An overview on the benefits and different applications of genetic engineering in agriculture. *Adv Genet Eng Perspect*, *12*, 1–2.

Kodama, K., Xie, X. and Kyozuka, J.(2023). The D14 and KAI2 orthologs of gymnosperms sense strigolactones and KL mimics, respectively, and the signals are transduced to control downstream genes. *Plant Cell Physiol*, *64*, 1057–1065. https://doi.org/10.1093/pcp/pcad072

Kubis, A. and Bar-Even, A.(2019). Synthetic biology approaches for improving photosynthesis. *70*, 1425–1433. https://doi.org/10.1093/jxb/erz029

Leister, D. (2023). Enhancing the light reactions of photosynthesis: Strategies, controversies, and perspectives. *Mol Plant*, *16*, 4–22. https://doi.org/10.1016/j.molp.2022.08.005

Li, C., Du, X. and Liu, C. (2025). Enhancing crop yields to ensure food security by optimizing photosynthesis. *J Genet Genom*, *52*, 1082–1095. https://doi.org/10.1016/j.jgg.2025.01.002

Liu, Q., Yang, F., Zhang, J., Liu, H., Rahman, S., Islam, S.et al. (2021). Application of CRISPR/Cas9 in crop quality improvement. *Int J Mol Sci*, *22*, 1–16. https://doi.org/10.3390/ijms22084206

Locato, V., Cimini, S. and DeGara, L. (2013). Strategies to increase vitamin C in plants: From plant defense perspective to food biofortification. *Front Plant Sci*, *4*, 1–12. https://doi.org/10.3389/fpls.2013.00152

Ma, M., Liu, Y., Bai, C., Yang, Y., Sun, Z., Liu, X.et al. (2021). The physiological functionality of PGR5/PGRL1-dependentcyclicelectrontransport in sustaining photosynthesis. *Front Plant Sci*, *12*. https://doi.org/10.3389/fpls.2021.702196

Mao, S., Jiang, J., Xiong, K., Chen, Y., Yao, Y., Liu, L.et al. (2024). Enzyme engineering: Performance optimization, novel sources, and applications in the food industry. *Foods*, *13*(23), 3846. https://doi.org/10.3390/foods13233846

Miranda-Astudillo, H., Ostolga-Chavarría, M., Cardol, P. and González-Halphen, D. (2022). Beyond being an energy supplier, ATP synthase is a sculptor of mitochondrial cristae. *Biochim Biophys Acta – Bioenerg*, *1863*. https://doi.org/10.1016/j.bbabio.2022.148569

Mlinarić, S., Begović, L., Tripić, N., Piškor, A. and Cesar, V. (2021). Evaluation of light-dependent photosynthetic reactions in *Reynoutria japonica* Houtt. leaves grown at different light conditions. *Front Plant Sci*, *12*, 1–13. https://doi.org/10.3389/fpls.2021.612702

Nowicka, B., Ciura, J., Szymańska, R. and Kruk, J. (2018). Improving photosynthesis, plant productivity and abiotic stress tolerance – Current trends and future perspectives. *J Plant Physiol*, *231*, 415–433. https://doi.org/10.1016/j.jplph.2018.10.022

Oliveira-Filho, E. R., Voiniciuc, C. and Hanson, A. D. (2023). Adapting enzymes to improve their functionality in plants: Why and how. *Biochem Soc Trans*, *51*, 1957–1966. https://doi.org/10.1042/BST20230532

Page, M. T. and Parry, M. A. J.2025. Shining light on photosynthesis. *J Exp Bot*,*76*(12), 3243–3246. https://doi.org/10.1093/jxb/eraf096

Pesaresi, P., Bono, P., Corn, S., Crosatti, C., Daniotti, S., Jensen, J. D. et al. (2025). Boosting photosynthesis opens new opportunities for agriculture sustainability and circular economy: The BEST-CROP research and innovation action. *Plant J*, *121*, 1–22. https://doi.org/10.1111/tpj.17264

Simkin, A. J., López-Calcagno, P. E. and Raines, C. A. (2019). Feeding the world: Improving photosynthetic efficiency for sustainable crop production. *J Exp Bot*, *70*, 1119–1140. https://doi.org/10.1093/jxb/ery445

Storti, M., Segalla, A., Mellon, M., Alboresi, A. and Morosinotto, T. (2020). Regulation of electron transport is essential for photosystem I stability and plant growth. *New Phytol*, *228*, 1316–1326. https://doi.org/10.1111/nph.16643

Strand, D. D. and Walker, B. J. (2023). Energetic considerations for engineering novel biochemistries in photosynthetic organisms. *Front Plant Sci*, *14*, 1–9. https://doi.org/10.3389/fpls.2023.1116812

Swift, J., Luginbuehl, L. H., Hua, L., Schreier, T. B., Donald, R. M., Stanley, S. et al. (2024). Exaptation of ancestral cell-identity networks enables C4 photosynthesis. *Nature*, *636*. https://doi.org/10.1038/s41586-024-08204-3

Tan, B. and Chen, S. (2023). Defining mechanisms of C3 to CAM photosynthes is transition toward enhancing crop stress resilience. *Int J Mol Sci*, *24*. https://doi.org/10.3390/ijms241713072

Taylor, A., Heyes, D. J. and Scrutton, N. S. (2022). Catalysis by nature's photoenzymes. *Curr Opin Struct Biol*, *77*, 102491. https://doi.org/10.1016/j.sbi.2022.102491

Tikkanen, M., Nurmi, M., Suorsa, M., Danielsson, R., Mamedov, F., Styring, S. et al. (2008). Phosphorylation-dependent regulation of excitation energy distribution between the two photosystems in higher plants. *Biochim Biophys Acta – Bioenerg*, *1777*, 425–432. https://doi.org/10.1016/j.bbabio.2008.02.001

Uchida, H., Kato, K., Suzuki, K., Yokota, A., Kawano, S., Matsunaga, S. and Okada, S. (2018). Algal genes encoding enzymes for photosynthesis and hydrocarbon biosynthesis as candidates for genetic engineering. *Cytologia (Tokyo)*, *83*, 7–17. https://doi.org/10.1508/cytologia.83.7

von Caemmerer, S. and Furbank, R. T. (2016). Strategies for improving C4 photosynthesis. *Curr Opin Plant Biol*, *31*, 125–134. https://doi.org/10.1016/j.pbi.2016.04.003

Wang, P., Khoshravesh, R., Karki, S., Tapia, R., Balahadia, C. P., Bandyopadhyay, A. et al. (2017). Re-creation of a key step in the evolutionary switch from C3 to C4 leaf anatomy. *Curr Biol*, *27*, 3278–3287.e6. https://doi.org/10.1016/j.cub.2017.09.040

Wang, Y., Phelps, A., Godbehere, A., Evans, B., Takizawa, C., Chinen, G. et al. (2025). Revolutionizing agriculture with CRISPR technology: Applications, challenges, and future perspectives. *Biotechnol J*, *20*. https://doi.org/10.1002/biot.70113

Weis, W. I. and Kobilka, B. K. (2018). The molecular basis of g protein-coupled receptor activation. *Annu Rev Biochem*, *87*, 897–919. https://doi.org/10.1146/annurev-biochem-060614-033910

Weissmann, S. and Brutnell, T. P. (2012). Engineering C4 photosynthetic regulatory networks. *Curr Opin Biotechnol*, *23*, 298–304. https://doi.org/10.1016/j.copbio.2011.12.018

Yang, W., and Lin, R. (2025). Advances in photosynthesis research: Unlocking the potential for food security, renewable energy, and climate resilience. *Journal of Integrative Plant Biology*. https://doi.org/10.1111/jipb.13908

Yang, Z., Cao, Y., Shi, Y., Qin, F., Jiang, C. and Yang, S. (2023). Genetic and molecular exploration of maize environmental stress resilience: Toward sustainable agriculture. *Mol Plant*, *16*, 1496–1517. https://doi.org/10.1016/j.molp.2023.07.005

Yin, X. R., Chen, K. S., Allan, A. C., Wu, R. M., Zhang, B., Lallu, N. and Ferguson, I. B. (2008). Ethylene-induced modulation of genes associated with the ethylene signalling pathway in ripening kiwifruit. *J Exp Bot*, *59*, 2097–2108. https://doi.org/10.1093/jxb/ern067

Yu, A., Xie, Y., Pan, X., Zhang, H., Cao, P., Su, X. et al. (2020). Photosynthetic phosphoribulokinase structures: Enzymatic mechanisms and the redox regulation of the Calvin-Benson-Bassham cycle. *Plant Cell*, *32*, 1556–1573. https://doi.org/10.1105/tpc.19.00642

Zhang, J., Ge, J., Dayananda, B. and Li, J. (2022). Effect of light intensities on the photosynthesis, growth and physiological performances of two maple species. *Front Plant Sci*, *13*, 1–10. https://doi.org/10.3389/fpls.2022.999026

4 Respiratory Enzymes Energy Metabolism and Stress Responses

Hamzeh Amiri

4.1 INTRODUCTION

Respiration stands as a fundamental biological process that transcends mere energy production, serving as a critical interface between cellular metabolism and stress adaptation in living organisms. In plants, the respiratory pathway—encompassing glycolysis, the tricarboxylic acid (TCA) cycle, and mitochondrial electron transport chain (ETC)—represents not only an energy-generating system but also a dynamic regulatory network that responds to environmental challenges (Millar et al., 2011). The enzymatic machinery of respiration exhibits remarkable plasticity, modulating its activity and organization to maintain metabolic homeostasis under diverse stress conditions, including drought, salinity, temperature extremes, and pathogen attack (Van Aken et al., 2009). This metabolic flexibility is achieved through intricate regulation of key enzymes such as alternative oxidase (AOX), uncoupling proteins (UCPs), and various dehydrogenases that collectively orchestrate the balance between energy production and stress protection (Arnholdt-Schmitt et al., 2006).

Recent advances in plant biochemistry have revealed that respiratory enzymes serve dual roles in both core metabolism and stress signaling networks. The mitochondrial ETC, traditionally viewed as simply an ATP-generating system, has emerged as a central hub for stress perception and response integration (Vanlerberghe, 2013). Enzymes such as succinate dehydrogenase (SDH) (Complex II) and cytochrome c oxidase (Complex IV) not only participate in energy transduction but also function as redox sensors that modulate stress-responsive gene expression (Gleason et al., 2021). Similarly, the AOX pathway, once considered a wasteful electron sink, is now recognized as a crucial component of plant defense systems against both abiotic and biotic stresses (Cvetkovska & Vanlerberghe, 2012).

The interplay between respiratory enzymes and stress hormones further underscores the integrative nature of plant stress responses. Abscisic acid (ABA)-mediated stomatal closure, for instance, directly impacts mitochondrial respiration by altering cellular CO_2/O_2 ratios, while simultaneously inducing specific isoforms of respiratory enzymes optimized for stress conditions (Giraud et al., 2008). Similarly, salicylic acid and jasmonic acid pathways intersect with mitochondrial function during pathogen responses, creating a complex web of metabolic and defense signaling (Amirsadeghi et al., 2007). These interactions highlight how respiratory enzymes have evolved to serve as metabolic rheostats, fine-tuning energy metabolism while coordinating defense mechanisms.

This chapter aims to provide a comprehensive synthesis of current knowledge regarding the multifaceted roles of respiratory enzymes in plant energy metabolism and stress responses. By examining the molecular mechanisms of enzyme regulation, their integration with stress signaling pathways, and their potential applications in crop improvement, we seek to bridge the gap between basic biochemical understanding and practical agricultural applications. The discussion will encompass both traditional biochemical perspectives and emerging concepts in metabolic signaling, offering readers a holistic view of respiration's role in plant stress adaptation.

4.2 ENZYMES IN AEROBIC RESPIRATION IN PLANTS

Aerobic respiration in plants involves a sophisticated network of enzymatic reactions that convert photosynthetic products into cellular energy through four interconnected metabolic stages: glycolysis, the oxidative decarboxylation of pyruvate, the TCA cycle, and oxidative phosphorylation. This process exhibits unique plant-specific adaptations that distinguish it from aerobic respiration in other eukaryotes, particularly in terms of enzyme isoforms, regulatory mechanisms, and metabolic flexibility (Plaxton & Podestá, 2006). Glycolytic enzymes in plants display remarkable compartmentalization, with parallel cytosolic and plastidic isoforms that allow metabolic crosstalk between respiration and photosynthesis (Giegé et al., 2003). Hexokinase (EC 2.7.1.1) not only initiates glycolysis but also functions as a sugar sensor, linking carbohydrate metabolism to stress signaling (Rolland et al., 2002). Plant phosphofructokinase (PFK; EC 2.7.1.11) is uniquely regulated by inorganic phosphate (Pi) and phosphoenolpyruvate (PEP), reflecting adaptations to fluctuating energy demands (Turner et al., 2005). The final glycolytic enzyme, pyruvate kinase (PK; EC 2.7.1.40) exists as multiple isozymes with distinct kinetic properties, including a plastid-localized form that facilitates carbon partitioning between respiration and biosynthetic pathways (Banakar et al., 2022; Andre et al., 2007).

The mitochondrial pyruvate dehydrogenase complex (mtPDC; EC 1.2.4.1) in plants contains additional regulatory subunits that confer sensitivity to light-dark transitions and redox status, allowing tight coordination with photosynthetic activity (Tovar-Méndez et al., 2003). Plant TCA cycle enzymes exhibit unusual plasticity, with several steps potentially bypassed through alternative pathways like the

DOI: 10.1201/9781003669920-4

γ-aminobutyrate (GABA) shunt during stress conditions (Fait et al., 2008). Citrate synthase (CS; EC 2.3.3.1) activity is modulated by organic acids that accumulate during photorespiration, while plant-specific isoforms of isocitrate dehydrogenase (IDH; EC 1.1.1.41) show differential sensitivity to NADPH/$NADP^+$ ratios (Lemaitre et al., 2007). The mitochondrial malic enzyme (EC 1.1.1.39) provides a critical anaplerotic function, replenishing TCA cycle intermediates while generating NAD(P)H for biosynthetic processes (Tronconi et al., 2008).

The plant mitochondrial ETC features several unique components, including alternative NAD(P)H dehydrogenases (NDs; EC 1.6.5.3) on both sides of the inner membrane that bypass proton-pumping complex I (Rasmusson et al., 2004). These enzymes allow flexible electron input from cytosolic and matrix NAD(P)H pools, particularly important during stress when redox balance is critical (Møller, 2001). The (AOX; EC 1.10.3.11) pathway, a plant-specific branch point at ubiquinone, uncouples carbon metabolism from ATP synthesis while reducing reactive oxygen species (ROS) production (Vanlerberghe, 2013). Plant cytochrome c oxidase (Complex IV; EC 1.9.3.1) contains additional subunits that may modulate its activity in response to nitric oxide (NO), a key signaling molecule in stress responses (Gupta et al., 2018). The ATP synthase (Complex V; EC 3.6.3.14) in plant mitochondria exhibits altered inhibitor sensitivity compared to mammalian systems, with unique ε and δ subunits that may confer regulatory flexibility (Hahn et al., 2018).

Post-translational modifications play a particularly important role in regulating plant respiratory enzymes. Reversible protein phosphorylation dynamically modulates pyruvate dehydrogenase complex (PDC), IDH, and several ETC components in response to light conditions and cellular energy status (Schwarzländer et al., 2012). Redox-sensitive cysteine residues in glycolytic and TCA cycle enzymes allow direct sensing of the cellular redox environment, linking respiration to antioxidant defense systems (Schmidtmann et al., 2014). These regulatory mechanisms enable plants to maintain energy homeostasis while responding to environmental fluctuations, demonstrating how respiratory enzymes have evolved specialized functions to meet the unique metabolic demands of photosynthetic organisms (Millar et al., 2011).

4.3 ENZYMES IN ANAEROBIC RESPIRATION AND FERMENTATION IN PLANTS

Plants have evolved sophisticated enzymatic systems to maintain energy metabolism during oxygen limitation, employing both anaerobic respiration and fermentation pathways that differ fundamentally from their aerobic counterparts. When oxygen becomes limiting due to flooding, waterlogging, or high tissue respiration rates, plants rapidly activate these alternative pathways to sustain ATP production and prevent cytoplasmic acidosis (Bailey-Serres & Voesenek, 2008). The shift from aerobic to anaerobic metabolism begins with the activation of pyruvate dehydrogenase complex (PDC; EC 4.1.1.1), which catalyzes the irreversible decarboxylation of pyruvate to acetaldehyde, representing the first committed step of ethanolic fermentation (Mustroph et al., 2010). This enzyme exists as multiple isoforms in plants with differing kinetic properties and stress responsiveness, with PDC1 and PDC2 showing particularly strong induction under hypoxia (Kürsteiner et al., 2003). The subsequent reduction of acetaldehyde to ethanol by alcohol dehydrogenase (ADH; EC 1.1.1.1) serves the critical function of regenerating NAD^+ to maintain glycolytic flux, with plant ADH enzymes exhibiting unique substrate specificities and regulatory properties compared to their microbial counterparts (Strommer, 2011). Arabidopsis contains seven ADH genes, with ADH1 being the primary hypoxia-responsive isoform that shows allosteric regulation by cellular pH and redox state (Amiri, 2025; Dolferus et al., 2008).

Parallel to ethanolic fermentation, plants activate lactate production through lactate dehydrogenase (LDH; EC 1.1.1.27), which converts pyruvate to lactate while oxidizing nicotinamide adenine dinucleotide (NADH) (Rivoal & Hanson, 1994). Plant LDH enzymes display distinctive kinetic properties, including higher Km values for pyruvate and greater sensitivity to pH changes compared to mammalian isoforms, reflecting adaptation to fluctuating stress conditions (Germain et al., 1997). The relative flux through these competing fermentation pathways is tightly regulated by cytoplasmic pH, with acidic conditions inhibiting LDH and promoting ethanolic fermentation through pyruvate dehydrogenase complex (PDC) activation (Ricard et al., 1994). This pH-dependent switch prevents excessive lactate accumulation that could lead to cytoplasmic acidosis, demonstrating the sophisticated metabolic control mechanisms in plants (Greenway & Gibbs, 2003). Additional fermentation enzymes include alanine aminotransferase (AlaAT; EC 2.6.1.2), which converts pyruvate to alanine while consuming glutamate, serving as an important alternative route for pyruvate metabolism and nitrogen balance during oxygen deprivation (Miyashita et al., 2007). The GABA shunt pathway, involving glutamate decarboxylase (GAD; EC 4.1.1.15) and GABA transaminase (GABA-T; EC 2.6.1.19), becomes particularly active during hypoxia, consuming protons and generating succinate that can feed into alternative mitochondrial respiration (Shelp et al., 2012).

Mitochondrial enzymes adapt to low oxygen by shifting from oxidative phosphorylation to alternative electron transport pathways. The (AOX; EC 1.10.3.11), while not unique to anaerobic conditions, becomes increasingly important during oxygen limitation by accepting electrons from ubiquinol without proton pumping, preventing over-reduction of respiratory chains (Vanlerberghe, 2013). Plant mitochondria also employ alternative NADH dehydrogenases (NDs; EC 1.6.5.3) that bypass Complex I, allowing continued NADH oxidation when the standard ETC is compromised (Rasmusson et al., 2004). The SDH complex (SDH; EC 1.3.5.1) operates in reverse during anoxia, functioning as fumarate reductase to maintain minimal electron flow and generate ATP through substrate-level phosphorylation (Gupta et al., 2009). These adaptations are complemented by post-translational modifications of respiratory enzymes, including phosphorylation of PDC by hypoxia-induced kinases and redox modifications of critical cysteine residues in fermentation enzymes (Licausi et al., 2011).

The regulation of anaerobic enzymes occurs at multiple levels, beginning with oxygen-sensing mechanisms involving the N-end rule pathway of protein degradation that controls stability of transcription factors like VII ETHYLENE RESPONSE FACTORS (ERF-VIIs) proteins (Gibbs et al., 2011). Hypoxia triggers rapid transcriptional activation of fermentation genes through promoter elements like the anaerobic response element (ARE) and GC-rich motifs that bind MYB and bHLH transcription factors (Lasanthi-Kudahettige et al., 2007). Post-transcriptional regulation includes alternative splicing of pre-mRNAs and selective translation of stress-responsive transcripts through mechanisms involving the 5′ untranslated region (5′UTR) of fermentation enzyme mRNAs (Branco-Price et al., 2008). Metabolic regulation occurs through allosteric control of PDC by AMP and pyruvate, pH-dependent modulation of LDH activity, and redox-sensitive switching between fermentation pathways (Zabalza et al., 2009). These multilayered regulatory mechanisms allow plants to fine-tune their metabolic responses according to the severity and duration of oxygen deprivation, with different enzymatic strategies dominating during transient hypoxia versus prolonged anoxia (Bailey-Serres et al., 2012). The evolutionary conservation of these anaerobic enzymes across plant species, along with their unique kinetic and regulatory properties compared to non-plant homologs, underscores their critical role in plant stress adaptation and survival in fluctuating environments (Mustroph, 2018).

4.4 REGULATION OF RESPIRATORY ENZYMES IN PLANTS

The regulation of respiratory enzymes in plants represents a complex, multilayered control system that integrates metabolic demands with environmental signals through transcriptional, post-translational, and allosteric mechanisms. At the transcriptional level, nuclear-encoded mitochondrial proteins are regulated by suites of transcription factors that respond to cellular energy status, with the *Arabidopsis thaliana* transcription factor ABSCISIC ACID INSENSITIVE 4 (ABI4) emerging as a key regulator that coordinates expression of AOX (AOX1a) and cytochrome c oxidase (COX5b-1) genes in response to sugar availability. The circadian clock exerts another layer of control, with core clock components like CIRCADIAN CLOCK ASSOCIATED 1 (CCA1) and TIMING OF CAB EXPRESSION 1 (TOC1) directly regulating the expression of TCA cycle enzymes such as IDH and malate dehydrogenase (MDH), creating diurnal oscillations in respiratory capacity (Haydon et al., 2011). Hypoxia-responsive transcription factors including group ERF-VIIs activate anaerobic genes through specific promoter motifs, simultaneously repressing genes encoding aerobic respiration components like cytochrome c oxidase while inducing pyruvate dehydrogenase complex (PDC) and ADH (Gibbs et al., 2011). Light quality and intensity modulate respiratory gene expression through phytochrome-interacting factors (PIFs) that regulate nuclear-encoded mitochondrial proteins, creating a direct link between photosynthetic electron transport and mitochondrial gene expression (Van Aken et al., 2016; Amiri, 2025).

Post-translational modifications provide rapid adjustment of respiratory enzyme activity, with reversible protein phosphorylation playing a particularly prominent role through mitochondrial-targeted protein kinases and phosphatases. The PDC is regulated by a dedicated mitochondrial pyruvate dehydrogenase kinase (PDK) that phosphorylates and inactivates the E1 subunit in response to high NADH/NAD^+ ratios, with plant-specific PDK isoforms showing differential sensitivity to metabolic effectors (Thelen et al., 2000). S-nitrosylation of cysteine residues provides redox-sensitive control over multiple respiratory enzymes, including aconitase and SDH, with NO concentrations fluctuating in response to environmental stress (Gupta et al., 2018). Lysine acetylation has emerged as a widespread modification affecting over 60% of mitochondrial proteins, including numerous TCA cycle enzymes whose acetylation status changes dramatically during the day-night cycle, suggesting a direct link between photosynthetic activity and mitochondrial protein modification (Finkemeier et al., 2011). Ubiquitination and subsequent proteasomal degradation regulate the turnover of ETC components, with the mitochondrial E3 ubiquitin ligase PROHIBITIN3 (PHB3) controlling Complex I abundance in response to ROS levels (Van Aken et al., 2007). Glycosylation modifications of respiratory enzymes, particularly O-GlcNAcylation, provide nutrient-sensing capabilities that link hexose availability to mitochondrial function (Zheng et al., 2019).

Allosteric regulation of plant respiratory enzymes exhibits unique features compared to other eukaryotes, reflecting the need to accommodate photosynthetic carbon flux and respond to environmental variability. Plant PFK shows atypical regulation by PEP and Pi, creating a sensitive switch that coordinates glycolytic flux with photosynthetic carbon partitioning (Plaxton, 1996). The TCA cycle enzyme CS is inhibited by NADH in most organisms but shows reduced sensitivity in plants, allowing continued operation under conditions of high reduction potential (Sienkiewicz-Porzucek et al., 2008). Alternative NDs located on both sides of the mitochondrial inner membrane provide bypass routes for electron transfer that are allosterically activated by Ca^{2+} and calmodulin, creating a direct link between cellular signaling and respiratory electron flow (Rasmusson et al., 2008). The AOX is activated by α-keto acids (particularly pyruvate) through reduction of intermolecular disulfide bonds, providing a metabolic sensor that enhances alternative electron flow when carbon substrates accumulate. UCPs in plant mitochondria are activated by fatty acids and inhibited by purine nucleotides but show distinct kinetic properties from their mammalian counterparts, suggesting specialized roles in thermogenesis and ROS management (Vercesi et al., 2006).

Metabolic regulation extends beyond individual enzymes to pathway-level control through substrate channeling and enzyme supercomplex formation. Evidence suggests that glycolytic enzymes associate with the mitochondrial outer membrane to form metabolons that direct carbon flow toward respiration, with these associations being dynamically regulated by cellular energy status (Giegé et al., 2003). Respiratory supercomplexes (respirasomes) containing varying stoichiometries of Complex I, III, and IV reorganize in response to light conditions and stress, optimizing electron transfer efficiency and minimizing ROS production (Eubel et al., 2004). The discovery of substrate-specific TCA cycle "modules" that operate under different physiological conditions reveals an unprecedented flexibility in plant mitochondrial metabolism, with enzymes like fumarase and malic enzyme participating in multiple metabolic configurations depending on cellular demands (Sweetlove et al., 2010). These regulatory mechanisms collectively enable plants to maintain energy homeostasis while responding to environmental fluctuations, demonstrating how respiratory control in photosynthetic organisms has evolved unique solutions to the challenges of sessile growth in variable environments (Millar et al., 2011). The integration of these regulatory layers—from gene expression to protein modification to metabolic organization—provides plants with the plasticity required to balance energy production with stress resilience, making plant respiratory regulation a paradigm of metabolic adaptability.

4.5 RESPIRATORY ENZYMES IN ENERGY METABOLISM DURING STRESS

Plants undergo significant metabolic reprogramming under environmental stress to sustain energy production and survival. Respiratory enzymes play a central role in this process by modulating glycolysis, the TCA cycle, and mitochondrial electron transport to maintain ATP supply while minimizing oxidative damage. Under hypoxia caused by flooding or waterlogging, plants shift from aerobic respiration to fermentation to regenerate NAD^+ and avoid metabolic arrest. Key enzymes such as LDH and ADH convert pyruvate into lactate and ethanol, respectively, preventing cytoplasmic acidosis and sustaining limited ATP production (Amiri, 2025; Bailey-Serres & Voesenek, 2008; Gibbs & Greenway, 2003).

During drought or heat stress, mitochondrial respiration is often suppressed due to reduced water availability or enzyme denaturation. In response, plants enhance glycolytic flux through the upregulation of hexokinase, PFK, and PK to compensate for decreased TCA cycle activity (Rizhsky et al., 2004). The TCA cycle itself undergoes modifications, with alternative pathways such as the GABA shunt becoming prominent to maintain carbon flow and support nitrogen metabolism (Fait et al., 2008). SDH, a key TCA cycle enzyme and part of mitochondrial Complex II, also plays a dual role in energy metabolism and ROS generation, linking respiration to oxidative stress responses (Millar et al., 2011) (Table 4.1).

The mitochondrial ETC is particularly sensitive to stress-induced damage. Under conditions such as cold, drought, or pathogen attack, plants activate the AOX pathway, which bypasses proton-pumping Complexes III and

TABLE 4.1
Role of Various Enzymes in Stress Condition

Enzyme	Stress Condition	Role in Stress Response	Regulatory Mechanism	References
Alternative oxidase (AOX)	Drought, cold, pathogen attack	Reduces ROS by diverting electrons from ETC, maintains respiration under stress	Activated by pyruvate, redox-sensitive disulfide bonds	Vanlerberghe (2013); Amiri (2025)
pyruvate dehydrogenase complex (PDC)	Hypoxia/Flooding	Shifts metabolism to ethanolic fermentation, regenerates NAD^+	Induced by low O_2 (ERF-VII transcription factors)	Mustroph et al. (2010)
Alcohol dehydrogenase (ADH)	Hypoxia/Flooding	Converts acetaldehyde to ethanol, prevents acidosis	Allosteric regulation by pH and NADH	Dolferus et al. (2008)
Lactate dehydrogenase (LDH)	Hypoxia/Flooding	Produces lactate as temporary fermentative product	Inhibited by low pH to favor ethanolic fermentation	Rivoal and Hanson (1994)
Succinate dehydrogenase (SDH, Complex II)	Oxidative stress	Acts as ROS sensor; reverse activity (fumarate reductase) under anoxia	Inhibited by NO, S-nitrosylation	Gupta et al. (2018)
Uncoupling proteins (UCPs)	Heat, oxidative stress	Dissipates proton gradient, reduces mtROS production	Activated by fatty acids, inhibited by ATP	Vercesi et al. (2006)
Glycolytic enzymes (e.g., PFK, PK)	Drought, salt stress	Maintains ATP supply when mitochondrial respiration is limited	Allosteric regulation by PEP, Pi, and Ca^{2+}	Plaxton (1996)
GABA Shunt Enzymes (GAD, GABA-T)	Drought, cold	Accumulates GABA, buffers pH, provides succinate for TCA cycle	Induced by Ca^{2+}/calmodulin signaling	Shelp et al. (2012)
Alternative NADH dehydrogenases (NDs)	Oxidative stress	Bypasses Complex I to reduce ROS, accepts electrons from cytosolic NADH	Activated by Ca^{2+}	Rasmusson et al. (2004)

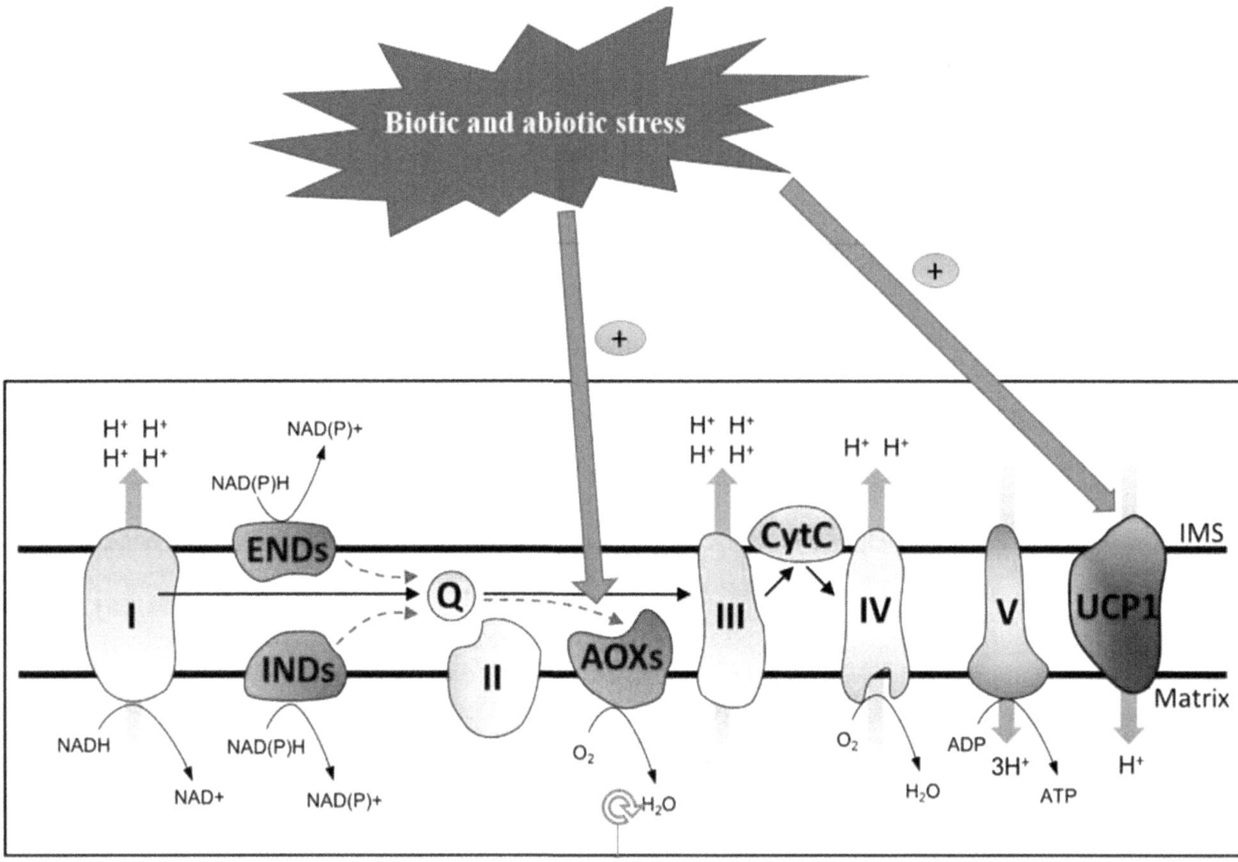

FIGURE 4.1 The effect of biotic and abiotic stresses on some inner mitochondrial membrane complexes in plants. Under stress, plants activate the alternative oxidase (AOX) pathway, which bypasses proton-pumping complexes III and IV, reducing electron leakage and subsequent ROS production. Additionally, the activation of uncoupling proteins (UCPs) dissipates the proton gradient to mitigate mitochondrial dysfunction caused by heat stress.

IV, reducing electron leakage and subsequent ROS production (Vanlerberghe, 2013). AOX serves as a critical safety valve, allowing continued respiration without excessive oxidative damage, while UCPs dissipate the proton gradient to mitigate heat-induced mitochondrial dysfunction (Vercesi et al., 2006). These adaptations ensure that energy metabolism remains flexible under stress, balancing ATP demand with cellular protection mechanisms (Figure 4.1).

4.6 CONCLUSION AND FUTURE PERSPECTIVES

The study of respiratory enzymes in plants has revealed a remarkably dynamic and adaptable system that integrates energy production with stress responses, ensuring survival under fluctuating environmental conditions. This chapter has highlighted how glycolysis, the TCA cycle, and mitochondrial electron transport are not merely metabolic pathways but sophisticated regulatory networks that respond to cellular energy demands, redox status, and external stressors. Enzymes such as AOX, UCPs, and fermentation-specific isoforms (e.g., PDC, ADH, LDH) exemplify the plasticity of plant respiration, allowing rapid metabolic reprogramming under hypoxia, drought, temperature extremes, and pathogen attack. Post-translational modifications, allosteric regulation, and enzyme supercomplex formation further fine-tune respiratory flux, demonstrating that plants have evolved intricate mechanisms to balance ATP synthesis with stress resilience. The dual roles of respiratory enzymes—both in core metabolism and stress signaling—underscore their centrality in plant adaptation, making them key targets for improving crop performance under climate change. Looking ahead, several promising research directions could deepen our understanding of respiratory regulation and its applications in agriculture. First, the development of advanced imaging techniques, such as fluorescent biosensors for real-time monitoring of mitochondrial NADH/NAD$^+$ ratios or ATP dynamics, could provide unprecedented spatial and temporal resolution of metabolic shifts during stress. Second, the exploration of enzyme moonlighting—where respiratory proteins like GAPDH or enolase assume non-catalytic roles in stress signaling or RNA binding—may reveal novel regulatory nodes linking metabolism to gene expression. Third, synthetic biology approaches could engineer stress-responsive promoters or allosteric regulation into respiratory enzymes to create "smart" metabolic pathways that automatically adjust to environmental cues. For instance, modifying

AOX expression or PDC kinetics in crops like rice or wheat could enhance flood tolerance without compromising yield. Another critical frontier is elucidating the role of respiratory enzymes in systemic signaling, particularly how mitochondrial ROS (mtROS) and retrograde signals coordinate whole-plant responses to stress. The interaction between respiratory enzymes and phytohormones (e.g., ABA, jasmonates) also warrants deeper investigation, as these pathways likely mediate trade-offs between growth and stress defense. Finally, integrating multi-omics data (transcriptomics, proteomics, metabolomics) with computational modeling could predict how manipulating specific enzymes might reshape metabolic networks under stress, guiding precision breeding or gene-editing strategies. In conclusion, respiratory enzymes represent both a fundamental framework for energy metabolism and a versatile toolkit for stress adaptation in plants. By unraveling their regulatory complexities and harnessing their adaptive potential, we can develop innovative solutions to enhance crop resilience, ensuring food security in an era of climate uncertainty. Future research should bridge fundamental discoveries with translational applications, transforming our knowledge of plant respiration into sustainable agricultural advances.

REFERENCES

Amiri, H. (2025). Sources of ROS Production in Plants Under Normal and Stressed Conditions. In *Plant Signaling Molecules in Regulation of ROS-Scavenging System: Adapting to Climate Change* (pp. 203–221). Singapore: Springer Nature Singapore.

Amirsadeghi, S., Robson, C. A., McDonald, A. E., & Vanlerberghe, G. C. (2007). Changes in plant mitochondrial electron transport alter cellular levels of reactive oxygen species and susceptibility to cell death signaling molecules. *Plant & Cell Physiology*, *48*(10), 1509–1519. https://doi.org/10.1093/pcp/pcm132

Andre, C., Froehlich, J. E., Moll, M. R., & Benning, C. (2007). A heteromeric plastidic pyruvate kinase complex involved in seed oil biosynthesis in Arabidopsis. *The Plant Cell*, *19*(6), 2006–2022. https://doi.org/10.1105/tpc.106.048629

Arnholdt-Schmitt, B., Costa, J. H., & de Melo, D. F. (2006). AOX – A functional marker for efficient cell reprogramming under stress? *Trends in Plant Science*, *11*(6), 281–287. https://doi.org/10.1016/j.tplants.2006.05.001

Bailey-Serres, J., Fukao, T., Gibbs, D. J., Holdsworth, M. J., Lee, S. C., Licausi, F., …, & Van Dongen, J. T. (2012). Making sense of low oxygen sensing. *Trends in Plant Science*, *17*(3), 129–138. https://doi.org/10.1016/j.tplants.2011.12.004

Bailey-Serres, J., & Voesenek, L. A. C. J. (2008). Flooding stress: Acclimations and genetic diversity. *Annual Review of Plant Biology*, *59*, 313–339. https://doi.org/10.1146/annurev.arplant.59.032607.092752

Banakar, M. H., Amiri, H., Ardakani, M. R. S., & Ranjbar, G. H. (2022). Susceptibility and tolerance of fenugreek (*Trigonella foenum-graceum* L.) to salt stress: Physiological and biochemical inspections. *Environmental and Experimental Botany*, *194*, 104748.

Branco-Price, C., Kaiser, K. A., Jang, C. J., Larive, C. K., & Bailey-Serres, J. (2008). Selective mRNA translation coordinates energetic and metabolic adjustments to cellular oxygen deprivation and reoxygenation in *Arabidopsis thaliana*. *The Plant Journal*, *56*(5), 743–755. https://doi.org/10.1111/j.1365-313X.2008.03642.x

Cvetkovska, M., & Vanlerberghe, G. C. (2012). Alternative oxidase modulates leaf mitochondrial concentrations of superoxide and nitric oxide. *New Phytologist*, *195*(1), 32–39. https://doi.org/10.1111/j.1469-8137.2012.04166.x

Dolferus, R., Klok, E. J., Delessert, C., Wilson, S., Ismond, K. P., Good, A. G., …, & Jacobs, M. (2008). Enhancing the anaerobic response. *Annals of Botany*, *91*(2), 111–117. https://doi.org/10.1093/aob/mcf048

Eubel, H., Heinemeyer, J., & Braun, H. P. (2004). Identification and characterization of respirasomes in potato mitochondria. *Plant Physiology*, *134*(4), 1450–1459. https://doi.org/10.1104/pp.103.038018

Fait, A., Fromm, H., Walter, D., Galili, G., & Fernie, A. R. (2008). Highway or byway: The metabolic role of the GABA shunt in plants. *Trends in Plant Science*, *13*(1), 14–19. https://doi.org/10.1016/j.tplants.2007.10.005

Finkemeier, I., Laxa, M., Miguet, L., Howden, A. J., & Sweetlove, L. J. (2011). Proteins of diverse function and subcellular location are lysine acetylated in Arabidopsis. *Plant Physiology*, *155*(4), 1779–1790. https://doi.org/10.1104/pp.110.171595

Germain, V., Ricard, B., Raymond, P., & Saglio, P. H. (1997). The role of sugars, hexokinase, and sucrose synthase in the determination of hypoxically induced tolerance to anoxia in tomato roots. *Plant Physiology*, *114*(1), 167–174. https://doi.org/10.1104/pp.114.1.167

Gibbs, D. J., Lee, S. C., Isa, N. M., Gramuglia, S., Fukao, T., Bassel, G. W., …, & Holdsworth, M. J. (2011). Homeostatic response to hypoxia is regulated by the N-end rule pathway in plants. *Nature*, *479*(7373), 415–418. https://doi.org/10.1038/nature10534

Gibbs, J., & Greenway, H. (2003). Mechanisms of anoxia tolerance in plants. I. Growth, survival and anaerobic catabolism. *Functional Plant Biology*, *30*(1), 1–47. https://doi.org/10.1071/PP98095

Giegé, P., Heazlewood, J. L., Roessner-Tunali, U., Millar, A. H., Fernie, A. R., Leaver, C. J., & Sweetlove, L. J. (2003). Enzymes of glycolysis are functionally associated with the mitochondrion in Arabidopsis cells. *The Plant Cell*, *15*(9), 2140–2151. https://doi.org/10.1105/tpc.012500

Giraud, E., Ho, L. H., Clifton, R., Carroll, A., Estavillo, G., Tan, Y. F., …, & Whelan, J. (2008). The absence of ALTERNATIVE OXIDASE1a in Arabidopsis results in acute sensitivity to combined light and drought stress. *Plant Physiology*, *147*(2), 595–610. https://doi.org/10.1104/pp.107.115121

Gleason, C., Huang, S., Thatcher, L. F., Foley, R. C., Anderson, C. R., Carroll, A. J., …, & Singh, K. B. (2021). Mitochondrial complex II has a key role in mitochondrial-derived reactive oxygen species influence on plant stress gene regulation and defense. *Proceedings of the National Academy of Sciences*, *118*(3), e2021492118.

Greenway, H., & Gibbs, J. (2003). Mechanisms of anoxia tolerance in plants. II. Energy requirements for maintenance and energy distribution to essential processes. *Functional Plant Biology*, *30*(10), 999–1036. https://doi.org/10.1071/PP98096

Gupta, K. J., Kumari, A., Florez-Sarasa, I., Fernie, A. R., & Igamberdiev, A. U. (2018). Interaction of nitric oxide with the components of the plant mitochondrial electron transport chain. *Journal of Experimental Botany*, *69*(14), 3413–3424. https://doi.org/10.1093/jxb/ery119

Gupta, K. J., Zabalza, A., & van Dongen, J. T. (2009). Regulation of respiration when the oxygen availability changes. *Physiologia Plantarum*, *137*(4), 383–391. https://doi.org/10.1111/j.1399-3054.2009.01253.x

Hahn, A., Parey, K., Bublitz, M., Mills, D. J., Zickermann, V., Vonck, J., …, & Meier, T. (2018). Structure of a complete ATP synthase dimer reveals the molecular basis of inner mitochondrial membrane morphology. *Molecular Cell*, *63*(3), 445–456. https://doi.org/10.1016/j.molcel.2016.06.025

Haydon, M. J., Mielczarek, O., Robertson, F. C., Hubbard, K. E., & Webb, A. A. (2011). Photosynthetic entrainment of the *Arabidopsis thaliana* circadian clock. *Nature*, *502*(7473), 689–692. https://doi.org/10.1038/nature12603

Kürsteiner, O., Dupuis, I., & Kuhlemeier, C. (2003). The pyruvate decarboxylase1 gene of Arabidopsis is required during anoxia but not other environmental stresses. *Plant Physiology*, *132*(2), 968–978. https://doi.org/10.1104/pp.102.016907

Lasanthi-Kudahettige, R., Magneschi, L., Loreti, E., Gonzali, S., Licausi, F., Novi, G., …, & Perata, P. (2007). Transcript profiling of the anoxic rice coleoptile. *Plant Physiology*, *144*(1), 218–231. https://doi.org/10.1104/pp.106.093997

Lemaitre, T., Urbanczyk-Wochniak, E., Flesch, V., Bismuth, E., Fernie, A. R., & Hodges, M. (2007). NAD-dependent isocitrate dehydrogenase mutants of Arabidopsis suggest the enzyme is not limiting for nitrogen assimilation. *Plant Physiology*, *144*(3), 1546–1558. https://doi.org/10.1104/pp.107.100677

Licausi, F., Kosmacz, M., Weits, D. A., Giuntoli, B., Giorgi, F. M., Voesenek, L. A., …, & Perata, P. (2011). Oxygen sensing in plants is mediated by an N-end rule pathway for protein destabilization. *Nature*, *479*(7373), 419–422. https://doi.org/10.1038/nature10536

Millar, A. H., Whelan, J., Soole, K. L., & Day, D. A. (2011). Organization and regulation of mitochondrial respiration in plants. *Annual Review of Plant Biology*, *62*, 79–104. https://doi.org/10.1146/annurev-arplant-042110-103857

Miyashita, Y., Dolferus, R., Ismond, K. P., & Good, A. G. (2007). Alanine aminotransferase catalyses the breakdown of alanine after hypoxia in *Arabidopsis thaliana*. *The Plant Journal*, *49*(6), 1108–1121. https://doi.org/10.1111/j.1365-313X.2006.03023.x

Møller, I. M. (2001). Plant mitochondria and oxidative stress: Electron transport, NADPH turnover, and metabolism of reactive oxygen species. *Annual Review of Plant Physiology and Plant Molecular Biology*, *52*, 561–591. https://doi.org/10.1146/annurev.arplant.52.1.561

Mustroph, A. (2018). Improving flooding tolerance of crop plants. *Agronomy*, *8*(9), 160. https://doi.org/10.3390/agronomy8090160

Mustroph, A., Lee, S. C., Oosumi, T., Zanetti, M. E., Yang, H., Ma, K., …, & Bailey-Serres, J. (2010). Cross-kingdom comparison of transcriptomic adjustments to low-oxygen stress highlights conserved and plant-specific responses. *Plant Physiology*, *152*(3), 1484–1500. https://doi.org/10.1104/pp.109.151845

Plaxton, W. C. (1996). The organization and regulation of plant glycolysis. *Annual Review of Plant Physiology and Plant Molecular Biology*, *47*, 185–214. https://doi.org/10.1146/annurev.arplant.47.1.185

Plaxton, W. C., & Podestá, F. E. (2006). The functional organization and control of plant respiration. *Critical Reviews in Plant Sciences*, *25*(2), 159–198. https://doi.org/10.1080/07352680600563876

Rasmusson, A. G., Geisler, D. A., & Møller, I. M. (2008). The multiplicity of dehydrogenases in the electron transport chain of plant mitochondria. *Mitochondrion*, *8*(1), 47–60. https://doi.org/10.1016/j.mito.2007.10.004

Rasmusson, A. G., Soole, K. L., & Elthon, T. E. (2004). Alternative NAD(P)H dehydrogenases of plant mitochondria. *Annual Review of Plant Biology*, *55*, 23–39. https://doi.org/10.1146/annurev.arplant.55.031903.141720

Ricard, B., Couée, I., Raymond, P., Saglio, P. H., Saint-Ges, V., & Pradet, A. (1994). Plant metabolism under hypoxia and anoxia. *Plant Physiology and Biochemistry*, *32*(1), 1–10.

Rivoal, J., & Hanson, A. D. (1994). Metabolic control of anaerobic glycolysis: Overexpression of lactate dehydrogenase in transgenic tomato roots supports the Davies-Roberts hypothesis and points to a critical role for lactate secretion. *Plant Physiology*, *106*(3), 1179–1185. https://doi.org/10.1104/pp.106.3.1179

Rizhsky, L., Liang, H., Shuman, J., Shulaev, V., Davletova, S., & Mittler, R. (2004). When defense pathways collide: The response of Arabidopsis to a combination of drought and heat stress. *Plant Physiology*, *134*(4), 1683–1696. https://doi.org/10.1104/pp.103.033431

Rolland, F., Moore, B., & Sheen, J. (2002). Sugar sensing and signaling in plants. *The Plant Cell*, *14*(suppl 1), S185–S205. https://doi.org/10.1105/tpc.010455

Schmidtmann, E., König, A. C., Orwat, A., Leister, D., Hartl, M., & Finkemeier, I. (2014). Redox regulation of Arabidopsis mitochondrial citrate synthase. *Molecular Plant*, *7*(1), 156–169. https://doi.org/10.1093/mp/sst144

Schwarzländer, M., König, A. C., Sweetlove, L. J., & Finkemeier, I. (2012). The impact of impaired mitochondrial function on retrograde signalling: A meta-analysis of transcriptomic responses. *Journal of Experimental Botany*, *63*(4), 1735–1750. https://doi.org/10.1093/jxb/err374

Shelp, B. J., Bozzo, G. G., Trobacher, C. P., Zarei, A., Deyman, K. L., & Brikis, C. J. (2012). Hypothesis/review: Contribution of putrescine to 4-aminobutyrate (GABA) production in response to abiotic stress. *Plant Science*, *193*, 130–135. https://doi.org/10.1016/j.plantsci.2012.06.001

Sienkiewicz-Porzucek, A., Nunes-Nesi, A., Sulpice, R., Lisec, J., Centeno, D. C., Carillo, P., …, & Fernie, A. R. (2008). Mild reductions in mitochondrial citrate synthase activity result in a compromised nitrate assimilation and reduced leaf pigmentation but have no effect on photosynthetic performance or growth. *Plant Physiology*, *147*(1), 115–127. https://doi.org/10.1104/pp.108.117978

Strommer, J. (2011). The plant ADH gene family. *The Plant Journal*, *66*(1), 128–142. https://doi.org/10.1111/j.1365-313X.2010.04458.x

Sweetlove, L. J., Beard, K. F., Nunes-Nesi, A., Fernie, A. R., & Ratcliffe, R. G. (2010). Not just a circle: Flux modes in the plant TCA cycle. *Trends in Plant Science*, *15*(8), 462–470. https://doi.org/10.1016/j.tplants.2010.05.006

Thelen, J. J., Miernyk, J. A., & Randall, D. D. (2000). Pyruvate dehydrogenase kinase from *Arabidopsis thaliana*: A protein histidine kinase that phosphorylates serine residues. *Biochemical Journal*, *349*(1), 195–201. https://doi.org/10.1042/bj3490195

Tovar-Méndez, A., Miernyk, J. A., & Randall, D. D. (2003). Regulation of pyruvate dehydrogenase complex activity in plant cells. *European Journal of Biochemistry*, *270*(6), 1043–1049. https://doi.org/10.1046/j.1432-1033.2003.03469.x

Tronconi, M. A., Fahnenstich, H., Gerrard Weehler, M. C., Andreo, C. S., Flügge, U. I., Drincovich, M. F., & Maurino, V. G. (2008). Arabidopsis NAD-malic enzyme functions as a homodimer and heterodimer and has a major impact

on nocturnal metabolism. *Plant Physiology*, *146*(4), 1540–1552. https://doi.org/10.1104/pp.107.114975
Turner, W. L., Plaxton, W. C., & Podestá, F. E. (2005). Plant cytosolic pyruvate kinase: A regulatory enzyme for glycolysis. *Phytochemistry Reviews*, *4*(2–3), 157–166.
Van Aken, O., Giraud, E., Clifton, R., & Whelan, J. (2009). Alternative oxidase: A target and regulator of stress responses. *Physiologia Plantarum*, *137*(4), 354–361. https://doi.org/10.1111/j.1399-3054.2009.01240.x
Van Aken, O., Pecenková, T., van de Cotte, B., De Rycke, R., Eeckhout, D., Fromm, H., …, & Van Breusegem, F. (2007). Mitochondrial type-I prohibitins of *Arabidopsis thaliana* are required for supporting proficient meristem development. *The Plant Journal*, *52*(5), 850–864. https://doi.org/10.1111/j.1365-313X.2007.03276.x
Van Aken, O., Zhang, B., Carrie, C., Uggalla, V., Paynter, E., Giraud, E., & Whelan, J. (2016). Defining the mitochondrial stress response in *Arabidopsis thaliana*. *Molecular Plant*, *9*(3), 440–450. https://doi.org/10.1016/j.molp.2015.12.013
Vanlerberghe, G. C. (2013). Alternative oxidase: A mitochondrial respiratory pathway to maintain metabolic and signaling homeostasis during abiotic and biotic stress in plants. *International Journal of Molecular Sciences*, *14*(4), 6805–6847. https://doi.org/10.3390/ijms14046805
Vercesi, A. E., Borecký, J., Maia, I. G., Arruda, P., Cuccovia, I. M., & Chaimovich, H. (2006). Plant uncoupling mitochondrial proteins. *Annual Review of Plant Biology*, *57*, 383–404. https://doi.org/10.1146/annurev.arplant.57.032905.105335
Zabalza, A., van Dongen, J. T., Froehlich, A., Oliver, S. N., Faix, B., Gupta, K. J., …, & Royuela, M. (2009). Regulation of respiration and fermentation to control the plant internal oxygen concentration. *Plant Physiology*, *149*(2), 1087–1098. https://doi.org/10.1104/pp.108.129288
Zheng, Q., Omanska-Klusek, A., Wiley, S. E., Robertson, I. M., Barua, B., & Shacter, E. (2019). O-GlcNAcylation regulates phosphorylation of tau: A mechanism involved in Alzheimer's disease. *Proceedings of the National Academy of Sciences*, *116*(12), 5451–5460.

5 Plant Enzymes and Their Role in Abiotic Stress Tolerance

Shifa Shaffique, Anis Ali Shah, Muhammad Imran, Muhammad Aaqil Khan, Javier A. Trujillo-Rugamas, Sang-Mo Kang, and In-Jung Lee

5.1 INTRODUCTION

Plants are basic providers of food, energy, and security. Unfortunately, they are threatened by adverse environmental challenges such as global climatic change conditions, global warming, and several kinds of abiotic stress, such as salinity, heat, drought, cold, flooding, and heavy metal toxicity (Ashraf & Harris, 2005; Chaffai et al., 2024; Costa & Farrant, 2019; Oshunsanya et al., 2019). Exposure to these stresses significantly causes oxidative stress in plants by the production of reactive oxygen species (ROS). This oxidative stress damages the biomolecules such as carbohydrates, proteins, and nucleic acids in plant cells, and damages the cell membrane's functional integrity (Banerjee & Roychoudhury, 2017; Choudhury et al., 2013, 2017; Miller et al., 2008). However, plants are sophisticated enough that they have an endogenous complex enzymatic and non-enzymatic regulatory network to regulate stress. Among these, enzymatic tolerance is the first line of defense that facilitates biochemical and metabolic reprogramming (Hasanuzzaman et al., 2011; Sharma et al., 2019).

5.2 ABIOTIC STRESS AND AGRICULTURE

Agricultural land is facing dual pressure: environmental stress and increasing demand for plant production. It is predicted that in 2050, the human population will increase to 10 billion. Moreover, it is also predicted that abiotic stress will be more prevalent all over the world (Ghazala, 2024; Junaid & Gokce, 2024; Minhas et al., 2017; Oshunsanya et al., 2019). Among all the plant stresses, abiotic stress is more prevalent, and it reduces the global crop production by 50–70%. However, with increasing global warming, abiotic stress can worsen soon, which will threaten food security (Chaffai et al., 2024; Yadav et al., 2023).

In response to stress, there is rapid generation of ROS such as hydrogen peroxide, superoxide radical, singlet oxygen, and hydroxyl ion. Mainly, they are generated in the chloroplasts, mitochondria, and peroxisomes of plant cells under stress conditions. When its concentration is low, it acts as a secondary messenger, while when its concentration is too high, it causes stress to the plant (Baxter et al., 2014; Demidchik, 2012; Gechev et al., 2006; Shaffique et al., 2025). It affects the plant's macro- and micro-biological molecules, such as carbohydrates, lipids, nucleic acids, cellular components, and their functional integrity, which negatively affect the plant's physiological, biochemical, and metabolic function (Apel & Hirt, 2004; Huang et al., 2019; Suzuki et al., 2012). However, in plants, there is a complex enzymatic system that plays a central role in ROS regulation, biochemical and metabolic regulation to sustain normal physio-biochemical processes. Enzymatic stress response depends upon several factors such as the type of abiotic stress, exposure duration, and its intensity (Dinakar et al., 2010; Gill & Tuteja, 2010; Hasanuzzaman et al., 2020; Mansoor et al., 2022; Navrot et al., 2007). For example, in response to drought stress, there is production of antioxidant and osmo-protectant enzymes that enhance stress tolerance and mitigate the negative effects (Iqbal, 2018; Ozturk et al., 2021). Similarly, in response to salt stress, there is production of enzymes that enhance ion regulation and facilitate osmolyte biosynthesis (Ahmad et al., 2019; Kamińska et al., 2022).

5.3 ANTIOXIDANT ENZYMATIC DEFENSE SYSTEM

Plants must face abiotic stress throughout their life cycle, from germination to harvesting. Exposure to stress causes disruption in biochemical metabolism in plants, and long-term stress can lead to several symptoms, which sometimes may result in plant death (Koyro et al., 2011; Shanker & Venkateswarlu, 2011). However, if the duration is no longer than, the plant stimulates the antioxidant defense system via activation of antioxidant defense molecules. Among antioxidant molecules, they are categorized into two forms: the enzymatic and non-enzymatic antioxidant defense molecules (Ahmad et al., 2010; Hasanuzzaman et al., 2011; Rajput et al., 2021). In this chapter, we will discuss the enzymatic defense mechanism. The activation of enzymatic defense molecules is a biochemical programming of cellular events that ultimately causes stress tolerance, stress signaling, and homeostasis of redox. Most importantly, seven important enzymatic molecules are involved in abiotic stress tolerance, as shown in Figure 5.1 and Table 5.1.

These enzymatic antioxidant molecules have the potential to enable physiological plant growth under marginal stress conditions. If the duration of stress becomes longer or is accompanied by other stressors, it may derange their metabolic adjustment and lead to metabolic reprogramming (Irato & Santovito, 2021; Rao et al., 2025). To ensure metabolic reprogramming, several enzymes facilitate the plant's potential adjustment in response to abiotic stress. For

DOI: 10.1201/9781003669920-5

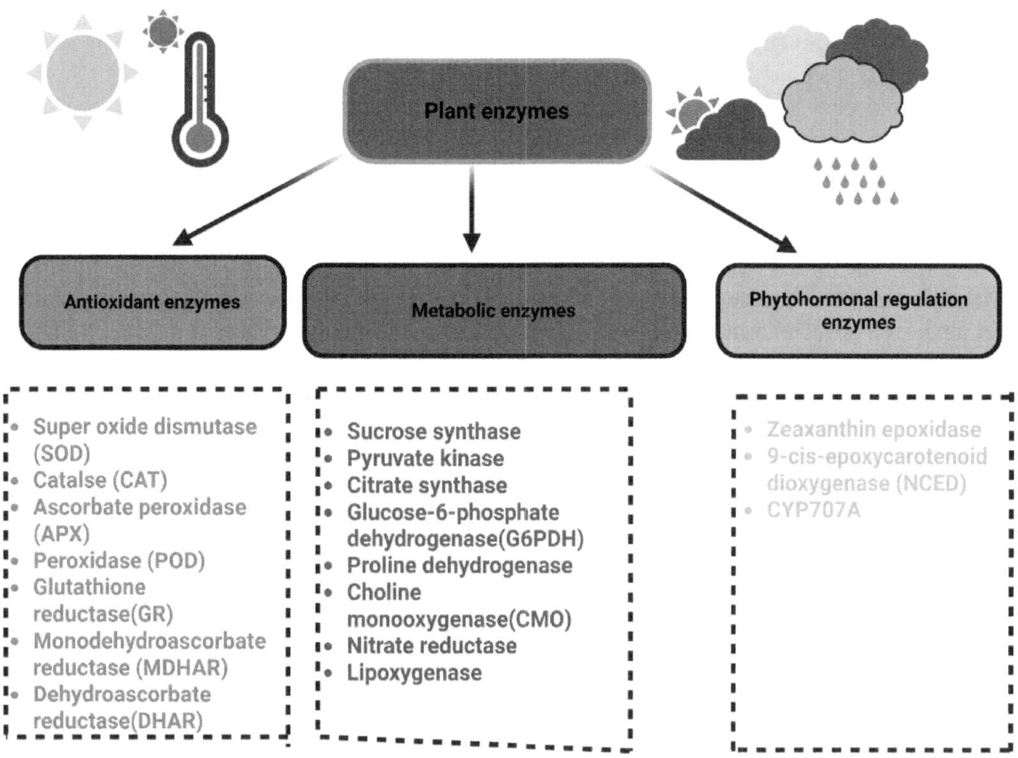

FIGURE 5.1 Plant enzymes and their function in regulating abiotic stress tolerance. The figure represents three important phytohormones based on their function: Antioxidant, metabolic, and phytohormonal regulatory enzymes.

instance, if antioxidant enzymes fail to reprogram cellular events, they further activate the sugar, organic acid, and amino acid synthesis and stimulate the Krebs cycle (Araujo et al., 2012; Zhang & Fernie, 2018, 2023). Several enzymatic molecules facilitate the Krebs cycle participation, such as sucrose synthase (SuSy), pyruvate kinase (PK), citrate synthase, glucose-6-phosphate dehydrogenase (G6PDH), proline dehydrogenase, choline monooxygenase (CMO), nitrate reductase (NR), and lipoxygenase (Araujo et al., 2012; Dhal & Pal, 2022; Scheible et al., 2000). The

TABLE 5.1
Antioxidant Defense Enzymatic Molecules, Their Presence in Plant Cells, and Their Role in Abiotic Stress Tolerance

Antioxidant Enzymatic Molecules	Biochemical Function	Presence	Role in Abiotic Stress Tolerance
Catalase (CAT)	Convert H_2O_2 into H_2O and O_2	Peroxisome	Protect peroxisomes from stress
Superoxide dismutase (SOD)	Catalyzes the removal of superoxide O^{2-} to H_2O_2 and O_2	Chloroplast Mitochondria Cytosol	Mitigate oxidative stress
Peroxidase (POD)	Catalyze H_2O_2-dependent oxidation of phenolic and non-phenolic substrates	Apoplast Cell wall vacuoles	Protect the cell wall, minimize the oxidation stress
Ascorbate peroxidase (APX)	Reduce H_2O_2 to water	Mitochondria Chloroplast Cytosol	Control line of ROS, protect photosynthetic machinery
Glutathione reductase (GR)	Reduction of oxidized glutathione (GSSG) to its reduced form, glutathione (GSH)	Chloroplast Cytosol Mitochondria	Facilitate the ascorbate glutathione cycle
Monodehydroascorbate reductase (MDHAR)	Regenerate ascorbate	Chloroplast Cytosol	Redox maintenance
Dehydroascorbate reductase (DHAR)	Regenerate ascorbate using glutathione	Mitochondria Chloroplast Cytosol	Sustain the ascorbate glutathione cycle

metabolic enzymes modulate carbohydrate, lipid, and nitrogen metabolism to adapt to the stress episode and facilitate stress tolerance. Several studies are in favor that SuSy plays a significant role in plant sugar metabolism, present mainly in the cytosol and plasma membrane of plant cells. It is involved in different metabolic pathways, such as cellulose and starch biosynthesis. It also facilitates the signaling process for stress tolerance and vascular tissue functioning (Stein & Granot, 2019). The futile cycles are connected and regulated by the SuSy to form a regulatory network. It tends to modulate in abiotic stress conditions to support carbon and energy demands (Nguyen-Quoc & Foyer, 2001). PK is an enzyme of glycolysis that tends to supply energy under unfavorable conditions and ensures the supply and demand of energy during stress conditions. It connects glycolysis to the tricarboxylic acid (TCA) cycles to provide energy in stressful situations, also known as regulatory enzymes in plants (Duggleby & Dennis, 1973; Tomlinson & Turner, 1973). G6PDH is an oxidoreductase enzyme that is primarily present in the cytosol and plastid of plant cells. Its main function is to provide the reducing power (NADPH) for maintaining carbon metabolism under stress conditions. Moreover, it has the potential to mediate redox homeostasis in abiotic stress conditions. Interestingly, it has a physiological impact on plant growth regulation from germination to stress alleviation (Jiang et al., 2022). Landi et al. (2016) conducted a study that revealed that G6PDH plays a fundamental role in modulating the drought stress response by triggering ROS in the experimental tomato plant. Proline dehydrogenase is mainly present in the mitochondrial inner membrane, which tends to facilitate the plant cells in post-stress recovery by maintaining the proline contents. Most importantly, it converts the stored form of proline into energy under stressful conditions via P5CS (Lehmann et al., 2010; Yoshiba et al., 1997). CMO is an foremost plant enzymes that help in glycine betaine biosynthesis in higher plants. It is located in the chloroplast stroma. Under abiotic stress conditions, it acts as an osmo-protector by converting choline into betaine aldehyde, which is oxidized by betaine. Under abiotic stress, it protects the photosynthetic machinery and regulates the enzymatic regulatory network (Hibino et al., 2002). NR is an oxidoreductase in nature that is found in the cytosol of plant cells. During nitrogen assimilation, it converts nitrate into nitrite. It is important that enzymes are involved in carbon–nitrogen interplay and induction of abiotic stress tolerance by nitric-oxide-mediated defense (Hoff et al., 1992; Meyer et al., 2005).

These enzymes provide substrates, reducing power, and energy for stimulation of the antioxidant defense activities (Zhang & Fernie, 2018). The integrated metabolic and antioxidant synergy leads to the upregulation of amino acids, osmo-protectant, organic acids, and membrane-assisted lipid that facilitate metabolic adjustment of the plant in stress (Nunes-Nesi et al., 2013; Omena-Garcia et al., 2017; Sweetlove et al., 2010; Zhang & Fernie, 2023) as shown in Figure 5.2 and Table 5.1.

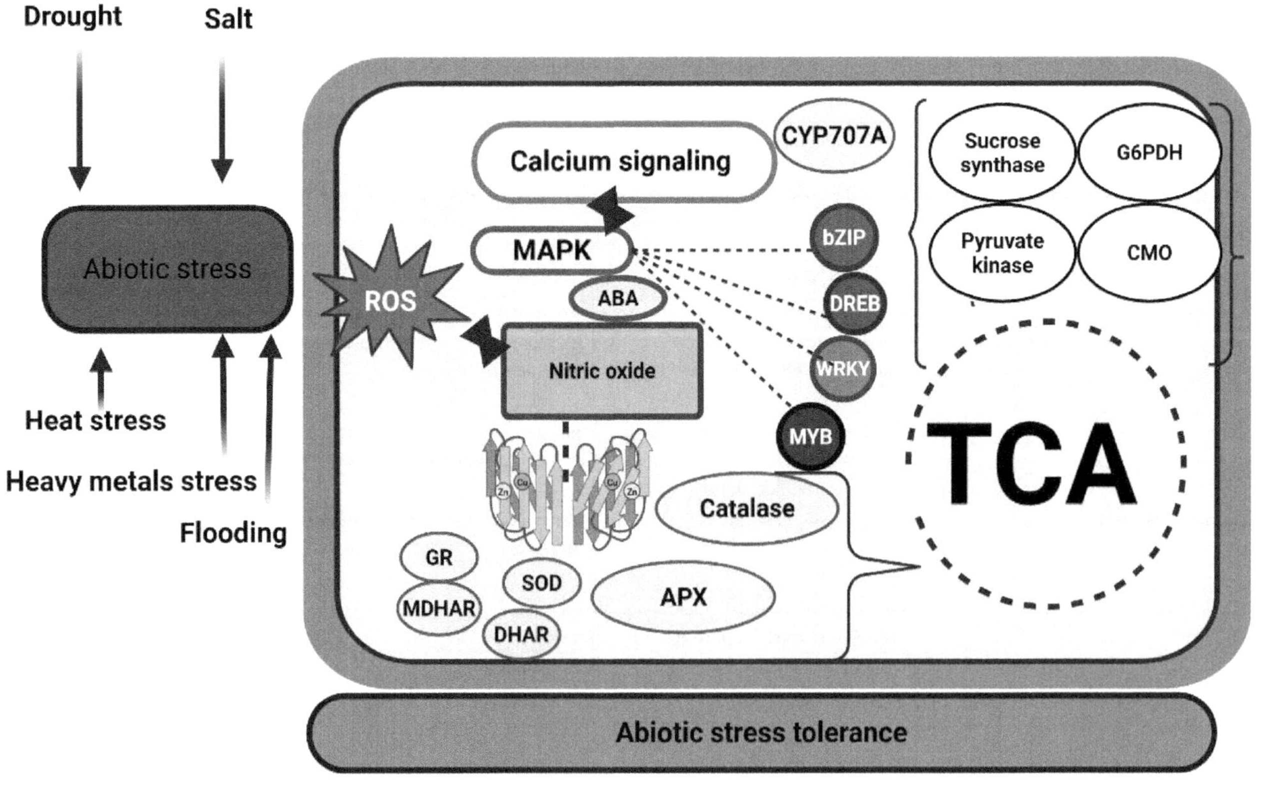

FIGURE 5.2 Effect of abiotic stress on plant cells and the generation of ROS, activation of phytohormones, antioxidant and metabolic enzymes, and their regulation in the mitigation of stress tolerance.

5.4 ENZYMES IN PHYTOHORMONAL REGULATION AND SIGNALING NETWORK

Plant physiology and biochemistry are not determined by only antioxidant defensive enzymes and metabolic enzymes adjustment but also by enzymes that trigger phytohormonal regulation. These enzymes are implicated in the biosynthesis and catabolism of phytohormones and their derivative compounds to induce environmental protection (Ciura & Kruk, 2018; EL Sabagh et al., 2022). They also contribute to the stress signaling initiation and biosynthesis, regulation, and catabolism. For example, in response to stress, the abscisic acid (ABA), also known as a stress phytohormone, rapidly accumulates, and its contents in plant cells are raised to activate further antioxidant defense molecules. However, to activate the biosynthesis of the ABA, several enzymes are involved, such as zeaxanthin epoxidase, abscisic aldehyde oxidase (AAO3), 9-cis-epoxycarotenoid dioxygenase (NCED), and, for catabolism, CYP707A (Endo et al., 2014; Jia et al., 2022; Nambara & Marion-Poll, 2005; Zeevaart & Yang, 2005). Rapid accumulation of ABA content leads to activation of the TCA cycles to regulate the stress condition (Dhal & Pal, 2022). In summary, it is stated that in the plant system, the enzymes that are responsible for biosynthesis, regulation, and catabolism of phytohormones are also important in triggering tolerance. Similarly, for jasmonic acid lipoxygenase, beta-oxidation enzymes are responsible. These enzymes modulate the synthesis of secondary metabolites and stimulate the antioxidant defense system.

5.5 MOLECULAR REGULATION OF STRESS-RELATED ENZYMES

Stress-related enzymatic regulation is governed by some transcriptional factors such as DREB, WRKY, and BZIP. These transcriptional factors bind with promoter motifs, regions such as HSE and DRE, to encode enzymatic genes (Aslam et al., 2019; Li et al., 2020). The signal transduction pathways, such as calcium signaling or MAPK, reactive oxygen and nitric species regulation, and ensure defense-related enzymatic activities (Jiang et al., 2017; Roychoudhury & Banerjee, 2017).

5.6 CONCLUSION AND FUTURE PROSPECTIVE

Enzymes are an important part of plant stress tolerance and adaptation under abiotic stress conditions. They act as dual functional molecules, biocatalysts as well as regulatory molecules. These enzymes enhance the metabolic regulation of carbon, nitrogen, and lipid, which improves the energy resource and membrane stability under stress conditions. Moreover, enzymes also contribute to the biosynthesis of phytohormones and signal transduction pathways that coordinate with the antioxidant defense system. These enzymes collectively orchestrate the multilayered defense regulatory network. At the molecular level, the expression pattern of these enzymes is regulated by transcriptional and post-transcriptional modification. Future studies on genetic engineering or biotechnology approaches to design plants with enzymes that improve plant protection under abiotic stress are required.

REFERENCES

Ahmad, P., Jaleel, C. A., Salem, M. A., Nabi, G., & Sharma, S. (2010). Roles of enzymatic and nonenzymatic antioxidants in plants during abiotic stress. *Critical Reviews in Biotechnology*, *30*(3), 161–175.

Ahmad, R., Hussain, S., Anjum, M. A., Khalid, M. F., Saqib, M., Zakir, I., Hassan, A., Fahad, S., & Ahmad, S. (2019). Oxidative stress and antioxidant defense mechanisms in plants under salt stress. In: *Plant abiotic stress tolerance: Agronomic, molecular and biotechnological approaches* (pp. 191–205). Springer.

Apel, K., & Hirt, H. (2004). Reactive oxygen species: Metabolism, oxidative stress, and signal transduction. *Annual Review of Plant Biology*, *55*(1), 373–399.

Araujo, W. L., NUNES-NESI, A., Nikoloski, Z., Sweetlove, L. J., & Fernie, A. R. (2012). Metabolic control and regulation of the tricarboxylic acid cycle in photosynthetic and heterotrophic plant tissues. *Plant, Cell & Environment*, *35*(1), 1–21.

Ashraf, M., & Harris, P. (2005). Abiotic stresses. In: *Plant resistance through breeding and molecular approaches.*

Aslam, R., Wali, Q., Sarwar, M., Naeem, M., & Zia, M. (2019). A comprehensive overview of transcription factors (WRKY, NAC and BZIP) in plants. *Journal of Biosciences*, *14*, 495–509.

Banerjee, A., & Roychoudhury, A. (2017). Abiotic stress, generation of reactive oxygen species, and their consequences: An overview. In: *Reactive oxygen species in plants: Boon or bane-revisiting the role of ROS* (pp. 23–50). Springer.

Baxter, A., Mittler, R., & Suzuki, N. (2014). ROS as key players in plant stress signalling. *Journal of Experimental Botany*, *65*(5), 1229–1240.

Chaffai, R., Ganesan, M., & Cherif, A. (2024). Abiotic stress in plants: Challenges and strategies for enhancing plant growth and development. In: *Plant adaptation to abiotic stress: From signaling pathways and microbiomes to molecular mechanisms* (pp. 1–30). Springer.

Choudhury, F. K., Rivero, R. M., Blumwald, E., & Mittler, R. (2017). Reactive oxygen species, abiotic stress and stress combination. *The Plant Journal*, *90*(5), 856–867.

Choudhury, S., Panda, P., Sahoo, L., & Panda, S. K. (2013). Reactive oxygen species signaling in plants under abiotic stress. *Plant Signaling & Behavior*, *8*(4), e23681.

Ciura, J., & Kruk, J. (2018). Phytohormones as targets for improving plant productivity and stress tolerance. *Journal of Plant Physiology*, *229*, 32–40.

Costa, M.-C. D., & Farrant, J. M. (2019). Plant resistance to abiotic stresses. In (Vol. 8, pp. 553): MDPI.

Demidchik, V. (2012). Reactive oxygen species and oxidative stress in plants. In: *Plant stress physiology* (pp. 24–58). CABI Wallingford UK.

Dhal, S., & Pal, H. (2022). Regulation of glycolysis and Krebs cycle during biotic and abiotic stresses. In: *Photosynthesis and respiratory cycles during environmental stress response in plants* (pp. 263–308). Apple Academic Press.

Dinakar, C., Abhaypratap, V., Yearla, S. R., Raghavendra, A. S., & Padmasree, K. (2010). Importance of ROS and antioxidant system during the beneficial interactions of mitochondrial metabolism with photosynthetic carbon assimilation. *Planta*, *231*(2), 461–474.

Duggleby, R. G., & Dennis, D. T. (1973). Pyruvate kinase, a possible regulatory enzyme in higher plants. *Plant Physiology*, *52*(4), 312–317.

EL Sabagh, A., Islam, M. S., Hossain, A., Iqbal, M. A., Mubeen, M., Waleed, M., Reginato, M., Battaglia, M., Ahmed, S., & Rehman, A. (2022). Phytohormones as growth regulators during abiotic stress tolerance in plants. *Frontiers in Agronomy*, *4*, 765068.

Endo, A., Okamoto, M., & Koshiba, T. (2014). ABA biosynthetic and catabolic pathways. In: *Abscisic acid: Metabolism, transport and signaling* (pp. 21–45). Springer.

Gechev, T. S., Van Breusegem, F., Stone, J. M., Denev, I., & Laloi, C. (2006). Reactive oxygen species as signals that modulate plant stress responses and programmed cell death. *Bioessays*, *28*(11), 1091–1101.

Ghazala, A. S. M. (2024). Abiotic threats of agriculture. In: *Achieving food security through sustainable agriculture* (pp. 103–142). IGI Global.

Gill, S. S., & Tuteja, N. (2010). Reactive oxygen species and antioxidant machinery in abiotic stress tolerance in crop plants. *Plant Physiology and Biochemistry*, *48*(12), 909–930.

Hasanuzzaman, M., Bhuyan, M. B., Parvin, K., Bhuiyan, T. F., Anee, T. I., Nahar, K., Hossen, M. S., Zulfiqar, F., Alam, M. M., & Fujita, M. (2020). Regulation of ROS metabolism in plants under environmental stress: A review of recent experimental evidence. *International Journal of Molecular Sciences*, *21*(22), 8695.

Hasanuzzaman, M., Hossain, M. A., da Silva, J. A. T., & Fujita, M. (2011). Plant response and tolerance to abiotic oxidative stress: Antioxidant defense is a key factor. In: *Crop stress and its management: Perspectives and strategies* (pp. 261–315). Springer.

Hibino, T., Waditee, R., Araki, E., Ishikawa, H., Aoki, K., Tanaka, Y., & Takabe, T. (2002). Functional characterization of choline monooxygenase, an enzyme for betaine synthesis in plants. *Journal of Biological Chemistry*, *277*(44), 41352–41360.

Hoff, T., Stummann, B. M., & Henningsen, K. W. (1992). Structure, function and regulation of nitrate reductase in higher plants. *Physiologia Plantarum*, *84*(4), 616–624.

Huang, H., Ullah, F., Zhou, D.-X., Yi, M., & Zhao, Y. (2019). Mechanisms of ROS regulation of plant development and stress responses. *Frontiers in Plant Science*, *10*, 800.

Iqbal, M. J. (2018). Role of osmolytes and antioxidant enzymes for drought tolerance in wheat. *Global Wheat Production*, *51*, 51–66.

Irato, P., & Santovito, G. (2021). Enzymatic and non-enzymatic molecules with antioxidant function. In (Vol. 10, pp. 579): MDPI.

Jia, K.-P., Mi, J., Ali, S., Ohyanagi, H., Moreno, J. C., Ablazov, A., Balakrishna, A., Berqdar, L., Fiore, A., & Diretto, G. (2022). An alternative, zeaxanthin epoxidase-independent abscisic acid biosynthetic pathway in plants. *Molecular Plant*, *15*(1), 151–166.

Jiang, J., Ma, S., Ye, N., Jiang, M., Cao, J., & Zhang, J. (2017). WRKY transcription factors in plant responses to stresses. *Journal of Integrative Plant Biology*, *59*(2), 86–101.

Jiang, Z., Wang, M., Nicolas, M., Ogé, L., Pérez-Garcia, M.-D., Crespel, L., Li, G., Ding, Y., Le Gourrierec, J., Grappin, P., & Sakr, S. (2022). Glucose-6-phosphate dehydrogenases: The hidden players of plant physiology. *International Journal of Molecular Sciences*, *23*(24), 16128. https://www.mdpi.com/1422-0067/23/24/16128

Junaid, M., & Gokce, A. (2024). Global agricultural losses and their causes. *Bulletin of Biological and Allied Sciences Research*, *2024*(1), 66–66.

Kamińska, I., Lukasiewicz, A., Klimek-Chodacka, M., Długosz-Grochowska, O., Rutkowska, J., Szymonik, K., & Baranski, R. (2022). Antioxidative and osmoprotecting mechanisms in carrot plants tolerant to soil salinity. *Scientific Reports*, *12*(1), 7266.

Koyro, H.-W., Ahmad, P., & Geissler, N. (2011). Abiotic stress responses in plants: An overview. In: *Environmental adaptations and stress tolerance of plants in the era of climate change* (pp. 1–28). Springer.

Landi, S., Nurcato, R., De Lillo, A., Lentini, M., Grillo, S., & Esposito, S. (2016). Glucose-6-phosphate dehydrogenase plays a central role in the response of tomato (*Solanum lycopersicum*) plants to short and long-term drought. *Plant Physiology and Biochemistry*, *105*, 79–89. https://doi.org/10.1016/j.plaphy.2016.04.013

Lehmann, S., Funck, D., Szabados, L., & Rentsch, D. (2010). Proline metabolism and transport in plant development. *Amino Acids*, *39*(4), 949–962.

Li, W., Pang, S., Lu, Z., & Jin, B. (2020). Function and mechanism of WRKY transcription factors in abiotic stress responses of plants. *Plants*, *9*(11), 1515.

Mansoor, S., Ali Wani, O., Lone, J. K., Manhas, S., Kour, N., Alam, P., Ahmad, A., & Ahmad, P. (2022). Reactive oxygen species in plants: From source to sink. *Antioxidants*, *11*(2), 225.

Meyer, C., Lea, U. S., Provan, F., Kaiser, W. M., & Lillo, C. (2005). Is nitrate reductase a major player in the plant NO (nitric oxide) game? *Photosynthesis Research*, *83*(2), 181–189.

Miller, G., Shulaev, V., & Mittler, R. (2008). Reactive oxygen signaling and abiotic stress. *Physiologia Plantarum*, *133*(3), 481–489.

Minhas, P. S., Rane, J., & Pasala, R. K. (2017). Abiotic stresses in agriculture: An overview. In: *Abiotic stress management for resilient agriculture* (3–8). Springer.

Nambara, E., & Marion-Poll, A. (2005). Abscisic acid biosynthesis and catabolism. *Annual Review of Plant Biology*, *56*(1), 165–185.

Navrot, N., Rouhier, N., Gelhaye, E., & Jacquot, J. P. (2007). Reactive oxygen species generation and antioxidant systems in plant mitochondria. *Physiologia Plantarum*, *129*(1), 185–195.

Nguyen-Quoc, B., & Foyer, C. H. (2001). A role for 'futile cycles' involving invertase and sucrose synthase in sucrose metabolism of tomato fruit. *Journal of Experimental Botany*, *52*(358), 881–889.

Nunes-Nesi, A., Araújo, W. L., Obata, T., & Fernie, A. R. (2013). Regulation of the mitochondrial tricarboxylic acid cycle. *Current Opinion in Plant Biology*, *16*(3), 335–343.

Omena-Garcia, R. P., Araújo, W. L., Gibon, Y., Fernie, A. R., & Nunes-Nesi, A. (2017). Measurement of tricarboxylic acid cycle enzyme activities in plants. In: *Plant respiration and internal oxygen: Methods and protocols* (pp. 167–182). Springer.

Oshunsanya, S. O., Nwosu, N. J., & Li, Y. (2019). Abiotic stress in agricultural crops under climatic conditions. In: *Sustainable agriculture, forest and environmental management* (pp. 71–100). Springer.

Ozturk, M., Turkyilmaz Unal, B., García-Caparrós, P., Khursheed, A., Gul, A., & Hasanuzzaman, M. (2021). Osmoregulation and its actions during the drought stress in plants. *Physiologia Plantarum*, *172*(2), 1321–1335.

Rajput, V. D., Harish, Singh, R. K., Verma, K. K., Sharma, L., Quiroz-Figueroa, F. R., Meena, M., Gour, V. S., Minkina, T., & Sushkova, S. (2021). Recent developments in enzymatic antioxidant defence mechanism in plants with special reference to abiotic stress. *Biology*, *10*(4), 267.

Rao, M. J., Duan, M., Zhou, C., Jiao, J., Cheng, P., Yang, L., Wei, W., Shen, Q., Ji, P., & Yang, Y. (2025). Antioxidant defense system in plants: Reactive oxygen species production, signaling, and scavenging during abiotic stress-induced oxidative damage. *Horticulturae*, *11*(5), 477.

Roychoudhury, A., & Banerjee, A. (2017). Abscisic acid signaling and involvement of mitogen activated protein kinases and calcium-dependent protein kinases during plant abiotic stress. *Mechanism of Plant Hormone Signaling Under Stress*, *1*, 197–241.

Scheible, W. R., Krapp, A., & Stitt, M. (2000). Reciprocal diurnal changes of phosphoenolpyruvate carboxylase expression and cytosolic pyruvate kinase, citrate synthase and NADP-isocitrate dehydrogenase expression regulate organic acid metabolism during nitrate assimilation in tobacco leaves. *Plant, Cell & Environment*, *23*(11), 1155–1167.

Shaffique, S., Aaqil Khan, M., Shahzad, R., Kang, S.-M., & Lee, I.-J. (2025). Chapter 4 – interplay of reactive oxygen species and reactive nitrogen species: Unrevealing their signaling and metabolic mechanism in response to biotic and abiotic stresses. In A. Husen (Ed.), *Role of antioxidants in mitigating plant stress* (pp. 63–74). Academic Press. https://doi.org/10.1016/B978-0-443-26799-4.00015-3

Shanker, A., & Venkateswarlu, B. (2011). *Abiotic stress response in plants: Physiological, biochemical and genetic perspectives*. BoD–Books on Demand.

Sharma, P., Jha, A. B., & Dubey, R. S. (2019). Oxidative stress and antioxidative defense system in plants growing under abiotic stresses. In: *Handbook of plant and crop stress, fourth edition* (pp. 93–136). CRC press.

Stein, O., & Granot, D. (2019). An overview of sucrose synthases in plants. *Frontiers in Plant Science*, *10*, 95.

Suzuki, N., Koussevitzky, S., Mittler, R., & Miller, G. (2012). ROS and redox signalling in the response of plants to abiotic stress. *Plant, Cell & Environment*, *35*(2), 259–270.

Sweetlove, L. J., Beard, K. F., Nunes-Nesi, A., Fernie, A. R., & Ratcliffe, R. G. (2010). Not just a circle: Flux modes in the plant TCA cycle. *Trends in Plant Science*, *15*(8), 462–470.

Tomlinson, J. D., & Turner, J. F. (1973). Pyruvate kinase of higher plants. *Biochimica et Biophysica Acta (BBA)-General Subjects*, *329*(1), 128–139.

Yadav, R. K., Tripathi, M. K., Tiwari, S., Tripathi, N., Asati, R., Chauhan, S., Tiwari, P. N., & Payasi, D. K. (2023). Genome editing and improvement of abiotic stress tolerance in crop plants. *Life*, *13*(7), 1456.

Yoshiba, Y., Kiyosue, T., Nakashima, K., Yamaguchi-Shinozaki, K., & Shinozaki, K. (1997). Regulation of levels of proline as an osmolyte in plants under water stress. *Plant and Cell Physiology*, *38*(10), 1095–1102.

Zeevaart, J. A., & Yang, S.-H. (2005). Abscisic acid metabolism. *Annual Review of Plant Biology*, *56*, 165–185.

Zhang, Y., & Fernie, A. R. (2018). On the role of the tricarboxylic acid cycle in plant productivity. *Journal of Integrative Plant Biology*, *60*(12), 1199–1216.

Zhang, Y., & Fernie, A. R. (2023). The role of TCA cycle enzymes in plants. *Advanced Biology*, *7*(8), 2200238.

6 Enzyme Systems in Plant–Water Relations and Drought Tolerance

Umair Rasheed, Abdul Sattar, Ahmad Sher, and Muhammad Ijaz

6.1 INTRODUCTION

Plant–water relations are the processes through which plants absorb, transport, and lose water, which plays a central role in the existence, growth, and metabolism of plants and creatures (Farooq et al. 2008). Water is essential to ensure turgor in plant cells, nutrient mobility, and photosynthesis. Water scarcity, a characteristic of drought stress, disrupts these processes and has been found to be one of the critical reasons limiting plant productivity at a global scale (Sun et al. 2020). Drought stress causes a decrease in leaf size, stem lengthening, and root growth and disrupts water and nutrient balances, which results in a decrease in water-use effectiveness and photosynthetic potential (Seleiman et al. 2021). Stomatal closure is among the major physiological activities to conserve water, but it also restricts CO_2 uptake, a process that decreases photosynthesis and leads to the formation of reactive oxygen species (ROS) that might harm cellular components (Sun et al. 2020; Wu et al. 2022). Antioxidant defense mechanisms, osmotic accumulation of proline and glycine betaine, and changes in hormone concentration, specifically the amounts of abscisic acid (ABA), are the various ways plants respond to counteract these effects (Wahab et al. 2022).

The ways to adapt to drought include morphological and molecular adaptation, such as formation of deeper or larger root systems to enhance water uptake, decreased leaf area to reduce transpiration, and synthesis of defensive proteins and compatible solutes, which improve cell turgor and cell structures (Farooq et al. 2008; Seleiman et al. 2021). Immediately, drought stress evokes the activity of certain genes and transcription factors, including dehydration-responsive element-binding proteins, aquaporins, and late developmental abundant proteins, to increase the capacity of a plant to endure water lack (Wu et al. 2022). According to recent studies, the involvement of epigenetic changes in the activation of gene expression during the drought phase and thereby the ability of plants to adjust and even impart stress resistance to their offspring generation may be considered (Aanniz et al. 2025). The severity and duration of drought, as well as the plant's capacity for recovery, depend on the efficiency of these physiological and molecular responses, with recovery of water relations and gas exchange often lagging behind the restoration of leaf water potential after rehydration (Ruehr et al. 2019). Understanding these complex water relations and drought responses is crucial for developing strategies to improve crop resilience, including breeding for drought-tolerant varieties, optimizing irrigation, and employing biotechnological approaches to enhance stress adaptation (Seleiman et al. 2021).

Enzyme systems are important attributes of adaptation to variable and changing environmental conditions by plants, having an impact on growth, development, and stress responsiveness (Oliveira-Filho et al. 2023). Adaptation is particularly important when transferring non-native enzymes to plants since these proteins will need to adapt to the intracellular environment, such as redox potential and metabolites, in order to work efficiently in crop improvement (Oliveira-Filho et al. 2023). Among key examples, the cytochrome P450 enzymes, by way of gene duplication and neofunctionalization, contribute to incredible plasticity across evolutionary history, rendering the ability of both genes and pathways to structure metabolism emerging lineage-specific and denoting adaptation to new conceptions needed in a new ecological niche (Hansen et al. 2021). Enzymes are also involved in the regulation of hormone biosynthesis and modification that influence plant perception and responsiveness to environmental cues like drought, salinity, and pathogens (Westfall et al. 2013) (Figure 6.1). Advances in directed evolution are enhancing enzyme functioning, resulting in enzyme versions with enhanced catalytic efficiency and stability, which facilitates quicker plant adaptation and aids in metabolic engineering (García-García et al. 2021). Enzyme-assisted extraction techniques also validate their relevance as they would release bioactive compounds which not only enhance the crop yield and quality but also contribute to human and animal health (Štreimikytė et al. 2022). Diversification of enzymes like coumaroyl conjugate 3-hydroxylases has been key to the synthesis of lignin or to defend against predators, representing relevant evolutionary innovations in land plants (Alber et al. 2019). Studies related to enzyme catalysis reveal how activity and stability are fine-tuned in response to environmental conditions, providing insights into the molecular mechanisms that enable the adaptation of plants and challenge the standard models linking activity and stability to enzyme engineering utilises naturally occurring stress-resistant enzymes and extremozymes to relocate desirable catalytic functions to crop plants, endowing them with enhanced stress-resistance characteristics and deserve (Chatterjee et al. 2023). Collectively, these findings highlight the wider significance of enzyme systems, not only

DOI: 10.1201/9781003669920-6

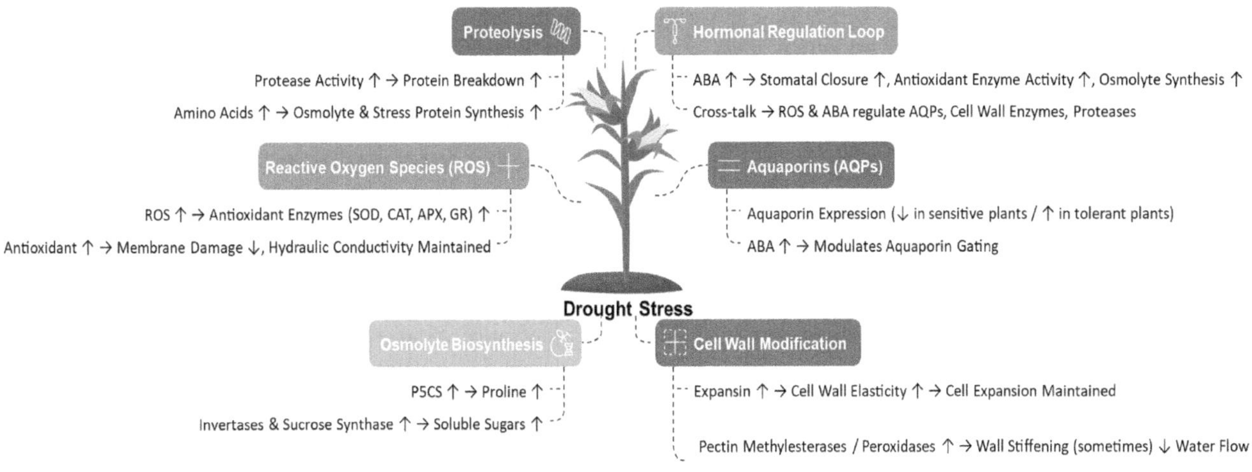

FIGURE 6.1 Enzymatic regulation of plant–water relations under drought stress.

for understanding plant biology but also for crop improvement, sustainable agriculture, and the development of new biotech innovations.

6.2 PLANT–WATER RELATIONS: A PHYSIOLOGICAL OVERVIEW

6.2.1 Water Uptake, Transport, and Loss in Plants

The process of water uptake, transfer, and evaporation in plants is regulated by the physiological processes and the surrounding conditions. Roots absorb the water in the soil depending on the soil texture, moisture, and resistance. The root structure is developed to maximize water uptake (Scharwies and Dinneny 2019). Water moves in roots through two ways, i.e., through the cell wall (apoplastic) or the intercellular way (symplastic). This flow is controlled by root conductivity, which is determined by the structure and proteins of aquaporins. Aquaporins increase water transport across membranes (Steudle 2000). Water subsequently flows upwards through xylem tissue that is pushed upwards by gradients formed during the transpiration of leaves. The updated technology of fast neutron tomography is showing the dynamics of water movement across roots in real-time (Tötzke et al. 2021).

Water uptake directly through foliar absorption of water by plants, which forms an important water subsidy where soil is dry (Berry et al. 2018). Water absorption and water loss are controlled by leaf cuticles and stomata, and their efficiency depends on the humidity, rainfall, and atmospheric particles (Fernández et al. 2020; Koch et al. 2025). The loss of water is driven by transpiration with stomatal regulation depending on atmospheric demand but is not exceeded by available root supply rendering the soil–plant–water pathway the bottleneck (Jackson et al. 2000). Silicon enhances water status even during stressed conditions but makes root uptake more and the activity of aquaporins better instead of decreasing loss (Chen et al. 2018). It will cause plants at the ecosystem levels to regulate the flow of water by using root uptake and transpiration strategies, by species-specific root uptake and transpiration arrangement, in response to environmental factors (Dubbert and Werner 2018). This complex root–foliar uptake, transportation, and controlled loss enables plants to balance water during different conditions and in their interactions with diverse stresses (Wang et al. 2021).

6.2.2 Role of Roots, Xylem, Stomata, and Aquaporins

Plant–water relations depend on interlinked operations of roots, xylem, stomata, and aquaporins and on phases of water absorption, transport, and regulation. The main organ of absorption is the root system, the efficiency of which depends on the architecture of the root system and the modulation of the hydraulic conductivity by aquaporins sensitive to soil moisture and stress conditions (Javot and Maurel 2002; Gambetta et al. 2017). Once absorbed, water is transported through the xylem, longer-distance conduction of which is dependent on the structure of vessels; the xylem parenchyma and the related aquaporins. These aquaporins support the repair of an embolism, along with stabilizing the transport with cycles of drought and rehydration (Secchi et al. 2017). At the leaf surface, stomata control transpiration and the exchange of gases. Their opening and closing are known to be sensitive to environmental stimuli but also to internal ones, with aquaporins present in guard cells being responsible for water flux and thus opening and closing the stomata, directly modulating transpiration efficiency (Pawlowicz and Masajada, 2019). Beyond the role of transporting water, some forms of aquaporin also transport small molecules such as CO_2, providing yet another connection between water regulation, photosynthesis, and thermal adaptation (Kaldenhoff et al. 2008). Aquaporin function is tightly regulated by mechanisms such as phosphorylation, changes in the pH conditions, and ROS, allowing plants to adjust to biotic or abiotic stress conditions (Chaumont and Tyerman 2014). They further play their roles in root system establishment, nutrient uptake, photosynthetic performance, and post-drought recovery, contributing toward plant strength (Patel and Mishra 2021). Collectively, roots, xylem, stomata, and aquaporins form an integrated system for maintenance of water balance, improvement of growth, and increase in survival under fluctuating environmental conditions (Wang et al. 2020).

6.3 IMPACT OF DROUGHT STRESS ON WATER BALANCE

Drought stress largely interferes with water balance in plants as it causes a decrease in the water supply and results in a cascade of physiological, biochemical, and morphological shifts that hamper growth and productivity (Farooq et al. 2008). Decreasing the surface area of the leaves, extending the stems, and increasing roots is one of the first lines of restoring moisture to plants and helps reduce water loss to the environment and preserve water supplies within the plant (Farooq et al. 2008). There is also a significant decrease in stomatal conductance in drought stress as plants close their stomata in order to reduce evaporation, but the effect of it also limits CO_2 uptake and reduces photosynthetic velocity, further influencing plant energy balance and growth (Marchin et al. 2020). Stomatal closure, in turn, leading to reduced transpiration, may finally result in leaves reaching an elevated temperature, exposing plants to increased thermal damage and oxidative stress through ROS build-up (Sun et al. 2020). Plants prevent wilting by depositing osmolytes, such as proline, glycine betaine, and soluble sugars, and maintain cells' water supply and ensure the stability of membranes and proteins when prone to drought (Pamungkas et al. 2022). ABA is another phytohormone synthesized during drought, potentially leading to stomatal closure and activation of stress genes, but at the cost of late embryogenesis abundance proteins and dehydrins (Mukarram et al. 2021). At molecular scales, drought stress activates transcription of certain genes and transcription factors, including dehydration-responsive element binding proteins and aquaporins that could control water uptake and withstand drought (Yang et al. 2021). Additionally, the effects of drought stress can include impairment of chloroplast structures and a reduction in photosynthetic capabilities, changes in pigment structure, and an induction of the formation of ROS, causing cell injury when antioxidant protection is not sufficient (Razi and Muneer 2021). Another way water adapts plants to drought is through the course of altering the proportion of root to shoot, such as lengthening and deepening roots, to get to deeper soil moisture and improving water-taking efficiency (Seleiman et al. 2021). Silicon release was eventually demonstrated to positively affect the water status of the plants in drought by improving water uptake, stomatal regulation, aiding osmotic adjustment, and protecting water balance in the entire plant (Wang et al. 2021).

6.4 ENZYMATIC REGULATION IN DROUGHT RESPONSE

6.4.1 Overview of Enzymatic Functions Under Abiotic Stress

Enzymes are crucial to plant processes and tolerance to a wide range of abiotic stresses, such as drought, salinity, temperature extremes, and exposure to heavy metals, mainly by maintaining cell integrity and preventing oxidative damage (Rajput et al. 2021). In abiotic stress, the excess accumulation of ROS can cause protein, lipid, and nucleic acid damage unless it is properly detoxified via antioxidant enzymes (Mishra et al. 2023). The major enzyme-based antioxidants that reduce the cellular toxicity include superoxide dismutases (SOD), which catalyze the breaking of the superoxide radicals into hydrogen peroxide and oxygen, as well as catalase (CAT), which helps in the dissociation of the hydrogen peroxide into oxygen and water (Anjum et al. 2016; Rajput et al. 2021). The role of ascorbate and glutathione is also irreplaceable, with ascorbate peroxidase (APX) and glutathione peroxidase (GPX) breaking down hydrogen peroxide and lipid peroxides like ascorbate and glutathione, respectively, in maintenance of redox homeostasis in various cellular compartments (Anjum et al. 2016). The glutathione–ascorbate cycle that includes glutathione reductase (GR), monodehydroascorbate reductase (MDHAR), and dehydroascorbate reductase (DHAR) is crucial in the restoration of reduced ascorbate and glutathione, in turn, maintaining ROS scavenging during unfavorable environmental circumstances (Rajput et al. 2021). Enzymatic regulation of ROS detoxification is not isolated, and enzymes are involved in the synthesis and breakdown of osmoprotectants such as proline that accumulates during stress to protect proteins and membranes, stabilize osmotic balance, and elevate other antioxidant enzyme activities (Hosseinifard et al. 2022). Abiotic stress provides a dynamically adjusted amino acid metabolism, where there is a higher rate of compatible solute synthesis (proline, arginine, and asparagine) released, and other amino acids are accumulated as a result of elevated protein cycling to contribute to stress adaptation and nitrogen storage (Hildebrandt 2018). In addition to the ROS regulation and osmolyte regulation, enzyme activities are also involved in lipid signaling pathways, whereby enzymes produce lipid-derived signaling molecules that mediate stress responses in cells and regulate the expression of gene transcription (Liang et al. 2023). Protective functions are also attributed to enzymes that participate in the biosynthesis of signaling molecules such as hydrogen sulfide (H_2S) and nitric oxide (NO) that mediate post-translational modifications and the opening of stress-response pathways, which further improve tolerance of plants to detrimental conditions (Saini et al. 2024). The enzyme-based antioxidant systems are not merely important as they act as targets of immediate response to stress but also of genetic engineering and breeding technologies to create stress-resistant crops (Mishra et al. 2023). These enzymatic defenses can be very effective when spatially and temporally controlled and can in addition collaborate with other physiologic and biochemical processes, so a balanced action is essential to the survival of plants during abiotic stress (You and Chan 2015).

6.4.2 General Classification of Stress-Responsive Enzymes

Stress-responsive enzymes are a heterogeneous group, without which organisms, in particular plants, are unable to adapt in a variety of severe environments, including drought, salinity, and heat, as well as oxidative stress. The most dominant of them are antioxidant enzymes, such as superoxide dismutase (SOD), catalase (CAT), peroxidases (POX), glutathione peroxidase (GPX), glutathione reductase (GR), ascorbate peroxidase (APX), monodehydroascorbate reductase (MDHAR), and dehydroascorbate reductase (DHAR). These enzymes work together to neutralize and detoxify ROS, thus averting oxidative injury to other cellular parts and re-establishing redox homeostasis in stressful situations (Rajput et al. 2021). Indicatively, SOD helps to transform superoxide radicals into hydrogen peroxide, which may be decomposed by CAT and POX, and hydrogen peroxide is further reduced by two more (APX and GPX) with ascorbate and glutathione as electron donors (Lee et al. 2001; Rajput et al. 2021). As shown in Table 6.1.

One more significant family is mitogen-activated protein kinases (MAPKs), which are c-Jun N-terminal kinases (JNKs) and p38-MAPKs. All these stress signals activate these kinases, which are primary minor transducers in signaling pathways that regulate gene expression, reorganization of the cytoskeleton, and cell survival or apoptosis to stress (Sugden and Clerk 1998). There are also new protein kinases, e.g., Krs-1 and Krs-2, that are directly triggered by specific stresses or apoptotic stimuli, indicating further involvement in early stress-specific phosphorylation events (Taylor et al. 1996).

Stress adaptation is also strongly associated with DNA repair enzymes. DNA glycosylases, such as those involved in the base excision repair pathway base excision repair, localize and repair the damage that happens in the DNA due to environmental stress, and maintaining genome stability and promoting the growth of plants in unfavorable conditions (Zhao et al. 2025). Also, glutaredoxins are oxidoreductase enzymes, the presence of which is essential to the operation of oxidative stress responses, particularly to delay oxidative damage by the reduction of ROS and maintaining thiol homeostasis (Malik et al. 2020).

6.5 KEY ENZYME SYSTEMS IN DROUGHT TOLERANCE

6.5.1 Antioxidant Enzymes

ROS are produced during aerobic metabolism and aggravated under stressful conditions and represent an important threat to cell integrity (Dvořák et al. 2021). The main antioxidant enzymes include superoxide dismutase (SOD), catalase (CAT), ascorbate peroxidase (APX), and glutathione reductase (GR), which all work together to counter redox imbalance and restore redox balance within plant and animal systems (Dey et al. 2016). These findings are summarized in Table 6.2a and Table 6.2b.

6.5.1.1 Superoxide Dismutase (SOD)

The initial defense against ROS is SOD that performs the reaction of converting superoxide anion ($O_2^{\bullet -}$) into hydrogen peroxide (H_2O_2) and molecular oxygen (O_2), thus minimizing chances of superoxide causing cellular damage (Mates 2000). SOD has several isoforms, including Cu/Zn-SOD, which is found in the cytosol, Mn-SOD, which is found in mitochondria, and Fe-SOD, which is found in chloroplasts, giving each a specific compartmental protection (Vives-Bauza et al. 2007). Environmental stresses, including drought, salinity, and pathogen attack, induce the rapid upregulation of SOD activity, which underscores its critical role in coping with stress (Berwal et al. 2021). Besides its protective role, SOD also distinctly affects cell

TABLE 6.1
General Classification of Stress-Responsive Enzymes and Their Main Functions

Enzyme Class	Main Function Under Stress	Reference
Superoxide dismutase (SOD)	Converts superoxide radicals to H_2O_2 and O_2	Rajput et al. (2021)
Catalase (CAT)	Breaks down H_2O_2 to water and O_2	Rajput et al. (2021)
Peroxidases (POX)	Scavenges H_2O_2, cell wall modification, signaling	Rajput et al. (2021)
Ascorbate peroxidase (APX)	Uses ascorbate to reduce H_2O_2 to water	Lee et al. (2001)
Glutathione peroxidase (GPX)	Reduces H_2O_2 and lipid peroxides using glutathione	Rajput et al. (2021)
Glutathione reductase (GR)	Regenerates reduced glutathione (GSH)	Rajput et al. (2021)
MAPKs (JNKs, p38-MAPKs)	Signal transduction, gene regulation	Sugden and Clerk (1998)
Krs-1, Krs-2 protein kinases	Early phosphorylation in stress response	Taylor et al. (1996)
DNA glycosylases	DNA repair, genome protection	Zhao et al. (2025)
Glutaredoxins	Redox regulation, ROS detoxification	Malik et al. (2020)

TABLE 6.2(a)
Major Antioxidant Enzymes and Their Functions in ROS Scavenging

Enzyme	Main Function in ROS Scavenging	Cellular Location	Reference
Superoxide dismutase	Converts superoxide anion ($O_2\bullet^-$) to H_2O_2 and O_2	Cytosol, mitochondria, chloroplasts	Mates (2000)
Catalase	Decomposes H_2O_2 to water and O_2	Peroxisomes	Janku et al. (2019)
Ascorbate peroxidase	Reduces H_2O_2 to water using ascorbate as electron donor	Chloroplasts, cytosol	Apak et al. (2021)
Glutathione reductase	Regenerates reduced glutathione (GSH) from oxidized glutathione (GSSG)	Cytosol, mitochondria	An et al. (2020)

signaling by readjusting the levels of ROS, which serve as peripheral messengers in physiological mechanisms (Nam et al. 2024). Insufficient or disabling mutations in SOD genes have been associated with neurodegenerative pathologies and other diseases, highlighting the importance of the enzyme in the physiological state of humans (Gamero-Sandemetrio et al. 2013).

6.5.1.2 Catalase (CAT)

Catalytic action CAT is a metalloheme enzyme, decomposing hydrogen peroxide to water and superoxide instead of allowing H_2O_2 build up to an extremophilic reaction, causing the formation of a highly reactive hydroxyl radical (Janku et al. 2019). It is mainly found in the peroxisomes, whereby it eliminates H_2O_2 produced in the process of fatty acid b-oxidation and photorespiration (Apak et al. 2016). CAT is also characterized by high turnover rates, which enables the radiation to be used effectively when ROS are generated rapidly, as is attacked by a pathogen or environmental stress (Ahmad et al. 2008). The enzyme activity can usually be adopted as a biomarker of oxidative stress and may be involved in aging and other disease prevention activities (Nordberg and Arnér 2001). Other antioxidant enzymes also react with CAT to create a coordinated system in which the enzyme successfully de-oxidizes the ROS (Joorabloo and Liu 2024).

6.5.1.3 Ascorbate Peroxidase (APX)

APX is a crucial enzyme of the ascorbate–glutathione loop since it utilizes ascorbate as an e-donor to convert H_2O_2 to water, hence the primary ROS detoxifier, particularly in the cytosol and chloroplasts (Apak et al. 2021). Transparency to cellular redox status APX is quickly activated by oxidative stress, plus APX is sensitized by the presence of H_2O_2 during photosynthesis and other metabolic functions (Sies and Jones 2020). Several APX isoforms are available with specific subcellular localization patterns and can be selectively targeted in compartments to control ROS scavenging (Bhushan and Gopinath 2015). The photosynthetic function and photooxidative defense of plants require APX activity (Mates,2000). The activity of the enzyme is directly related to the presence of ascorbate, which is replenished by other agents of the antioxidant system (Li et al. 2021).

6.5.1.4 Glutathione Reductase (GR)

GR preserves the presence of the major cellular antioxidant reduced glutathione (GSH) by catalyzing the reduction of oxidized glutathione (GSSG) to GSH conducted by NADPH (An et al. 2020). This response is essential to the synthesis of GSH, the raw material of glutathione peroxidase and other enzymes in the process of neutralizing ROS (Noctor et al. 2012). GR activity has a strong association with the ascorbate glutathione cycle to maintain the continual clearing out of H_2O_2 and other peroxides (Foyer and Noctor 2011). The improved stress tolerance of plants and the resistance to oxidative injury of animal cells, under the conditions of improved GR activity, are mentioned (Gill and Tuteja 2010). The indirect control of the enzyme is extensively governed in all aspects, which include transcriptional, post-transcriptional, and post-translational; all responsive to the cellular redox status (Mullineaux and Rausch 2005).

6.5.2 Role in ROS Scavenging

SOD, CAT, APX, and GR collaborate with each other to create a powerful enzymatic cascade that effectively removes ROS and oxidative damage (Mittler 2002). The initial action involves SOD transforming superoxide radicals into H_2O_2, detoxified by CAT and APX, and GSH regenerated through GR in order to counterattack the ROS

TABLE 6.2(b)
Sequential Action of Antioxidant Enzymes in ROS Detoxification

	ROS Species	Enzyme Involved	Reaction Catalyzed	Reference
1	Superoxide ($O_2\bullet -$)	Superoxide dismutase	$2O_2\bullet - + 2H+ \rightarrow H_2O_2 + O_2$	Vives-Bauzá et al. (2007)
2	Hydrogen peroxide	Catalase	$2H_2O_2 \rightarrow 2H_2O + O_2$	Apak et al. (2016)
3	Hydrogen peroxide	Ascorbate peroxidase	H_2O_2 + ascorbate $\rightarrow 2H_2O$ + dehydroascorbate	Sies and Jones (2020)
4	Oxidized glutathione	Glutathione reductase	GSSG + NADPH + H+ $\rightarrow$ 2GSH + NADP+	Foyer and Noctor (2011)

once again (Foyer and Noctor 2005). This multidimensional system of cellular defense is important to survive in normal and stressful conditions because it preserves healthy redox and does not cause ROS of harmful nature to accumulate (Apel and Hirt 2004). This antioxidant network may be disrupted, contributing to oxidative stress and dysfunction of cells and the development of several diseases, such as cancer, neurodegeneration, and cardiovascular conditions (Valko et al. 2007).

6.5.3 Osmolyte Biosynthesis Enzyme

Plants are exposed to different abiotic stresses (drought, salinity, extreme temperatures) which may cause disruption of cellular homeostasis leading to survival damage. To combat these stresses, plants synthesize compatible solutes referred to as osmolytes, such as proline, trehalose, glycine betaine, and polyols.

6.5.3.1 Biosynthesis and Degradation of Proline

Proline is a key osmolyte, which builds up in plant tissue in response to osmotic stress and is part of osmotic adaptation, protein and membrane stabilization, and ROS scavenging (Meena et al. 2019). Production of proline by plants takes place via the two enzymes that transform glutamate into proline: Delta 1-pyrroline-5-carboxylate synthase (P5CS) and Delta 1-pyrroline-5-carboxylate reductase (P5CR) (Hu et al. 1992). P5CS is a bifunctional enzyme that carries out the initial phosphorylation of glutamate to 5-carboxyl gamma-glutamyl phosphate and a reduction of glutamic semialdehyde, which, under physiological conditions, spontaneously cyclizes to give Delta-1-pyrroline-5-carboxyl gamma-glutamyl phosphate (P5C) (Hu et al. 1992). P5CR, in turn, catalyses the reduction of P5C to proline (Meena et al. 2019). The production of proline is highly controlled, and the rate-limiting pathway is the P5CS; this step is inhibited by feedback with proline. This negative feedback is reduced in stressful conditions to permit increased proline to accumulate (Hong et al. 2000). This can be increased through genetic manipulation to alleviate the inhibition that osmotic stress can impose to raise proline concentrations and improve osmotic stress tolerance (Hong et al. 2000). In addition to its role as an osmolyte, proline is a stabilizer of subcellular architecture, redox homeostasis, and a signaling molecule in response to stress (Escalante-Magaña et al. 2019).

6.5.3.2 Trehalose Phosphate Synthase, Trehalose Biosynthesis

Trehalose is a non-reducing disaccharide that is an osmoprotectant in plants, but natural concentrations of endogenous trehalose are generally relatively low (Hassan et al. 2022). It is biosynthesized through a two-step enzymatic cascade: TPS initially transforms UDP-glucose and glucose-6-phosphate into trehalose-6-phosphate phosphatase (TPP) then dephosphorylates trehalose-6-phosphate to free trehalose. It has been reported that transgenic expression of TPS genes can increase stress tolerance, but the constitutive accumulation of trehalose can negatively affect plant growth (Hassan et al. 2022). Trehalose helps to osmotically adjust cells by stabilizing proteins and membranes, protecting cellular structures against dehydration, and enhancing antioxidant defenses (Hassan et al. 2022). Besides, it synergistically interacts with other osmolytes like proline and glycine betaine to provide stress tolerance (Hassan et al. 2022).

6.5.3.3 Enzymes Dealing with Glycine Betaine, Polyol Synthesis

Glycine betaine (GB) is a quaternary ammonium ion, which occurs in elevated concentrations during abiotic stress, particularly in halophytes and in select crop species (Annunziata et al. 2019). The oxidative pathway of GB biosynthesis includes two steps (i.e., first oxidation of choline to betaine aldehyde (carried out by choline monooxygenase (CMO)) and the secondary oxidation of betaine aldehyde to GB (carried out by betaine aldehyde dehydrogenase (BADH)) (Annunziata et al. 2019). The stressful conditions that initiate the transcriptional upregulation of CMO and BADH, in turn, improve GB accumulation (Annunziata et al. 2019).GB serves its function as an osmolyte and stabilizes proteins and membranes, protects photosynthetic structures, and scavenges ROS (Annunziata et al. 2019). It also regulates gene expression and enzyme activity, which is part of stress adaptation (Hernandez-Leon and Valenzuela-Soto 2022). Exogenous application of GB can make stress-responsive genes active and also improve stress tolerance in non-accumulating species (Ashraf and Foolad 2007). Polyols include mannitol and sorbitol, which are sugar alcohols synthesized by enzymes like mannitol-1-phosphate dehydrogenase and sorbitol-6-phosphate dehydrogenase (Hong et al. 2000). These compounds serve as osmoprotectants, ROS hunters, and stabilizers of macromolecules under stress (Hong et al. 2000).

6.5.4 Contribution to the Osmotic Adjustment

Osmotic adjustment is the mechanism that allows plants to accumulate compatible solute materials to ensure cell turgor pressure and take up water under stress conditions (Ashraf and Foolad 2007). Proline, trehalose, glycine betaine, and polyols play a major role in this process, and each has unique and overlapping roles (Escalante-Magaña et al. 2019). Proline is a rapid and significant buildup under stress that has immediate effects terms of osmotic balance and protection against oxidative damage (Meena et al. 2019). GB accumulation is more significant for young tissue and in prolonged stress - long-term protection and stabilization of cellular structures (Annunziata et al. 2019). Though there is a lesser quantity of trehalose, it improves the stress tolerance by stabilizing proteins and membranes along with disrupting the stress-responsive signals (Hassan et al. 2022).

TABLE 6.3
Key Enzymes in Osmolyte Biosynthesis and Their Roles in Osmotic Adjustment

Osmolyte	Key Enzymes	Biosynthetic Pathway	Role in Osmotic Adjustment	Reference
Proline	P5CS, P5CR	Glutamate → P5C → Proline	Osmolyte, ROS scavenger, stabilizer	Hu et al. (1992)
Trehalose	TPS, TPP	UDP-glucose + Glucose-6-P → Trehalose	Osmoprotectant, stabilizes proteins	Hassan et al. (2022)
Glycine Betaine	CMO, BADH	Choline → Betaine aldehyde → GB	Osmolyte, stabilizes membranes, ROS scav.	Annunziata et al. (2019)
Polyols	Mannitol-1-P dehydrogenase	Fructose-6-P → Mannitol	Osmoprotectant, ROS scavenger	Hong et al. (2000)

Polyols participate in osmotic adjustment and ROS detoxification, and more so in plant survival in unfavorable conditions (Hong et al. 2000). The interaction between these osmolytes and their biosynthetic enzymes is very important for effective osmotic adjustment and stress resistance. Genetic engineering to optimize the expression of these enzymes has overall proven to be a promising strategy to boost the ability of plants to tolerate abiotic stresses (Ashraf and Foolad 2007). As Table 6.3 summarizes all osmolyte biosynthesis and their roles.

6.5.5 Cell-Wall-Modifying Enzymes

Plant cell walls are dynamic assemblages that need to balance tough support and the flexible range of lifestyle change, particularly in cases of drought stress. Two important classes of cell wall-acting enzymes, expansins and peroxidases, are key players in coordinating the loosening and hardening of the wall in response to environmental stimuli.

6.5.5.1 Expansins: Structure, Function, and Drought Response

Expansins are a family of non-enzymatic proteins that loosely bind cellulose microfibrils and the hemicellulose, especially xyloglucans, to promote turgor-driven expansion in cells (Balkova et al. 2025). The expansin gene family could be divided into four subgroups: 4-alpha- and 4-beta-expansins, expansin-like A, and expansin-like B, which function autonomously in plant development and adaptation to stress (Dabravolski and Isayenkov 2025). Expansins are largely active in cells that are undergoing rapid elongation or shortening (root and apical meristem of the plant's stems and leaves), yet they are also highly developmentally and environmentally regulated in their expression patterns (Cosgrove 2016). Stomatal closure, osmolyte buildup, and expression of stress-responsive genes are among the physiological and molecular adaptations that plants invoke in drought response (Gall et al. 2015). Expansin expression is commonly upregulated in growing tissues such as roots and young shoots under water deficit, supporting continued cell elongation and proliferation (Yang et al. 2020). In *Arabidopsis*, some expansins, such as EXPA1, have been overexpressed, which led to increased drought tolerance through increased photosynthetic efficiency, indicating faster senescence (Wang et al. 2024). Likewise, transgenic lines of wheat and soybean with superexpression of expansin proteins are characterized by enhanced root elongation, ion homeostasis, and osmotic homeostasis, which leads to enhanced drought tolerance (Narayan et al. 2020). Expansins were also shown to interact with phytohormones – such as ABA, auxins, and ethylene to create a mechanistic and hormonal nexus regulating cell wall plasticity during stress (Zhao et al. 2012). In particular, direct activation of the TaEXPA2 gene in wheat is mediated by a MYB transcription factor, and TaEXPA2 overexpression improves wheat resistance to drought through an increase in the growth of lateral roots and the antioxidative potential (Han et al. 2015). Similar circumstances have been observed in sugarcane, where EaEXPA1 expansionary corresponds with high relative water content, high photosynthetic capacity, and high drought resistance (Chen et al. 2016).

6.5.5.2 Peroxidases: Dual Roles in Cell Wall Stiffening and Loosening

Another class of cell-wall-modifying enzymes of significance is the peroxidases, which carry out the oxidation of a wide range of substrates in the presence of hydrogen peroxide. Peroxidase enzymes can play complementary roles in cell wall structural engineering: first, to confer mechanical rigidity and oxidative cross-linking of phenolic compounds and glycoproteins, they can cause cell wall stiffening, which is intricately important to maintain cell integrity in stress responses (Tenhaken 2015). However, during stress or prolonged drought, ROS begin to dominate the homeostatic cellular balance, and pathological concentrations of hydrogen peroxide have been shown to cause the formation of hydroxyl radicals. These radicals cleave the cell wall polymers, causing loosening of the wall and allowing the stressed organs to grow continuously (Hoffmann et al. 2020). It is also known that peroxidase activity is very strongly controlled and often is synchronized with expansins and other cell-wall-modifying enzymes. Whereas in soybean, overexpression of the expansin gene GsEXLB14 results in overexpression of both expansin and peroxidase genes, which increase tolerance to drought by increasing root growth and stress-responsive gene expression (Wang et al. 2024). In wheat, overexpression of TaEXPB23 may result not only in oxidative stress tolerance but also in increased cell wall-bound peroxidase activity, further validating the role activity of peroxidases in protection against the effects of stress (Han et al. 2015).

6.5.5.3 Enzymatic Modulation of Cell Wall Plasticity Under Drought

Drought stress causes a group of physiological, biochemical, and molecular changes in plants, many of which are mediated by cell-wall alterations (Gall et al. 2015). One important adaptation is the differential regulation of cell-wall-modifying enzymes that ensures growth in essential tissues while limiting it in others. Expansins are expressed at high levels in cells that are actively growing, and elongation of root tips is a prerequisite for water foraging (Yang et al. 2020). This is usually followed by upregulation of xyloglucan endotransglucosylase/hydrolase (XTH) and other remodeling enzymes, which in cooperation maintain the extensibility of the cell wall under water deficit (Ganie and Ahammed 2021). Peroxidases on the other hand can play dual roles in wall stiffening and loosening depending on the availability of ROS and electrons (Tenhaken 2015). Early in drought, peroxidase-driven cross-linking of cell walls prevents excessive water loss and mechanical damage to cells. As pressure from the tension continues, ROS production results in radical-mediated cleavage of wall polymers, allowing cell wall loosening and continued growth in certain tissues (Hoffmann et al. 2020). Expansin and peroxidase activities are further regulated by cell wall remodeling, such as by elevated branching of pectic polymers, impaired methylesterification of homogalacturonan, influencing wall hydration and plasticity (Gall et al. 2015). These changes are therefore greater in stress-tolerant cultivars, indicating that cell wall plasticity becomes an important factor in stress adaptation (Samalova et al. 2022).

6.5.6 Hydrolytic Enzymes and Aquaporin-Mediated Water Regulation

Hydrolytic enzymes like amylases and proteases play a crucial role in the mobilization of stored reserves during germination and early development of seeds. Amylases play an important role in the breakdown of starch – the main storage carbohydrate – into soluble sugars, which can be used for energy and biosynthesis processes. Similarly, storage proteins are broken down by proteases into amino acids, which are used in the synthesis of new proteins, or, in a state of stress or rapid growth, these can be used as alternate energy substrates. The expression of these enzymes is controlled by hormonal and environmental signals, thus guaranteeing the proper and temporally appropriate use of reserves (Kapilan et al. 2018).

Aquaporins (AQPs) form an important subgroup of the major intrinsic protein (MIP) superfamily, and they are essential in the translocation of water and small neutral solutes across biological membranes, to maintain water homeostasis, cellular turgor, and adaptability to stress. They can be grouped as plasma membrane intrinsic proteins (PIPs), tonoplast intrinsic proteins (TIPs), nodulin-like intrinsic proteins (NIPs), small basic intrinsic proteins (SIPs), and X intrinsic proteins (XIPs); the majority are localized at the plasma membrane and tonoplast (Kapilan et al. 2018). Dynamically, phosphorylation, tetramerization, pH, ions, and ROS are able to modulate aquaporin functionality. Conformational changes induced by phosphorylation on selected serines on PIPs increase water permeability, and dephosphorylation reduces flux. The pore-open conformation is maintained by interaction with other proteins, including 14-3-3 proteins, and enables further transport (Prado et al. 2019). Such regulatory mechanisms allow rapid adaptation to drought, flooding, or changes in light conditions. Phosphorylation state is also under circadian control, as shown in *Arabidopsis* where in silico prediction identified phosphorylation of AtPIP2;1 at Ser280 and Ser283, coordinately regulating its hydraulic conductivity along the diurnal environmental rhythm (Nesverova and Törnroth-Horsefield 2019; Prado et al. 2019).

6.5.7 Influence on Hydraulic Conductivity

Among other factors, aquaporins form a key influence on plant hydraulic conductivity, in other words, the relative ease of water going through plant tissues (Chaumont and Tyerman 2014). High expression and satisfactory activity of aquaporins in the plasma membranes of root and leaf tissue have a direct effect on the rate of water uptake in the soil and its transportation to the aerial portions of the organism (Gal et al. 2023). PIPs found in roots allow water to flow radially out of the soil into the xylem to set the overall genosomatic state of the plant (Gal et al. 2023). Certain aquaporins within leaves facilitate water entering the xylem into the mesophyll cells, prompting changes in transpiration and photosynthetic organization (Shekoofa and Sinclair 2018). Drought, salinity, low root temperature, and other environmental stresses can rapidly cause downregulation of water transport through changes in aquaporin expression and activity (Kapilan et al. 2018). Considering the example, in tobacco, the lower concentration of electrolytes in the root, which was a consequence of a lower expression level of the gene NtAQP1, a family of the PIP1, decreased the root hydraulic conductivity and water stress resistance (Kapilan et al. 2018). Prityl phosphate synthase 8;6, Pir2 and PIP2 were also involved; moreover, the knockout of PIP1;6 in *Setaria viridis* led to a modified root hydraulic conductivity and gas exchange, hence showing the essential contribution of PIPs toward whole plant–water homeostasis (Gal et al. 2023). There are also other factors that could regulate aquaporin gating, such as pH, cations, and ROS, thus offering numerous control levels of water transport (Kapilan et al. 2018). The circadian clock and environmental signals, like light and darkness, are able to regulate aquaporin phosphorylation, resulting in daily variations of leaf hydraulic conductivity (Prado et al. 2019). In wheat, phosphorylation of TaPIP2;10 acts on specific serine residues, which increases plant growth and defense, showing that aquaporins present multifunctional roles in plant physiology (Lu et al. 2022).

6.6 ENZYME-MEDIATED SIGNAL TRANSDUCTION PATHWAYS

The MAPK cascade is a key element of the drought signal transduction in plants, which integrates environmental cues and regulates ABA accumulation and stomatal closure, which are key players in drought tolerance. In cotton, the GhMAP3K62-GhMKK16-GhMPK32 kinase cascade phosphorylates the transcription factor GhEDT1, which upregulates the expression of GhNCED3, a key enzyme in ABA biosynthesis, to ultimately increase the ABA content and enhance the drought tolerance (Chen et al. 2022). Overexpression of GhMKK16 in cotton plants leads to the accumulation of more ABA and increases drought tolerance, whereas knockdown of GhMKK16 or other cascade members decreases ABA content and drought tolerance, which shows that kinases play critical roles in this pathway (Chen et al. 2022). In soybean, the WRKY transcription factor GmWRKY54 directly activates genes in ABA and calcium (Ca^{2+}) signaling pathways, including ABA receptors and SnRK2 kinases, which increase stomatal closure and drought tolerance (Wei et al. 2019). For example, gmWRKY54 is targeted to the promoters of PYL8 (an ABA receptor), SRK2A (a SnRK2 kinase), CIPK11, and CPK3 and stimulates their expression, thereby placing itself in an upstream role within the ABA signaling cascade, as opposed to numerous WRKY proteins that have been found to exert action downstream (Wei et al. 2019). The *Arabidopsis* calcium-dependent protein kinase (CPK) 6 is ABA and dehydration-inducible and positively regulates ABA signaling and drought tolerance through phosphorylation of ABA-responsive element binding factor (ABF/AREB), thereby improving its transcriptional activity (Zhang et al. 2019). Overexpression of CPK6 resulted in enhanced tolerance to drought, and loss-of-function mutants had reduced sensitivity to ABA, demonstrating that kinase-mediated phosphorylation plays a key role in ABA-involved stress responses for tolerance to drought stress (Zhang et al. 2019). At14a-like1 (AFL1), a gene encoding a stress-induced membrane protein associated with endocytosis, signal transduction, and accumulation of proline in creeping bentgrass, overexpressing it improved drought tolerance capacity by regulating the genes related to phytohormone biosynthesis, redox homeostasis, and secondary metabolites (Tan et al. 2025). AFL1-mediated drought stress tolerance is correlated with transcriptional regulation of ABA, auxin, and strigolactone pathways and enzymes involved in lignin, cutin, suberin, and wax biosynthesis, suggesting a broad sphere of crosstalk between enzymatic and hormonal signaling (Tan et al. 2025). Protein phosphatases, including clade F PP2C ZmPP84 in maize, are known as negative regulators of drought tolerance because they dephosphorylate the MAPK pathway members, suppressing their ABA-induced responses by inhibiting stomatal closure (Guo et al. 2022).

6.7 GENETIC AND BIOTECHNOLOGICAL INSIGHTS

Drought tolerance is a complex trait, and it involves activation of numerous signaling mechanisms and differentially expressed molecular responses such as upregulation of stress-responsive kinases, phosphatases, and transcription factors (Mahmood et al. 2019). Genetic engineering of soybean, including overexpression of GmWRKY54, has shown that functional manipulation of the significant transcription factors is suitable to improve drought tolerance via the activation of both ABA and Ca^{2+} signaling (Wei et al. 2019). CRISPR/Cas9 as well as other genome editing technologies are now increasingly being employed to characterize and modify the function of genes involved in drought signaling, allowing it to precisely modify key candidate genes identified from transcriptomics, proteomics, and epigenomics approaches (Mahmood et al. 2019). Lignin biosynthesis and salicylic acid (SA) signaling were long recognized in providing drought stress tolerance, and a phenylalanine ammonia lyase (PAL) from *Fritillaria unibracteata* supports stress tolerance by regulating processes of secondary metabolism and hormone crosstalk (Qin et al. 2022). Wheat Aux/IAA protein TaIAA15-1A promotes drought resistance in *Brachypodium* by modulating ABA signaling pathway, ABA sensitivity, and antioxidant enzyme activities, resulting in less accumulation of ROS (Su et al. 2022). Cao et al. (2024) have identified a transcription factor ZmDof22 in maize upon CRISPR/Cas9 and gene co-expression network analysis, which boosts drought tolerance by stomatal closure and antioxidant enzyme activity through ABA pathways.

6.8 ENVIRONMENTAL AND DEVELOPMENTAL MODULATION OF ENZYMES

In *Arabidopsis*, the Gigantea-Enhanced EM function and diurnal antioxidant response (GIGANTEA-Enhanced EM LEVEL, EEL) complex improves the response to drought conditions by controlling diurnal transcription of NCED3, which is known to be a rate-limiting enzyme in ABA biosynthesis, thus connecting circadian clock signaling to response to drought (Baek et al. 2020). In potato, the StbHLH47 transcription factor negatively regulates drought stress tolerance by regulating genes involved in the ABA signaling pathway and phenylpropanoid (PPP) pathway and antioxidant enzymes, while overexpression of StbHLH47 results in more sensitivity to drought conditions (Wang et al. 2025). PtoERF15 in poplar, an AP2 homeodomain transcription factor, imparts Blattinger's cellulosic water-stress resistance through jasmonic acid (JA) signaling at the posttranscriptional level to control xylem maturation and facilitate metabolic maintenance of stem water potential under water stress (Kong et al. 2023). OsERF71 in rice has been demonstrated to be ABA- and drought-induced, and its overexpression has been shown to enhance drought tolerance, via induction of ABA-responsive and proline biosynthesis genes leading to accumulation of proline and the retention of water (Li et al. 2018). The interaction of ABF4 and ABR1 forms a regulatory cascade for the enhanced drought tolerance in trifoliate orange by synergistically regulating amylase-mediated starch catabolism for the supply of soluble sugars for drought tolerance (Zhang et al. 2022). Epigenetic reversion such as lysine deacetylation by TaSRT1 in wheat could regulate drought resistance through able to influence the stability level and transcriptional activity of some important drought-associated protein (Chang et al. 2025) In wheat, TaERF87 and TaAKS1 synergistically transactivate genes for proline biosynthesis catalyzing the proline accumulation and ABA-responsive element binding factors (Du et al. 2022), which improved drought tolerance. A novel flavin-containing monooxygenase protein (FMO1) from tomato was found to interact with catalase 2 (CAT2) to negatively regulate drought tolerance through ROS homeostasis and the ABA signaling pathway, demonstrating the integration of enzymatic and hormonal mechanisms in drought responses (Wang et al. 2023). A wheat R2R3 MYB gene TaMpc1-D4 negatively regulates drought tolerance through the downregulation of stress and antioxidant genes, and its silencing improves drought resistance and enzyme activity (Li et al. 2020).

6.9 CONCLUSION

Plant–water relations underpin growth, productivity, and survival, especially under drought stress, which disrupts turgor, nutrient flow, and photosynthesis. Enzyme systems form the core of plant adaptive responses, supporting osmotic balance, redox homeostasis, structural flexibility, and efficient water transport. Antioxidant, biosynthetic, and cell-wall-modifying enzymes, along with aquaporins and key signaling proteins, collectively maintain metabolic stability during water deficit. Modern genetic engineering and multi-omics tools now enable precise improvement of these pathways, offering new opportunities to develop drought-resilient cultivars. Together, these insights strengthen efforts to enhance water-use efficiency, crop performance, and agricultural sustainability under changing climatic conditions.

REFERENCES

Aanniz, T., El Baaboua, A., Aboulaghras, S., Bouyahya, A., Benali, T., Balahbib, A., El Omari, N., Butnariu, M., Muzammil, K., Yadav, K., Al Abdulmonem, W., Lee, L. H., Zengin, G. and Chamkhi, I. 2025. Impact of water stress to plant epigenetic mechanisms in stress and adaptation. *Physiol. Plant. 177*, e70058.

Ahmad, P., Sarwat, M. and Sharma, S. 2008. Reactive oxygen species, antioxidants and signaling in plants. *J. Plant Biol. 51*, 167–173.

Alber, A., Renault, H., Basilio-Lopes, A., Bassard, J. E., Liu, Z., Ullmann, P., Lesot, A., Bihel, F., Schmitt, M., Werck-Reichhart, D. and Ehlting, J. 2019. Evolution of coumaroyl conjugate 3-hydroxylases in land plants: Lignin biosynthesis and defense. *Plant J. 99*, 873–892.

An, Z., Yan, J., Zhang, Y. and Pei, R. 2020. Applications of nanomaterials for scavenging reactive oxygen species in the treatment of central nervous system diseases. *J. Mater. Chem. B. 8*, 8231–8246.

Anjum, N., Sharma, P., Gill, S., Hasanuzzaman, M., Khan, E. A., Kachhap, K., Mohamed, A., Thangavel, P., Devi, G. D., Vasudhevan, P., Sofo, A., Khan, N., Misra, A. N., Lukatkin, A., Singh, H., Pereira, E. and Tuteja, N. 2016. Catalase and ascorbate peroxidase—Representative H_2O_2-detoxifying heme enzymes in plants. *Environ. Sci. Pollut. Res. 23*, 19002–19029

Annunziata, M., Ciarmiello, L., Woodrow, P., dell'Aversana, E. and Carillo, P. 2019. Spatial and temporal profile of glycine betaine accumulation in plants under abiotic stresses. *Front. Plant Sci. 10*, 230.

Apak, R., Calokerinos, A., Gorinstein, S., Segundo, M., Hibbert, D., Gülçin, I., Çekiç, S., Güçlü, K., Özyürek, M., Çelik, S., Magalhães, L. and Arancibia-Avila, P. 2021. Methods to evaluate the scavenging activity of antioxidants toward reactive oxygen and nitrogen species (IUPAC technical report). *Pure Appl. Chem. 93*. 1279–1296.

Apak, R., Özyürek, M., Güçlü, K. and Çapanoğlu, E. 2016. Antioxidant activity/capacity measurement. 3. Reactive oxygen and nitrogen species (ROS/RNS) scavenging assays, oxidative stress biomarkers, and chromatographic/chemometric assays. *J. Agric. Food Chem. 64*, 1046–1070.

Ashraf, M. and Foolad, M. 2007. Roles of glycine betaine and proline in improving plant abiotic stress resistance. *Environ. Exp. Bot. 59*, 206–216.

Baek, D., Kim, W. Y., Cha, J., Park, H., Shin, G., Park, J., Lim, C. J., Chun, H. J., Li, N. Y., Kim, D., Lee, S. Y., Pardo, J. M., Kim, M. C. and Yun, D. 2020. The GIGANTEA-ENHANCED EM LEVEL complex enhances drought tolerance via regulation of abscisic acid synthesis. *Plant Physiol. 183*, 1–15.

Balkova, D., Mala, K., Hejátko, J., Panzarová, K., Abdelhakim, L., Pleskačová, B. and Samalova, M. 2025. Differential expression and localization of expansins in *Arabidopsis* shoots: Implications for cell wall dynamics and drought tolerance. *Front. Plant Sci. 16*, 1546819

Berry, Z., Emery, N., Gotsch, S. and Goldsmith, G. 2018. Foliar water uptake: Processes, pathways, and integration into plant water budgets. *Plant Cell Environ. 42*, 410–423.

Berwal, M. K., Kumar, R., Prakash, K., Rai, G. K., and Hebbar, K. B. 2021. Antioxidant defense system in plants against abiotic stress. In *Abiotic stress tolerance mechanisms in plants*. Eds. Rai, G. K., Kumar, R. R., and Bagati, S. (CRC Press), 175–202.

Bhushan, B. and Gopinath, P. 2015. Antioxidant nanozyme: A facile synthesis and evaluation of the reactive oxygen species scavenging potential of nanoceria encapsulated albumin nanoparticles. *J. Mater. Chem. B. 3*, 4843–4852.

Cao, L., Ye, F., Fahim, A. M., Ma, C., Pang, Y., Zhang, X., Zhang, Q. and Lu, X. 2024. Transcription factor ZmDof22 enhances drought tolerance by regulating stomatal movement and antioxidant enzymes activities in maize (*Zea mays* L.). *Theor. Appl. Genet. 137*, 1–15.

Chang, S., Yang, Q., Chu, W., Liu, X., Li, J., Liu, Z., Lin, J., Liu, D., Zhao, D., Peng, X., Xin, M., Yao, Y., Xie, X., Peng, H., Ni, Z., Sun, Q. and Hu, Z. 2025. Lysine deacetylase TaSRT1 mediates wheat drought tolerance by deacetylating TaDT-A to reduce its protein stability and transcriptional activity. *Plant Biotechnol. J. 23*, 1–15.

Chatterjee, A., Puri, S., Sharma, P., Deepa, P. R. and Chowdhury, S. 2023. Nature-inspired enzyme engineering and sustainable catalysis: Biochemical clues from the world of plants and extremophiles. *Front. Bioeng. Biotechnol. 11*, 1203456.

Chaumont, F. and Tyerman, S. 2014. Aquaporins: Highly regulated channels controlling plant water relations. *Plant Physiol. 164*, 1600–1618

Chen, D., Wang, S., Yin, L. and Deng, X. 2018. How does silicon mediate plant water uptake and loss under water deficiency? *Front. Plant Sci. 9*, 281.

Chen, L., Zhang, B., Xia, L., Yue, D., Han, B., Sun, W., Wang, F., Lindsey, K., Zhang, X. and Yang, X. 2022. The GhMAP3K62-GhMKK16-GhMPK32 kinase cascade regulates drought tolerance by activating GhEDT1-mediated ABA accumulation in cotton. *J. Adv. Res. 38*, 1–15.

Chen, Y., Han, Y. Y., Zhang, M., Zhou, S. S., Kong, X. and Wang, W. 2016. Overexpression of the wheat expansin gene TaEXPA2 improved seed production and drought tolerance in transgenic tobacco plants. *PLoS One. 11*, e0153494.

Cosgrove, D. J. 2016. Catalysts of plant cell wall loosening. *F1000Res. 5*, F1000

Dabravolski, S. and Isayenkov, S. V. 2025. Expansins in salt and drought stress adaptation: From genome-wide identification to functional characterisation in crops. *Plants. 14*, 1327.

Dey, S., Sidor, A. and O'Rourke, B. 2016. Compartment-specific control of reactive oxygen species scavenging by antioxidant pathway enzymes. *J. Biol. Chem. 291*, 7868–7880.

Du, L., Huang, X., Ding, L., Wang, Z., Tang, D., Chen, B., Ao, L., Liu, Y., Kang, Z. and Mao, H. 2022. TaERF87 and TaAKS1 synergistically regulate TaP5CS1/TaP5CR1-mediated proline biosynthesis to enhance drought tolerance in wheat. *New Phytol. 236*, 1–15.

Dubbert, M. and Werner, C. 2018. Water fluxes mediated by vegetation: Emerging isotopic insights at the soil and atmosphere interfaces. *New Phytol. 221*, 1754–1763.

Dvořák, P., Krasylenko, Y., Zeiner, A., Šamaj, J. and Takáč, T. 2021. Signaling toward reactive oxygen species-scavenging enzymes in plants. *Front. Plant Sci. 12*, 628992.

Escalante-Magaña, C., Aguilar-Caamal, L. F., Echevarría-Machado, I., Medina-Lara, F., Sánchez Cach, L. and Martínez-Estévez, M. 2019. Contribution of glycine betaine and proline to water deficit tolerance in pepper plants. *HortScience. 54*, 1031–1038.

Farooq, M., Wahid, A., Kobayashi, N., Fujita, D. and Basra, S. M. 2008. Plant drought stress: Effects, mechanisms and management. *Agron. Sustain. Dev.* 153–188.

Fernández, V., Gil-Pelegrín, E. and Eichert, T. 2020. Foliar water and solute absorption: An update. *Plant J. 105*, 870–883.

Foyer, C. H. and Noctor, G. 2005. Redox homeostasis and antioxidant signaling: A metabolic interface between stress perception and physiological responses. *Plant Cell. 17*, 1866–1875.

Foyer, C. H. and Noctor, G. 2011. Ascorbate and glutathione: The heart of the redox hub. *Plant Physiol. 155*, 2–18.

Gal, A., Dalal, A., Anfang, M., Sharma, D., Binenbaum, J., Muchaki, P., Kumar, R., Egbaria, A., Duarte, K. E., Kelly, G., de Souza, W. R. and Sade, N. 2023. Plasma membrane aquaporins regulate root hydraulic conductivity in the model plant *Setaria viridis*. *Plant Physiol. 192*, 123–137.

Gall, H., Philippe, F., Domon, J. M., Gillet, F., Pelloux, J. and Rayon, C. 2015. Cell wall metabolism in response to abiotic stress. *Plants.*

Gambetta, G., Knipfer, T., Fricke, W. and McElrone, A. 2017. Aquaporins and root water uptake.

Gamero-Sandemetrio, E., Gómez-Pastor, R. and Matallana, E. 2013. Zymogram profiling of superoxide dismutase and catalase activities allows Saccharomyces and non-Saccharomyces species differentiation and correlates to their fermentation performance. *Appl. Microbiol. Biotechnol. 97*, 1757–1768.

Ganie, S. and Ahammed, G. 2021. Dynamics of cell wall structure and related genomic resources for drought tolerance in rice. *Plant Cell Rep. 40*, 437–459.

García-García, J. D., Van Gelder, K., Joshi, J., Bathe, U., Leong, B. J., Bruner, S., Liu, C. C. and Hanson, A. 2021. Using continuous directed evolution to improve enzymes for plant applications. *Plant Physiol. 187*, 1–13.

Gill, S. S. and Tuteja, N. 2010. Reactive oxygen species and antioxidant machinery in abiotic stress tolerance in crop plants. *Plant Physiol. Biochem. 48*, 909–930.

Guo, Y., Shi, Y., Wang, Y., Liu, F., Li, Z., Qi, J., Wang, Y., Zhang, J., Yang, S., Wang, Y. and Gong, Z. 2022. The clade F PP2C phosphatase ZmPP84 negatively regulates drought tolerance by repressing stomatal closure in maize. *New Phytol. 236*, 678–692.

Han, Y. Y., Chen, Y., Yin, S., Zhang, M. and Wang, W. 2015. Overexpression of TaEXPB23, a wheat expansin gene, improves oxidative stress tolerance in transgenic tobacco plants. *J. Plant Physiol. 173*, 62–71.

Hansen, C. C., Nelson, D., Møller, B. and Werck-Reichhart, D. 2021. Plant cytochrome P450 plasticity and evolution. *Mol. Plant. 14*, 1244–1265.

Hassan, M., Nawaz, M., Shah, A., Raza, A., Barbanti, L., Skalický, M., Hashem, M., Brestič, M., Pandey, S., Alamri, S., Mostafa, Y., Sabagh, A. E. and Qari, S. 2022. Trehalose: A key player in plant growth regulation and tolerance to abiotic stresses. *J. Plant Growth Regul. 41*, 1–18.

Hernandez-Leon, S. G. and Valenzuela-Soto, E. 2022. Glycine betaine is a phytohormone-like plant growth and development regulator under stress conditions. *J. Plant Growth Regul. 41*, 1–13.

Hildebrandt, T. M. 2018. Synthesis versus degradation: Directions of amino acid metabolism during *Arabidopsis* abiotic stress response. *Plant Mol. Biol. 98*, 121–135.

Hoffmann, N., Benske, A., Betz, H., Schuetz, M. and Samuels, A. 2020. Laccases and peroxidases co-localize in lignified secondary cell walls throughout stem development. *Plant Physiol. 184*, 806–822.

Hong, Z., Lakkineni, K., Zhang, Z. and Verma, D. 2000. Removal of feedback inhibition of delta(1)-pyrroline-5-carboxylate synthetase results in increased proline accumulation and protection of plants from osmotic stress. *Plant Physiol. 122*, 1129–1136.

Hosseinifard, M., Stefaniak, S., Ghorbani Javid, M., Soltani, E., Wojtyla, Ł. and Garnczarska, M. 2022. Contribution of exogenous proline to abiotic stresses tolerance in plants: A review. *Int. J. Mol. Sci. 23*, 5186.

Hu, C. A. A., Delauney, A. and Verma, D. 1992. A bifunctional enzyme (delta 1-pyrroline-5-carboxylate synthetase) catalyzes the first two steps in proline biosynthesis in plants. *Proc. Natl. Acad. Sci. USA. 89*, 9354–9358.

Jackson, R. B., Sperry, J. and Dawson, T. 2000. Root water uptake and transport: Using physiological processes in global predictions. *Trends Plant Sci. 5*, 482–488.

Janku, M., Luhová, L. and Petřivalský, M. 2019. On the origin and fate of reactive oxygen species in plant cell compartments. *Antioxidants. 8*, 105.

Javot, H. and Maurel, C. 2002. The role of aquaporins in root water uptake. *Ann. Bot.* 90, 301–313.

Joorabloo, A. and Liu, T. 2024. Recent advances in reactive oxygen species scavenging nanomaterials for wound healing. *Exploration. 4*, 20230066.

Kaldenhoff, R., Ribas-Carbó, M., Flexas Sans, J., Lovisolo, C., Heckwolf, M. and Uehlein, N. 2008. Aquaporins and plant water balance. *Plant Cell Environ*. 31, 658–666.

Kapilan, R., Vaziri, M. and Zwiazek, J. 2018. Regulation of aquaporins in plants under stress. *Biol. Res. 51*, 4.

Koch, I., Kahmen, A. and Burkhardt, J. 2025. Aerosol deposition affects water uptake and water loss of beech leaves. *Tree Physiol. 45*, 055

Kong, L., Song, Q., Wei, H., Wang, Y., Lin, M., Sun, K., Zhang, Y., Yang, J., Li, C. and Luo, K. 2023. The AP2/ERF transcription factor PtoERF15 confers drought tolerance via JA-mediated signaling in Populus. *New Phytol. 239*, 1–15.

Lee, D. H., Kim, Y. and Lee, C. B. 2001. The inductive responses of the antioxidant enzymes by salt stress in the rice (Oryza sativa L.). *J. Plant Physiol. 158*, 737–745.

Li, D., Ding, Z., Du, K., Ye, X. and Cheng, S. 2021. Reactive oxygen species as a link between antioxidant pathways and autophagy. *Oxid. Med. Cell. Longev. 2021*, 5583215.

Li, J., Guo, X., Zhang, M., Wang, X., Zhao, Y., Yin, Z., Zhang, Z., Wang, Y., Xiong, H., Zhang, H., Todorovska, E. and Li, Z. 2018. OsERF71 confers drought tolerance via modulating ABA signaling and proline biosynthesis. *Plant Sci. 274*, 1–10.

Li, X., Tang, Y., Li, H., Luo, W., Zhou, C., Zhang, L. and Lv, J. 2020. A wheat R2R3 MYB gene TaMpc1-D4 negatively regulates drought tolerance in transgenic *Arabidopsis* and wheat. *Plant Sci. 299*, 110603.

Liang, Y., Huang, Y., Liu, C., Chen, K. and Li, M. 2023. Functions and interaction of plant lipid signaling under abiotic stresses. *Plant Biol. 25*, 361–378.

Lu, K., Chen, X., Yao, X., An, Y., Wang, X., Qin, L., Li, X., Wang, Z., Liu, S., Sun, Z., Zhang, L., Chen, L., Li, B., Liu, B., Wang, W., Ding, X., Yang, Y., Zhang, M., Zou, S. and Dong, H. 2022. Phosphorylation of a wheat aquaporin at two sites enhances both plant growth and defense. *Mol. Plant. 15*, 1601–1617.

Mahmood, T., Khalid, S., Abdullah, M., Ahmed, Z., Shah, M., Ghafoor, A. and Du, X. 2019. Insights into drought stress signaling in plants and the molecular genetic basis of cotton drought tolerance. *Cells. 9*(1), 25.

Malik, W., Wang, X., Wang, X., Shu, N., Cui, R., Chen, X., Wang, D., Lu, X., Yin, Z., Wang, J. and Ye, W. 2020. Genome-wide expression analysis suggests glutaredoxin genes response to various stresses in cotton. *Int. J. Biol. Macromol. 146*, 1–12.

Marchin, R., Ossola, A., Leishman, M. and Ellsworth, D. 2020. A simple method for simulating drought effects on plants. *Front. Plant Sci. 10*, 1715.

Mates, J. 2000. Effects of antioxidant enzymes in the molecular control of reactive oxygen species toxicology. *Toxicol. 153*, 83–104.

Meena, M., Divyanshu, K., Kumar, S., Swapnil, P., Zehra, A., Shukla, V., Yadav, M. and Upadhyay, R. 2019. Regulation of L-proline biosynthesis, signal transduction, transport, accumulation and its vital role in plants during variable environmental conditions. *Heliyon. 5*, e02952.

Mishra, N., Jiang, C., Chen, L., Paul, A., Chatterjee, A. and Shen, G. 2023. Achieving abiotic stress tolerance in plants through antioxidative defense mechanisms. *Front. Plant Sci. 14*, 1110622.

Mittler, R. 2002. Oxidative stress, antioxidants and stress tolerance. *Trends Plant Sci. 7*, 405–410.

Mukarram, M., Choudhary, S., Kurjak, D., Petek, A. and Khan, M. A. 2021. Drought: Sensing, signalling, effects and tolerance in higher plants. *Physiol. Plant*. 172, 1291–1300.

Mullineaux, P. M. and Rausch, T. 2005. Glutathione, photosynthesis and the redox regulation of stress-responsive gene expression. *Photosynth. Res. 86*, 459–474.

Nam, Y. E., Kim, H. J. and Kwon, O. 2024. Acute and prolonged effects of *Bacillus amyloliquefaciens* GF424-derived SOD on antioxidant defense in healthy individuals challenged with intense aerobic exercise. *Free Radic. Biol. Med. 215*, 1–12.

Narayan, J., Chakravarthi, M., Nerkar, G., Manoj, V. M., Dharshini, S., Subramonian, N., Premachandran, M. N., Kumar, R. A., Surendar, K., Hemaprabha, G., Ram, B. and Appunu, C. 2020. Overexpression of expansin EaEXPA1, a cell wall loosening protein enhances drought tolerance in sugarcane. *Ind. Crops Prod. 159*, 113035.

Nesverova, V. and Törnroth-Horsefield, S. 2019. Phosphorylation-dependent regulation of mammalian aquaporins. *Cells. 8*, 30.

Noctor, G., Mhamdi, A. and Foyer, C. H. 2012. The roles of reactive oxygen metabolism in drought: Not so cut and dried. *Plant Physiol. 164*, 1636–1648.

Nordberg, J. and Arnér, E. S. J. 2001. Reactive oxygen species, antioxidants, and the mammalian thioredoxin system. *Free Radic. Biol. Med. 31*, 1287–1312.

Oliveira-Filho, E. R., Voiniciuc, C. and Hanson, A. D. 2023. Adapting enzymes to improve their functionality in plants: Why and how. *Biochem. Soc. Trans. 51*, 1801–1812.

Pamungkas, S. S. T., Suwarto, Suprayogi, N. and Farid. 2022. Drought stress: Responses and mechanism in plants. *Rev. Agric. Sci. 10*, 168–185.

Patel, J. and Mishra, A. 2021. Plant aquaporins alleviate drought tolerance in plants by modulating cellular biochemistry, root architecture and photosynthesis. Physiol. Plant. *172*, 1030–1044.

Pawlowicz, I. and Masajada, K. 2019. Aquaporins as a link between water relations and photosynthetic pathway in abiotic stress tolerance in plants. *Gene. 687*, 166–172.

Prado, K., Cotelle, V., Li, G., Bellati, J., Tang, N., Tournaire-Roux, C., Martinière, A., Santoni, V. and Maurel, C. 2019. Oscillating aquaporin phosphorylation and 14-3-3 proteins mediate the circadian regulation of leaf hydraulics. *Plant Cell. 31*, 417–429.

Qin, Y., Li, Q., An, Q., Li, D. B., Huang, S., Zhao, Y., Chen, W., Zhou, J. and Liao, H. 2022. A phenylalanine ammonia lyase from *Fritillaria unibracteata* promotes drought tolerance by regulating lignin biosynthesis and SA signaling pathway. *Int. J. Biol. Macromol. 206*, 1–12.

Rajput, V., Harish, R., Singh, K., Verma, L., Sharma, F., Quiroz-Figueroa, M., Meena, V., Gour, T., Minkina, S., Sushkova, S. and Mandzhieva. 2021. Recent developments in enzymatic antioxidant defence mechanism in plants with special reference to abiotic stress. *Biology. 10*, 267.

Razi, K. and Muneer, S. 2021. Drought stress-induced physiological mechanisms, signaling pathways and molecular response of chloroplasts in common vegetable crops. *Crit. Rev. Biotechnol*. 41, 669–691.

Ruehr, N., Grote, R., Mayr, S. and Arneth, A. 2019. Beyond the extreme: Recovery of carbon and water relations in woody plants following heat and drought stress. *Tree Physiol. 39*, 1285–1299.

Saini, N., Modolo, L., Deswal, R., Sehrawat, A., Yadav, N. and Sangwan, N. S. 2024. Expanding roles of cross-talk between hydrogen sulfide and nitric oxide under abiotic stress in plants. *Plant Physiol. Biochem. 214*, 108852.

Samalova, M., Gahurova, E. and Hejátko, J. 2022. Expansin-mediated developmental and adaptive responses: A matter of cell wall biomechanics? Quant. *Plant Biol. 3*, e11.

Scharwies, J. D. and Dinneny, J. 2019. Water transport, perception, and response in plants. *J. Plant Res. 132*, 311–324.

Secchi, F., Pagliarani, C. and Zwieniecki, M. 2017. The functional role of xylem parenchyma cells and aquaporins during recovery from severe water stress. *Plant Cell Environ. 40*, 858–871.

Seleiman, M., Al-Suhaibani, N., Ali, N., Akmal, M., Alotaibi, M. A., Refay, Y., Dindaroglu, T., Abdul-Wajid, H. H. and Battaglia, M. 2021. Drought stress impacts on plants and different approaches to alleviate its adverse effects. *Plants. 10*, 259.

Shekoofa, A. and Sinclair, T. 2018. Aquaporin activity to improve crop drought tolerance. *Cells. 7*, 123.

Sies, H. and Jones, D. P. 2020. Reactive oxygen species (ROS) as pleiotropic physiological signalling agents. *Nat. Rev. Mol. Cell Biol. 21*, 363–383.

Steudle, E. 2000. Water uptake by plant roots: An integration of views. *Plant Soil. 226*, 45–56.

Štreimikytė, P., Viškelis, P. and Viškelis, J. 2022. Enzymes-assisted extraction of plants for sustainable and functional applications. *Int. J. Mol. Sci. 23*, 1234.

Su, P., Sui, C., Li, J., Wan, K., Sun, H., Wang, S., Liu, X. and Guo, S. 2022. The Aux/IAA protein TaIAA15-1A confers drought tolerance in *Brachypodium* by regulating abscisic acid signal pathway. *Plant Cell Rep. 41*, 2467–2482.

Sugden, P. and Clerk, A. 1998. "Stress-responsive" mitogen-activated protein kinases (c-Jun N-terminal kinases and p38 mitogen-activated protein kinases) in the myocardium. *Circ. Res. 83*, 345–352.

Sun, Y., Wang, C., Chen, H. Y. H. and Ruan, H. 2020. Response of plants to water stress: A meta-analysis. *Front. Plant Sci. 11*, 978.

Tan, Z., Wang, Y., Liu, Y., Jiang, H., Li, Y., Zhong, X., Zhuang, L., Yang, Z., Zhang, X. and Huang, B. 2025. Transcriptional regulation mechanisms in AsAFL1-mediated drought tolerance for creeping bentgrass (*Agrostis stolonifera*). *Physiol. Plant. 173*, 1–15.

Taylor, L. K., Wang, H. C. R. and Erikson, R. 1996. Newly identified stress-responsive protein kinases, Krs-1 and Krs-2. *Proc. Natl. Acad. Sci. USA. 93*, 10305–10310.

Tenhaken, R. 2015. Cell wall remodeling under abiotic stress. *Front. Plant Sci. 5*, 771.

Tötzke, C., Kardjilov, N., Hilger, A., Rudolph-Mohr, N., Manke, I. and Oswald, S. 2021. Three-dimensional in vivo analysis of water uptake and translocation in maize roots by fast neutron tomography. *Sci. Rep. 11*, 10578.

Valko, M., Leibfritz, D., Moncol, J., Cronin, M. T., Mazur, M. and Telser, J. 2007. Free radicals and antioxidants in normal physiological functions and human disease. *Int. J. Biochem. Cell Biol. 39*, 44–84.

Vives-Bauza, C., Starkov, A. and García-Arumí, E. 2007. Measurements of the antioxidant enzyme activities of superoxide dismutase, catalase, and glutathione peroxidase. *Methods Cell Biol. 80*, 379–393.

Wahab, A., Abdi, G., Saleem, M., Ali, B., Ullah, S., Shah, W., Mumtaz, S., Yasin, G., Mureșan, C. and Marc, R. 2022. Plants' physio-biochemical and phyto-hormonal responses to alleviate the adverse effects of drought stress: A comprehensive review. *Plants. 11*, 1620.

Wang, D., Hou, L., Zhang, L. and Liu, P. 2021. The mechanisms of silicon on maintaining water balance under water deficit stress. *Physiol. Plant. 173*, 1253–1262.

Wang, L., Zhang, T., Li, C., Zhou, C., Liu, B., Wu, Y., He, F., Xu, Y., Li, F. and Feng, X. 2024. Overexpression of wild soybean expansin gene GsEXLB14 enhanced the tolerance of transgenic soybean hairy roots to salt and drought stresses. *Plants. 13*, 1656.

Wang, L., Zhou, Y., Ding, Y., Chen, C., Chen, X., Su, N., Zhang, X., Pan, Y. and Li, J. 2023. Novel flavin-containing monooxygenase protein FMO1 interacts with CAT2 to negatively regulate drought tolerance through ROS homeostasis and ABA signaling pathway in tomato. *Hortic. Res. 10*, 1–15.

Wang, M., Wang, R., Mur, L., Ruan, J., Shen, Q. and Guo, S. 2021. Functions of silicon in plant drought stress responses. *Hortic. Res. 8*, 254.

Wang, P., Wu, X., Li, N., Nie, H., Ma, Y., Wu, J., Zhang, Z. and Ma, Y. 2025. The StbHLH47 transcription factor negatively regulates drought tolerance in potato (*Solanum tuberosum* L.). *BMC Plant Biol. 25*(1), 15.

Wang, Y., Zhao, Z., Liu, F., Sun, L. and Hao, F. 2020. Versatile roles of aquaporins in plant growth and development. *Int. J. Mol. Sci. 21*, 9485.

Wei, W., Liang, D., Bian, X. H., Shen, M., Xiao, J., Zhang, W. K., Ma, B., Lin, Q., Lv, J., Chen, X., Chen, S. and Zhang, J. S. 2019. GmWRKY54 improves drought tolerance through activating genes in ABA and Ca^{2+} signaling pathways in transgenic soybean. *Plant J. 100*, 327–343.

Westfall, C., Muehler, A. M. and Jez, J. 2013. Enzyme action in the regulation of plant hormone responses. *J. Biol. Chem. 288*, 17647–17655.

Wu, J., Wang, J., Hui, W., Zhao, F., Wang, P., Su, C. and Gong, W. 2022. Physiology of plant responses to water stress and related genes: A review. *Forests. 13*, 324.

Yang, J., Zhang, G., An, J., Li, Q., Chen, Y., Zhao, X., Wu, J., Wang, Y., Hao, Q., Wang, W. and Wang, W. 2020. Expansin gene TaEXPA2 positively regulates drought tolerance in transgenic wheat (*Triticum aestivum* L.). *Plant Sci. 298*, 110596.

Yang, X., Lu, M., Wang, Y., Wang, Y., Liu, Z. and Chen, S. 2021. Response mechanism of plants to drought stress. *Horticulturae. 7*, 50.

You, J. and Chan, Z. 2015. ROS regulation during abiotic stress responses in crop plants. *Front. Plant Sci. 6*, 1092.

Zhang, H., Liu, D., Yang, B., Liu, W., Mu, B., Song, H., Chen, B., Li, Y., Ren, D., Deng, H. and Jiang, Y. Q. 2019. *Arabidopsis* CPK6 positively regulates ABA signaling and drought tolerance through phosphorylating ABA-responsive element binding factors. *J. Exp. Bot. 70*, 5371–5384.

Zhang, Y., Zhu, J., Khan, M. L., Wang, Y., Xiao, W., Fang, T., Qu, J., Xiao, P., Li, C. and Liu, J. H. 2022. Transcription factors ABF4 and ABR1 synergistically regulate amylase-mediated starch catabolism in drought tolerance. *Plant Physiol. 190*, 1–15.

Zhao, M. R., Han, Y. Y., Feng, Y., Li, F. and Wang, W. 2012. Expansins are involved in cell growth mediated by abscisic acid and indole-3-acetic acid under drought stress in wheat. *Plant Cell Rep. 31*, 671–685.

Zhao, Y., Chen, D., Grin, I. R., Zharkov, D. O. and Yu, B. 2025. Developing plant-derived DNA repair enzyme resources through studying the involvement of base excision repair DNA glycosylases in stress responses of plants. *Physiol. Plant. 173*, e14012.

7 Impact of Soil Nutrients on Enzyme Activity in Plants

Farwa Basit, Elshan Musazade, Wang Hao, Kebaichun Zhu, and Javaid Akhter Bhat

7.1 INTRODUCTION

Soil fertility is one of the most important factors influencing plant growth and productivity, as it affects both the regulation of plant biochemical processes and the availability of nutrients. Enzyme activity plays a crucial role in maintaining metabolic functions, promoting nutrient absorption, and helping plants adapt to environmental stressors (Tang, 2024). Various processes such as photosynthesis, respiration, nitrogen fixation, antioxidant defense, and hormone production are facilitated by plant enzymes. Therefore, the proper functioning of these enzyme systems is essential for the growth and resilience of plants (Basit et al., 2025; Gunal et al., 2025).

Soil-supplied nutrients are essential for regulating enzyme functions in plants. Macronutrients such as potassium, phosphorus, and nitrogen significantly influence enzyme synthesis, activity, and stability in primary metabolism (Shrivastav et al., 2020). For instance, nitrogen is crucial because it is a structural component of proteins and amino acids, determining the amount of enzymatic machinery available in plants (Tang, 2024). Similarly, potassium acts as a cofactor for many metabolic enzymes, helping to regulate osmotic balance and glucose metabolism. Phosphorus, on the other hand, is a key component of adenosine triphosphate (ATP) and nucleotides, playing a vital role in phosphorylation processes that activate enzymes (Wan et al., 2025).

Soil nutrients not only play essential roles in primary metabolic functions but also have indirect effects on enzyme activity. Reactive oxygen species (ROS), which can disrupt enzymatic processes, are often generated due to nutrient deficiencies or imbalances. This leads to increased production of antioxidant enzymes such as superoxide dismutase (SOD), catalase (CAT), and peroxidase (Ahmad et al., 2025). Additionally, the availability of nutrients and the regulation of enzymes are influenced by the soil's physicochemical properties, including pH, organic matter concentration, and microbial activity. In the rhizosphere, where plant roots and soil microorganisms interact, enzyme activity and nutrient cycling are closely interconnected (Lei et al., 2025).

Research indicates that nutrient management practices in agriculture significantly influence plant enzymatic activity and overall metabolic efficiency. For instance, balanced fertilization not only enhances the activity of enzymes associated with stress tolerance and nutrient absorption but also improves crop yield (Xing et al., 2025). Conversely, nutritional stress can hinder essential enzymatic processes, reducing growth and making plants more vulnerable to environmental changes (Nikitha et al., 2025). Understanding the relationship between soil nutrients and plant enzyme activity has become essential as global agriculture faces the challenge of maintaining soil health while increasing production.

This chapter examines how soil nutrients affect plant enzyme activity, emphasizing both direct and indirect regulatory processes. It examines how certain macro- and micronutrients function as cofactors and regulators of enzyme systems, talks about how nutrient management and soil characteristics impact enzymatic responses, and takes into account the wider ramifications for crop productivity, plant physiology, and sustainable agriculture.

7.2 ESSENTIAL PLANT NUTRIENTS

7.2.1 Criteria for Nutrient Essentiality

When all three of Arnon and Stout's (1939) requirements are satisfied, an element is deemed vital. Among them are the following:

- Insufficient amounts of the element prevent the plant from completing its life cycle.
- The deficiency is element-specific and may only be avoided or fixed by providing the relevant element.
- Because it is a component of an essential metabolite or necessary for the function of a certain enzyme system, the element directly contributes to the plant's nutrition.

The following elements are essential for the growth of higher plants, totaling 17 recognized nutrients. These essential nutrients include carbon (C), hydrogen (H), oxygen (O), nitrogen (N), phosphorus (P), potassium (K), calcium (Ca), magnesium (Mg), sulfur (S), iron (Fe), manganese (Mn), zinc (Zn), copper (Cu), boron (B), molybdenum (Mo), chlorine (Cl), and nickel (Ni). Carbon (C), hydrogen (H), and oxygen (O) are primarily obtained from water and carbon dioxide, and therefore, they are not classified as mineral nutrients (Taiz, 2002). Essential nutrients are crucial for plant growth and reproduction (Figure 7.1). For an element to be classified as an essential nutrient, it must meet certain criteria. These criteria include its necessity for a plant's life cycle, its importance for most plants, and the absence of a completely interchangeable alternative (Moini & Ferdowsi, 2024). However, in some plant species, certain elements

DOI: 10.1201/9781003669920-7

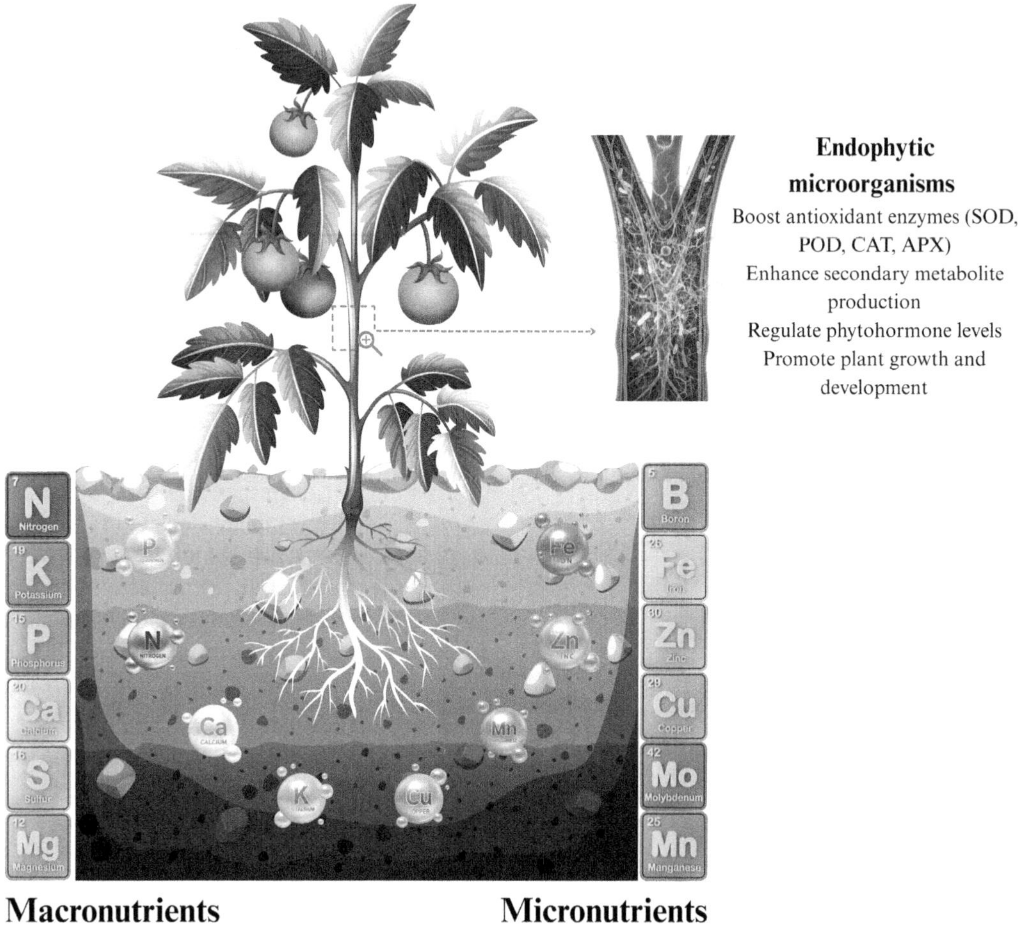

FIGURE 7.1 Plants regulate endophytes through nutrient uptake from the soil, enhancing growth and development.

may only fulfill specific roles or promote growth in particular ways. In addition to basic components, some beneficial components may only be required for particular plant species or may even have the ability to stimulate development. Notable positive aspects consist of sodium, cobalt, and silicon. Other elements have also been suggested as possible promoters of plant development, including vanadium, aluminum, and selenium (Moini & Ferdowsi, 2024).

7.3 CLASSIFICATION OF SOIL NUTRIENTS

7.3.1 Classification of Plant Nutrients Based on Quantity

The relative concentration of essential nutrients in plant tissue determines whether they are categorized as macronutrients or micronutrients. This is the classification that is based on numbers (Kirkby, 2023).

7.3.1.1 Macronutrients

Nutrients that plants require in large quantities are classified as macronutrients. The primary macronutrients include carbon (C), hydrogen (H), oxygen (O), nitrogen (N), phosphorus (P), potassium (K), calcium (Ca), magnesium (Mg), and sulfur (S). Among these, nitrogen, phosphorus, and potassium are referred to as primary nutrients because plants need them in greater amounts. Fertilizers containing N, P, and K can help correct deficiencies of these nutrients when they are identified. Calcium, magnesium, and sulfur are known as secondary nutrients, as they are needed in moderate amounts by plants and can sometimes lead to localized deficiencies (Monib et al., 2023).

7.3.1.2 Micronutrients

Nutrients that plants need in smaller quantities but are just as essential as macronutrients are classified as micronutrients. These include iron (Fe), manganese (Mn), zinc (Zn), copper (Cu), nickel (Ni), boron (B), molybdenum (Mo), and chlorine (Cl). Micronutrients can be categorized as cationic (Fe, Mn, Zn, Cu, Ni) or anionic (B, Mo, and Cl). Cationic micronutrients are absorbed as divalent cations, while anionic micronutrients are taken up in their anionic forms. Additionally, boron can be absorbed by plants as the neutral H_3BO_3 molecule (Akpojevwe Abafe et al., 2025).

There are also certain elements that promote plant growth and are essential for some species, but not for all higher plant species. These are known as beneficial elements and include silicon (Si), sodium (Na), cobalt (Co), and selenium (Se). The role of these elements and their optimal concentrations vary depending on the specific element

TABLE 7.1
Classification of Plant Nutrients Based on Biochemical Function

Biochemical Category	Nutrient Examples	Primary Functions in Plants	Key Enzymatic/ Biochemical Roles	Deficiency Symptoms	Key References
Structural nutrients	N, P, S, Ca, Mg	Form proteins, nucleic acids, phospholipids, chlorophyll, and cell walls	Protein synthesis, ATP formation, chlorophyll production	Stunted growth, chlorosis, poor root development, weak cell walls	Marschner (2011) and Mittag et al. (2015)
Enzymatic (catalytic) nutrients	Fe, Mn, Zn, Cu, Mo	Serve as cofactors or prosthetic groups in enzymatic reactions	Redox reactions, nitrogen assimilation, photosynthetic electron transport	Interveinal chlorosis, necrotic spots, poor photosynthesis, weak enzyme activity	Huang et al. (2025)
Regulatory nutrients	K, Ca, B	Modulate physiological processes, maintain osmotic balance, and activate enzymes	Secondary messengers, enzyme activation, and membrane stability	Leaf scorching, wilting, poor flowering, reduced fruit set	Marschner (2011) and Mittag et al. (2015)
Carbon, hydrogen, oxygen	C, H, O	Form water and organic molecules	Fundamental structural roles in carbohydrates, lipids, and nucleic acids	Reduced growth under CO_2-limited conditions	Marschner (2011) and Sharma et al. (2025a)

and plant species, as well as the environmental conditions in which the plants grow (Venkateswarlu et al., 2025; White & Zasoski, 1999).

7.3.2 Classification of Nutrients Based on Biochemical Function

A classification based on biochemical properties is preferred because the classification based on relative concentrations in plant tissues does not adequately consider the physiology of higher plants and the role of nutrients in plant physiology on a larger scale (Table 7.1). Therefore, this classification, as proposed by Mengel and Kirkby (1987), is adopted.

The classification is outlined in Table 7.2, where nutrients are grouped into four basic categories:

1. The first group consists of elements that are involved in the formation of organic compounds in plants. These elements are assimilated through biochemical reactions that involve oxidation and reduction (redox reactions).
2. The second group includes elements that play a role in energy storage and structural maintenance. Phosphorus, silicon, and boron are key contributors to these functions.
3. The third group encompasses nutrients that are present in ionic forms and serve various functions, such as acting as cofactors and aiding in osmoregulation.
4. The fourth group contains nutrients that are specifically involved in redox reactions.

7.4 ENZYME SYSTEM IN PLANTS

Enzymes are biological catalysts that play a crucial role in regulating nearly all biochemical processes in plants, such as metabolism, growth, stress responses, and nutrient assimilation. They speed up reactions without being consumed in the process and are highly specific to their substrates. Plant enzymes can be categorized based on their biochemical function, location, or metabolic role, with each type contributing significantly to plant physiology.

7.4.1 Photosynthetic Enzymes

Photosynthesis is a vital metabolic process in plants that depends on important enzymes for energy transfer and carbon fixation. Ribulose-1,5-bisphosphate carboxylase/oxygenase, or Rubisco, is the main enzyme in the Calvin cycle that fixes carbon dioxide. In C_4 and crassulacean acid metabolism (CAM) plants, phosphoenolpyruvate (PEP) carboxylase catalyzes the first fixation of CO_2 into molecules with four carbons (Tan & Chen, 2023). By enabling the quick interconversion of CO_2 and bicarbonate, carbonic anhydrase maximizes Rubisco's substrate availability (Dhatwalia, 2025). The availability of nutrients, including iron, magnesium, and nitrogen, essential cofactors, or structural elements, affects these enzymes.

7.4.2 Nitrogen Metabolism Enzymes

Specialized enzymes are essential for plants to effectively absorb nitrogen, which is a crucial macronutrient. The process of nitrate assimilation begins with nitrate reductase (NR), an enzyme that catalyzes the reduction of nitrate to nitrite. Subsequently, nitrite reductase (NiR) converts nitrite into ammonium. Following this, glutamate synthase (GOGAT) and glutamine synthetase (GS) incorporate ammonium into amino acids. The activity of these enzymes is tightly regulated by the availability of nitrogen, and they are also sensitive to deficiencies in iron, sulfur, and molybdenum (Jayakumar et al., 2025; Liu et al., 2025).

7.4.3 Phosphorus-Related Enzymes

Energy transfer reactions depend heavily on phosphorus, which is an essential component of ATP and nucleotides.

TABLE 7.2
Critical Roles of Essential Nutrients in Plant Growth and Enzyme Function

Nutrient	Type	Primary Functions in Plants	Key Enzymatic Roles	Deficiency Symptoms	Key References
Nitrogen (N)	Macronutrient	Component of amino acids, proteins, nucleic acids, chlorophyll	Nitrate reductase, glutamine synthetase	Chlorosis, stunted growth, poor leaf development	Ahmad et al. (2025) and Pravallika et al. (2025)
Phosphorus (P)	Macronutrient	ATP formation, nucleotides, phospholipids; energy transfer	Kinases, ATPases, phosphatases	Dark green/purplish leaves, weak roots, delayed maturity	Malhotra et al. (2018)
Potassium (K)	Macronutrient	Enzyme activation, osmotic regulation, photosynthesis, protein synthesis	Pyruvate kinase, starch synthase	Leaf scorching, weak stems, reduced yield	Ahmad et al. (2025) and Malhotra et al. (2018)
Calcium (Ca)	Macronutrient	Cell wall stabilization, signal transduction	Ca^{2+}-dependent protein kinases, pectin methylesterase	Necrosis of young leaves, blossom end rot, poor root development	Jeevanraj et al. (2025)
Magnesium (Mg)	Macronutrient	Chlorophyll center, ATP-dependent enzyme cofactor	Rubisco, ATPases	Interveinal chlorosis, weak stems, reduced photosynthesis	Chaudhry et al. (2021)
Sulfur (S)	Macronutrient	Component of cysteine, methionine, coenzymes, redox reactions	Glutathione reductase, sulfite reductase	Yellowing of young leaves, delayed growth	Zenda et al. (2021)
Iron (Fe)	Micronutrient	Electron transport, chlorophyll synthesis	Catalase, cytochromes, peroxidases	Interveinal chlorosis, poor growth, leaf necrosis	Roychoudhury et al. (2025b)
Zinc (Zn)	Micronutrient	Enzyme cofactor, protein synthesis, auxin metabolism	Carbonic anhydrase, alcohol dehydrogenase	Small leaves, shortened internodes, chlorosis	Sharma et al. (2025b)
Manganese (Mn)	Micronutrient	Photosystem II activity, enzyme activation	Superoxide dismutase (Mn-SOD), oxidoreductases	Interveinal chlorosis, necrotic leaf spots	Jiang et al. (2024) and Schmidt et al. (2016)
Copper (Cu)	Micronutrient	Oxidative enzymes, lignin synthesis	Polyphenol oxidase, laccase	Twisted young leaves, dieback, reduced flowering	Rather et al. (2025)
Molybdenum (Mo)	Micronutrient	Nitrogen assimilation	Nitrate reductase, nitrogenase	Pale leaves, poor nitrogen utilization	Roychoudhury and Chakraborty (2022)
Boron (B)	Micronutrient	Cell wall structure, sugar transport	Borate cross-linking in pectin	Deformed growing points, brittle leaves, poor fruit set	Day and Aasim (2020)
Chlorine (Cl)	Micronutrient	Photosynthesis, osmotic balance	Oxygen-evolving complex (PSII)	Wilting, chlorosis, reduced growth	Roychoudhury et al. (2025a)
Nickel (Ni)	Micronutrient	Cofactor of urease, nitrogen recycling	Urease	Accumulation of urea, poor seed germination, stunted growth	Lizcano Toledo et al. (2025)

Inorganic phosphate is released for plant absorption when organic phosphorus molecules are hydrolyzed by enzymes like acid and alkaline phosphatases. Phosphorylation processes that control metabolic pathways and activate or deactivate other enzymes are mediated by ATPases and kinases. Higher phosphatase activity is frequently induced by phosphorus deprivation as a compensatory mechanism (Cao et al., 2022).

7.4.4 Antioxidant and Stress-Related Enzymes

Due to environmental conditions, including drought, salt, and disease attack, plants are continuously subjected to oxidative stress. By scavenging ROS, enzymes such as glutathione reductase (GR), peroxidases (POD), CAT, and SOD shield cellular constituents. The availability of micronutrients, including iron, manganese, copper, and zinc, which act as cofactors in redox processes, affects the activity of these antioxidant enzymes (Cai et al., 2024).

7.4.5 Other Metabolic and Hydrolytic Enzymes

During germination and growth, hydrolytic enzymes like amylases, proteases, cellulases, and ureases help to release stored nutrients. For example, urease needs nickel as a cofactor in order to hydrolyze urea into ammonia. These enzymes' activity frequently indicates the nutritional quality of the soil and the general health of the plants (Neemisha & Sharma, 2022). As a result, plant enzyme activity is dynamic and nutrient-dependent, reacting to environmental factors as well as the availability of macro- and micronutrients. Enzyme activity monitoring is a useful technique in plant nutrition and agronomy research because it offers information on the physiological state of plants, their efficiency in using nutrients, and their reactions to stress (Schulz et al., 2023).

7.5 DIRECT OR INDIRECT INFLUENCE OF SOIL NUTRIENTS ON ENZYME ACTIVITY

Soil nutrients are crucial for regulating enzymatic activities in plants, which in turn affects their growth, metabolism, and responses to stress. Enzymes rely on the availability of both macro- and micronutrients, which serve as structural components, cofactors, or activators. This availability controls the rate and efficiency of biochemical reactions within the plant (Shrivastav et al., 2020; Wang et al., 2012).

7.5.1 Macronutrient Effects

7.5.1.1 Nitrogen (N)

Nitrogen, a significant component of proteins and amino acids, has a direct impact on the synthesis and function of enzymes like Rubisco, GS, and NR. Reduced enzyme concentrations, decreased photosynthetic capacity, and stunted development are the results of a nitrogen shortage.

7.5.1.2 Phosphorus (P)

Phosphorus is essential for ATP-dependent reactions. Its deficiency triggers phosphatases that hydrolyze organic P compounds, releasing inorganic phosphate, a crucial substrate for metabolic enzymes. Adequate phosphorus enhances the activities of kinases and ATPases, supporting energy transfer and metabolic regulation.

7.5.1.3 Potassium (K)

Potassium serves as a cofactor for enzymes that facilitate carbohydrate metabolism, protein synthesis, and osmotic regulation. A deficiency in potassium reduces the activity of enzymes like pyruvate kinase, which impairs energy metabolism and affects overall growth.

7.5.1.4 Calcium (Ca) and Magnesium (Mg)

Calcium regulates enzyme activity through Ca^{2+}-dependent signaling, while magnesium helps stabilize ATP and activates enzymes such as Rubisco. Deficiencies in these minerals can lead to impaired enzyme function and disrupted metabolic pathways.

7.5.2 Micronutrient Effects

Molybdenum (Mo), copper (Cu), manganese (Mn), zinc (Zn), iron (Fe), and nickel (Ni) are essential trace elements that act as cofactors in various enzymatic reactions. Manganese is necessary for the enzyme SOD (Mn-SOD); zinc is required for carbonic anhydrase; copper is crucial for polyphenol oxidase; molybdenum is needed for NR; nickel is essential for urease; and iron is important for CAT and cytochrome enzymes. Deficiencies or imbalances in these elements can lead to reduced enzymatic efficiency, compromised redox balance, and decreased nutrient absorption.

7.5.3 Indirect Effects of Nutrient Availability

Soil nutrients have indirect effects on enzyme activity, in addition to their primary biochemical functions, by influencing the following factors:

7.5.3.1 Microbial Activity

Extracellular enzymes produced by soil bacteria help mineralize organic matter, which enhances the absorption of nutrients by plants. Nutrient-rich soils typically stimulate microbial enzymatic activity (Wang et al., 2023).

7.5.3.2 Soil pH and Structure

These factors affect the solubility and availability of nutrients, which in turn influence the activation of enzymes in both plants and microbes (Chai et al., 2025).

7.5.3.3 Stress Responses

When ROS accumulate due to nutrient deficiencies, antioxidant enzymes such as SOD, CAT, and peroxidase are activated as a protective mechanism (Cai et al., 2024).

7.5.4 Agricultural Implications

Balanced nutrition management enhances enzymatic activity, which in turn improves photosynthesis, nitrogen uptake, and stress tolerance in plants. Conversely, nutritional imbalances can reduce plant productivity by inhibiting key enzyme pathways. Therefore, monitoring enzyme activity can serve as a valuable bio-indicator of soil fertility and plant nutrient status, guiding sustainable agricultural practices (Wang et al., 2023).

7.6 AGRICULTURAL AND ECOLOGICAL IMPLICATIONS OF NUTRIENT–ENZYME INTERACTIONS

Nutrient–enzyme interactions are essential for plant health, impacting crop yield, quality, stress resilience, and the functioning of soil ecosystems. Understanding these interactions is crucial for advancing sustainable agriculture and enhancing soil health (Solangi et al., 2024).

7.6.1 Impact of Nutrient-Enzyme Interaction on Crop Yield and Quality

The availability of essential nutrients directly affects the activity of plant enzymes involved in various metabolic processes. For example, nitrogen availability influences enzymes like NR and GS, which are vital for protein synthesis and overall plant growth. Sufficient nutrient supply ensures optimal enzyme function, ultimately leading to improved crop yield and quality (Sobhy et al., 2025).

7.6.2 Importance of Nutrient–Enzyme Interaction in Sustainable Agriculture and Nutrient Use Efficiency

Efficient nutrient management is critical for sustainable agriculture (Figure 7.2). Nutrient–enzyme interactions significantly contribute to nutrient uptake and utilization. Enzymes such as phosphatases and ureases help mobilize nutrients from soil organic matter, thereby enhancing nutrient use efficiency (Table 7.2). Sustainable practices that maintain or improve soil nutrient levels can optimize enzyme activity, leading to better resource utilization and a reduced environmental impact (Xing et al., 2025).

7.6.3 Role of Nutrient–Enzyme Interaction in Environmental Stress Tolerance

Nutrient availability influences the expression and activity of enzymes that are involved in stress responses. For instance, antioxidant enzymes like SOD and CAT protect plants from oxidative damage during drought or salinity stress and are regulated by nutrient levels (Bogati & Walczak, 2022; Xiao et al., 2023). Adequate nutrients support the synthesis of these enzymes, enhancing plant resilience to both abiotic and biotic stresses (Nurzhan et al., 2022).

7.6.4 Implications of Nutrient–Enzyme Interaction for Soil Health and Biogeochemical Cycling

Soil health is closely linked to microbial activity, which is influenced by nutrient availability. Nutrient-enzyme interactions affect microbial processes such as organic matter decomposition and nutrient cycling. Enzymes produced by soil microorganisms, like phosphatases and glucosidases, play a critical role in these processes. Maintaining balanced nutrient levels supports microbial health and activity, promoting soil fertility and ecosystem stability (Ma et al., 2017; Tian et al., 2021).

7.7 CONCLUSION AND FUTURE PERSPECTIVES

This book chapter highlights the physiological and biochemical roles of soil nutrients in plants. Nutrient deficiencies are increasingly common worldwide and significantly affect crop yields and agricultural productivity. Some nutrients that lead to impairments are integral components or prosthetic groups of enzymes, which in turn influence enzyme-catalyzed reactions involved in plant metabolism,

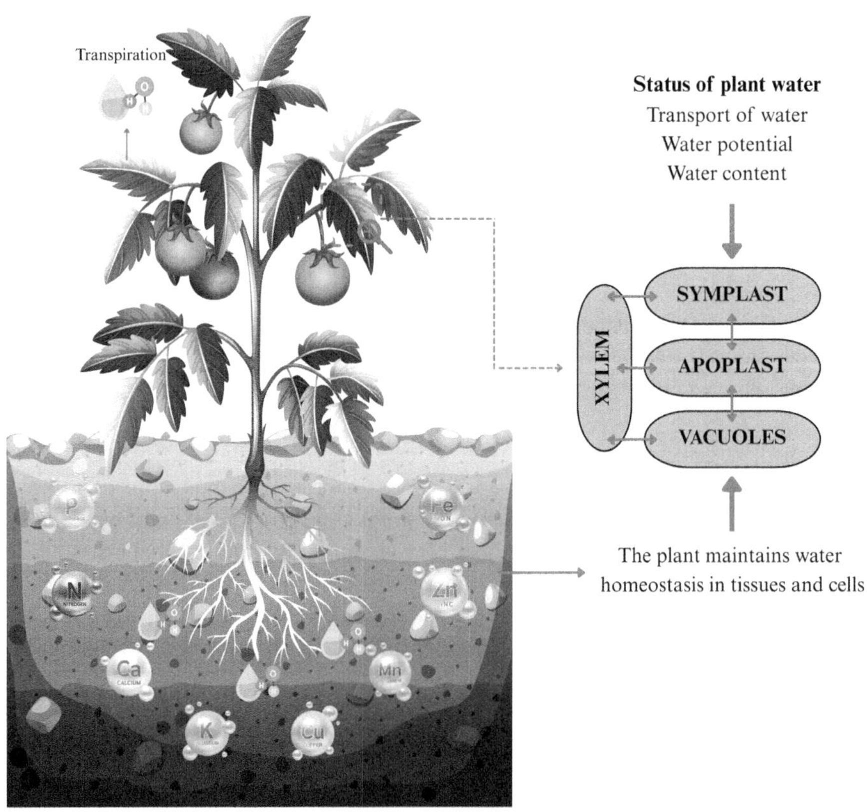

FIGURE 7.2 Plants absorb water and mineral nutrients from the soil, transport them through the xylem, and regulate water status via apoplast, symplast, and vacuoles to maintain cellular homeostasis and support growth.

including the synthesis of macromolecules and associated biochemical processes. Key nutrients such as nitrogen, phosphorus, calcium, and boron are structurally important for plants. Additionally, they play a vital role in maintaining the homeostasis of the plant system and ensuring an appropriate cellular redox status.

The extent of yield losses in plants is significantly influenced by the availability of nutrients, as well as their interaction with soil and environmental conditions. To achieve higher productivity, it is essential to study the mechanisms of nutrient uptake and translocation, particularly the role of transporters in maintaining nutrient homeostasis that favors yield. Although several studies on iron and zinc transporters have yielded positive results, further research is needed.

Additionally, research should focus on protective mechanisms against environmental stress and utilize molecular genetics and biotechnological tools to understand how plants adapt to low nutrient levels. This is particularly crucial as global food demand continues to rise each year. Moreover, nutrient biofortification, a technique aimed at increasing micronutrient levels to enhance seedling vigor and yield, is highly relevant. Genetic modification of crops to boost nutrient fortification for the betterment of human and animal nutrition would also be beneficial.

REFERENCES

Ahmad, A., Ahmad, S., Akram, T., Sehrish, A.K., Tabassam, R., Tayyeb, M., Ali, S., 2025. Addressing nutrient deficiency and sufficiency in plants, Role of antioxidants in abiotic stress management. Elsevier, pp. 185–199.

Akpojevwe Abafe, E., Smith, N.W., Maxwell, T.M., McNabb, W.C., 2025. Trends in micronutrient research since the SDGs: A global perspective. Critical Reviews in Food Science and Nutrition, 65(31), 7871–7882.

Arnon, D.I., Stout, P., 1939. The essentiality of certain elements in minute quantity for plants with special reference to copper. Plant Physiology 14, 371.

Basit, F., Khalid, M., Bahadur, A., Bhat, J.A., Ahmad, A., Ahmad, P., 2025. Alleviation of vanadium toxicity through individual and combined supplementation of spermine and spermidine in soybean: Insights into oxidative stress management, detoxification mechanisms, and root exudates. South African Journal of Botany 184, 468–479.

Bogati, K., Walczak, M., 2022. The impact of drought stress on soil microbial community, enzyme activities and plants. Agronomy 12, 189.

Cai, Q., Zou, Y., Zhong, N., Jia, G., 2024. Effects of nutritional stress on soil fertility and antioxidant enzymes of rice in different growth periods. Frontiers in Plant Science 15, 1471682.

Cao, N., Zhi, M., Zhao, W., Pang, J., Hu, W., Zhou, Z., Meng, Y., 2022. Straw retention combined with phosphorus fertilizer promotes soil phosphorus availability by enhancing soil P-related enzymes and the abundance of phoC and phoD genes. Soil and Tillage Research 220, 105390.

Chai, J., Yao, T., Li, H., Wang, X., Liu, X., Li, C., Li, S., Bai, J., 2025. Meadow degradation affects microbial community structure and vegetation characteristics by increasing soil pH. Land Degradation & Development 36, 790–801.

Chaudhry, A.H., Nayab, S., Hussain, S.B., Ali, M., Pan, Z., 2021. Current understandings on magnesium deficiency and future outlooks for sustainable agriculture. International Journal of Molecular Sciences 22, 1819.

Day, S., Aasim, M., 2020. Role of boron in growth and development of plant: Deficiency and toxicity perspective, Plant micronutrients: Deficiency and toxicity management. Springer, pp. 435–453.

Dhatwalia, D., 2025. 6 photosynthesis, productivity. Impact of climate change on medicinal and herbal plant microRNA, 95.

Gunal, H., Budak, M., Gunal, E., Kilic, M., Al Tawaha, A.R.M., Al-Tawaha, A.R., Karnwal, A., Nesterova, N., Singh, A., Rajput, V.D., 2025. Soil fertility in arid lands: Strategies for sustainable management and fertilization, sustainable agriculture under drought stress. Elsevier, 61–69. Academic Press. https://doi.org/10.1016/B978-0-443-23956-4.00006-5

Huang, K., Hu, C., Tan, Q., Wu, S., Shabala, S., Yu, M., Sun, X., 2025. Nanozymes as a tool to boost agricultural production: From preparation to application. Environmental Science: Nano 12, 98–120.

Jayakumar, K., Sakthilatha, D., Venkatesan, R., 2025. Advancements in bioelectrocatalysis for electrochemical ammonia production: Integrating enzyme-based catalysis. ChemistrySelect 10, e202500224.

Jeevanraj, R., Sivakumar, R., Boominathan, P., Kavitha, P., Sendhilvel, V., 2025. The essential role of macronutrients and micronutrients in improving crop resilience to biotic stress. Russian Journal of Plant Physiology 72, 126.

Jiang, Y., Sun, Y., Zhang, P., Zhou, P., Wang, Q., Zhu, G., Zhao, W., Li, Y., Wang, Q., Rui, Y., 2024. Investigation of the effects and mechanisms of manganese-based NMs on rice growth. Environmental Science and Pollution Research 31, 34368–34380.

Kirkby, E.A., 2023. Introduction, definition, and classification of nutrients, Marschner's mineral nutrition of plants. Elsevier, 3–9. Academic Press. https://doi.org/10.1016/B978-0-12-819773-8.00016-2

Lei, Y., Kuai, Y., Guo, M., Zhang, H., Yuan, Y., Hong, H., 2025. Phosphate-solubilizing microorganisms for soil health and ecosystem sustainability: A forty-year scientometric analysis (1984–2024). Frontiers in Microbiology 16, 1546852.

Liu, H., Gao, X., Fan, W., Fu, X., 2025. Optimizing carbon and nitrogen metabolism in plants: From fundamental principles to practical applications. Journal of Integrative Plant Biology 67(6), 1447–1466.

Lizcano Toledo, R., Melo, W.J.D., Prado, R.D.M., Olivera Viciedo, D., Peruca de Melo, G.M., de Araújo, A.S.F., Calero Hurtado, A., 2025. Morpho-physiological response to micronutrient deficiencies and nickel addition of passion fruit plants. Journal of Plant Nutrition 48, 670–689.

Ma, Y.-H., Fu, S.-L., Zhang, X.-P., Zhao, K., Chen, H.Y., 2017. Intercropping improves soil nutrient availability, soil enzyme activity and tea quantity and quality. Applied Soil Ecology 119, 171–178.

Malhotra, H., Vandana Sharma, S., Pandey, R., 2018. Phosphorus nutrition: Plant growth in response to deficiency and excess, Plant nutrients and abiotic stress tolerance. Springer, pp. 171–190.

Marschner, H., 2011. Marschner's mineral nutrition of higher plants. Academic Press.

Mittag, J., Gabrielyan, A., Ludwig-Müller, J., 2015. Knockout of GH3 genes in the moss physcomitrella patens leads to increased IAA levels at elevated temperature and in darkness. Plant Physiology and Biochemistry 97, 339–349.

Moini, J., Ferdowsi, K., 2024. Handbook of nutritional disorders. CRC Press.

Monib, A.W., Alimyar, O., Mohammad, M.U., Akhundzada, M.S., Niazi, P., 2023. Macronutrients for plants growth and

humans health. Journal for Research in Applied Sciences and Biotechnology 2, 268–279.
Neemisha Sharma, S., 2022. Soil enzymes and their role in nutrient cycling, Structure and functions of pedosphere. Springer, pp. 173–188.
Nikitha, S., Prabhanjan, S., Sathyanarayan, A., 2025. Plant nutritional deficiency detection: A survey of predictive analytics approaches. Iran Journal of Computer Science 8, 83–101.
Nurzhan, A., Tian, H., Nuralykyzy, B., He, W., 2022. Soil enzyme activities and enzyme activity indices in long-term arsenic-contaminated soils. Eurasian Soil Science 55, 1425–1435.
Pravallika, P.M., Kumar, K.M., Varalakshmi, D., 2025. Prediction of Nutrient Deficiency in Crops using Plant Health Data, 2025 6th International Conference on Intelligent Communication Technologies and Virtual Mobile Networks (ICICV). IEEE, pp. 203–209.
Rather, G.A., Manhas, A., Sharma, N., 2025. Mobility of nutrients and their metabolism in agriculture crops, agricultural biotechnology. Apple Academic Press, pp. 75–93.
Roychoudhury, A., Chakraborty, S., 2022. Cobalt and molybdenum: Deficiency, toxicity, and nutritional role in plant growth and development, Plant nutrition and food security in the era of climate change. In *Plant nutrition and food security in the era of climate change* (pp. 255–270). Academic Press. https://doi.org/10.1016/B978-0-12-822916-3.00021-4
Roychoudhury, A., Mukherjee, S., Ghosh, A., 2025a. 10 molecular and physiological. Agricultural Crop Improvement: Plant and Soil Relationships 135.
Roychoudhury, A., Mukherjee, S., Ghosh, A., 2025b. Molecular and physiological studies of the role of micronutrients in sustainable growth and development of crop plants, agricultural crop improvement. CRC Press, pp. 135–147.
Schmidt, S.B., Jensen, P.E., Husted, S., 2016. Manganese deficiency in plants: The impact on photosystem II. Trends in Plant Science 21, 622–632.
Schulz, C., Lemaire, Q., Berthier, A., Descat, A., Kouach, M., Vercoutter-Edouart, A.S., Yazidi-Belkoura, I.E., Hardivillé, S., Goossens, J.F., Lefebvre, P., 2023. Control of lipid metabolism by the dynamic and nutrient-dependent post-translational modification O-GlcNAcylation. Natural Sciences 3, e20220006.
Sharma, N., Chaudhary, A., Sharma, N., Shukla, S., Joshi, N.C., Vishwakarma, K., Singh, P.K., Mishra, A., 2025a. Plant nutrient requirements and nutrient homeostasis in plants, Soil health and nutrition management. CABI GB, pp. 1–20.
Sharma, V., Maurya, A., Kandhol, N., Singh, V.P., Sharma, S., Peralta-Videa, J., Tripathi, D.K., 2025b. Zinc deficiency in plants: An insight into fortification strategies, Zinc in plants. Elsevier, pp. 249–264.
Shrivastav, P., Prasad, M., Singh, T.B., Yadav, A., Goyal, D., Ali, A., Dantu, P.K., 2020. Role of nutrients in plant growth and development, Contaminants in agriculture: Sources, impacts and management. Springer, pp. 43–59.
Sobhy, D. M., & Anandhi, A. (2025). Soil Nutrient Monitoring Technologies for Sustainable Agriculture: A Systematic Review. Sustainability, 17(18), 8477.
Solangi, F., Zhu, X., Solangi, K.A., Iqbal, R., Elshikh, M.S., Alarjani, K.M., Elsalahy, H.H., 2024. Responses of soil enzymatic activities and microbial biomass phosphorus to improve nutrient accumulation abilities in leguminous species. Scientific Reports 14, 11139.
Taiz, L., 2002. Mineral nutrition. Plant Physiology, 67–86. https://cir.nii.ac.jp/crid/1571698600870516352
Tan, B., Chen, S., 2023. Defining mechanisms of C3 to CAM photosynthesis transition toward enhancing crop stress resilience. International Journal of Molecular Sciences 24, 13072.
Tang, L., 2024. Soil fertility, Plant nutrition and nutrient management. MDPI, p. 34.
Tian, J., Ge, F., Zhang, D., Deng, S., Liu, X., 2021. Roles of phosphate solubilizing microorganisms from managing soil phosphorus deficiency to mediating biogeochemical P cycle. Biology 10, 158.
Venkateswarlu, M., Rallapalli, S., Singh, A., Chalapathi, G.S.S., Kumar, S., Katpatal, Y.B., Sujatha, G., 2025. Macro and micronutrient based soil fertility zonation using fuzzy logic and geospatial techniques. Scientific Reports 15, 26772.
Wan, H., Hou, J., Wei, Z., Liu, F., 2025. Contrasting maize responses to soil phosphorus and potassium availability driven by biochar under reduced irrigation. Plant and Soil 508, 677–700.
Wang, B., Xue, S., Liu, G.B., Zhang, G.H., Li, G., Ren, Z.P., 2012. Changes in soil nutrient and enzyme activities under different vegetations in the Loess Plateau area, Northwest China. Catena 92, 186–195.
White, J.G., Zasoski, R.J., 1999. Mapping soil micronutrients. Field Crops Research 60, 11–26.
Xiao, L., Min, X., Liu, G., Li, P., Xue, S., 2023. Effect of plant–plant interactions and drought stress on the response of soil nutrient contents, enzyme activities and microbial metabolic limitations. Applied Soil Ecology 181, 104666.
Xing, Y., Xie, Y., Wang, X., 2025. Enhancing soil health through balanced fertilization: A pathway to sustainable agriculture and food security. Frontiers in Microbiology 16, 1536524.
Zenda, T., Liu, S., Dong, A., Duan, H., 2021. Revisiting sulphur—The once neglected nutrient: It's roles in plant growth, metabolism, stress tolerance and crop production. Agriculture 11, 626.

8 Enzyme Mediated Adaptation to High Salinity Stress

Babar Iqbal, Khulood Fahad Alabbosh, Kashif, Qaiser Javed, and Amir Abdullah Khan

8.1 INTRODUCTION

8.1.1 Importance of Salinity Stress in Agriculture

Salinity constitutes a significant abiotic stressor, as it markedly impedes plant growth, development, and yield, thereby jeopardizing global food security (Khan et al., 2021; Mehmood et al., 2021; Verma et al., 2024). Excessive soil salinity causes osmotic stress, ion toxicity, nutritional deficiencies, and oxidative damage, all of which hinder plant metabolism and productivity ((El-Ramady et al., 2024). These effects diminish water absorption, hinder enzymatic activity, disturb nutritional equilibrium, and expedite the formation of reactive oxygen species (ROS), finally resulting in membrane damage and cellular dysfunction (Zhang, Hou, et al., 2025). To sustain elevated agricultural productivity in salinity-affected areas, the formulation of appropriate management techniques is necessary (Demo et al., 2025).

8.1.2 Overview of Plant Adaptation Strategies

Plants respond to salt by modifying their morphology, physiology, biochemistry, and molecular structure as shown in Figure 8.1 (Khan, Yongfeng, et al., 2025; Sharma, Tisarum, et al., 2024). Salinity, a significant abiotic stressor, impedes growth and yield; therefore, understanding plant adaptations is essential for cultivating salt-tolerant crops (Sackey et al., 2025). Morphological responses encompass inhibited

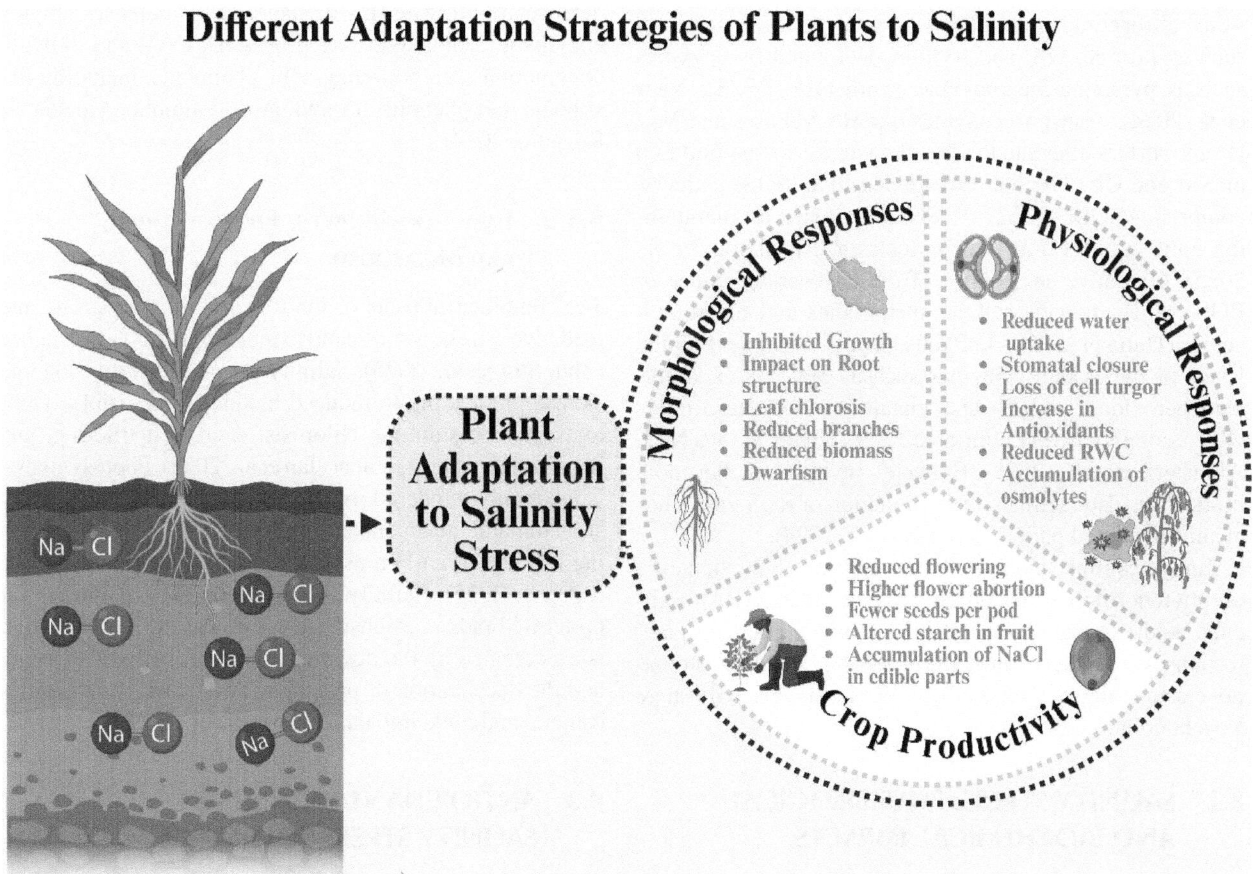

FIGURE 8.1 Plants under salinity adapt different strategies such as morphological, physiological, and in terms of crop productivity to overcome salinity-induced stress.

DOI: 10.1201/9781003669920-8

shoot and root development, along with alterations in leaf morphology, including reduced, thicker leaves or salt glands that excrete excess ions (Zhu, Zhou, et al., 2025). Plants manage salinity by osmotic adjustment, ion control, and antioxidant processes (Shafi et al., 2025). Cells absorb suitable solutes such as proline, glycine betaine (GB), and sorbitol to maintain osmotic equilibrium. These solutes preserve the integrity of proteins and membranes (Khan, Wang, Zeb, et al., 2025; Sevgi & Leblebici, 2025). Ion homeostasis relies on maintaining a high K^+/Na^+ ratio by the selective transit of Na^+ and its storage in vacuoles (Ikram et al., 2025a; Yuan, Zhou, et al., 2025). Plants enhance their antioxidant systems to combat oxidative stress. For example, they enhance the activity of enzymes such as superoxide dismutase (SOD), catalase (CAT), and peroxidases (PODs), facilitating the elimination of ROS (Venkatachalam, Mohan, & Seralathan, 2025).

8.1.3 Role of Enzymes in Biochemical and Physiological Adaptation

Plant enzymes are crucial for the biochemical and physiological alterations that facilitate salinity tolerance (Sharma, Tisarum et al., 2024). Salinity induces osmotic, ionic, and oxidative stressors, initiating a cascade of enzymatic activities (Zhang, Zhang, et al., 2025). Osmotic stress restricts water absorption, prompting plants to produce osmolytes such as proline, GB, and sorbitol, facilitated by enzymes such as pyrroline-5-carboxylate synthetase (P5CS) (Kaur et al., 2024). Transport enzymes like H^+-ATPases and Na^+/H^+ antiporters alleviate ion toxicity caused by the build-up of Na^+ and Cl^-. The Salt Overly Sensitive (SOS) pathway, comprising SOS1, SOS2, and SOS3, is crucial for maintaining ionic equilibrium in the cytoplasm (Abdehpour et al., 2025). Oxidative stress arises from an overabundance of ROS, which damage cellular membranes and macromolecules (Dutta et al., 2024). Plants respond by increasing the levels of antioxidant enzymes such as SOD, CAT, ascorbate peroxidase (APX), and glutathione reductase (GR), which collaboratively eliminate ROS (Khan, Wang, Nisa, Alnusairi, & Shi, 2022). Enzymes involved in flavonoid production also facilitate the elimination of ROS via anthocyanin-mediated pathways (Altaf et al., 2024).

Enzymes that modulate the synthesis and signaling of phytohormones, including abscisic acid (ABA), salicylic acid (SA), and gibberellins (GA), can augment plant responses to stress through processes such as stomatal closure and the release of defense signals (Ghosh & Roychoudhury, 2024).

8.2 SALINITY STRESS: PHYSIOLOGICAL AND BIOCHEMICAL IMPACTS

Salinity stress impairs plant physiology and metabolism by causing ion toxicity, osmotic imbalance, and oxidative damage, hence limiting growth and photosynthetic efficiency (Alghamdi, 2024).

8.2.1 Ion Toxicity

Salinity stress is characterized by an excessive accumulation of Na^+ and Cl^-, which disrupts ion homeostasis and impairs cellular function (Ikram et al., 2025b). Elevated salt concentrations impede plants' nutrient absorption by competing with essential ions such as K^+, Ca^{2+}, and Mg^{2+} (Khan, Wang, Akbar, & Alhoqail, 2025). They inhibit crucial metabolic enzymes and diminish biochemical efficiency (Khan, Wang, Zeb, et al., 2025). Ultimately, they render membranes less stable, hence altering the movement of substances across them (Khan et al., 2021).

8.2.2 Osmotic Stress and Water Deficit

Salinity diminishes the availability of soil water, hence impeding root absorption of water. This induces osmotic stress, resulting in decreased turgor, cellular dehydration, stomatal closure, and inhibited cell development. These variables combined hinder growth and photosynthesis (Haghpanah, Hashemipetroudi, Arzani, & Araniti, 2024). Simultaneously, salinity elevates the concentrations of ROS, such as superoxide, hydrogen peroxide, and hydroxyl radicals, which damage lipids, proteins, and nucleic acids, adversely affecting cellular health and metabolism (Saleem et al., 2025). Plants initiate their antioxidant defenses to mitigate this damage. These defenses employ enzymatic mechanisms, such as SOD, CAT, and POD, in conjunction with non-enzymatic chemicals, including flavonoids and phenolics (Vicidomini, Palumbo, Moccia, & Roviello, 2024).

8.2.3 Impact on Growth, Photosynthesis, and Metabolism

The combined impacts of ion toxicity, osmotic stress, and oxidative stress significantly reduce plant performance (Zhu, Sun, et al., 2025). Salinity impedes cell division and elongation, leading to reduced biomass, and visible symptoms include stunting, chlorosis, decreased tillering, and impaired root development (Jan et al., 2025). Photosynthesis is markedly hindered by decreased chlorophyll synthesis, stomatal conductance, and RuBisCO efficacy, leading to lowered carbon assimilation (Salesse-Smith, Wang, & Long, 2025). Salinity disrupts metabolic pathways for proteins, lipids, carbohydrates, and energy. The situation is exacerbated by the downregulation of RuBisCO activity and the deceleration of protein synthesis, which intensifies osmotic and ionic imbalances (Verma et al., 2024).

8.3 ANTIOXIDANT ENZYMES IN SALINITY STRESS TOLERANCE

Excessive production of ROS by cells disrupts their structures and functioning. This phenomenon is referred to as oxidative damage. Plants activate antioxidant enzymes and utilize additional mechanisms, such as osmolyte storage

TABLE 8.1
Different Plant Enzymes and Their Physiological Functions under Saline Regime

Enzyme/ Pathway	Main Function	Physiological Role Under Salinity Stress	Examples/Notes	Key references
Antioxidant enzymes	Detoxification of reactive oxygen species (ROS)	Protect cells from oxidative damage, maintain redox balance	Superoxide dismutase (SOD), catalase (CAT), ascorbate peroxidase (APX), glutathione reductase (GR)	Hasanuzzaman et al. (2020) and Singh et al. (2020)
Ion transport enzymes	Regulation of Na^+ and K^+ flux	Maintain ionic balance, prevent Na^+ toxicity, support enzyme activity	Na^+/H^+ antiporters (SOS1, NHX), H^+-ATPases, vacuolar H^+-pyrophosphatase	Gupta and Huang (2014) and Zhu (2016)
Osmolyte biosynthesis enzymes	Synthesis of compatible solutes	Osmotic adjustment, stabilization of proteins and membranes, ROS detoxification	P5CS (proline), BADH (glycine betaine), trehalose biosynthesis enzymes	Szabados and Savouré (2010) and Rai et al. (2011)
Secondary metabolism enzymes	Production of phenolics and flavonoids	Strengthen antioxidant capacity, cell wall modification, signaling	Phenylalanine ammonia-lyase (PAL), chalcone synthase (CHS)	Sharma et al. (2019)
Microbial enzyme interactions	Modulation of plant enzymatic activity by rhizobacteria/ mycorrhizae	Enhance antioxidant defense, ion balance, osmolyte accumulation	PGPR-induced SOD, CAT, APX activity	Etesami and Beattie (2018)

and ion transport regulation, to enhance their defenses against salt stress (Table 8.1).

8.3.1 The Role of Antioxidant Enzymes

SOD, CAT, POD, APX, and GR are critical antioxidant enzymes that neutralize ROS and preserve cellular redox balance as shown in Figure 8.2.

8.3.1.1 Superoxide Dismutase (SOD)

SOD enzymes are crucial for plant salinity tolerance, mitigating oxidative damage caused by ROS that accumulate under salt stress as shown in Figure 8.2 (Han et al., 2025). Salinity stress disrupts cellular homeostasis, increasing the production of ROS, which subsequently causes damage to lipids, proteins, and DNA (Dildar, Cui, Ikhwanuddin, & Ma, 2025). SODs, a class of metalloenzymes, catalyze the

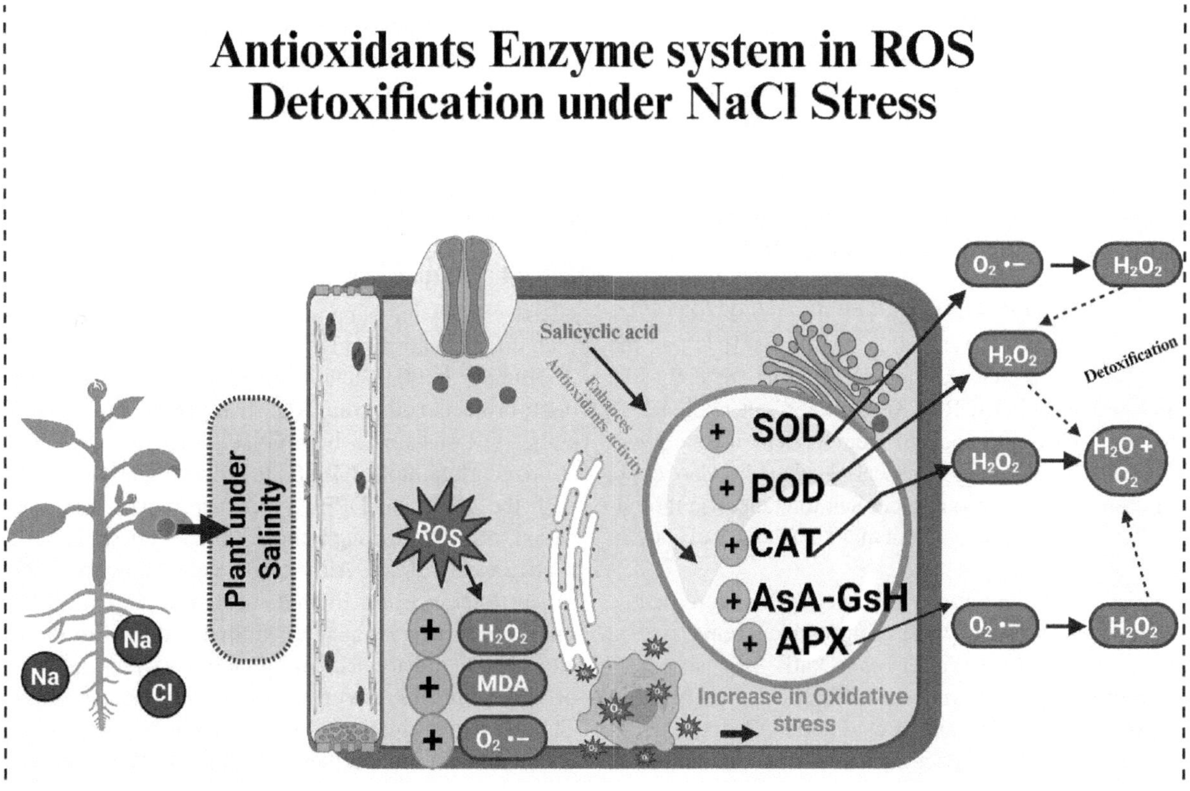

FIGURE 8.2 The role of antioxidants in alleviation of salinity stress through detoxification of reactive oxygen species.

conversion of superoxide radicals into hydrogen peroxide (H_2O_2) and oxygen (O_2). They are the primary mechanism by which cells safeguard themselves against ROS (Farnum & Goldsmith, 2024).

8.3.1.1.1 Mechanisms of SOD in Salinity Tolerance

Antioxidant Défense: SODs protect cells from oxidative harm by transforming superoxide radicals into hydrogen peroxide (H_2O_2), thereby preventing the adverse effects of ROS on cellular components (Kozlov, Javadov, & Sommer, 2024).

- *Isozymes and Their Localization*: Distinct SOD isozymes in chloroplasts, mitochondria, and the cytosol mitigate ROS at their origin. Plants have enhanced salt tolerance when they possess increased levels of Cu/Zn-SOD derived from the mangrove *Kandelia candel* in tobacco or mitochondrial Mn-SOD in rice chloroplasts. These enzymes reduce ROS (Sinha et al., 2025).
- *Nitric Oxide (NO) Regulation*: NO modulates SOD activity and mitigates oxidative damage caused by salinity; hence, it enhances salt tolerance in plants (Jumaah et al., 2024).
- *Genetic Manipulation*: Transgenic plants exhibiting increased SOD gene expression demonstrate improved salt tolerance. Cytoplasmic SOD (cytSOD) in plum and PagSOD2a in poplar augment SOD activity, diminish lipid peroxidation, and boost salinity tolerance (Yao et al., 2024).

8.3.1.2 Catalase (CAT)

CAT is essential for plant salinity tolerance, alleviating oxidative damage induced by ROS under salt stress (Rao et al., 2025). Salinity frequently induces the overproduction of ROS, resulting in cellular damage (Singh, Rajput, et al., 2024). CAT preserves redox equilibrium by catalyzing the decomposition of hydrogen peroxide (H_2O_2) into water and oxygen (Singh, Guha, Pathak, & Kumar, 2025).

8.3.1.2.1 Mechanisms of Catalase in Salinity Tolerance

CAT 2 (CAT2), an enzyme that scavenges H_2O_2, is the mechanism by which calmodulin-binding receptor-like cytoplasmic kinase 3 (CRCK3) regulates salt tolerance in Arabidopsis. CRCK3 knockout mutants demonstrate sensitivity to salt, highlighting its essential role (Liu, Gao, et al., 2025). Jasmonic acid (JA) reduces salt tolerance via inhibiting CAT2 in a MYC2-dependent manner (Zeng et al., 2025).

- *Genetic Manipulation*: Salt tolerance is increased by overexpressing CAT genes. Transgenic rice expressing *Escherichia coli* katE has shown improved growth and seed output in high NaCl environments (Nagamiya et al., 2007; Padmavathi et al., 2023). Osmolytes such as sorbitol, mannitol, or sucrose enhance CAT activity under salt stress (Alhudhaibi et al., 2024).
- Exogenous applications of SA augment CAT activity and alleviate salt-induced damage in *Capsicum annuum* by balancing ROS and antioxidant defenses (Rao et al., 2025).
- *Microbial and Rootstock Interactions*: In heterografted chrysanthemums (*Artemisia annua* rootstock), quinoa cultivars, and *Suaeda salsa* infected with *Glomus mosseae*, increased CAT activity improves salt tolerance (Li, Meng, et al., 2022).

8.3.1.3 Ascorbate Peroxidase (APX)

APX is crucial for plant salinity tolerance, alleviating oxidative stress induced by the accumulation of ROS during salt stress as described in Figure 8.2 (Omar et al., 2024; Rao et al., 2025). Salt stress increases ROS, which negatively impact lipids, proteins, and DNA (Singh, Rajput, et al., 2024). APX employs ascorbate as an electron donor to eliminate hydrogen peroxide (H_2O_2), so safeguarding cells from oxidative damage (Corpas, González-Gordo, & Palma, 2024).

8.3.1.3.1 APX Function and the Ascorbate–Glutathione Cycle

APX utilizes ascorbate to decompose H_2O_2, hence maintaining enzyme stability and ensuring its availability for the degradation of ROS. APX is an essential component of the plant's antioxidant defense mechanism, serving as a crucial element of the ascorbate–glutathione cycle (Anjum et al., 2016; Li, 2023).

- The activity of APX varies across different tissues and genotypes under conditions of salt stress. Diverse sweet potato cultivars exhibiting both salt tolerance and sensitivity demonstrate differing levels of APX expression. Salt stress elevates the activity of APX and GR in the leaves of *Atriplex hortensis* (Kachout, Hamza, Bouraoui, Leclerc, & Ouerghi, 2013).

8.3.1.4 Glutathione Reductase (GR)

GR is crucial for plant salinity tolerance by maintaining glutathione balance and mitigating oxidative damage (Basu & Kumar, 2024). Glutathione (GSH), a primary antioxidant, protects cells from ROS that accumulate during salt stress. GR sustains a high GSH/GSSG ratio by reducing oxidized glutathione (GSSG) to reduced glutathione (GSH) with the aid of NADPH, thereby promoting the detoxification of ROS through the ascorbate–glutathione cycle (Patwa et al., 2024). Although direct examinations of GR in salt tolerance are limited, related studies highlight its importance in stress adaptation. In tomatoes, salinity stress alters sugar metabolism and protein N-glycosylation, both of which are linked to fruit quality (Wolters et al., 2024). Thorough investigation of maize, sorghum, and pearl millet under water stress clarifies GR's role in resilience (Aher & Palakolanu, 2025). Augmented gibberellin receptor activity and gene upregulation are associated with improved redox balance, enhancing plant defense against salinity and other stresses (Samanta & Roychoudhury, 2025).

8.3.2 Enzyme Crosstalk in ROS Detoxification

The effective neutralization of ROS necessitates substantial interaction among antioxidant enzymes (Table 8.1) (Rao et al., 2025). The interaction between SOD and CAT/APX involves SOD-converting superoxide radicals into hydrogen peroxide (H_2O_2), which is then detoxified by CAT and APX. This sequential mechanism inhibits the build-up of both superoxide and H_2O_2, hence mitigating oxidative damage (Thiruvengadam et al., 2024). The Ascorbate–Glutathione (AsA–GSH) Cycle: APX is an essential element of the AsA–GSH cycle, a redox mechanism that alleviates H_2O_2. APX employs ascorbate to diminish H_2O_2, resulting in the formation of monodehydroascorbate (MDHA). MDHA reductase (MDHAR) converts MDHA back into ascorbate. Dehydroascorbate reductase (DHAR) facilitates the transformation of dehydroascorbate (DHA) into ascorbate, utilizing glutathione (GSH), which is subsequently restored by GR. This cycle preserves the redox equilibrium of ascorbate and glutathione, facilitating the ongoing removal of ROS (Rao et al., 2025). The detoxification of ROS is regulated by signaling molecules like calcium ions (Ca^{2+}) and NO. Calcium ions (Ca^{2+}) activate antioxidant enzymes and modulate the expression of stress-responsive genes, whereas NO interacts with hydrogen peroxide (H_2O_2) to affect antioxidant defense mechanisms and subsequent signaling pathways (Hussan et al., 2025; Jiang et al., 2024). Phytohormones, especially SA, augment the antioxidant defense mechanism. SA therapy increases the synthesis and function of antioxidant enzymes, hence improving resistance to salt-induced oxidative stress (Ghosh & Roychoudhury, 2024; Song et al., 2025). Halophytes, inherently adapted to elevated salinity, typically exhibit enhanced constitutive or salt-induced antioxidant enzyme activity compared to glycophytes. This underscores the essential function of effective ROS removal in promoting salt tolerance (Chaffai, Ganesan, & Cherif, 2024).

8.4 ION HOMEOSTASIS AND MEMBRANE TRANSPORT ENZYMES

8.4.1 Role of Na^+/H^+ Antiporters (NHX Family)

Na^+/H^+ antiporters (NHXs) are integral membrane proteins in the tonoplast, plasma membrane, and endomembrane compartments (Table 8.1). They facilitate the interchange of Na^+ and H^+ to uphold ionic homeostasis and stabilize intracellular pH. Under conditions of elevated salinity, NHXs sequester excess Na^+ in vacuoles or expel it from the cytosol (Bui et al., 2025). This inhibits ions from becoming poisonous and preserves osmotic equilibrium. NHX activity functions alongside the SOS pathway to enable the transfer of cations and protons across membranes. Their function promotes cellular proliferation, enlargement, and stress regulation (Xie et al., 2025).

NHX genes are found in multiple plant species, with varying copy numbers that suggest unique roles (for example, 30 in wheat, 25 in potatoes, 11 in Rosa multiflora, and 7 in common beans) (Gulzar et al., 2025). In reaction to salt stress, NHXs are often activated, promoting Na^+ sequestration and regulating ROS levels, thereby linking ion transport with antioxidant defense mechanisms (Bui et al., 2025; Jiang et al., 2024).

8.4.2 H^+-ATPases and Vacuolar Pumps

Plasma membrane H^+-ATPases (PM H^+-ATPases) establish an electrochemical proton gradient by transporting H^+ from the cytosol to the apoplast (Li, Grauschopf, Hedrich, Dreyer, & Konrad, 2024). This gradient facilitates the movement of ions across the plasma membrane and tonoplast by secondary active transport, contributing to growth, development, and stress responses. An autoinhibitory domain at the C-terminus regulates plasma membrane H^+-ATPases. Phosphorylation and interaction with the 14-3-3 protein activate the pump. The overexpression of PM H^+-ATPases or activators like PeREM6.5 enhances salt tolerance (Li, Wan, et al., 2024).

8.4.3 Vacuolar H^+-ATPases and H^+-Pyrophosphatases

V-ATPases and V-PPases, located in the tonoplast, acidify the vacuole, facilitating vacuolar transport and maintaining stable ion concentrations (Table 8.1) (Cha, Min, & Seo, 2024; Savoi et al., 2024). They facilitate Na^+ sequestration, K^+ accumulation, and pH regulation, all of which are crucial for adaptation to salinity levels. V-ATPases are proton pumps composed of many homologous subunits (Salazar et al., 2024). For instance, the overexpression of MdVHA-B1 or LmVHA-E1 enhances salt tolerance (Kotula et al., 2020). V-PPases collaborate with V-ATPases by supplying energy to the vacuolar membrane. Both pumps are dynamically modulated in response to salt stress, initially exhibiting increased activity followed by a subsequent decline over time (Mansour, 2023).

8.5 OSMOLYTE BIOSYNTHESIS ENZYMES

Osmolytes, such as proline, GB, and soluble carbohydrates, are essential for plant adaptation to salinity stress. They facilitate osmotic adjustment, stabilize proteins and membranes, and mitigate oxidative damage by eliminating ROS (Kumar, Kumar, et al., 2024). Specific enzymes facilitate the synthesis of these compounds, and their activity is typically augmented in the presence of salt (Hussain et al., 2024; Khan, Alnusaire, et al., 2025). Comprehending the regulation of these enzyme pathways provides valuable insights for enhancing plant resistance to salinity.

8.5.1 Proline Biosynthesis (Δ^1-Pyrroline-5-Carboxylate Synthetase, P5CS)

Proline biosynthesis is a prevalent response of plants to saline stress, significantly contributing to osmotic pressure regulation and aiding in stress mitigation (Renzetti, Funck, & Trovato, 2024). Salinity-induced osmotic stress can hinder plant development and yield, while proline accumulation facilitates the preservation of cellular homeostasis (Fatima et al., 2024). Proline functions as an osmoprotectant, regulating osmotic pressure in cells and preventing excessive water loss in saline environments (Koc, Aycan, & Mitsui, 2024). The increase of proline diminishes the cell's osmotic potential, hence aiding osmotic correction and stabilization (Boussora et al., 2024). Proline acts as an antioxidant by removing ROS that build up during salt stress in plants (Wang, Wan, Wu, Yang, & Chen, 2024). Proline functions as an antioxidant, protecting cells from damage inflicted by free radicals (Hareem et al., 2024). Proline stabilizes proteins and cell membranes by maintaining their structural integrity under stress (Li, Wan, et al., 2025).

8.5.2 Glycine Betaine Metabolism

GB is an essential osmoprotectant that enhances plant resilience to salinity stress (Table 8.1) (Chen et al., 2025). Salinity, an essential abiotic factor, constrains plant development and productivity worldwide (Mansinhos et al., 2024). The accumulation of GB is an essential adaptation mechanism that allows plants to sustain physiological stability in saline conditions (Chen et al., 2025). GB is a zwitterionic molecule that functions as an effective solute, maintaining osmotic equilibrium and cellular turgor, promoting water uptake and averting desiccation (Xia et al., 2025). Besides osmotic adjustment, GB safeguards proteins and enzymes from denaturation and activity loss at high salt concentrations, while maintaining their functional structures (Chen et al., 2025). It also maintains the membrane, protecting it from damage caused by ion toxicity and oxidative stress (Kumar, Paul, et al., 2024). GB governs the selective transport of solutes crucial for cellular homeostasis by preserving membrane stability (Rasheed et al., 2024). Salinity stress is often linked to the overproduction of ROS, resulting in oxidative damage (Singh, Rajput, et al., 2024). GB alleviates this damage by bolstering the body's defenses against free radicals, specifically via the activation of SOD, CAT, and POD, which jointly diminish the amounts of ROS in the body (Chanthini et al., 2024; Ma & Li, 2025).

8.5.3 Trehalose and Polyol-Related Enzymes

Trehalose, a non-reducing disaccharide, in conjunction with polyols such as mannitol and sorbitol, is essential for plant adaptation to saline environments. These molecules function as compatible solutes, preserving osmotic equilibrium and safeguarding cells from ionic toxicity in high-salinity conditions (Bhattacharya & Kundu, 2020; Nawaz et al., 2022). Trehalose specifically stabilizes proteins and membranes, hence protecting cellular structures during osmotic stress (Kuczyńska-Wiśnik, Stojowska-Swędrzyńska, & Laskowska, 2024). Trehalose metabolism is intricately associated with responses to salt stress, as trehalose and its intermediates influence signaling networks that regulate tolerance mechanisms (Bai et al., 2025). Trehalose, whether endogenous or exogenous, has demonstrated the ability to augment plant tolerance, with exogenous application facilitating growth and mitigating salinity-induced damage (Shah, Parrey, Barwal, Mohammad, & Siddiqui, 2024). Trehalose not only safeguards cells against osmotic stress but also diminishes oxidative damage. It maintains the function of antioxidant enzymes and may facilitate the removal of ROS, hence mitigating oxidative damage (Kuczyńska-Wiśnik et al., 2024). Trehalose and polyols together enhance the structural and metabolic stability of plants, hence reducing their susceptibility to salt damage.

8.6 ENZYMES IN HORMONAL REGULATION UNDER SALINITY STRESS

Salinity stress significantly impacts plant physiology, and enzymes play a crucial role in regulating hormones to mitigate these effects (Table 8.1) (Khan et al., 2021). Plants require hormones to detect and react to salinity. Enzymes participate in the synthesis, degradation, and signaling of several plant hormones (Aizaz et al., 2024; Hussain et al., 2024).

8.6.1 Abscisic Acid (ABA) Biosynthesis Enzymes

Salt stress in plants induces alterations in their hormonal makeup (Ahmadzai, Hu, Zhang, & Li, 2025). ABA levels frequently rise in response to salinity stress, resulting in stomatal closure to minimize water loss (Wang, Ma, et al., 2025). Under these circumstances, enzymes such as 9-cis-epoxycarotenoid dioxygenase (NCED), implicated in ABA production, are elevated (Kasbi et al., 2024). ABA activates signaling pathways, including protein kinases and transcription factors (TFs), hence inducing the expression of stress-responsive genes (Ma, Hu, & Jiang, 2024).

8.6.2 Ethylene and Jasmonic Acid Modulation

Salinity influences the production of ethylene (Gieniec et al., 2024). The enzymes 1-aminocyclopropane-1-carboxylic acid (ACC) synthase and ACC oxidase regulate ethylene production (Houben et al., 2024). Ethylene influences various aspects of plant growth and their responses to stress (Wang, Wen, Suprun, & Zhu, 2025). JA is a hormone that assists plants in managing stress. Salinity influences the enzymes responsible for the synthesis of JA and signal transduction (Ahmad et al., 2025). The interplay of several hormones is crucial for regulating plant responses to salt (Sultana et al., 2025). SA and gibberellins (GA) are key components of the intricate hormonal network that enables plants to adapt to salinity (Elsisi et al., 2024).

8.6.3 Role of Auxin- and Gibberellin-Related Enzymes

Salt induces stress in plants, resulting in a disordered interaction of their hormones. This encompasses auxin and gibberellin, along with the enzymes that facilitate their degradation and signal transduction (Thilakarathne, Liu, & Zou, 2025). These hormones are essential for regulating plant growth, development, and stress response (Kumar, Ashique, et al., 2024). Auxin, namely indole-3-acetic acid (IAA), is a crucial phytohormone that enables plant responses to saline stress (Zhang, Li, Gao, Liu, & Fu, 2024). It performs numerous functions, such as altering root growth and transmitting stress signals (Su et al., 2024). TAA1 (Tryptophan Aminotransferase of Arabidopsis 1) and YUC (Yucca) family flavin-containing monooxygenases are enzymes that collaboratively synthesize auxin (Tang et al., 2024).

8.7 MOLECULAR AND GENETIC REGULATION OF STRESS-RESPONSIVE ENZYMES

Salinity stress profoundly affects plant growth and productivity. It commences complex biochemical and genetic regulatory mechanisms. These methods encompass several stress-responsive enzymes, modifications in gene expression, and complex transcriptional regulation (Khan et al., 2021). Comprehending these processes is essential for developing salt-tolerant crops (Li, Grauschopf, et al., 2024; Omar et al., 2024; Padmavathi et al., 2023).

8.7.1 Gene Expression and Transcriptional Control

TFs are essential for modulating gene expression in response to salt stress (Aizaz et al., 2024). This control encompasses several transcription factor families, including WRKY, NAC, MYB, bZIP, and ERF (Li, Khoso, et al., 2024). Transgenic Arabidopsis that overexpresses HcSCL13, a GRAS transcription factor from *Halostachys caspica*, exhibits accelerated development and enhanced tolerance to salt stress (Zhang, Li, Fan, Zhou, & Wang, 2020). The Arabidopsis NF-YC1 gene is crucial for salt tolerance (Ao et al., 2023). Post-transcriptional mechanisms are equally significant, in conjunction with transcriptional regulation. SKIP, a component of the spliceosome, regulates alternative gene splicing in Arabidopsis during salt stress (Liu, Wang, et al., 2025). The regulation of mRNA degradation in plant responses to saline and osmotic stress underscores the significance of mRNA metabolism (Prasetyaningrum et al., 2023). Small RNAs, like microRNAs (miRNAs), are crucial epigenetic regulators of gene expression in response to and adaptation to salt stress (Rashid et al., 2022; Zhang, Zhang, et al., 2024). For example, zma-miR164 is suppressed in maize under saline stress conditions. This alters the expression of its target genes, which belong to the NAC transcription factor family (Yang et al., 2022).

8.8 BIOTECHNOLOGICAL APPROACHES TO ENHANCE ENZYME-MEDIATED SALINITY TOLERANCE

Biotechnological techniques offer significant potential for enhancing plants' salt tolerance via enzymes. This constitutes a significant issue for global agriculture. Salinity stress adversely affects plant growth, development, and productivity by inducing oxidative damage, osmotic stress, and ionic imbalance. Enhancing the activity and stability of enzymes under saline conditions can significantly augment the resilience of plants (Chaudhary et al., 2024; Salma Santhosh et al., 2025).

8.8.1 Genetic Engineering of Key Enzymes

Genetic engineering offers a feasible method to improve plant salinity tolerance by targeting enzymes that govern ion homeostasis, osmotic adjustment, and oxidative stress response (Aamir et al., 2025). Globally, soil salinity significantly impedes plant growth and productivity (Nurbekova et al., 2024). Enhancing ion control is a significant method to do this. Overexpression of Na^+/H^+ antiporters, such as SOS1, facilitates the maintenance of K^+/Na^+ equilibrium by extruding Na^+ from the cell (Cebrián et al., 2021). The SOS pathway, comprising SOS3/SCaBP8, SOS2, and SOS1, is integral to this control (Ali, Petrov, Yun, & Gechev, 2023). An alternative method involves the engineering of osmoprotectant synthesis. Elevating proline concentrations via P5CS or enhancing GB production through betaine aldehyde dehydrogenase (BADH) aids plants in adapting to saline stress (Goszcz et al., 2025; Hossain et al., 2022). Detoxifying ROS is equally essential. The overexpression of antioxidant enzymes such as SOD, CAT, and glutathione S-transferase (GST) increases the elimination of ROS and reduces oxidative damage (Udomsak et al., 2024). Numerous SOD isoforms have been identified in pak choi, highlighting their importance in growth and stress tolerance (Gantait, Masih, Addesso, Maxton, & Sofo, 2024).

8.8.2 CRISPR-Cas Applications in Modifying Enzyme Regulation

CRISPR-Cas systems provide a powerful method for modifying enzyme regulation in plants to enhance salt stress tolerance (Sharma, Sidhu, et al., 2024; Younas et al., 2024). CRISPR-Cas can precisely alter genes associated with ionic homeostasis, osmoprotectant synthesis, and ROS detoxification (Cannea & Padiglia, 2025). A significant application involves altering the functionality of ion transporters. For example, modifying Na^+/H^+ antiporters such as SOS1 using CRISPR-Cas might enhance sodium efflux, hence facilitating K^+/Na^+ balance in conditions of elevated salinity (Pabuayon, Jiang, Qian, Chung, & Shi, 2021; Shang et al., 2024). Enzymes participating in osmoprotectant synthesis, specifically within the proline and GB pathways, may be altered to augment the accumulation of these solutes and maintain cellular turgor (Rasheed et al., 2024). CRISPR-Cas9 has been employed in rice to enable targeted mutagenesis of certain genes, hence improving salinity tolerance responses (Chakraborty & Wylie, 2025; Rasheed et al., 2024). Genome-wide association studies (GWAS) have identified genes associated with salt tolerance in maize seedlings, offering compelling targets for CRISPR-mediated editing (Amombo et al., 2025). The functional validation of the RING-H2 type E3 ubiquitin ligase gene MdATL16 indicated that its overexpression enhances salt tolerance in transgenic tomato and apple (Yuan, Wei, et al., 2024). Besides salinity, CRISPR-Cas-based genome editing can improve tolerance to many abiotic and biotic challenges in crop plants (Mohamed, Khan, & Basit, 2024).

8.9 FUTURE PERSPECTIVES AND CHALLENGES

8.9.1 Gaps in Knowledge

- The ability of organisms to regulate salinity is governed by a complex interaction of enzymatic, molecular, and physiological systems. Presently, our understanding is insufficient since most research concentrates exclusively on particular enzymes or pathways, neglecting the interaction among osmolyte production, antioxidant defense, ion homeostasis, and hormone regulation.
- Numerous conclusions derived from controlled laboratory trials do not fully reflect the intricacies of multi-stress field conditions, which encompass the interplay of salinity with drought, heat, and nutrient deficits.
- The function of little researched enzymes, particularly those associated with secondary metabolism or epigenetic control, remains inadequately examined.
- The restricted transfer of knowledge from model plants to crops and from glycophytes to halophytes hinders practical applications.

8.9.2 Integration of Omics Approaches

- Future study should employ multi-omics approaches (genomics, transcriptomics, proteomics, metabolomics, ionomics) to comprehensively elucidate the complexity of plant responses to salinity.
- Systems biology and computational modeling can enhance our understanding of enzyme functionality, reveal regulatory hubs, and elucidate the operation of networks that may not be immediately apparent.
- Machine learning and AI can be employed to integrate extensive data and uncover concealed patterns that facilitate the development of targeted medicines.
- Comparative omics of salt-tolerant halophytes and sensitive glycophytes can uncover novel enzymatic pathways for the engineering of tolerance.
- To comprehend the regulation of stress-responsive enzymes post-transcriptionally and across generations, it is essential to integrate epigenomics with small RNA research.

8.9.3 Potential Applications in Crop Improvement

- Genetic engineering and genome editing (CRISPR-Cas) can enhance key enzymatic functions in ion transport (e.g., Na^+/H^+ antiporters), osmolyte biosynthesis (e.g., P5CS for proline, BADH for GB), and ROS detoxification (e.g., SOD, CAT, APX).
- Marker-assisted selection and speed breeding can accelerate the introgression of enzyme-related traits into elite cultivars.
- Exploiting beneficial microbes that enhance enzymatic activities (antioxidant capacity, ion regulation, osmolyte accumulation) represents a sustainable, eco-friendly strategy.
- Precision agriculture practices (enzyme priming with osmoprotectants or hormones, smart irrigation, salt-tolerant rootstocks) can complement molecular approaches.
- Ultimately, combining molecular innovations with ecological and agronomic practices will be crucial for developing salt-resilient, high-yielding crops to ensure food security under expanding soil salinization.

8.10 CONCLUSION

Plants utilize many enzymes to mitigate salt stress. SOD, CAT, APX, and GR are antioxidant enzymes that maintain redox homeostasis and safeguard cells from oxidative stress-induced damage. Ion transport enzymes, including as Na^+/H^+ antiporters, H^+-ATPases, and vacuolar pumps, manage ionic equilibrium by eliminating or sequestering excess Na^+ while maintaining constant K^+ concentrations. Enzymes involved in the synthesis of osmolytes, such as P5CS for proline and BADH for GB, contribute to protein stability, osmotic regulation, and the elimination of ROS. Enzymes that govern the production and signaling of phytohormones like as ABA, ethylene, jasmonates, auxins, and gibberellins also affect stomatal behavior, growth alterations, and stress signaling. These enzymatic activities function collaboratively within a network that assists plants in managing the osmotic, ionic, and oxidative effects of salt stress. Biotechnology, molecular breeding, advantageous microorganisms, and precision agronomy centered on enzymes can enhance crop resilience under adverse conditions and promote sustainable agriculture. To ensure that farmers derive authentic advantages, strategies must attain equilibrium between stress tolerance and production stability. We can apply our knowledge of enzymes in breeding programs, genome editing, and agricultural practices to address salinity issues with minimal environmental impact. This will create farming systems that are more robust and sustainable.

REFERENCES

Aamir, M., Kashyap, S. P., Krishna, R., Mishra, P., Singh, P., Ansari, W. A., …, & Rai, N. (2025). Advances in salinity tolerance breeding in tomato: From conventional breeding tools to genome editing. *Environmental Sustainability*, *8*(2), 139–157.

Abdehpour, Z., Ehsanzadeh, P., & Ghanati, F. (2025). Expression level of NHX1 and SOS1 genes is key to high salt tolerance of ancient emmer wheat: Implications to tackling salt stress penalties of wheat yield. *Functional Plant Biology*, *52*(8).

Ahmad, J., Hayat, F., Khan, U., Altaf, M. A., Ahmad, M., Akram, M. T., …, & Alyafei, M. (2025). Jasmonic acid as a regulator in plant stress tolerance: Mechanisms and agricultural implications. *Journal of Crop Health*, *77*(4), 118.

Ahmadzai, A. S., Hu, C., Zhang, C., & Li, Y. (2025). Mechanisms of anthocyanin-mediated salt stress alleviation and cellular homeostasis in plants. *Plant Growth Regulation*, *105*(3), 655–673.

Aizaz, M., Lubna, Jan, R., Asaf, S., Bilal, S., Kim, K.-M., & AL-Harrasi, A. (2024). Regulatory dynamics of plant hormones and transcription factors under salt stress. *Biology*, *13*(9), 673.

Alghamdi, S. (2024). Drought and salinity effects on plant growth: A comprehensive review. *SABRAO Journal of Breeding and Genetics*, *56*(6), 2331–2340.

Alhudhaibi, A. M., Ibrahim, M. A., Abd-Elaziz, S. M., Farag, H. R., Elsayed, S. M., Ibrahim, H. A., …, & Elkelish, A. (2024). Enhancing salt stress tolerance in wheat (*Triticum aestivum*) seedlings: Insights from trehalose and mannitol. *BMC Plant Biology*, *24*(1), 472.

Ali, A., Petrov, V., Yun, D.-J., & Gechev, T. (2023). Revisiting plant salt tolerance: Novel components of the SOS pathway. *Trends in Plant Science*, *28*(9), 1060–1069.

Altaf, M. M., Ashraf, S., Khalofah, A., Khan, M. Q. N., & Anwar, M. (2024). Melatonin foliar application promoted anthocyanin accumulation boosted reactive oxygen scavenging by upregulating the key enzymes activities and genes expression of anthocyanin biosynthesis pathway in green and purple mustard under vanadium stress. *Scientia Horticulturae*, *331*, 113112.

Amombo, E., Mumm, R. H., Nilahyane, A., Hirich, A., Hijri, M., & Ntui, V. O. (2025). Enhancing crop nutrition in arid and semiarid regions of Africa through genome editing using CRISPR/Cas. *Food and Energy Security*, *14*(3), e70068.

Anjum, N. A., Sharma, P., Gill, S. S., Hasanuzzaman, M., Khan, E. A., Kachhap, K., …, & Vasudhevan, P. (2016). Catalase and ascorbate peroxidase—Representative H_2O_2-detoxifying heme enzymes in plants. *Environmental Science and Pollution Research*, *23*(19), 19002–19029.

Ao, Y., Wu, Q., Zhang, C., Zheng, J., Zhou, G., Yang, S., …, & Yang, M. (2023). Transcriptome profiles reveal NF-YC1-regulated pathways in adaption to salt stress in Arabidopsis thaliana. *Environmental and Experimental Botany*, *216*, 105515.

Bai, G., Zhang, H., Li, Y., Yang, D.-H., Fei, M., Pang, T., …, & Gu, J. (2025). Integration of metabolic and transcriptomic analyses for revealing the galactose metabolism of tobacco (*Nicotiana tabacum*) under salt stress. *Frontiers in Plant Science*, *16*, 1614515.

Basu, S., & Kumar, G. (2024). Regulation of nitro-oxidative homeostasis: An effective approach to enhance salinity tolerance in plants. *Plant Cell Reports*, *43*(8), 193.

Bhattacharya, S., & Kundu, A. (2020). Sugars and sugar polyols in overcoming environmental stresses. In A. Roychoudhury & D.K. Tripathi (Eds.), *Protective chemical agents in the amelioration of plant abiotic stress: Biochemical and molecular perspectives* (pp. 71–101). https://doi.org/10.1002/9781119552154.

Boussora, F., Triki, T., Bennani, L., Bagues, M., Ben Ali, S., Ferchichi, A., …, & Guasmi, F. (2024). Mineral accumulation, relative water content and gas exchange are the main physiological regulating mechanisms to cope with salt stress in barley. *Scientific Reports*, *14*(1), 14931.

Bui, H. H. T., Nguyen, D. H., Dinh, L. T. T., Trinh, H. T. T., Vu, T. K., & Bui, V. N. (2025). Comprehensive In Silico analysis of the NHX (Na^+/H^+ antiporter) gene In Rice (*Oryza sativa* L.). *International Journal of Plant Biology*, *16*(1), 6.

Cannea, F. B., & Padiglia, A. (2025). Antioxidant defense systems in plants: Mechanisms, regulation, and biotechnological strategies for enhanced oxidative stress tolerance. *Life*, *15*(8), 1293.

Cebrián, G., Iglesias-Moya, J., García, A., Martínez, J., Romero, J., Regalado, J. J., …, & Jamilena, M. (2021). Involvement of ethylene receptors in the salt tolerance response of *Cucurbita pepo*. *Horticulture Research*, *8*.

Cha, S., Min, W. K., & Seo, H. S. (2024). Arabidopsis COP1 guides stomatal response in guard cells through pH regulation. *Communications Biology*, *7*(1), 150.

Chaffai, R., Ganesan, M., & Cherif, A. (2024). Plant adaptations and molecular responses to salt stress. In *Plant adaptation to abiotic stress: From signaling pathways and microbiomes to molecular mechanisms*(pp. 137–176). Springer.

Chakraborty, A., & Wylie, S. J. (2025). CRISPR/Cas9 for heat stress tolerance in rice: A review. *Plant Molecular Biology Reporter*, *43*(3), 1047–1056.

Chanthini, K. M.-P., Pavithra, G.-S., Senthil-Nathan, S., & Malafaia, G. (2024). An in-depth review on the mechanistic insights of marine macroalgal compounds in enhancing plant tolerance to stress induced by saline soil conditions. *Toxin Reviews*, *43*(4), 651–669.

Chaudhary, M. T., Majeed, S., Rana, I. A., Ali, Z., Jia, Y., Du, X., …, & Azhar, M. T. (2024). Impact of salinity stress on cotton and opportunities for improvement through conventional and biotechnological approaches. *BMC Plant Biology*, *24*(1), 20.

Chen, J., Zhang, J., Liu, Y., Zhang, K., Zhu, F., & Xie, Y. (2025). Advances in the biosynthetic regulation and functional mechanisms of Glycine betaine for enhancing plant stress resilience. *International Journal of Molecular Sciences*, *26*(16), 7971.

Corpas, F. J., González-Gordo, S., & Palma, J. M. (2024). Ascorbate peroxidase in fruits and modulation of its activity by reactive species. *Journal of Experimental Botany*, *75*(9), 2716–2732.

Demo, A. H., Gemeda, M. K., Abdo, D. R., Guluma, T. N., & Adugna, D. B. (2025). Impact of soil salinity, sodicity, and irrigation water salinity on crop production and coping mechanism in areas of dryland farming. *Agrosystems, Geosciences & Environment*, *8*(1), e70072.

Dildar, T., Cui, W., Ikhwanuddin, M., & Ma, H. (2025). Aquatic organisms in response to salinity stress: Ecological impacts, adaptive mechanisms, and resilience strategies. *Biology*, *14*(6), 667.

Dutta, S., Sengupta, P., Izuka, E., Menuba, I., & Nwagha, U. (2024). Oxidative and nitrosative stress and female reproduction: Roles of oxidants and antioxidants. *Journal of Integrated Science and Technology*, *12*(3), 754–754.

El-Ramady, H., Prokisch, J., Mansour, H., Bayoumi, Y. A., Shalaby, T. A., Veres, S., & Brevik, E. C. (2024). Review of crop response to soil salinity stress: Possible approaches from leaching to nano-management. *Soil Systems*, *8*(1), 11.

Elsisi, M., Elshiekh, M., Sabry, N., Aziz, M., Attia, K., Islam, F., …, & Abdelrahman, M. (2024). The genetic orchestra of salicylic acid in plant resilience to climate change induced abiotic stress: Critical review. *Stress Biology*, *4*(1), 31.

Etesami, H., & Beattie, G. A. (2018). Mining halophytes for plant growth-promoting halotolerant bacteria to enhance the salinity tolerance of non-halophytic crops. *Frontiers in Microbiology*, *9*, 148.

Farnum, B. H., & Goldsmith, C. R. (2024). Use of intramolecular quinol redox couples to facilitate the catalytic transformation of O_2 and O_2-derived species. *Accounts of Chemical Research*, *58*(1), 101–112.

Fatima, N., Ashraf, M. A., Rasheed, R., Shad, M. I., & Ali, S. (2024). Exogenous Myo-inositol weakens salinity effects by improving photosynthesis, ROS homeostasis, osmotic adjustment, and nutrient acquisition in maize (*Zea mays* L.). *Journal of Plant Growth Regulation*, *43*(12), 4556–4576.

Gantait, A., Masih, S. A., Addesso, R., Maxton, A., & Sofo, A. (2024). Glucosinolates mediated regulation of enzymatic activity in response to oxidative stress in Brassica spp. *Plants*, *13*(23), 3422.

Ghosh, P., & Roychoudhury, A. (2024). Molecular basis of salicylic acid–phytohormone crosstalk in regulating stress tolerance in plants. *Brazilian Journal of Botany*, *47*(3), 735–750.

Gieniec, M., Miszalski, Z., Rozpądek, P., Jędrzejczyk, R. J., Czernicka, M., & Nosek, M. (2024). How the ethylene biosynthesis pathway of semi-halophytes is modified with prolonged salinity stress occurrence? *International Journal of Molecular Sciences*, *25*(9), 4777.

Goszcz, A., Furtak, K., Stasiuk, R., Wójtowicz, J., Musiałowski, M., Schiavon, M., & Dębiec-Andrzejewska, K. (2025). Bacterial osmoprotectants-a way to survive in saline conditions and potential crop allies. *FEMS Microbiology Reviews*, 49, fuaf020.

Gulzar, R. M. A., Munir, R., Yi, Z. R., Jan, N., Rauf, A., Kaleem, Z., …, & Sheteiwy, M. S. (2025). Genome-wide identification of NHX gene family and effects of salicylic acid in regulating antioxidant activity as well as nutrients deposition under salt stress in three Brassica species. *BMC Plant Biology*, *25*(1), 740.

Haghpanah, M., Hashemipetroudi, S., Arzani, A., & Araniti, F. (2024). Drought tolerance in plants: Physiological and molecular responses. *Plants*, *13*(21), 2962.

Han, T., Sun, X., He, S., Li, J., Tian, X., Sun, Q., …, & Chen, L. (2025). Cu-doped ZnO nanozyme with citric acid modification alleviate saline-alkaline stress in wheat by scavenging reactive oxygen species. *Plant Physiology and Biochemistry*, 110365.

Hareem, M., Danish, S., Pervez, M., Irshad, U., Fahad, S., Dawar, K., …, & Datta, R. (2024). Optimizing chili production in drought stress: Combining Zn-quantum dot biochar and proline for improved growth and yield. *Scientific Reports*, *14*(1), 6627.

Hasanuzzaman, M., Bhuyan, M. B., Zulfiqar, F., Raza, A., Mohsin, S. M., Mahmud, J. A., ... & Fotopoulos, V. (2020). Reactive oxygen species and antioxidant defense in plants under abiotic stress: Revisiting the crucial role of a universal defense regulator. *Antioxidants*, *9*(8), 681.

Hossain, A., Syed, M. A., Maitra, S., Garai, S., Mondal, M., Akter, B., …, & Skalicky, M. (2022). Genetic regulation, biosynthesis, and the roles of osmoprotective compounds in abiotic stress tolerance in plants. In *Plant abiotic stress physiology* (pp. 61–114). Apple Academic Press.

Houben, M., Vaughan-Hirsch, J., Pattyn, J., Mou, W., Roden, S., Martinez Roig, A., …, & De Coninck, B. (2024). 1-Aminocyclopropane-1-carboxylic acid oxidase determines the fate of ethylene biosynthesis in a tissue-specific way to fine-tune development and stress resilience. *bioRxiv*, 2024-02.

Hussain, S., Hafeez, M. B., Azam, R., Mehmood, K., Aziz, M., Ercisli, S., …, & Hussain, S. (2024). Deciphering the role of phytohormones and osmolytes in plant tolerance against salt stress: Implications, possible cross-talk, and prospects. *Journal of Plant Growth Regulation*, *43*(1), 38–59.

Hussan, M. U., Mahmood, U., Zarbakhsh, S., Hussain, S., Hafeez, M. B., Zahra, N., …, & Yan, M. (2025). Mitigating Cd stress in alfalfa: The role of melatonin and nano-calcium oxide in enhancing photosynthesis and antioxidant defense. *BMC Plant Biology*, *25*(1), 1–17.

Ikram, M., Batool, M., Ullah, M., Khalid, B., El-Badri, A. M., Alfagham, A. T., …, & Xu, Z. (2025a). Sulforaphane modulates Na^+/K^+ homeostasis and hormonal balance in rice under salt stress by regulating OsHKT1 and OsHKT5 expression. *International Journal of Biological Macromolecules*, *309*, 142783.

Ikram, M., Batool, M., Ullah, M., Khalid, B., El-Badri, A. M., Mohamed, I. A., …, & Zhao, J. (2025b). Molecular alchemy: Converting stress into resilience via secondary metabolites and calcium signaling in rice. *Rice*, *18*(1), 32.

Jan, S., Ashraf Bhat, M., Kumar, A., Altaf Wani, M., Ahmad Bhat, F., Kanth, R. H., …, & Altaf, H. (2025). Plants response to different abiotic stresses. *Journal of Crop Health*, *77*(5), 1–27.

Jiang, S., Lan, Z., Zhang, Y., Kang, X., Zhao, L., Wu, X., & Gao, H. (2024). Mechanisms by which exogenous substances enhance plant salt tolerance through the modulation of ion membrane transport and reactive oxygen species metabolism. *Antioxidants*, *13*(9), 1050.

Jumaah, W. N., Nabi, R. B. S., Rolly, N. K., Adamu, T. A., Tayade, R., Aye, N. C., …, & Yun, B.-W. (2024). Nitric oxide mediated growth enhancement of tomato under salinity stress. *Plant Biotechnology Reports*, *18*(3), 437–446.

Kachout, S. S., Hamza, K. J., Bouraoui, N. K., Leclerc, J. C., & Ouerghi, Z. (2013). Salt-induced changes in antioxidative enzyme activities in shoot tissues of two Atriplex varieties. *Notulae Botanicae Horti Agrobotanici Cluj-Napoca*, *41*(1), 115–121.

Kasbi, E. A., Taleei, A., & Amiri, R. M. (2024). Effect of drought stress on the expression pattern of genes involved in ABA biosynthesis in desi-type chickpea (*Cicer arietinum* L.). *Molecular Biology Reports*, *51*(1), 469.

Kaur, G., Sanwal, S. K., Kumar, A., Pundir, R. K., Yadav, M., & Sehrawat, N. (2024). Role of osmolytes dynamics in plant metabolism to cope with salinity induced osmotic stress. *Discover Agriculture*, *2*(1), 59.

Khan, A. Wang, A., Hussain, T., Amna, T., Ali, F., Shi, F., …, & Mehmood, S. (2021). Halotolerant-Koccuria rhizophila (14asp)-induced amendment of salt stress in pea plants by limiting Na^+ uptake and elevating production of antioxidants. *Agronomy*, *11*(10), 1907.

Khan, A. A., Alnusaire, T. S., Akbar, R., Iqbal, B., Zeb, A., & Soliman, M. H. (2025). Use of biostimulants to enhance temperature tolerance in oilseed crops. In *Oilseed crops under abiotic stress: Mitigation strategies and future perspectives* (pp. 123–146). Springer.

Khan, A. A., Alabbosh, K. F., Kashif, Iqbal, B., Manan, S., Alhoqail, W. A., ... & Wang, Y. F. (2025). Ion homeostasis, osmotic adjustment, and ROS detoxification underlie pea salinity tolerance induced by *Pseudomonas putida* RT12. *Microbiology Research, 16*(11), 227.

Khan, A. A., Wang, T., Nisa, Z. U., Alnusairi, G. S., & Shi, F. (2022). Insights into cadmium-induced morphophysiological disorders in *Althea rosea* Cavan and its phytoremediation through the exogeneous citric acid. *Agronomy, 12*(11), 2776.

Khan, A. A., Wang, Y.-F., Akbar, R., & Alhoqail, W. A. (2025). Mechanistic insights and future perspectives of drought stress management in staple crops. *Frontiers in Plant Science, 16*, 1547452.

Khan, A. A., Wang, Y.-F., Zeb, A., Hayat, K., Alhoqail, W. A., & Soliman, M. H. (2025). Beneficial elements and their roles against soil pollution. In *Beneficial elements for remediation of heavy metals in polluted soil* (pp. 1–32). Elsevier.

Khan, A. A., Yongfeng, W., Soliman, M. H., Alhoqail, W. A., ALHaithloul, H. S., Alghanem, S. M., & Latef, A. A. H. A. (2025). PGPR in agriculture: A sustainable approach to increasing climate change resilience. In *Plant-microbe interactions for environmental and agricultural sustainability* (pp. 129–154). Springer.

Koc, Y. E., Aycan, M., & Mitsui, T. (2024). Self-defense mechanism in rice to salinity: Proline. *J, 7*(1), 103–115.

Kotula, L., Garcia Caparros, P., Zörb, C., Colmer, T. D., & Flowers, T. J. (2020). Improving crop salt tolerance using transgenic approaches: An update and physiological analysis. *Plant, Cell & Environment, 43*(12), 2932–2956.

Kozlov, A. V., Javadov, S., & Sommer, N. (2024). *Cellular ROS and antioxidants: Physiological and pathological role* (Vol. 13, pp. 602). MDPI.

Kuczyńska-Wiśnik, D., Stojowska-Swędrzyńska, K., & Laskowska, E. (2024). Intracellular protective functions and therapeutical potential of trehalose. *Molecules, 29*(9), 2088.

Kumar, P., Ashique, S., Kumar, N., Jain, A., Sharma, H., Pandey, S. N., & Singh, A. (2024). Regulation of plant hormones under abiotic stress conditions in plants. *Plant Secondary Metabolites and Abiotic Stress*, 243–276. https://doi.org/10.1002/9781394186457

Kumar, P., Kumar, A., Agrawal, A., Bhusan, R., Kumar, S., Rehman, S., & Kumar, R. (2024). Osmolytes as stress sensors in plants: Acclimatizing plants under stress conditions. In *Cellular osmolytes: From chaperoning protein folding to clinical perspectives* (pp. 199–210). Springer.

Kumar, P., Paul, D., Jhajhriya, S., Kumar, R., Dutta, S., Siwach, P., & Das, S. (2024). Understanding heat-shock proteins' abundance and pivotal function under multiple abiotic stresses. *Journal of Plant Biochemistry and Biotechnology, 33*(4), 492–513.

Li, K., Grauschopf, C., Hedrich, R., Dreyer, I., & Konrad, K. R. (2024). K^+ and pH homeostasis in plant cells is controlled by a synchronized K^+/H^+ antiport at the plasma and vacuolar membrane. *New Phytologist, 241*(4), 1525–1542.

Li, M., Guo, P., Nan, N., Ma, A., Liu, W., Wang, T.-J., …, & Xu, Z.-Y. (2024). Plasma membrane-localized H^+-ATPase OsAHA3 functions in saline–alkaline stress tolerance in rice. *Plant Cell Reports, 43*(1), 9.

Li, M., Wan, Y., Fan, H., Liu, J., Cai, W., Yuan, J., …, & Fu, G. (2025). Integrated transcriptomic and physiological and biochemical analysis reveal the molecular mechanism of proline regulating the resistance of *Saccharomyces cerevisiae* to tannic acid stress. *Food Bioscience, 7*, 107083.

Li, S. (2023). Novel insight into functions of ascorbate peroxidase in higher plants: More than a simple antioxidant enzyme. *Redox Biology, 64*, 102789.

Li, S., Khoso, M. A., Xu, H., Zhang, C., Liu, Z., Wagan, S., …, & Liu, L. (2024). WRKY transcription factors (TFs) as key regulators of plant resilience to environmental stresses: Current perspective. *Agronomy, 14*(10), 2421.

Li, W., Meng, R., Liu, Y., Chen, S., Jiang, J., Wang, L., …, & Chen, F. (2022). Heterografted chrysanthemums enhance salt stress tolerance by integrating reactive oxygen species, soluble sugar, and proline. *Horticulture Research, 9*, uhac073.

Liu, G., Wang, H., Gao, H., Yu, S., Liu, C., Wang, Y., …, & Zhang, D. (2025). Alternative splicing of functional genes in plant growth, development, and stress responses. *International Journal of Molecular Sciences, 26*(12), 5864.

Liu, X., Gao, L., Yun, Q., Wang, W., Liu, X., Tang, W., …, & Shang, J.-X. (2025). Protein tyrosine phosphatase IBR5 positively affects salt stress response by modulating CAT2 activity. *Plant Science, 359*, 112615.

Ma, Y., & Li, X. (2025). Plant growth-promoting bacteria-mediated climatic stress tolerance in plants. In *Microorganisms resilience to climate change* (pp. 113–174). Springer.

Ma, Z., Hu, L., & Jiang, W. (2024). Understanding AP2/ERF transcription factor responses and tolerance to various abiotic stresses in plants: A comprehensive review. *International Journal of Molecular Sciences, 25*(2), 893.

Mansinhos, I., Gonçalves, S., & Romano, A. (2024). How climate change-related abiotic factors affect the production of industrial valuable compounds in Lamiaceae plant species: A review. *Frontiers in Plant Science, 15*, 1370810.

Mansour, M. M. F. (2023). Role of vacuolar membrane transport systems in plant salinity tolerance. *Journal of Plant Growth Regulation, 42*(3), 1364–1401.

Mehmood, S., Khan, A. A., Shi, F., Tahir, M., Sultan, T., Munis, M. F. H., …, & Chaudhary, H. J. (2021). Alleviation of salt stress in wheat seedlings via multifunctional *Bacillus aryabhattai* PM34: An in-vitro study. *Sustainability, 13*(14), 8030.

Mohamed, H. I., Khan, A., & Basit, A. (2024). CRISPR-Cas9 system mediated genome editing technology: An ultimate tool to enhance abiotic stress in crop plants. *Journal of Soil Science and Plant Nutrition, 24*(2), 1799–1822.

Nagamiya, K., Motohashi, T., Nakao, K., Prodhan, S. H., Hattori, E., Hirose, S., …, & Takabe, T. (2007). Enhancement of salt tolerance in transgenic rice expressing an *Escherichia coli* catalase gene, kat E. *Plant Biotechnology Reports, 1*(1), 49–55.

Nawaz, M., Hassan, M. U., Chattha, M. U., Mahmood, A., Shah, A. N., Hashem, M., …, & Thabit, M. A. (2022). Trehalose: A promising osmo-protectant against salinity stress—Physiological and molecular mechanisms and future prospective. *Molecular Biology Reports, 49*(12), 11255–11271.

Nurbekova, Z., Satkanov, M., Beisekova, M., Akbassova, A., Ualiyeva, R., Cui, J., …, & Zhangazin, S. (2024). Strategies for achieving high and sustainable plant productivity in saline soil conditions. *Horticulturae, 10*(8), 878.

Omar, A. A. A., Zaman, M. M. U., Abedin, M. T., Alam, M. S., Khatun, M. A., Akter, N., & Sayed, M. A. (2024). Unraveling the dynamics of salt stress response in wheat cultivars: Insights into growth inhibition, antioxidant enzyme activity, and AOX gene expression. *Journal of Crop Science and Biotechnology, 27*(4), 439–448.

Pabuayon, I. C. M., Jiang, J., Qian, H., Chung, J.-S., & Shi, H. (2021). Gain-of-function mutations of AtNHX1 suppress SOS1 salt sensitivity and improve salt tolerance in Arabidopsis. *Stress Biology, 1*(1), 14.

Padmavathi, G., Singh, R., Arun, M., Umakanth, B., Prasad, G., & Muralidharan, K. (2023). *Genetic enhancement for salt tolerance in rice Molecular and Physiological Insights into*

Plant Stress Tolerance and Applications in Agriculture (pp. 40–84). Bentham Science Publishers Sharjah.

Patwa, N., Pandey, V., Gupta, O. P., Yadav, A., Meena, M. R., Ram, S., & Singh, G. (2024). Unravelling wheat genotypic responses: Insights into salinity stress tolerance in relation to oxidative stress, antioxidant mechanisms, osmolyte accumulation and grain quality parameters. *BMC Plant Biology*, *24*(1), 875.

Prasetyaningrum, P., Litthauer, S., Vegliani, F., Battle, M. W., Wood, M. W., Liu, X., …, & Jones, M. A. (2023). Inhibition of RNA degradation integrates the metabolic signals induced by osmotic stress into the Arabidopsis circadian system. *Journal of Experimental Botany*, *74*(18), 5805–5819.

Rai, B., Kaur, J., Catalina, M., Anand, S. C., Jacobs, R., & Teughels, W. (2011). Effect of simulated microgravity on salivary and serum oxidants, antioxidants, and periodontal status. *Journal of Periodontology*, *82*(10), 1478–1482.

Rao, M. J., Duan, M., Zhou, C., Jiao, J., Cheng, P., Yang, L., …, & Yang, Y. (2025). Antioxidant defense system in plants: Reactive oxygen species production, signaling, and scavenging during abiotic stress-induced oxidative damage. *Horticulturae*, *11*(5), 477.

Rasheed, Y., Khalid, F., Ashraf, H., Asif, K., Maqsood, M. F., Naz, N., …, & Rana, S. (2024). Enhancing plant stress resilience with osmolytes and nanoparticles. *Journal of Soil Science and Plant Nutrition*, *24*(2), 1871–1906.

Rashid, M. M., Vaishnav, A., Verma, R. K., Sharma, P., Suprasanna, P., & Gaur, R. (2022). Epigenetic regulation of salinity stress responses in cereals. *Molecular Biology Reports*, *49*(1), 761–772.

Renzetti, M., Funck, D., & Trovato, M. (2024). Proline and ROS: A unified mechanism in plant development and stress response? *Plants*, *14*(1), 2.

Sackey, O. K., Feng, N., Mohammed, Y. Z., Dzou, C. F., Zheng, D., Zhao, L., & Shen, X. (2025). A comprehensive review on rice responses and tolerance to salt stress. *Frontiers in Plant Science*, *16*, 1561280.

Salazar, O. R., Chen, K., Melino, V. J., Reddy, M. P., Hřibová, E., Čížková, J., …, & Aranda, M. (2024). SOS1 tonoplast neo-localization and the RGG protein SALTY are important in the extreme salinity tolerance of *Salicornia bigelovii*. *Nature Communications*, *15*(1), 4279.

Saleem, M. A., Khan, A., Tu, J., Huang, W., Liu, Y., Feng, N., …, & Xue, Y. (2025). Salinity stress in rice: Multilayered approaches for sustainable tolerance. *International Journal of Molecular Sciences*, *26*(13), 6025.

Salesse-Smith, C. E., Wang, Y., & Long, S. P. (2025). Increasing Rubisco as a simple means to enhance photosynthesis and productivity now without lowering nitrogen use efficiency. *New Phytologist*, *245*(3), 951–965.

Salma Santhosh, S., Meena, S., Baskar, M., Karthikeyan, S., Vanniarajan, C., & Ramesh, T. (2025). Transformative strategies for saline soil restoration: Harnessing halotolerant microorganisms and advanced technologies. *World Journal of Microbiology and Biotechnology*, *41*(5), 1–41.

Samanta, S., & Roychoudhury, A. (2025). Molecular crosstalk of jasmonate with major phytohormones and plant growth regulators during diverse stress responses. *Journal of Plant Growth Regulation*, *44*(1), 62–88.

Sharma, P., Sharma, P., Arora, P., Verma, V., Khanna, K., Saini, P., & Bhardwaj, R. (2019). Role and regulation of ROS and antioxidants as signaling molecules in response to abiotic stresses. In *Plant signaling molecules* (pp. 141–156). Woodhead Publishing.

Szabados, L., & Savouré, A. (2010). Proline: a multifunctional amino acid. *Trends in Plant Science*, *15*(2), 89–97.

Savoi, S., Shi, M., Sarah, G., Weber, A., Torregrosa, L., & Romieu, C. (2024). Time-resolved transcriptomics of single *Vitis vinifera* fruits: Membrane transporters as switches of the double sigmoidal growth. *Journal of Experimental Botany*, *76*(11), 3105–3124.

Sevgi, B., & Leblebici, S. (2025). Exogenous sucrose alleviates salt stress in sunflower (*Helianthus annuus* L.) and canola (*Brassica napus* L.) by modulating osmotic adjustment and antioxidant defense system. *Physiology and Molecular Biology of Plants*, *31*(3), 405–418.

Shafi, Z., Shahid, M., Ilyas, T., Pandey, V. K., Aijaz, S. A., Singh, R., & Sahu, P. K. (2025). Unveiling the plant growth regulators crosstalk in agricultural crop response to salinity-stress: A concise review. *Physiologia Plantarum*, *177*(4), e70402.

Shah, S. H., Parrey, Z. A., Barwal, S. K., Mohammad, F., & Siddiqui, M. H. (2024). Deciphering the mechanism of action and crosstalk of brassinosteroids with other plant growth regulators in orchestrating physio-biochemical responses in plants under salt stress. *Plant Growth Regulation*, *104*(3), 1285–1306.

Shang, C., Sihui, L., Li, C., Hussain, Q., Chen, P., Hussain, M. A., & Nkoh Nkoh, J. (2024). SOS1 gene family in mangrove (*Kandelia obovata*): Genome-wide identification, characterization, and expression analyses under salt and copper stress. *BMC Plant Biology*, *24*(1), 805.

Sharma, M., Sidhu, A. K., Samota, M. K., Shah, P., Pandey, M. K., & Gangurde, S. S. (2024). Technological advancements in the CRISPR toolbox for improving plant salt tolerance. *Discover Agriculture*, *2*(1), 102.

Sharma, M., Tisarum, R., Kohli, R. K., Batish, D. R., Cha-Um, S., & Singh, H. P. (2024). Inroads into saline-alkaline stress response in plants: Unravelling morphological, physiological, biochemical, and molecular mechanisms. *Planta*, *259*(6), 130.

Singh, A., Rajput, V. D., Agrawal, S., Ghazaryan, K., Minkina, T., Al Tawaha, A. R. M., …, & Papadakis, M. (2024). Nanoparticles mediated salt stress resilience: A holistic exploration of physiological, biochemical, and Nano-omics approaches. *Reviews of Environmental Contamination and Toxicology*, *262*(1), 19.

Singh, N., Guha, L., Pathak, Z., & Kumar, H. (2025). Nanoshields: Exploring the role of antioxidant mimicking nanoparticles as regenerative therapy in spinal cord injury. *Biomaterials Advances*, *178*, 214484.

Sinha, D., Dey, A., Maurya, A. K., Abdi, G., Sen, S., Nath, S., …, & Sohail, M. (2025). Oxidative responses, antioxidative system, and redox regulation during water stress. In *Water stress in crop plants and its management* (pp. 77–129). CRC Press.

Song, X., Chen, J., Xu, C., Cai, X., Song, W., Chang, A., …, & Luo, C. (2025). Physiological and molecular mechanisms of exogenous salicylic acid in enhancing salt tolerance in tobacco seedlings by regulating antioxidant defence system and gene expression. *Frontiers in Plant Science*, *16*, 1545865.

Su, J., Liu, Y., Han, F., Gao, F., Gan, F., Huang, K., & Li, Z. (2024). ROS, an important plant growth regulator in root growth and development: Functional genes and mechanism. *Biology*, *13*(12), 1033.

Sultana, R., Imam, Z., Kumar, R. R., Banu, V. S., Nahakpam, S., Bharti, R., …, & Singh, D. R. (2025). Signaling and defence mechanism of jasmonic and salicylic acid response in pulse crops: Role of WRKY transcription factors in stress response. *Journal of Plant Growth Regulation*, *44*(1), 5–21.

Tang, C., Zhang, Y., Liu, X., Zhang, B., Si, J., Xia, H., …, & Kong, L. (2024). Nitrate starvation induces lateral root organogenesis in *Triticum aestivum* via auxin signaling. *International Journal of Molecular Sciences*, *25*(17), 9566.

Thilakarathne, A. S., Liu, F., & Zou, Z. (2025). Plant signaling hormones and transcription factors: Key regulators of plant responses to growth, development, and stress. *Plants*, *14*(7), 1070.

Thiruvengadam, R., Venkidasamy, B., Easwaran, M., Chi, H. Y., Thiruvengadam, M., & Kim, S.-H. (2024). Dynamic interplay of reactive oxygen and nitrogen species (ROS and RNS) in plant resilience: Unveiling the signaling pathways and metabolic responses to biotic and abiotic stresses. *Plant Cell Reports*, *43*(8), 198.

Udomsak, W., Kucinska, M., Pospieszna, J., Dams-Kozlowska, H., Chatuphonprasert, W., & Murias, M. (2024). Antioxidant enzymes in cancer cells: Their role in photodynamic therapy resistance and potential as targets for improved treatment outcomes. *International Journal of Molecular Sciences*, *25*(6), 3164.

Venkatachalam, J., Mohan, H., & Seralathan, K.-K. (2025). Role of plant growth promoting bacteria in alleviating heavy metals induced oxidative stress in plant systems: A review. *Environmental Science and Pollution Research*, *32*(33), 19619–19638.

Verma, O., Sharma, S., Kumar, V., Singh, T., Kumar, R., & Auji, R. (2024). Salinity stress effect on staple food crops and novel mitigation strategies. *Biologia*, *79*(8), 2359–2374.

Vicidomini, C., Palumbo, R., Moccia, M., & Roviello, G. N. (2024). Oxidative processes and xenobiotic metabolism in plants: Mechanisms of defense and potential therapeutic implications. *Journal of Xenobiotics*, *14*(4), 1541–1569.

Wang, F., Wan, C., Wu, W., Yang, S., & Chen, X. (2024). Exogenous melatonin (MT) enhances salt tolerance of okra (*Abelmoschus esculentus* L.) plants by regulating proline, photosynthesis, ion homeostasis and ROS pathways. *Vegetos*, *37*(1), 224–238.

Wang, X., Ma, G., Feng, N., Zheng, D., Zhou, H., Li, J., …, & Huang, Y. (2025). Abscisic acid can improve the salt tolerance and yield of rice by improving its physiological characteristics. *Agronomy*, *15*(2), 309.

Wang, X., Wen, H., Suprun, A., & Zhu, H. (2025). Ethylene signaling in regulating plant growth, development, and stress responses. *Plants*, *14*(3), 309.

Wolters, S. M., Laibach, N., Riekötter, J., Roelfs, K.-U., Müller, B., Eirich, J., …, & Schulze Gronover, C. (2024). The interaction networks of small rubber particle proteins in the latex of Taraxacum koksaghyz reveal diverse functions in stress responses and secondary metabolism. *Frontiers in Plant Science*, *15*, 1498737.

Xia, Y., Chen, X., Jiang, X., & Shen, J. (2025). Enhanced denitrification under saline Conditions: Glycine betaine as a key osmoprotectant. *Bioresource Technology*, *429*, 132517.

Xie, W., Liu, H., Ren, D., Wei, Y., Liu, Y., Tang, L., …, & Rao, Y. (2025). A K^+-efflux antiporter is vital for tolerance to salt stress in rice. *Rice*, *18*(1), 57.

Yang, S., Zhao, D., Li, M., Cai, D., Zhang, Y., Ding, D., …, & Duan, P. (2022). Conserved Zma-miR164 family is responsible for maize heterosis by negatively regulating NAC transcription factor during internode elongation. *South African Journal of Botany*, *150*, 806–812.

Yao, P., Zhang, C., Bi, Z., Liu, Y., Liu, Z., Wei, J., …, & Sun, C. (2024). Overexpression of potato PYL16 gene in tobacco enhances the transgenic plant tolerance to drought stress. *International Journal of Molecular Sciences*, *25*(16), 8644.

Younas, A., Riaz, N., Rashid, M., Tufail, A., Hyder, S., & Noreen, Z. (2024). CRISPR/Cas system for achieving abiotic stress tolerance. In: S. Fiaz & C. S. Prakash (Eds.), *OMICs-based techniques for global food security* (pp. 213–231). https://doi.org/10.1002/9781394209156.ch11

Yuan, P., Wei, Y., Gao, X., Song, C., Jiao, J., Wang, M., …, & Bai, T. (2024). The RING-H2 type E3 ubiquitin ligase gene MdATL16 positively regulates salt tolerance in transgenic tomato and apple. *Environmental and Experimental Botany*, *220*, 105689.

Yuan, Z., Zhou, X., Zhang, Y., Wang, Y., Yan, H., Sun, W., …, & Wu, C. (2025). Stage-dependent mineral element dynamics in 'Junzao' Jujube: Ionic homeostasis and selective transport under graduated saline-alkali stress. *Horticulturae*, *11*(7), 726.

Zeng, T., Su, H., Wang, M., He, J., Gu, L., Wang, H., …, & Zhu, B. (2025). The role of MYC2 transcription factors in plant secondary metabolism and stress response mechanisms. *Plants*, *14*(8), 1255.

Zhang, D., Zhang, D., Zhang, Y., Li, G., Sun, D., Zhou, B., & Li, J. (2024). Insights into the epigenetic basis of plant salt tolerance. *International Journal of Molecular Sciences*, *25*(21), 11698.

Zhang, J., Li, S., Gao, X., Liu, Y., & Fu, B. (2024). Genome-wide identification and expression pattern analysis of the Aux/IAA (auxin/indole-3-acetic acid) gene family in alfalfa (*Medicago sativa*) and the potential functions under drought stress. *BMC Genomics*, *25*(1), 382.

Zhang, S., Li, X., Fan, S., Zhou, L., & Wang, Y. (2020). Overexpression of HcSCL13, a *Halostachys caspica* GRAS transcription factor, enhances plant growth and salt stress tolerance in transgenic Arabidopsis. *Plant Physiology and Biochemistry*, *151*, 243–254.

Zhang, Y., Hou, S., Athanassiou, C. G., Guo, L., Zhao, S., Hu, Q., …, & Liu, Q. (2025). Regulatory role and mechanisms of reactive oxygen species in grains during post-harvest storage: The vanguard sentinel of quality deterioration. *Food Reviews International*, *41*(5), 1259–1289.

Zhang, Z., Zhang, Y., Cao, Y., Chen, S., Zhu, Y., Wang, Y., …, & Wu, Y. (2025b). Integrated physiological and transcriptomic analysis reveals osmoregulatory collapse, oxidative stress, and apoptosis drive hypo-salinity-induced mortality in Solen grandis. *International Journal of Biological Macromolecules*, *310*, 143495.

Zhu, D., Sánchez-Ferrer, A., & Nyström, L. (2016). Antioxidant activity of individual steryl ferulates from various cereal grain sources. *Journal of Natural Products*, *79*(2), 308–316.

Zhu, L., Sun, Y., Wang, R., Zeng, J., Li, J., Huang, M., …, & Guo, S. (2025). Applied potassium negates osmotic stress impacts on plant physiological processes: A meta-analysis. *Horticulture Research*, *12*(2), uhae318.

Zhu, Z., Zhou, Y., Liu, X., Meng, F., Xu, C., & Chen, M. (2025). Integrated transcriptomic and metabolomic analyses uncover the key pathways of *Limonium bicolor* in response to salt stress. *Plant Biotechnology Journal*, *23*(3), 715–730.

9 Oxidative Stress and the Role of Antioxidant Enzymes in Plants

Elshan Musazade, Nargiz Bayramova, Nargiz Sultanova, Javaid Akhter Bhat, and Xianzhong Feng

9.1 INTRODUCTION

Plants are constantly challenged by various environmental stresses that impede agricultural productivity and global food supply. The accelerating climate changes have further heightened the frequency and severity of abiotic stresses, including drought, salinity, temperature extremes, nutrient imbalance, flooding, and heavy metal toxicity, whose combined effects can cause >50% crop yield loss, severely affecting plant ecosystems and their ability to support productivity (Yang et al., 2024).

One of the most important early consequences of environmental stress is the overproduction of reactive oxygen species (ROS). ROS are inevitable products of aerobic metabolism that mainly arise via electron transfer in chloroplasts, mitochondria, and peroxisomes. These ROS, such as superoxide anions (O_2^-), hydrogen peroxide (H_2O_2), hydroxyl radicals (•OH), and singlet oxygen ($^1O_2$), are produced by the incomplete reduction of molecular O_2 in the presence of electron donors. Depending on their abundance, cellular compartmentalization, and redox status, ROS function as signaling molecules at low levels (where they tailor gene expression/stress responses), whereas at high levels they become harmful oxidants that mediate lipid peroxidation, protein denaturation, DNA damage, or programmed cell death (Waszczak et al., 2018).

When the production of ROS overwhelms antioxidant defenses, cells undergo an upset in their oxidation state, a condition called oxidative stress. It is the result of an imbalance between ROS production and accumulation and an organism's capacity to detoxify or neutralize these reactive intermediates. This leads to the loss of redox homeostasis, followed by oxidative damage of vital biomolecules, with a negative impact on cell integrity and physiological activity (Savchenko and Tikhonov, 2021) (Figure 9.1).

Plants have developed a sophisticated antioxidant defense system to alleviate ROS-induced damage and preserve redox homeostasis as well as sustain normal development in both optimal and stressful situations. This system is composed of enzymatic antioxidants that wipe out or neutralize ROS, and non-enzymatic antioxidants that complement each other to maintain redox homeostasis (Rajput et al., 2021). The enzymatic machinery is represented by superoxide dismutase (SOD), catalase (CAT), ascorbate peroxidase (APX), and glutathione reductase (GR), which are the main ROS-scavenging enzymes, constituting the first line of defense. Enzymes like glutathione S-transferases (GSTs), ferritin, polyphenol oxidase, and alternative oxidase (AOX) also participate in the protective mechanisms during stress. Ascorbic acid (AsA), glutathione (GSH), phenolic acids, flavonoids, anthocyanins, carotenoids, amino acids, vitamin E, and different mineral cofactors are non-enzymatic antioxidants that contribute to the redox buffering potential of cells (Irato and Santovito, 2021).

Plants have to maintain a balance between ROS production and scavenging for cellular homeostasis, which is crucial for developmental signaling, photosynthesis, and metabolism under regular physiologic conditions. When the balance is disturbed, triggered by unfavorable environmental conditions, excessive ROS are generated, and damage to biomolecules may end with the activation of stereotyped pathways leading to programmed cell death. To mitigate this imbalance, plants trigger their antioxidant armada involving the transcriptional re-programming as well as translational and post-translational regulation of defense-related genes to sustain redox balance during stress acclimation (Sachdev et al., 2021).

In this review, we discuss the concept of oxidative stress and its critical implications in plant physiology, as well as an extensive analysis of antioxidant defense systems devoted to maintaining cell integrity. Attention is explicitly focused on antioxidant enzymes, their regulation, and the importance of antioxidants in tolerating stress. We also discuss progress in transgenic expression of antioxidant genes to achieve stress tolerance and develop robust crop species. Together, this information provides an integrated view of how plants sense, tolerate, and adjust to oxidative stimuli as well as potential solutions for future crop improvement based on molecular and biotechnological interventions.

9.2 OXIDATIVE STRESS IN PLANTS

9.2.1 Definition and Nature of Oxidative Stress

Oxidative stress in plants can be described in two complementary ways: (1) as a physiological state where oxidant production overcomes the reductive capacity, leading to redox imbalance and molecular damage, and (2) as an abiotic stress, such as drought and salinity, that injures plant cells but also activates defense and signaling processes. In general, the source of oxidative stress is related to the activation of molecular oxygen (O_2) to intermediate reactive forms O_2^-, H_2O_2, $^1O_2$, and •OH, also known as ROS (Demidchik, 2015).

DOI: 10.1201/9781003669920-9

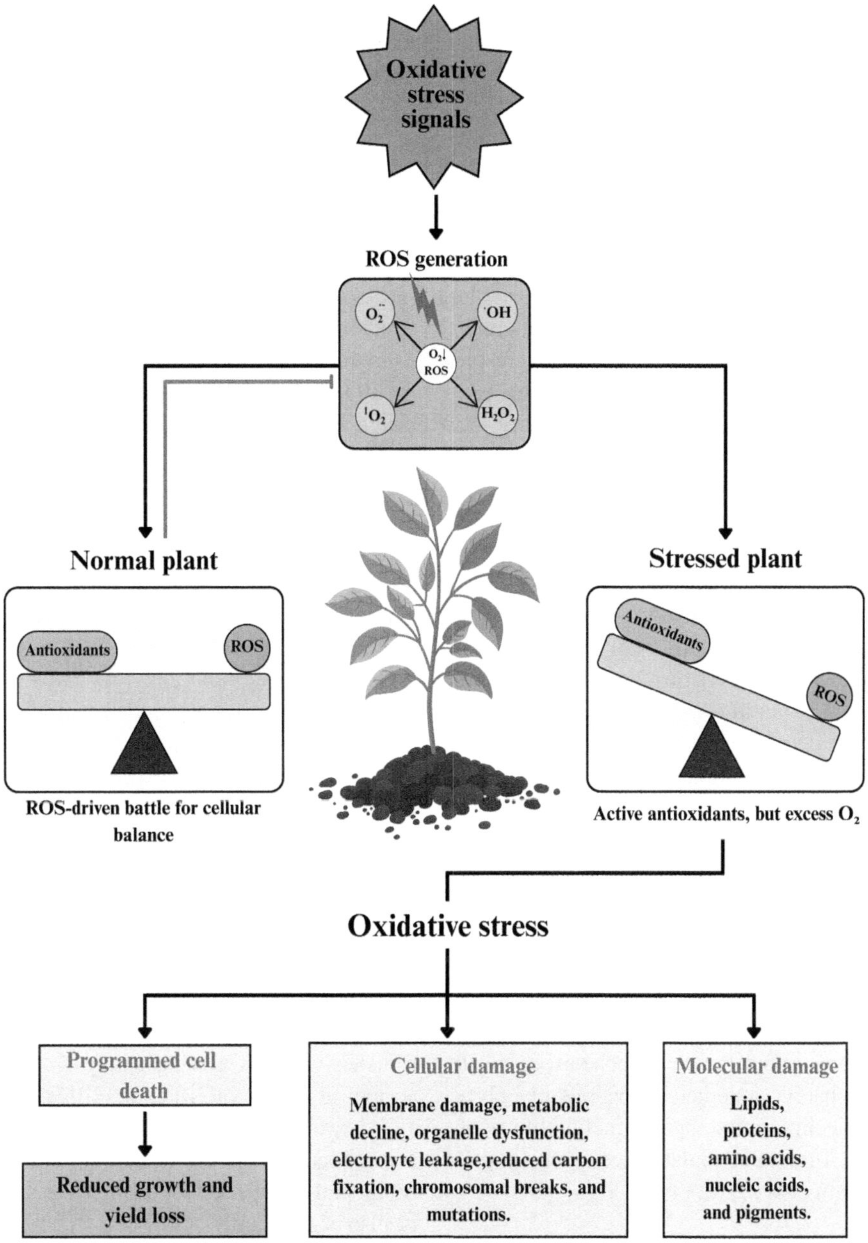

FIGURE 9.1 A diagrammatic presentation of oxidative stress responses and the ROS equilibrium in plants. Under unstressed conditions, a desired balance of ROS production and antioxidant defense is maintained in plants (redox homeostasis). Under stress conditions, however, excessive accumulation of reactive oxygen species (ROS) such as O_2^-, •OH, $^1O_2$, and H_2O_2 disrupts the balance between their production and scavenging, leading to oxidative stress. This condition causes extensive cellular and molecular damage, including membrane breakdown, organelle dysfunction, electrolyte leakage, and macromolecule oxidation, ultimately resulting in reduced growth and yield.

The antioxidant system prevents the cells from this potential damage by redox homeostasis, while daily and seasonal fluctuations in oxidative status are held in physiologically safe ranges. However, in adverse environmental circumstances, excess production of ROS disrupts this balance and causes oxidative stress. Because in specific ranges of concentrations, ROS are essential intracellular and intercellular messengers that translate environmental signals into acclimation responses. Outside these ranges, unfettered ROS build-up causes permanent damage to cells and can cause them to die (Mittler et al., 2022).

9.2.2 Cellular and Molecular Damage Under Oxidative Stress

Oxidative stress induces severe cellular and molecular damage, representing compromised membrane structure, nucleic acid protection, and metabolic homeostasis. Cell membranes are the first ROS targets. Lipid peroxidation, initiated by $^1O_2$

and •OH, attacks on polyunsaturated fatty acids, such as linoleic and linolenic acids, and generates lipid dimers and radicals that compromise membrane integrity and photosynthetic performance. Proteins, pigments, and cofactors of photosystem II (PSII) are well known for being susceptible to damage. Singlet oxygen is involved in lipid hydroperoxides (LOOH) formation and photoinhibition. Oxidation of membrane proteins and receptors also disturbs permeability, leading to an increased cellular breakdown (Hasanuzzaman et al., 2020). Lipid peroxidation (LPO) products, i.e., malondialdehyde (MDA) and α, β-unsaturated hydroxyalkenals, covalently modify amino acid residues (arginine, lysine, proline, threonine, and tryptophan), thus enhancing proteolytic processes and the damage of enzyme activity. D1/D2 reaction center proteins, Rubisco, and SOD are frequent oxidation or nitrosylation targets (Møller et al., 2007). As such, MDA is extensively applied in biochemical studies as a biomarker for oxidative stress.

ROS also damage nucleic acids. The •OH reacts with deoxyribose to form thymine glycol, hydroxymethyl urea, and strand breaks (Saha et al., 2015). Additional lesions, including 8-hydroxyguanine and dehydro-2′-deoxyguanosine, result from nucleotide oxidation and DNA–protein crosslinking. Such unrepaired modifications cause transcriptional errors, genomic instability, and impaired photosynthetic and signaling functions (Choudhury et al., 2013).

Metabolically, oxidative stress inhibits enzymes of glycolysis, the tricarboxylic-acid (TCA) cycle, and the pentose-phosphate (PP) pathway, such as glyceraldehyde-3-phosphate dehydrogenase and fructose-1,6-bisphosphate aldolase, which are particularly sensitive to oxidation (Banerjee and Roychoudhury, 2017). As compensation, stressed cells redirect carbon flux toward NADPH generation and PP pathway intermediates (ribose-5-phosphate and ribulose-5-phosphate). Nonetheless, overall glycolytic and TCA activities decline, reducing the metabolite pools of citrate, isocitrate, fumarate, malate, succinate, and 2-oxoglutarate (Lehmann et al., 2012).

9.3 MAJOR ANTIOXIDANT ENZYMES AND THEIR ROLES

Plants' antioxidant defense is a highly orchestrated and compartmentalized network that presents organ-, organelle-, and tissue-specific adaptations in response to the fluctuating environment. It involves enzymatic and non-enzymatic constituents, which act in concert to regulate redox homeostasis and limit oxidative stress. Non-enzymatic antioxidants are represented by AsA, the primary cellular redox buffer mainly located in the cytosol, and GSH (γ-glutamyl-cysteinyl-glycine), as well as tocopherols, carotenoids, phenolic compounds, amino acids, and other vitamins. This enzymatic machinery is composed of the main ROS-scavengers that are SOD, CAT, and GPOX (guaiacol peroxidase), and also enzymes of the AsA–GSH cycle that are APX, monodehydroascorbate reductase (MDHAR), dehydroascorbate reductase (DHAR), and GR. These enzymes, which are not only conserved among aerobic organisms but have also coevolved to work cooperatively in various subcellular organelles, permit plants to sense and effectively react to oxidative stress (Mittler et al., 2022).

9.3.1 SOD

SOD (EC 1.15.1.1) is an important antioxidant enzyme, which catalyzes O_2^- dismutation to H_2O_2 and O_2, a reaction that occurs at about 10,000 times the rate of spontaneous dismutation. By scavenging the •OH generated by metal-catalyzed reactions, SOD becomes the first-line defense for ROS in aerobic organisms (Bowler et al., 1992). SOD has three major isoforms based on the metal cofactor used: Cu/Zn-SOD, Mn-SOD, and Fe-SOD. The majority of the Cu/Zn-SOD, which is the most common form of these enzymes in plants, appears to be present in cytosol, chloroplast stroma, apoplast, and peroxisomes. Mn-SOD exists mainly in mitochondria and peroxisomes, whereas the less common Fe-SOD is found only within the chloroplasts of particular species (Mittler et al., 2004).

By structure, SOD enzymes have two calcium-binding α-helices and β-sheets, and β-strands primarily dominate Cu/Zn-SODs, whereas Mn- and Fe-SODs are enzymes with an α-helix-rich structure (Dehury et al., 2013). Mn- and Fe-SODs, however, are closely related in their evolutionary relationship, with Cu/Zn-SODs forming a structurally distinct clade (Dehury et al., 2013). The active site is characterized by highly conserved histidine and aspartate residues that are involved in metal chelation, where, for Fe-SOD, three histidines in conjunction with one aspartate provide the ligands for the metal center (Pilon et al., 2011). Metal cofactors (Cu^{2+}, Zn^{2+}, Mn^{2+}, and Fe^{2+}) assist in electron transfer, with Cu driving catalysis and Zn contributing to structural stability (Gill et al., 2015).

It is well known that SOD activity in plants may undergo a stress-induced modulation as an adaptive response to environmental stimuli. Under drought stress, the activity of SOD is also significantly enhanced in *Phaseolus vulgaris* and *Oryza sativa* (Zlatev et al., 2006), as well as under water-stressed irrigation in *Trifolium repens* (Chang-Quan and Rui-Chang, 2008). Salinity stress also increases SOD activity in *Cicer arietinum* and *Solanum lycopersicum* (Gapińska et al., 2007), and all isoforms mitigate oxidative stress in *C. arietinum* (Eyidogan and Öz, 2007). This is further confirmed by genetic manipulations, where the overexpression of Mn-SOD significantly increased salt tolerance in *Arabidopsis thaliana* (Wang et al., 2004). Furthermore, the SOD activity was increased under UV-B radiation in cultivars of *Triticum aestivum*, *Vigna radiata*, and *Glycine max* cultivars (Agrawal et al., 2009), highlighting its common role in combating oxidative stress due to different abiotic stresses.

9.3.2 CAT

CAT (EC 1.11.1.6) is a heme-containing antioxidant enzyme that catalyzes the decomposition of H_2O_2 into H_2O

and O_2 without the need for external reductants. The catalytic mechanism proceeds in two distinct steps: oxidation of heme iron (Fe^{3+}) by one molecule of H_2O_2, producing an iron–oxo intermediate identified as compound I (first step); and reduction by a second H_2O_2 molecule, which restores the enzyme to its resting state and results in the formation of O_2 and H_2O. Formation of compound I results in an apparent change in the absorption spectrum of the enzyme, which can be followed both *in vivo* and *in vitro* (Oshino et al., 1975).

Structurally, native CAT is a homotetramer of 240 kDa with subunits that range from 54 to 59 kDa. However, isoform variability exists. Bigger dimers (~320 kDa) were found in *Mesembryanthemum crystallinum* and smaller homodimers in *Capsicum annuum* (Rodríguez-Ruiz et al., 2019). CAT isoforms are expressed in specific tissues and at specific developmental stages. Three functional classes have been reported in *Nicotiana tabacum*: Class I (mainly in photosynthetic tissues and involved in H_2O_2 photorespiratory scavenging), Class II (vascular function, abscisic acid (ABA) response, lignification, senescence), and Class III (seed/reproductive organ function that is related to catabolism of fatty acids by glyoxysomes) (Černý et al., 2018). Likewise, *Zea mays* also possesses three genes, *CAT1*, *CAT2*, and *CAT3*, but they have identical coding sequences, different intron structures, and promoter regions (Fink and Scandalios, 2002). *A. thaliana* has three CAT genes, and the *Gossypium* species possesses seven. The copy number of the gene varies across different taxa. For instance, three copies were identified in *O. sativa*, *Cucurbita pepo*, and *Cucumis sativus*, whereas only one was found in *Ipomoea batatas*, *Ricinus communis*, and *S. lycopersicum* (Wang et al., 2019).

CATs are mainly present in peroxisomes and catabolize H_2O_2 that is produced by photorespiration and during β-oxidation of fatty acids. Each CAT tetramer harbors four heme groups, of which Fe^{2+} shuttles between oxidation states to mediate the reaction of two H_2O_2 molecules into water and O_2 (Arora et al., 2020). CATs, with millimolar K_m values, operate best under conditions of stress when H_2O_2 levels are raised, whereas peroxidase (PODs) have a greater affinity and peak activity at lower substrate concentrations (Černý et al., 2018). Therefore, CAT is a must-have high-capacity enzyme to remove H_2O_2, which always works together with PODs in the maintenance and defense of cellular redox homeostasis against oxidative injury.

CAT gene expression is tightly controlled at multiple levels, including transcriptional control by specific transcription factors and post-transcriptional regulation, such as alternative splicing and mRNA sponging (Wang et al., 2019). For example, the transgenic *N. tabacum* overexpressing *TaMIR1119* showed increased SOD, CAT, and POD activities under water deficit (Shi et al., 2018). Enhanced CAT activity has been reported in *T. aestivum* under drought stress (Simova-Stoilova et al., 2010), *C. arietinum* under salinity (Eyidogan and Öz, 2007), and *O. sativa* in heat-tolerant mutants of rice plants (Zafar et al., 2020). On the other hand, CAT activity in *Arabidopsis* is reduced under photoperiod stress (Abuelsoud et al., 2020). Furthermore, hormonal manipulation is a significant factor in CAT control. Melatonin pretreatment significantly increases the CAT activity under drought in *Brassica napus* roots, suggesting that there is a close cross-talk between the redox and hormonal homeostasis for stress tolerance (Dai et al., 2020).

9.3.3 Peroxidases (PODs, APX, GPOX)

Class III peroxidases (POXs) (EC 1.11.1.7), heme-binding enzymes, are plant heme proteins primarily located in the apoplast and vacuoles, where they scavenge H_2O_2 and engage in lignification, suberization, cell wall reinforcement, and stress protection. Contrary to the Class I POXs, for example, cytochrome c peroxidase, CAT-peroxidase, and APX, or Class II fungal-type POXs (Mn-dependent peroxidase and ligninase), plant Class III POXs are processed as secretory glycoproteins in the endoplasmic reticulum, and Golgi apparatus, before being released into the environment (Hiraga et al., 2001). These enzymes are integral to the plant antioxidant framework that serves as a bridge between ROS scavenging and structural remodeling during environmental acclimatization.

The POXs are usually 33–55 kDa glycoproteins, consisting of 300–350 amino acids. Nonetheless, multimeric forms have been reported, such as homotetramers in *Cocos nucifera* (Mujer et al., 1983), homodimers in *Brassica oleracea* (Regalado et al., 1999), and heterotrimers in *Leucaena leucocephala* (Pandey and Dwivedi, 2011). High diversity of isoforms correlates with functional specialization. Two peroxidase (POX) isoforms have been identified in *Fagopyrum esculentum*: POX I (46.1 kDa, optimum activity at 30°C) and POX II (58.1 kDa; optimum activity at 10°C) (Suzuki et al., 2006). Three distinct POX catalytic mechanisms have been identified in structural studies of horseradish peroxidase that accommodate a range of substrates (Veljović Jovanović et al., 2018), and this provides further evidence for the broad functional flexibility exhibited by this enzyme.

Concerning the Class I POXs, APX (EC 1.11.1.11) is a key enzyme of the Foyer–Halliwell–Asada cycle, metabolizing H_2O_2 into water with AsA as electron donor (Caverzan et al., 2012). Fe-dependent APX, which is non-storable in the absence of AsA, is localized in different subcellular compartments, including cytosol (Wang et al., 2005), mitochondria (Jiménez et al., 1998), chloroplasts (Yoshimura et al., 2002), peroxisomes (Mullen and Trelease, 2000), and glyoxysomes (Bunkelmann and Trelease, 1996). The calculated molecular weights range from 28 to 31 kDa in *C. pepo* (Yamaguchi et al., 1995) to 33 kDa in *Apium graveolens* (Liu et al., 2019b), and up to 34 kDa for the plastidic APX in *N. tabacum* (Madhusudhan et al., 2003) or even more, with a molecular weight of up to 110.6 kDa from *Actinidia deliciosa* (Liao et al., 2020). Promoter analyses have identified ABA-responsive (ABRE) and heat-shock (HSE) motifs (Saxena et al., 2020), consistent with its stress-inducible nature. The heme prosthetic group of the enzyme is required, and its activity is blocked by cyanide, azide, or sulfhydryl reagents like 5,5′-dithiobis-(2-nitrobenzoic acid) (DTNB). APX is an enzyme that reduces H_2O_2 with AsA as the electron donor, mediating the oxidation of AsA and H_2O_2 to create monodehydroascorbate (MDHA) and water. MDHA subsequently spontaneously dismutates into dehydroascorbate (DHA). Interestingly, 4-coumarate 3-hydroxylase can also function as a bifunctional cytosolic APX that oxidizes both AsA and 4-coumarate (Barros et al., 2019).

In detail, functional assays have emphasized APX as a key enzyme of ROS scavenging and redox balance. As AgAPX1 exhibited optimum activity at 55°C and was significantly induced by drought in *A. graveolens*, its expression led to increased tolerance of *Arabidopsis* plants to water deficit (Liu et al., 2019a). Likewise, *Dioscorea alata* APX provided better chilling and flooding tolerance, whereas H_2O_2 supplementation recovered enzymatic activity in yam lines with inhibited APX expression (Chen et al., 2019). Overexpression of *AtApx1* in *Brassica juncea* enhanced salinity tolerance (Saxena et al., 2020); and higher APX activity was observed under UV-B exposure (Rao et al., 1996) and lead (Pb) stress (Malar et al., 2016). In contrast, silencing of cytosolic APX by RNAi in *Gossypium* resulted in increased accumulation of ROS, compromised stomatal conductance activity, decreased CO_2 assimilation, and loss of yield. Signaling molecules and metabolic cross-talk can influence APX activity. Melatonin is reported to increase the activity of MeAPX2 in *Manihot esculenta* (Guo et al., 2020). In *Camellia sinensis*, CsAPX1 interacts with the nitrogen-regulatory protein CsGLB1 to synchronize AsA metabolism under nitrogen deficiency (Li et al., 2020). In *C. sativus*, APX isoenzymes are activated to different extents under low temperatures, which compensate for the inactivation of CAT (Lee and Lee, 2000). Together, these results define APX as a key node in the integration of ROS detoxification, stress signaling, and metabolic control.

GPOX (EC 1.11.1.7), a heme-containing enzyme, catalyzes the oxidation of guaiacol with H_2O_2 as an oxidant and so directly participates in the cellular H_2O_2 detoxification process (Mika and Lüthje, 2003). GPOXs play a role along with APXs and other Class III POXs in the Foyer–Halliwell–Asada pathway, where they work together with flavonoids and phenolic compounds to reinforce secondary antioxidant protection (Agati et al., 2012).

9.3.4 GR

GR (EC 1.8.1.7) is an NAD(P)H-dependent flavoprotein oxidoreductase. It plays a role in cellular redox homeostasis by catalyzing the reduction of oxidized glutathione (GSSG) to its reduced form, GSH, as one of the key steps in the AsA–GSH cycle. GR functions as a homodimeric enzyme of 100–150 kDa, each subunit containing one flavin adenine dinucleotide (FAD) cofactor and having two active-site cysteine residues (Sabeh et al., 1993). The enzyme acts through a ping-pong kinetic mechanism whereby NADPH initially reduces FAD, and the reduced form of flavin then donates electrons to cysteine thiols to fuel the thiol–disulfide exchange reaction concomitant with converting GSSG to GSH (Gill et al., 2013). In addition to its role in ROS scavenging, GR functions in sulfur assimilation by exporting the GSH pool to supply adenosine 5′-phosphosulfate (APS) reductase activity that creates a positive feedback loop for antioxidant metabolism (Cohen et al., 2020).

The vast majority (80%) of total GR activity is chloroplast-localized, with the remaining divided between cytosol, mitochondria, peroxisomes, and nucleus (Gill et al., 2013). In *A. thaliana*, GR1 localizes to the cytosol, peroxisomes, and nucleus, whereas GR2 (with an N-terminal transit peptide) targets chloroplasts and mitochondria (Trivedi et al., 2013). Functional analyses have shown that repression of GR2 in *Arabidopsis* results in H_2O_2 accumulation, increased sensitivity to high light stress, and defective PSII repair, indicating a fundamental function for this enzyme in safeguarding the photosynthetic apparatus and redox homeostasis (Marty et al., 2019).

9.4 ANTIOXIDANT ENZYME REGULATION AND CROSS-TALK

Plants use complex regulatory networks to counteract oxidative stress with enzymatic antioxidants closely integrated with signaling molecules and phytohormones. Upon sensing stress, key hormones, such as salicylic acid (SA), jasmonic acid (JA), ethylene (ET), and ABA, are synthesized rapidly to trigger downstream signaling pathways and transcriptional re-programming that elicit the expression of antioxidant enzymes (Apel and Hirt, 2004). These enzymes function synergistically to restore the cellular redox balance, limit ROS accumulation, and reinforce plant resilience under adverse environmental conditions.

ABA mediates a wide range of abiotic stress responses through mitogen-activated protein kinase (MAPK) cascades and Ca^{2+}/calmodulin (CaM)-dependent kinases, both of which interact with ROS and reactive nitrogen species (RNS) such as H_2O_2 and nitric oxide (NO). Cross-talk between ROS and RNS, generated mainly by NO, activates antioxidant defenses through the regulation of redox-sensitive enzymes and signaling molecules. Hydrogen sulfide (H_2S) also induces the activity of SOD, CAT, and APX and thus decreases oxidative damage under stress conditions as well. In addition, redox regulation is also supported by calcium signaling, since Ca^{2+} directly activates CAT and GR activity, whereas CaM activates CAT function (Chaki et al., 2020). Chloroplastic glutathione peroxidase (GPX) is considered a molecular interface between redox homeostasis and SA signaling. The disruption of this gene in *Arabidopsis* reduces tolerance to photooxidative stress. However, it increases resistance against virulent bacteria, as well as AsA and SA contents, affecting leaf development and immunity (Chang et al., 2009). Furthermore, hormones such as ABA, JA, and SA also control the GSH redox couple (GSH/GSSG), thus connecting hormonal and redox signaling pathways in an integrated network regulating plant stress adaptation and defense (Hasanuzzaman et al., 2017).

In addition to classic hormone interactions, melatonin plays a multifaceted role in plant stress signaling, interacting with numerous hormonal and redox pathways to induce SA-, gibberellic acid (GA)-, and ABA-dependent responses, which collectively contribute to increased tolerance. It induces NO biosynthesis by increasing the expression of nitrate reductase (NR) and NO synthase (NOS) genes. At the same time, NO also stimulates melatonin production in a parallel pathway, creating a double positive feedback loop. Melatonin and NO have been shown to synergistically stimulate expression of antioxidant enzymes through MAPK- and SA-mediated signaling cascades. However, melatonin directly scavenges excessive NO as well when the concentration exceeds in physiologic response (Zhang et al., 2020b). Moreover, a linkage between the δ-aminolevulinic acid (ALA) and antioxidant enzymes involving γ-aminobutyric acid (GABA), which mediates positive cross-talk, has been shown in tomatoes. Exogenous application of ALA increased the activities of SOD, CAT, APX, and GR, leading to enhanced cold tolerance, while knocking down GABA decreased the activity of these enzymes and stress resistance (Liu et al., 2020).

The total antioxidative activity in plants depends mainly on the redox states of AsA and GSH, and thiol-based switches counteract the enzyme activities through oxidoreductases (Meyer et al., 2021). Nucleoredoxin 1(NRX1) is a critical node in this network as it controls CATs, APX (APX1), GSTs, glutaredoxins, and methionine sulfoxide reductases (MSRs) for improved tolerance to oxidative stress (Kneeshaw et al., 2017). The thioredoxin-based regulation further includes targeting of mitochondrial manganese SOD (MSD1), peroxiredoxins, and also GPX6 (Yoshida et al., 2013), as well as plastidic TRX y1 that is associated with CSD2, CAT3, MDHAR6, and other ROS-scavenging enzymes there (Marchand et al., 2010). Concurrently, phosphorylation events coordinate antioxidant activity into global stress signaling cascades. For instance, CAT1 is induced by BAK1 under high light (Zhang et al., 2020a), whereas CAT3 is phosphorylated by the calcium-dependent protein kinase CPK8 in response to drought (Zou et al., 2015). Rice STRK1-mediated phosphorylation and activation of CATC confer salt and oxidative stress tolerance (Zhou et al., 2018). APX and MDAR are also targets for phosphorylation, as proved by WKS1-suppressed leaf thylakoid APX activity in wheat upon stripe rust infection, accompanied by enhanced ROS production and cell death (Gou et al., 2015).

9.5 ADVANCES IN ENHANCING ANTIOXIDANT DEFENSE IN PLANTS

Sustainable food supply is the cornerstone of global food security, but diverse environmental stresses hamper crop yields. As a result, the design of stress-tolerant plants has been one of the focal points in contemporary agricultural research. Functional analysis has clearly revealed that the overexpression of antioxidant-enzyme genes resulted in an increase in tolerance to a wide range of stresses (Table 9.1), and gene silencing or knockout experiments have also proved the essential roles of these genes in maintaining stress resistance. The vast majority of these studies have focused on abiotic stress factors, such as salinity, heat, drought, chilling, flooding, and heavy metals. In contrast, biotic stresses have been relatively less studied. With almost all of the former showing stronger constitutive or stress-inducible promoters and a higher tolerance, their great potential can be expected to be integrated into crop breeding programs for increased stress resistance and yield stability in crops (Li et al., 2025).

It was reported earlier that improved antioxidant enzyme activity in specific cellular compartments is followed by increased plant stress resistance. For example, chloroplast-targeted APX and SOD provide potato with drought tolerance (Ahmad et al., 2010), whereas downregulation of chloroplastic APX makes plants more sensitive to drought (Cunha et al., 2019). Cytosolic overexpression is also advantageous, since plum plants with elevated APX expression have been reported to be more drought-tolerant (Diaz-Vivancos et al., 2016), and wheat lines expressing higher levels of Cu/Zn-SOD are more resistant to salt and oxidative stress (Wang et al., 2016). In addition, the expression of Cu/Zn-SOD genes from *Avicennia marina* and *Arachis hypogaea* imparted stress tolerance in transgenic rice and tobacco (Negi et al., 2015). Peroxisomal targeting has also shown a similar role in osmotic and drought stress, as was the case with wheat CAT1 overexpressing *Arabidopsis*, which increased osmotic tolerance (Feki et al., 2015) or cytosolic Cu/Zn-SOD and peroxisomal CAT1 overexpression transgenic cassava lines presenting higher drought tolerance (Xu et al., 2013).

Despite encouraging progress in controlled environments, most studies on antioxidant-based stress tolerance remain confined to laboratory or greenhouse conditions, hindering their translation to realistic agricultural settings. Notably, *Arabidopsis DTP6* (*At1g68490*) expression in *Arabidopsis* and maize improved drought tolerance over three consecutive years under variable field conditions (Simmons et al., 2021), whereas *Medicago sativa* (alfalfa) lines overexpressing SOD demonstrated enhanced water-deficit tolerance in field trials (McKersie et al., 1996). However, despite these promising outcomes, no antioxidant-engineered crop has yet been commercially deployed, and conventional breeding approaches have struggled to confer stable, broad-spectrum stress tolerance solely through enhanced antioxidant capacity.

9.6 CONCLUSION AND FUTURE PERSPECTIVES

The plants, sessile organisms, are constantly exposed to various abiotic and biotic stresses such as drought, salinity, temperature extremes, and heavy metal toxicity that disrupt metabolic homeostasis, resulting in the generation of

TABLE 9.1
Transgenic Expression of Antioxidant Enzyme Genes and Improved Abiotic Stress Tolerance in Plants

Transgenic Plant(s)	Gene(s)/Source	Stress Condition	Key Findings	Reference
Arabidopsis (*Arabidopsis thaliana*)	*AgAPX1* from *Atriplex gmelinii*	Drought tolerance	Increased AsA and resistance to drought, along with increased photosynthesis efficiency.	Liu et al. (2019a)
Arabidopsis (*A. thaliana*)	*DaAPX* from *Dioscorea alata*	Flood/chilling stress	Transgenic lines retained more chlorophyll and a high antioxidant activity during co-stress.	Chen et al. (2019)
Arabidopsis (*A. thaliana*)	Cu-ZnSOD from *Cucurbita moschata* and AtSOD from *Arabidopsis* (under ubiquitin promoter)	Low temperature	Increased chilling resistance and activity of antioxidant enzymes, relieved oxidative damage under cold stress.	Lin et al. (2019)
Pomelo (*Citrus grandis*)	*PebHLH* from *Poncirus trifoliata*	Low temperature	PebHLH overexpression activated genes, including CsSOD and CsAPX, to enhance cold tolerance and antioxidative response.	Geng et al. (2019)
Irish potato (*Solanum tuberosum*)	*SOD1* gene under CAMV 35S promoter	Low temperature	The SOD activity is 1.3 times higher in transgenics, and POD and CAT are increased, with the capacity to scavenge ROS being enhanced.	Che et al. (2020)
Arabidopsis (*A. thaliana*)	*AtGPXL5*	Salt stress	Increased GSH pool and redox status, 12–14% increased growth relative to wild type on salt substrates.	Riyazuddin et al. (2019)
Chinese mustard (*Brassica juncea*)	*Apx1* from *Arabidopsis*	Salt stress	Increased GPX, CAT, and POD activities; enhanced photosynthesis, salt tolerance.	Qi et al. (2020)
Sweet potato (*Ipomoea batatas*)	*swpa1*	Salt stress	Transgenic lines had 2–3 times more peroxidase activity and greater salinity tolerance.	Kim et al. (2020)
Tobacco (*Nicotiana tabacum*)	*SlMDHAR* from *Solanum lycopersicum*	Salt stress	Higher AsA/DHA ratio and improved salt tolerance.	Li et al. (2021)
Tomato (*S. lycopersicum*)	FeSOD gene from *Arabidopsis*	Salt stress	The transgenic lines with overexpressed antioxidant enzymes also displayed reductions in salt stress damage and an enhanced structural organization.	Bogoutdinova et al. (2020)

excessively elevated ROS. An unbalanced ROS production and detoxification results in oxidative stress, a condition that disrupts redox balance, causing extensive damage to lipids, proteins, nucleic acids, and cellular membranes. ROS function in developmental and defense processes as essential signaling molecules at controlled levels but are highly reactive and toxic molecules upon overproduction. In order to counteract the dangers of these free radicals, plants have developed an intricate antioxidant defense system comprising both enzymatic and non-enzymatic components, which work together in concert to keep redox equilibrium and guard cellular structures.

The enzymatic antioxidants, including SOD, CAT, APX, GPX, and GR, are the primary defense against oxidative stress, metabolizing ROS and controlling redox signaling. The content of these enzymes is dynamically controlled on the level of transcription, translation, and post-translational modification, promoting adaptation to changing environmental conditions by the plant. Oxidative injuries caused by ROS are primarily determined by the balance between ROS generation and antioxidant defense, which varies between stress endurances and species. Thus, the discovery of the regulatory system mediating oxidative stress response and antioxidant is crucial for further increasing plant resistance and productivity.

The molecular and signaling mechanisms involved in the perception of oxidative stress, regulation of ROS-antioxidant balance, and redox homeostasis could be further studied in the future. Multi-omics approaches combined with sophisticated biotechnological tools, such as the CRISPR/Cas genome editing technique, would offer novel opportunities for manipulation of antioxidant enzyme activity and the construction of stress-tolerant genotypes. Activation of the antioxidant machinery by genetic manipulation has promising prospects for sustainable agriculture and global food security in the face of increasing climate change.

Funding: This research was supported by the Biological Breeding-National Science and Technology Major Project of China (2023ZD0403203) and the National Natural Science Foundation of China (U21A20215 and 32488102).

REFERENCES

Abuelsoud W, Cortleven A, Schmülling T. Photoperiod stress alters the cellular redox status and is associated with an increased peroxidase and decreased catalase activity 2020. https://doi.org/10.1101/2020.03.05.978270

Agati G, Azzarello E, Pollastri S, Tattini M. Flavonoids as antioxidants in plants: Location and functional significance. Plant Sci 2012;196:67–76. https://doi.org/10.1016/j.plantsci.2012.07.014

Agrawal SB, Singh S, Agrawal M. Chapter 3 ultraviolet-B induced changes in gene expression and antioxidants in plants. 2009;47–86. https://doi.org/10.1016/S0065-2296(10)52003-2

Ahmad R, Kim Y, Kim M, Kwon S, Cho K, Lee H, et al. Simultaneous expression of choline oxidase, superoxide dismutase and ascorbate peroxidase in potato plant chloroplasts provides synergistically enhanced protection against various abiotic stresses. Physiol Plant 2010;138:520–533. https://doi.org/10.1111/j.1399-3054.2010.01348.x

Apel K, Hirt H. REACTIVE OXYGEN SPECIES: Metabolism, oxidative stress, and signal transduction. Annu Rev Plant Biol 2004;55:373–399. https://doi.org/10.1146/annurev.arplant.55.031903.141701

Arora M, Saxena P, Abdin MZ, Varma A. Interaction between *Piriformospora indica* and *Azotobacter chroococcum* diminish the effect of salt stress in *Artemisia annua* L. by enhancing enzymatic and non-enzymatic antioxidants. Symbiosis 2020;80:61–73. https://doi.org/10.1007/s13199-019-00656-w

Banerjee A, Roychoudhury A. Abiotic Stress, Generation of Reactive Oxygen Species, and Their Consequences. React. Oxyg. Species Plants, Wiley; 2017, p. 23–50. https://doi.org/10.1002/9781119324928.ch2

Barros J, Escamilla-Trevino L, Song L, Rao X, Serrani-Yarce JC, Palacios MD, et al. 4-coumarate 3-hydroxylase in the lignin biosynthesis pathway is a cytosolic ascorbate peroxidase. Nat Commun 2019;10:1994. https://doi.org/10.1038/s41467-019-10082-7

Bogoutdinova LR, Lazareva EM, Chaban IA, Kononenko NV, Dilovarova T, Khaliluev MR, et al. Salt stress-induced structural changes are mitigated in transgenic tomato plants over-expressing superoxide dismutase. Biology (Basel) 2020;9:297. https://doi.org/10.3390/biology9090297

Bowler C, Montagu MV, Inze D. Superoxide dismutase and stress tolerance. Annu Rev Plant Biol 1992;43:83–116. https://doi.org/10.1146/annurev.pp.43.060192.000503

Bunkelmann JR, Trelease RN. Ascorbate peroxidase (a prominent membrane protein in oilseed glyoxysomes). Plant Physiol 1996;110:589–598. https://doi.org/10.1104/pp.110.2.589

Caverzan A, Passaia G, Rosa SB, Ribeiro CW, Lazzarotto F, Margis-Pinheiro M. Plant responses to stresses: Role of ascorbate peroxidase in the antioxidant protection. Genet Mol Biol 2012;35:1011–1019. https://doi.org/10.1590/S1415-47572012000600016

Černý M, Habánová H, Berka M, Luklová M, Brzobohatý B. Hydrogen peroxide: Its role in plant biology and cross-talk with signalling networks. Int J Mol Sci 2018;19:2812. https://doi.org/10.3390/ijms19092812

Chaki M, Begara-Morales JC, Barroso JB. Oxidative stress in plants. Antioxidants 2020;9:481. https://doi.org/10.3390/antiox9060481

Chang CCC, Ślesak I, Jordá L, Sotnikov A, Melzer M, Miszalski Z, et al. Arabidopsis chloroplastic glutathione peroxidases play a role in cross talk between photooxidative stress and immune responses. Plant Physiol 2009;150:670–683. https://doi.org/10.1104/pp.109.135566

Chang-Quan W, Rui-Chang L. Enhancement of superoxide dismutase activity in the leaves of white clover (*Trifolium repens* L.) in response to polyethylene glycol-induced water stress. Acta Physiol Plant 2008;30:841–847. https://doi.org/10.1007/s11738-008-0189-8

Che Y, Zhang N, Zhu X, Li S, Wang S, Si H. Enhanced tolerance of the transgenic potato plants overexpressing Cu/Zn superoxide dismutase to low temperature. Sci Hortic (Amsterdam) 2020;261:108949. https://doi.org/10.1016/j.scienta.2019.108949

Chen Z, Lu HH, Hua S, Lin KH, Chen N, Zhang Y, et al. Cloning and overexpression of the ascorbate peroxidase gene from the yam (*Dioscorea alata*) enhances chilling and flood tolerance in transgenic Arabidopsis. J Plant Res 2019;132:857–866. https://doi.org/10.1007/s10265-019-01136-4

Choudhury S, Panda P, Sahoo L, Panda SK. Reactive oxygen species signaling in plants under abiotic stress. Plant Signal Behav 2013;8:e23681. https://doi.org/10.4161/psb.23681

Cohen A, Hacham Y, Welfe Y, Khatib S, Avice J, Amir R. Evidence of a significant role of glutathione reductase in the sulfur assimilation pathway. Plant J 2020;102:246–261. https://doi.org/10.1111/tpj.14621

Cunha JR, Carvalho FEL, Lima-Neto MC, Jardim-Messeder D, Cerqueira JVA, Martins MO, et al. Proteomic and physiological approaches reveal new insights for uncover the role of rice thylakoidal APX in response to drought stress. J Proteomics 2019;192:125–136. https://doi.org/10.1016/j.jprot.2018.08.014

Dai L, Li J, Harmens H, Zheng X, Zhang C. Melatonin enhances drought resistance by regulating leaf stomatal behaviour, root growth and catalase activity in two contrasting rapeseed (*Brassica napus* L.) genotypes. Plant Physiol Biochem 2020;149:86–95. https://doi.org/10.1016/j.plaphy.2020.01.039

Dehury B, Sarma K, Sarmah R, Sahu J, Sahoo S, Sahu M, et al. In silico analyses of superoxide dismutases (SODs) of rice (*Oryza sativa* L.). J Plant Biochem Biotechnol 2013;22:150–156. https://doi.org/10.1007/s13562-012-0121-6

Demidchik V. Mechanisms of oxidative stress in plants: From classical chemistry to cell biology. Environ Exp Bot 2015;109:212–228. https://doi.org/10.1016/j.envexpbot.2014.06.021

Diaz-Vivancos P, Faize L, Nicolás E, Clemente-Moreno MJ, Bru-Martinez R, Burgos L, et al. Transformation of plum plants with a cytosolic ascorbate peroxidase transgene leads to enhanced water stress tolerance. Ann Bot 2016;117:1121–1131. https://doi.org/10.1093/aob/mcw045

Eyidogan F, Öz MT. Effect of salinity on antioxidant responses of chickpea seedlings. Acta Physiol Plant 2007;29:485–493. https://doi.org/10.1007/s11738-007-0059-9

Feki K, Kamoun Y, Ben Mahmoud R, Farhat-Khemakhem A, Gargouri A, Brini F. Multiple abiotic stress tolerance of the transformants yeast cells and the transgenic Arabidopsis plants expressing a novel durum wheat catalase. Plant Physiol Biochem 2015;97:420–431. https://doi.org/10.1016/j.plaphy.2015.10.034

Fink RC, Scandalios JG. Molecular evolution and structure–function relationships of the superoxide dismutase gene families in angiosperms and their relationship to other eukaryotic and prokaryotic superoxide dismutases. Arch Biochem Biophys 2002;399:19–36. https://doi.org/10.1006/abbi.2001.2739

Gapińska M, Skłodowska M, Gabara B. Effect of short- and long-term salinity on the activities of antioxidative enzymes and lipid peroxidation in tomato roots. Acta Physiol Plant 2007;30:11–18. https://doi.org/10.1007/s11738-007-0072-z

Geng J, Wei T, Wang Y, Huang X, Liu JH. Overexpression of *PtrbHLH*, a basic helix-loop-helix transcription factor from *Poncirus trifoliata*, confers enhanced cold tolerance in pummelo (*Citrus grandis*) by modulation of H_2O_2 level via regulating a CAT gene. Tree Physiol 2019. https://doi.org/10.1093/treephys/tpz081

Gill SS, Anjum NA, Gill R, Yadav S, Hasanuzzaman M, Fujita M, et al. Superoxide dismutase—Mentor of abiotic stress tolerance in crop plants. Environ Sci Pollut Res 2015;22:10375–10394. https://doi.org/10.1007/s11356-015-4532-5

Gill SS, Anjum NA, Hasanuzzaman M, Gill R, Trivedi DK, Ahmad I, et al. Glutathione and glutathione reductase: A boon in disguise for plant abiotic stress defense operations. Plant Physiol Biochem 2013;70:204–212. https://doi.org/10.1016/j.plaphy.2013.05.032

Gou JY, Li K, Wu K, Wang X, Lin H, Cantu D, et al. Wheat stripe rust resistance protein WKS1 reduces the ability of the thylakoid-associated ascorbate peroxidase to detoxify reactive oxygen species. Plant Cell 2015;27:1755–1770. https://doi.org/10.1105/tpc.114.134296

Guo K, Li Z, Tian H, Du X, Liu Z, Huang H, et al. Cytosolic ascorbate peroxidases plays a critical role in photosynthesis by modulating reactive oxygen species level in stomatal guard cell. Front Plant Sci 2020;11. https://doi.org/10.3389/fpls.2020.00446

Hasanuzzaman M, Bhuyan MHMBZulfiqar F, Raza A, Mohsin SM, Al Mahmud J, et al. Reactive oxygen species and antioxidant defense in plants under abiotic stress: Revisiting the crucial role of a universal defense regulator. Antioxidants 2020;9:1–52. https://doi.org/10.3390/antiox9080681

Hasanuzzaman M, Nahar K, Anee TI, Fujita M. Glutathione in plants: Biosynthesis and physiological role in environmental stress tolerance. Physiol Mol Biol Plants 2017;23:249–268. https://doi.org/10.1007/s12298-017-0422-2

Hiraga S, Sasaki K, Ito H, Ohashi Y, Matsui H. A large family of class III plant peroxidases. Plant Cell Physiol 2001;42:462–468. https://doi.org/10.1093/pcp/pce061

Irato P, Santovito G. Enzymatic and non-enzymatic molecules with antioxidant function. Antioxidants 2021;10:579. https://doi.org/10.3390/antiox10040579

Jiménez A, Hernández JA, Barceló AR, Sandalio LM, Del Río LA, Sevilla F. Mitochondrial and peroxisomal ascorbate peroxidase of pea leaves. Physiol Plant 1998;104:687–692. https://doi.org/10.1034/j.1399-3054.1998.1040424.x

Kim YH, Kim HS, Park SC, Ji CY, Yang J, Lee H, et al. Overexpression of swpa4 peroxidase enhances tolerance to hydrogen peroxide and high salinity-mediated oxidative stress in transgenic sweetpotato plants. Plant Biotechnol Rep 2020;14:301–307. https://doi.org/10.1007/s11816-020-00602-z

Kneeshaw S, Keyani R, Delorme-Hinoux V, Imrie L, Loake GJ, Le Bihan T, et al. Nucleoredoxin guards against oxidative stress by protecting antioxidant enzymes. Proc Natl Acad Sci 2017;114:8414–8419. https://doi.org/10.1073/pnas.1703344114

Lee DH, Lee CB. Chilling stress-induced changes of antioxidant enzymes in the leaves of cucumber: In gel enzyme activity assays. Plant Sci 2000;159:75–85. https://doi.org/10.1016/S0168-9452(00)00326-5

Lehmann M, Laxa M, Sweetlove LJ, Fernie AR, Obata T. Metabolic recovery of *Arabidopsis thaliana* roots following cessation of oxidative stress. Metabolomics 2012;8:143–153. https://doi.org/10.1007/s11306-011-0296-1

Li H, Liu H, Wang Y, Teng RM, Liu J, Lin S, et al. Cytosolic ascorbate peroxidase 1 modulates ascorbic acid metabolism through cooperating with nitrogen regulatory protein P-II in tea plant under nitrogen deficiency stress. Genomics 2020;112:3497–3503. https://doi.org/10.1016/j.ygeno.2020.06.025

Li J, Li H, Yang N, Jiang S, Ma C, Li H. Overexpression of a monodehydroascorbate reductase gene from sugar beet M14 increased salt stress tolerance. Sugar Tech 2021;23:45–56. https://doi.org/10.1007/s12355-020-00877-0

Li Y, Tan Z, Liu Y, Peng Y, Liu C. Root-specific overexpression of the CmDUF239-1 gene enhances heat tolerance in melon seedlings by upregulating antioxidant enzymes activities, proline content, and expression of heat shock protein-related genes. Horticulturae 2025;11:1198. https://doi.org/10.3390/horticulturae11101198

Liao GL, Liu Q, Li YQ, Zhong M, Huang CH, Jia DF, et al. Identification and expression profiling analysis of ascorbate peroxidase gene family in *Actinidia chinensis* (Hongyang). J Plant Res 2020;133:715–726. https://doi.org/10.1007/s10265-020-01206-y

Lin KH, Sei SC, Su YH, Chiang CM. Overexpression of the Arabidopsis and winter squash superoxide dismutase genes enhances chilling tolerance via ABA-sensitive transcriptional regulation in transgenic Arabidopsis. Plant Signal Behav 2019;14:1685728. https://doi.org/10.1080/15592324.2019.1685728

Liu JX, Feng K, Duan AQ, Li H, Yang QQ, Xu ZS, et al. Isolation, purification and characterization of an ascorbate peroxidase from celery and overexpression of the AgAPX1 gene enhanced ascorbate content and drought tolerance in Arabidopsis. BMC Plant Biol 2019a;19:488. https://doi.org/10.1186/s12870-019-2095-1

Liu T, Jiao X, Yang S, Zhang Z, Ye X, Li J, et al. Crosstalk between GABA and ALA to improve antioxidation and cell expansion of tomato seedling under cold stress. Environ Exp Bot 2020;180:104228. https://doi.org/10.1016/j.envexpbot.2020.104228

Liu Y, Jiang H, Zhao Y, Li X, Dai X, Zhuang J, et al. Three *Camellia sinensis* glutathione S-transferases are involved in the storage of anthocyanins, flavonols, and proanthocyanidins. Planta 2019b;250:1163–1175. https://doi.org/10.1007/s00425-019-03206-2

Madhusudhan R, Ishikawa T, Sawa Y, Shigeoka S, Shibata H. Characterization of an ascorbate peroxidase in plastids of tobacco BY-2 cells. Physiol Plant 2003;117:550–557. https://doi.org/10.1034/j.1399-3054.2003.00066.x

Malar S, Shivendra Vikram S, JC Favas P, Perumal V. Lead heavy metal toxicity induced changes on growth and antioxidative enzymes level in water hyacinths [*Eichhornia crassipes* (Mart.)]. Bot Stud 2016;55:54. https://doi.org/10.1186/s40529-014-0054-6

Marchand CH, Vanacker H, Collin V, Issakidis-Bourguet E, Maréchal PL, Decottignies P. Thioredoxin targets in Arabidopsis roots. Proteomics 2010;10:2418–2428. https://doi.org/10.1002/pmic.200900835

Marty L, Bausewein D, Müller C, Bangash SAK, Moseler A, Schwarzländer M, et al. Arabidopsis glutathione reductase 2 is indispensable in plastids, while mitochondrial glutathione is safeguarded by additional reduction and transport systems. New Phytol 2019;224:1569–1584. https://doi.org/10.1111/nph.16086

McKersie BD, Bowley SR, Harjanto E, Leprince O. Water-deficit tolerance and field performance of transgenic alfalfa overexpressing superoxide dismutase. Plant Physiol 1996;111:1177–1181. https://doi.org/10.1104/pp.111.4.1177

Meyer AJ, Dreyer A, Ugalde JM, Feitosa-Araujo E, Dietz KJ, Schwarzländer M. Shifting paradigms and novel players in Cys-based redox regulation and ROS signaling in plants – And where to go next. Biol Chem 2021;402:399–423. https://doi.org/10.1515/hsz-2020-0291

Mika A, Lüthje S. Properties of guaiacol peroxidase activities isolated from corn root plasma membranes. Plant Physiol 2003;132:1489–1498. https://doi.org/10.1104/pp.103.020396

Mittler R, Vanderauwera S, Gollery M, Van Breusegem F. Reactive oxygen gene network of plants. Trends Plant Sci 2004;9:490–498. https://doi.org/10.1016/j.tplants.2004.08.009

Mittler R, Zandalinas SI, Fichman Y, Van Breusegem F. Reactive oxygen species signalling in plant stress responses. Nat Rev

Mol Cell Biol 2022;23:663–679. https://doi.org/10.1038/s41580-022-00499-2

Møller IM, Jensen PE, Hansson A. Oxidative modifications to cellular components in plants. Annu Rev Plant Biol 2007;58:459–481. https://doi.org/10.1146/annurev.arplant.58.032806.103946

Mujer CV, Mendoza EMT, Ramirez DA. Coconut peroxidase isoenzymes: Isolation, partial purification and physicochemical properties. Phytochemistry 1983;22:1335–1340. https://doi.org/10.1016/S0031-9422(00)84011-X

Mullen RT, Trelease RN. The sorting signals for peroxisomal membrane-bound ascorbate peroxidase are within its C-terminal tail. J Biol Chem 2000;275:16337–16344. https://doi.org/10.1074/jbc.M001266200

Negi NP, Shrivastava DC, Sharma V, Sarin NB. Overexpression of CuZnSOD from *Arachis hypogaea* alleviates salinity and drought stress in tobacco. Plant Cell Rep 2015;34:1109–1126. https://doi.org/10.1007/s00299-015-1770-4

Oshino N, Jamieson D, Sugano T, Chance B. Optical measurement of the catalase-hydrogen peroxide intermediate (compound I) in the liver of anaesthetized rats and its implication to hydrogen peroxide production *in situ*. Biochem J 1975;146:67–77. https://doi.org/10.1042/bj1460067

Pandey VP, Dwivedi UN. Purification and characterization of peroxidase from *Leucaena leucocephala*, a tree legume. J Mol Catal B Enzym 2011;68:168–173. https://doi.org/10.1016/j.molcatb.2010.10.006

Pilon M, Ravet K, Tapken W. The biogenesis and physiological function of chloroplast superoxide dismutases. Biochim Biophys Acta – Bioenerg 2011;1807:989–998. https://doi.org/10.1016/j.bbabio.2010.11.002

Qi Q, Yanyan D, Yuanlin L, Kunzhi L, Huini X, Xudong S. Overexpression of *SlMDHAR* in transgenic tobacco increased salt stress tolerance involving *S*-nitrosylation regulation. Plant Sci 2020;299:110609. https://doi.org/10.1016/j.plantsci.2020.110609

Rajput VD, HarishSingh RK, Verma KK, Sharma L, Quiroz-Figueroa FR, et al. Recent developments in enzymatic antioxidant defence mechanism in plants with special reference to abiotic stress. Biology (Basel) 2021;10:267. https://doi.org/10.3390/biology10040267

Rao MV, Paliyath G, Ormrod DP. Ultraviolet-B- and ozone-induced biochemical changes in antioxidant enzymes of Arabidopsis thaliana. Plant Physiol 1996;110:125–136. https://doi.org/10.1104/pp.110.1.125

Regalado C, Arvizu OP, García-almendarez BE, Whitaker JR. Purification and properties of two acid peroxidases from brussels sprouts (*Brassica oleraceae* L.). J Food Biochem 1999;23:435–450. https://doi.org/10.1111/j.1745-4514.1999.tb00029.x

Riyazuddin R, Bela K, Horváth E, Rigó G, Gallé Á, Szabados L, et al. Overexpression of the Arabidopsis glutathione peroxidase-like 5 gene (AtGPXL5) resulted in altered plant development and redox status. Environ Exp Bot 2019;167:103849. https://doi.org/10.1016/j.envexpbot.2019.103849

Rodríguez-Ruiz M, González-Gordo S, Cañas A, Campos MJ, Paradela A, Corpas FJ, et al. Sweet pepper (*Capsicum annuum* L.) fruits contain an atypical peroxisomal catalase that is modulated by reactive oxygen and nitrogen species. Antioxidants 2019;8:374. https://doi.org/10.3390/antiox8090374

Sabeh F, Wright T, Norton SJ. Purification and characterization of a glutathione peroxidase from the Aloe vera plant. Enzym Protein 1993;47:92–98. https://doi.org/10.1159/000468662

Sachdev S, Ansari SA, Ansari MI, Fujita M, Hasanuzzaman M. Abiotic stress and reactive oxygen species: Generation, signaling, and defense mechanisms. Antioxidants 2021;10:277. https://doi.org/10.3390/antiox10020277

Saha P, Mukherjee A, Biswas AK. Modulation of NaCl induced DNA damage and oxidative stress in mungbean by pretreatment with sublethal dose. Biol Plant 2015;59:139–146. https://doi.org/10.1007/s10535-014-0460-3

Savchenko T, Tikhonov K. Oxidative stress-induced alteration of plant central metabolism. Life 2021;11:304. https://doi.org/10.3390/life11040304

Saxena SC, Salvi P, Kamble NU, Joshi PK, Majee M, Arora S. Ectopic overexpression of cytosolic ascorbate peroxidase gene (Apx1) improves salinity stress tolerance in *Brassica juncea* by strengthening antioxidative defense mechanism. Acta Physiol Plant 2020;42:45. https://doi.org/10.1007/s11738-020-3032-5

Shi G, Fu J, Rong L, Zhang P, Guo C, Xiao K. TaMIR1119, a miRNA family member of wheat (*Triticum aestivum*), is essential in the regulation of plant drought tolerance. J Integr Agric 2018;17:2369–2378. https://doi.org/10.1016/S2095-3119(17)61879-3

Simmons CR, Lafitte HR, Reimann KS, Brugière N, Roesler K, Albertsen MC, et al. Successes and insights of an industry biotech program to enhance maize agronomic traits. Plant Sci 2021;307:110899. https://doi.org/10.1016/j.plantsci.2021.110899

Simova-Stoilova L, Vaseva I, Grigorova B, Demirevska K, Feller U. Proteolytic activity and cysteine protease expression in wheat leaves under severe soil drought and recovery. Plant Physiol Biochem 2010;48:200–206. https://doi.org/10.1016/j.plaphy.2009.11.003

Suzuki T, Honda Y, Mukasa Y, Kim S. Characterization of peroxidase in buckwheat seed. Phytochemistry 2006;67:219–224. https://doi.org/10.1016/j.phytochem.2005.11.014

Trivedi DK, Gill SS, Yadav S, Tuteja N. Genome-wide analysis of glutathione reductase (GR) genes from rice and Arabidopsis. Plant Signal Behav 2013;8:e23021. https://doi.org/10.4161/psb.23021

Veljović Jovanović S, Kukavica B, Vidović M, Morina F, Menckhoff L. Class III Peroxidases: Functions, Localization and Redox Regulation of Isoenzymes. Antioxidants Antioxid. Enzym. High, Springer International Publishing; 2018, p. 269–300. https://doi.org/10.1007/978-3-319-75088-0_13

Wang M, Zhao X, Xiao Z, Yin X, Xing T, Xia G. A wheat superoxide dismutase gene TaSOD2 enhances salt resistance through modulating redox homeostasis by promoting NADPH oxidase activity. Plant Mol Biol 2016;91:115–130. https://doi.org/10.1007/s11103-016-0446-y

Wang W, Cheng Y, Chen D, Liu D, Hu M, Dong J, et al. The catalase gene family in cotton: Genome-wide characterization and bioinformatics analysis. Cells 2019;8:86. https://doi.org/10.3390/cells8020086

Wang Y, Wisniewski M, Meilan R, Cui M, Webb R, Fuchigami L. Overexpression of cytosolic ascorbate peroxidase in tomato confers tolerance to chilling and salt stress. J Am Soc Hortic Sci 2005;130:167–173. https://doi.org/10.21273/JASHS.130.2.167

Wang Y, Ying Y, Chen J, Wang X. Transgenic *Arabidopsis* overexpressing Mn-SOD enhanced salt-tolerance. Plant Sci 2004;167:671–677. https://doi.org/10.1016/j.plantsci.2004.03.032

Waszczak C, Carmody M, Kangasjärvi J. Reactive oxygen species in plant signaling. Annu Rev Plant Biol 2018;69:209–236. https://doi.org/10.1146/annurev-arplant-042817-040322

Xu J, Duan X, Yang J, Beeching JR, Zhang P. Coupled expression of Cu/Zn-superoxide dismutase and catalase in cassava improves tolerance against cold and drought stresses. Plant Signal Behav 2013;8:e24525. https://doi.org/10.4161/psb.24525

Yamaguchi K, Mori H, Nishimura M. A novel isoenzyme of ascorbate peroxidase localized on glyoxysomal and leaf peroxisomal membranes in pumpkin. Plant Cell Physiol 1995;36:1157–1162. https://doi.org/10.1093/oxfordjournals.pcp.a078862

Yang Y, Tilman D, Jin Z, Smith P, Barrett CB, Zhu YG, et al. Climate change exacerbates the environmental impacts of agriculture. Science 2024;385. https://doi.org/10.1126/science.adn3747

Yoshida K, Noguchi K, Motohashi K, Hisabori T. Systematic exploration of thioredoxin target proteins in plant mitochondria. Plant Cell Physiol 2013;54:875–892. https://doi.org/10.1093/pcp/pct037

Yoshimura K, Yabuta Y, Ishikawa T, Shigeoka S. Identification of a *cis* element for tissue-specific alternative splicing of chloroplast ascorbate peroxidase pre-mRNA in higher plants. J Biol Chem 2002;277:40623–40632. https://doi.org/10.1074/jbc.M201531200

Zafar SA, Hameed A, Ashraf M, Khan AS, Qamar Z, Li X, et al. Agronomic, physiological and molecular characterisation of rice mutants revealed the key role of reactive oxygen species and catalase in high-temperature stress tolerance. Funct Plant Biol 2020;47:440. https://doi.org/10.1071/FP19246

Zhang S, Li C, Ren H, Zhao T, Li Q, Wang S, et al. BAK1 mediates light intensity to phosphorylate and activate catalases to regulate plant growth and development. Int J Mol Sci 2020a;21:1437. https://doi.org/10.3390/ijms21041437

Zhang W, Cao J, Fan X, Jiang W. Applications of nitric oxide and melatonin in improving postharvest fruit quality and the separate and crosstalk biochemical mechanisms. Trends Food Sci Technol 2020b;99:531–541. https://doi.org/10.1016/j.tifs.2020.03.024

Zhou YB, Liu C, Tang DY, Yan L, Wang D, Yang YZ, et al. The receptor-like cytoplasmic kinase STRK1 phosphorylates and activates CatC, thereby regulating H_2O_2 homeostasis and improving salt tolerance in rice. Plant Cell 2018;30:1100–1118. https://doi.org/10.1105/tpc.17.01000

Zlatev ZS, Lidon FC, Ramalho JC, Yordanov IT. Comparison of resistance to drought of three bean cultivars. Biol Plant 2006;50:389–394. https://doi.org/10.1007/s10535-006-0054-9

Zou JJ, Li XD, Ratnasekera D, Wang C, Liu WX, Song LF, et al. Arabidopsis CALCIUM-DEPENDENT PROTEIN KINASE8 and CATALASE3 function in abscisic acid-mediated signaling and H_2O_2 homeostasis in stomatal guard cells under drought stress. Plant Cell 2015;27:1445–1460. https://doi.org/10.1105/tpc.15.00144

10 Role of Plant Enzymes in Plant–Microbial Interaction and Symbiosis

Shifa Shaffique and In-Jung Lee

10.1 INTRODUCTION

Plant lives closely with a diverse range of soil microbes, such as fungi, bacteria, and parasites. Communication between microbes and plants is maintained via complicated molecular interactions. For the mutual advantage of both the kingdom, plants, and microbes, they might have a symbiotic or non-symbiotic relationship (Harvey et al., 2009; Kawaguchi & Minamisawa, 2010; Larimer et al., 2010). In symbiotic association, their relationship confers the induction of secondary metabolites such as sugar, organic acid, amino acid, and certain phytohormones that tend to enhance the growth and production of plants (Barrow et al., 2008; Kawaguchi & Minamisawa, 2010; Porter & Sachs, 2020). While in a non-symbiotic relationship, they are pathogenic in nature and cause disease to plants and reducing plant production and yield (Lamichhane & Venturi, 2015; Wheeler & Luke, 1963).

Plant–microbial interaction is known as cross-kingdom communication, where the two different kingdoms communicate with each other in an interactive environment (Duan et al., 2024; Rodriguez & Roossinck, 2012). In symbiotic interplay, beneficial microbes play an important role in plant growth and production as shown in Figure 10.1. They enhance the nutrient uptake and also help plants to tolerate certain kinds of stress conditions (Duan et al., 2024; Lee et al., 2022). The beneficial microbes are known as plant growth-promoting rhizobacteria (PGPRs). These microbes have the potential to process certain kinds of distinctive features, such as the production of exopolysaccharide (EPS), indole acetic acid (IAA), siderophore, and

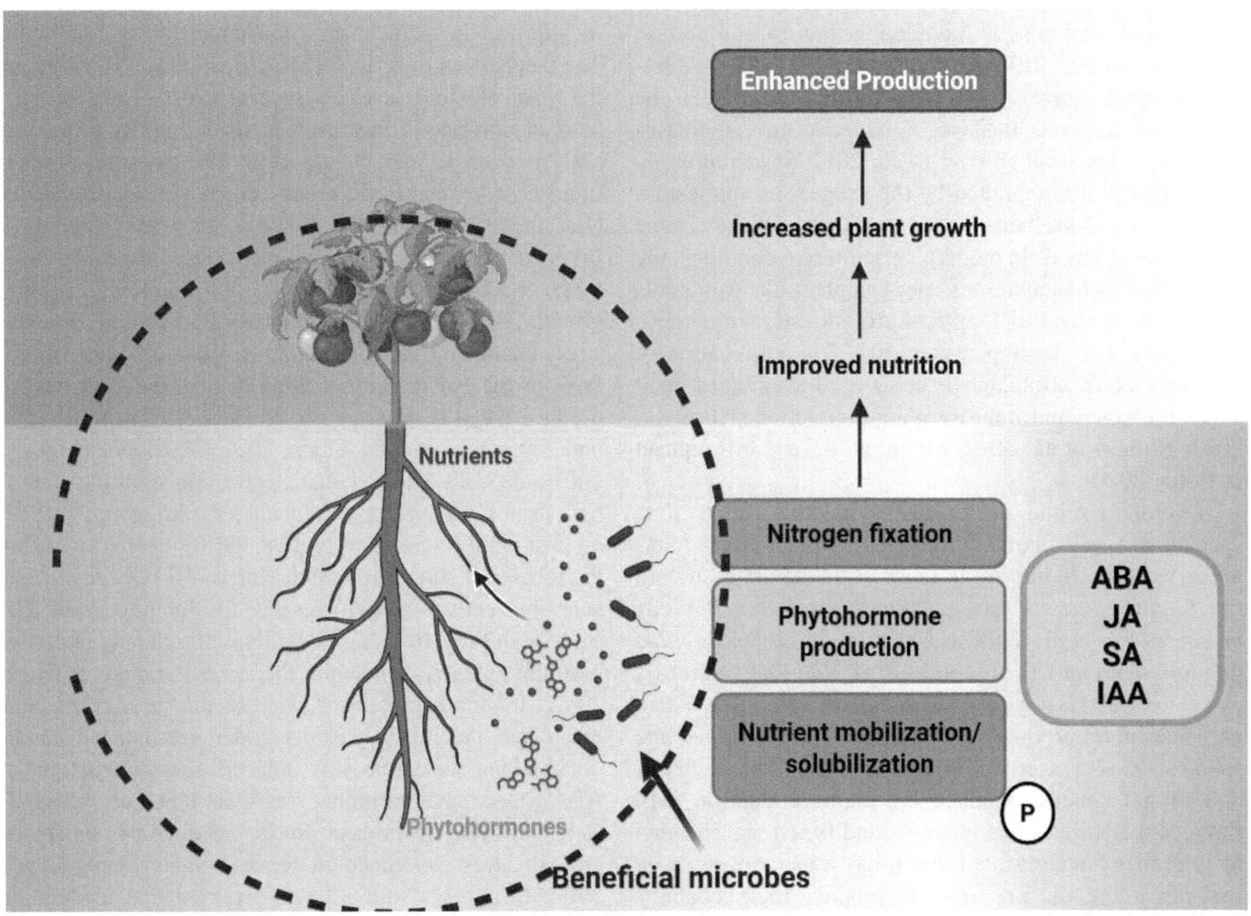

FIGURE 10.1 Plant–microbial interaction enhanced the plant growth and production via nitrogen fixation, production of phytohormones, and nutrient mobilization.

DOI: 10.1201/9781003669920-10

have a tendency to solubilize the phosphate. Interestingly, most microbes have the potential to produce phytohormones such as IAA, abscisic acid (ABA), jasmonic acid (JA) and salicylic acid (SA), and melatonin (MT) (Arshad & Frankenberger, 1991; Baca & Elmerich, 2007; Costacurta & Vanderleyden, 1995). These phytohormones make them unique in enhancing the plant stress tolerance and improving the plant production (Shaffique et al., 2022; Shaffique et al., 2023). Several studies are in favor that microbes tend to improve plant growth and yield in several aspects. Moreover, it is also reported that inoculation of these microbes also enhances the plant production and works as a bio-stimulant in enhancing the plant stress tolerance (De Zelicourt et al., 2013; Misu et al., 2025). The beneficial microbes induce a symbiotic relationship with plants and stimulate the defense-related components, such as to support the induced systemic resistance (ISR) (Jakubowska et al., 2025). This causes the upregulation of antioxidant defense activities and leads to a reduction in the oxidative stress in plants (González et al., 2024; Singh et al., 2021). In plants with exposure to external stimuli, there is production of reactive oxygen species (ROS). When plants are in contact with external stimuli, it enhances the oxidation stress in plants; however, plant tries to fight the adverse conditions in several ways, such as plant activation of endogenous defense activity, phytohormones, and genetic modulation, but up to a certain limit (Gull et al., 2019; Shaffique et al., 2024). If the duration of stress increases and the plant fails to combat the stress, then in plants, the level of oxidative stress increases to an alarming level (Baxter et al., 2014; Sewelam et al., 2016). If it is not regulated by the exogenous application of a bio-stimulant, then plants may die or may show some symptoms of stress. In modern agriculture technology, the application of beneficial microbes as plant bio-stimulants has revolutionized the agriculture industry in several developing and developed countries. The application of microbes as bio-stimulants is being used to enhance plant stress tolerance and improve plant production (Ali et al., 2022; Kumari et al., 2022; Melini et al., 2023; Rouphael & Colla, 2020).

Several microbial species, such as bacteria, the *Bacillus* species, are well known for enhancing plant stress tolerance. Various studies are in favor of the application ooid the *Bacillus strains* such as *Bacillus subtilis*, *Bacillus mycoides* (Yi et al., 2018, Hashem et al., 2019), *Bacillus amyloliquefaciens* (Liu et al., 2016), *Bacillus velezensis* (Sun et al., 2022) not only tolerate plant stress, but it also enhances plant production. Also, with this *Pseudomonas* species (Zainurin et al., 2025), several other bacterial and fungal species enhance the plant production with the recent advances that genomic and functional insights of microbes have revealed that it has a metabolic region that has a specific precursor to produce the secondary metabolites have advances in scientific studies (Shaffique et al., 2025).

10.2 ROLE OF ENZYMES IN PLANT–MICROBIAL INTERACTION AND SYMBIOSIS

When beneficial microbes come across the plant to start the symbiotic relationship, the plant cell uses receptors like kinases (RLKs) and to identify or detect the microbial molecules (Saijo et al., 2018) (Antolín-Llovera et al., 2012; Bhat & Haney, 2025; Zipfel, 2014). The recognition stimulates signal transduction enzymes such as mitogen-activated protein kinases (MAPKs) and calcium-dependent protein kinases (CDPKs). These enzymes modulate and activate the expression pattern of genes to transmit the message to the nucleus to induce he plant–microbial symbiotic interaction (Romeis & Herde, 2014; Van Gerrewey & Chung, 2024). Several other enzymes also participate in the symbiotic relationship of plant–microbial interaction: these are phosphatases, cellulases, phytases, pectinases, ACC-deaminase, nitrate reductases (NR), and glutamine synthetases (GS). The details are given below and in Figure 10.2.

10.3 ACC-DEAMINASE

The 1,aminocyclopropane-1-carboxylate (ACC) deaminase is an important enzyme that is produced by microbes, specifically the PGPRs (Glick, Todorovic, et al., 2007). It tends to improve the plant production under stress conditions. The mechanism of action of ACC-deaminase is to regulate the plant ethylene level. In several stress conditions, the level of ethylene is increased in several stress conditions such as drought and metal stress. The potential production of ACC-deaminase breaks down ethylene precursor (1-aminocyclopropane-1-carboxylic acid) and converts it into α-ketobutyrate and ammonia (Gamalero et al., 2023; Singh et al., 2015). Interestingly, it also improves the plant growth and helps the plant–microbial communication and stress tolerance (Naing et al., 2021). Several studies are in favor of the potential application of microbes that tend to produce the ACC-deaminase, enhancing plant production under stress conditions (Glick, 2014; Glick, Cheng, et al., 2007). A study showed that application of three microbial strains, *Enterobacter cloacae*, *Pseudomonas putida*, on post-flood recovery in tomato, the results declared that microbes that show expression pattern of ACC-deaminase genes present substantial tolerance for flooding stress. The study is in favor that ACC-deaminase producing microbes have the potential to tolerate the stress (Grichko & Glick, 2001). In another study, *Leclercia adecarboxylata* was inoculated into tomato plants under salt-induced conditions. When salt stress was induced, the upregulation of ABA (stress phytohormone) was seen; however, the application of ACC-deaminase producing microbes enhanced the salt stress tolerance in tomato plants (Kang et al., 2019). In another study, it was reported that inoculation of endophytic fungi that have the potential to produce the ACC-deaminase enhances the drought stress tolerance and

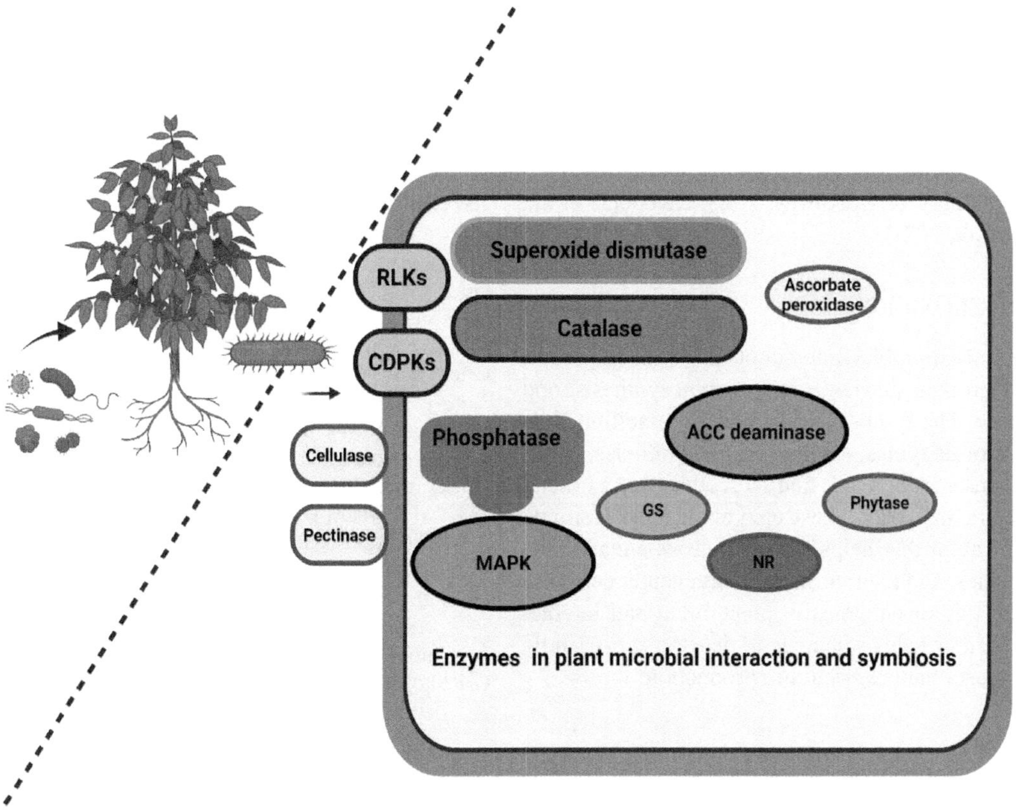

FIGURE 10.2 Role of enzymes in plant–microbial interaction and symbiotic interplay. The figure represents that the application of beneficial microbes activates a series of enzymes to facilitate communication.

reduces the plant stress markers such as hydrogen peroxide and ethylene in *Moringa oleifera* (Rehman et al., 2022).

10.4 PHOSPHATASES

Phosphorus (P) is an important nutrient that improves plant growth and production, as it is important for several important functions such as nitrogen fixation, photosynthesis, and protein and nucleic acid formation. In the natural environment, the P is available in the form of the organic form; however, the phosphatase enzyme converts the organic P into the inorganic P (Pi), which plants can absorb easily. This enzyme is produced by plants and beneficial microbes that enhance the availability of P for plants in nutrient-deficient environments (Nannipieri et al., 2010). There are four kinds of phosphatases: alkaline phosphatases, acid phosphatases, phytase, and phosphodiesterase. The phytase converts the phytic acid into the organic P form, while phosphodiesterase hydrolyses the DNA and RNA, and phospholipid to release the P in stressful conditions. When plants are in a symbiotic relationship with microbes, they enhance the phosphatase activity to ensure the availability of P under stressful circumstances (Yadav & Tarafdar, 2003). Beneficial microbes play a significant role in P solubilization, P–plant interaction, and biogeochemical cycling of P in the plant and soil system (Ingle & Padole, 2017; Li et al., 2021; Rawat et al., 2021). Beneficial microbes, along with enhancing the P availability, also improve he nutrient uptake and biosynthesis of phytohormones with synergistic interaction of plants (Awasthi et al., 2011).

10.5 CELLULASE AND PECTINASE

Cellulase is a hydrolytic enzyme that is secreted in plant–microbial interactions by microbes and enzymes. As microbes has the adhesive potential and ability to form biofilm, so by secreting these enzymes microbes secrete this enzyme and establish colonization without harming the host. However, some pathogenic microorganisms also produce, so the nature of the relationship, either symbiotic or non-symbiotic, depends upon the nature of the microbe (Augimeri et al., 2015; Lu et al., 2006). Pectinase is also a hydrolytic enzyme that is important in plant–microbial interactions. Beneficial microbes produce the optimal amount of pectinase to penetrate the root surface to establish the symbiotic relationship between plant and microbe (Prade et al., 1999).

10.6 ANTIOXIDANT ENZYMATIC MOLECULES

During plant–microbial interaction, the plant cell may experience a change in redox state. The modulation in redox state causes the production of ROS; however, the optimal amount of ROS acts as a signaling molecule, while the overproduction of ROS causes stress in plants. However, to deal with this situation, plant produces the antioxidant defense enzymes such as superoxide dismutase SOD, catalase CAT ascorbate peroxidase APX. In symbiotic plant–microbial interaction, they act as a signaling molecule to overcome the stress conditions and provide symbiotic stability (Choudhary & Upadhyay, 2023; Helepciuc et al., 2014; Singh et al., 2021).

10.7 CONCLUSION

Symbiotic plant–microbial interaction plays an important role in plant growth, nutrient uptake, photosynthesis, and stress tolerance. The symbiotic relationship is facilitated by several kinds of enzymes, such as ACC-deaminase, cellulase, phosphatase, SOD, CAT, and APX, that play a crucial role in symbiotic interplay. These enzymatic plays facilitate the communication that helps plants to grow, enhance the defense activities, and tolerate stress. This chapter certainly addresses the role of enzymes in plant–microbial interaction, which further helps scientists to develop eco-friendly solutions for sustainable agriculture production.

REFERENCES

Ali, S., Moon, Y.-S., Hamayun, M., Khan, M. A., Bibi, K., & Lee, I.-J. (2022). Pragmatic role of microbial plant biostimulants in abiotic stress relief in crop plants. *Journal of Plant Interactions*, *17*(1), 705–718.

Antolín-Llovera, M., Ried, M. K., Binder, A., & Parniske, M. (2012). Receptor kinase signaling pathways in plant-microbe interactions. *Annual Review of Phytopathology*, *50*(1), 451–473.

Arshad, M., & Frankenberger, W., Jr. (1991). Microbial production of plant hormones. *Plant and Soil*, *133*(1), 1–8.

Augimeri, R. V., Varley, A. J., & Strap, J. L. (2015). Establishing a role for bacterial cellulose in environmental interactions: Lessons learned from diverse biofilm-producing *Proteobacteria*. *Frontiers in Microbiology*, *6*, 1282.

Awasthi, R., Tewari, R., & Nayyar, H. (2011). Synergy between plants and P-solubilizing microbes in soils: Effects on growth and physiology of crops. *International Research Journal of Microbiology*, *2*(12), 484–503.

Baca, B., & Elmerich, C. (2007). Microbial production of plant hormones. In: *Associative and endophytic nitrogen-fixing bacteria and cyanobacterial associations* (pp. 113–143). Springer.

Barrow, J. R., Lucero, M. E., Reyes-Vera, I., & Havstad, K. M. (2008). Do symbiotic microbes have a role in regulating plant performance and response to stress? *Communicative & Integrative Biology*, *1*(1), 69–73.

Baxter, A., Mittler, R., & Suzuki, N. (2014). ROS as key players in plant stress signalling. *Journal of Experimental Botany*, *65*(5), 1229–1240.

Bhat, A., & Haney, C. H. (2025). The role of plant receptor-like kinases in sensing extrinsic and host-derived signals and shaping the microbiome. *Cell Host & Microbe*, *33*(8), 1233–1240.

Choudhary, P., & Upadhyay, P. (2023). Relevance of the antioxidative mechanism during plant-microbe interaction. In: *Plant-microbe interaction-recent advances in molecular and biochemical approaches* (pp. 123–140). Elsevier.

Costacurta, A., & Vanderleyden, J. (1995). Synthesis of phytohormones by plant-associated bacteria. *Critical Reviews in Microbiology*, *21*(1), 1–18.

De Zelicourt, A., Al-Yousif, M., & Hirt, H. (2013). Rhizosphere microbes as essential partners for plant stress tolerance. *Molecular Plant*, *6*(2), 242–245.

Duan, S., Feng, G., Limpens, E., Bonfante, P., Xie, X., & Zhang, L. (2024). Cross-kingdom nutrient exchange in the plant–arbuscular mycorrhizal fungus–bacterium continuum. *Nature Reviews Microbiology*, *22*(12), 773–790.

Gamalero, E., Lingua, G., & Glick, B. R. (2023). Ethylene, ACC, and the plant growth-promoting enzyme ACC deaminase. *Biology*, *12*(8), 1043.

Glick, B. R. (2014). Bacteria with ACC deaminase can promote plant growth and help to feed the world. *Microbiological Research*, *169*(1), 30–39.

Glick, B. R., Cheng, Z., Czarny, J., & Duan, J. (2007). Promotion of plant growth by ACC deaminase-producing soil bacteria. In: *New perspectives and approaches in plant growth-promoting rhizobacteria research* (329–339). Springer.

Glick, B. R., Todorovic, B., Czarny, J., Cheng, Z., Duan, J., & McConkey, B. (2007). Promotion of plant growth by bacterial ACC deaminase. *Critical Reviews in Plant Sciences*, *26*(5-6), 227–242.

González, M.-C., Roitsch, T., & Pandey, C. (2024). Antioxidant responses and redox regulation within plant-beneficial microbe interaction. *Antioxidants*, *13*(12), 1553.

Grichko, V. P., & Glick, B. R. (2001). Amelioration of flooding stress by ACC deaminase-containingplant growth-promoting bacteria. *Plant Physiology and Biochemistry*, *39*(1), 11–17.

Gull, A., Lone, A. A., & Wani, N. U. I. (2019). Biotic and abiotic stresses in plants. In *Abiotic and biotic stress in plants*. IntechOpen.

Harvey, P., Warren, R., & Wakelin, S. (2009). Potential to improve root access to phosphorus: The role of non-symbiotic microbial inoculants in the rhizosphere. *Crop and Pasture Science*, *60*(2), 144–151.

Hashem, A., Tabassum, B., & Abd_Allah, E. F. (2019). *Bacillus subtilis*: A plant-growth promoting rhizobacterium that also impacts biotic stress. *Saudi Journal of Biological Sciences*, *26*(6), 1291–1297.

Helepciuc, F. E., Mitoi, M. E., Manole-Paunescu, A., Aldea, F., Brezeanu, A., & Cornea, C. P. (2014). Induction of plant antioxidant system by interaction with beneficial and/or pathogenic microorganisms. *Romanian Biotechnological Letters*, *19*(3), 9366–9375.

Ingle, K. P., & Padole, D. A. (2017). Phosphate solubilizing microbes: An overview. *International Journal of Current Microbiology and Applied Sciences*, *6*(1), 844–852.

Jakubowska, Z., Gradowski, M., & Dobrzyński, J. (2025). Role of plant growth-promoting bacteria (PGPB) in enhancing phenolic compounds biosynthesis and its relevance to abiotic stress tolerance in plants: A review. *Antonie van Leeuwenhoek*, *118*(9), 1–17.

Kang, S.-M., Shahzad, R., Bilal, S., Khan, A. L., Park, Y.-G., Lee, K.-E., Asaf, S., Khan, M. A., & Lee, I.-J. (2019). Indole-3-acetic-acid and ACC deaminase producing *Leclercia adecarboxylata* MO1 improves *Solanum lycopersicum* L. growth and salinity stress tolerance by endogenous secondary metabolites regulation. *BMC Microbiology*, *19*(1), 80. https://doi.org/10.1186/s12866-019-1450-6

Kawaguchi, M., & Minamisawa, K. (2010). Plant–microbe communications for symbiosis. *Plant and Cell Physiology*, *51*(9), 1377–1380.

Kumari, M., Swarupa, P., Kesari, K. K., & Kumar, A. (2022). Microbial inoculants as plant biostimulants: A review on risk status. *Life*, *13*(1), 12.

Lamichhane, J. R., & Venturi, V. (2015). Synergisms between microbial pathogens in plant disease complexes: A growing trend. *Frontiers in Plant Science*, *6*, 385.

Larimer, A. L., Bever, J. D., & Clay, K. (2010). The interactive effects of plant microbial symbionts: A review and meta-analysis. *Symbiosis*, *51*(2), 139–148.

Lee, K. K., Kim, H., & Lee, Y.-H. (2022). Cross-kingdom co-occurrence networks in the plant microbiome: Importance and ecological interpretations. *Frontiers in Microbiology*, *13*, 953300.

Li, J. t., Lu, J. l., Wang, H. y., Fang, Z., Wang, X. j., Feng, S. w., Wang, Z., Yuan, T., Zhang, S. c.,, & Ou, S. n. (2021). A comprehensive synthesis unveils the mysteries of phosphate-solubilizing microbes. *Biological Reviews*, *96*(6), 2771–2793.

Liu, Y., Chen, L., Zhang, N., Li, Z., Zhang, G., Xu, Y., Shen, Q., & Zhang, R. (2016). Plant-microbe communication enhances auxin biosynthesis by a root-associated bacterium, *Bacillus amyloliquefaciens* SQR9. *Molecular Plant-Microbe Interactions*, *29*(4), 324–330.

Lu, Y., Zhang, Y.-H. P., & Lynd, L. R. (2006). Enzyme–microbe synergy during cellulose hydrolysis by *Clostridium thermocellum*. *Proceedings of the National Academy of Sciences*, *103*(44), 16165–16169.

Melini, F., Melini, V., Luziatelli, F., Abou Jaoudé, R., Ficca, A. G., & Ruzzi, M. (2023). Effect of microbial plant biostimulants on fruit and vegetable quality: Current research lines and future perspectives. *Frontiers in Plant Science*, *14*, 1251544.

Misu, I. J., Kayess, M. O., Siddiqui, M. N., Gupta, D. R., Islam, M. N., & Islam, T. (2025). Microbiome engineering for sustainable rice production: Strategies for biofertilization, stress tolerance, and climate resilience. *Microorganisms*, *13*(2), 233.

Naing, A. H., Maung, T. T., & Kim, C. K. (2021). The ACC deaminase-producing plant growth-promoting bacteria: Influences of bacterial strains and ACC deaminase activities in plant tolerance to abiotic stress. *Physiologia Plantarum*, *173*(4), 1992–2012.

Nannipieri, P., Giagnoni, L., Landi, L., & Renella, G. (2010). Role of phosphatase enzymes in soil. In: *Phosphorus in action: Biological processes in soil phosphorus cycling* (pp. 215–243). Springer.

Porter, S. S., & Sachs, J. L. (2020). Agriculture and the disruption of plant–microbial symbiosis. *Trends in Ecology & Evolution*, *35*(5), 426–439.

Prade, R. A., Zhan, D., Ayoubi, P., & Mort, A. J. (1999). Pectins, pectinases and plant-microbe interactions. *Biotechnology and Genetic Engineering Reviews*, *16*(1), 361–392.

Rawat, P., Das, S., Shankhdhar, D., & Shankhdhar, S. (2021). Phosphate-solubilizing microorganisms: Mechanism and their role in phosphate solubilization and uptake. *Journal of Soil Science and Plant Nutrition*, *21*(1), 49–68.

Rehman, B., Javed, J., Rauf, M., Khan, S. A., Arif, M., Hamayun, M., Gul, H., Khilji, S. A., Sajid, Z. A., & Kim, W.-C. (2022). ACC deaminase-producing endophytic fungal consortia promotes drought stress tolerance in *M. oleifera* by mitigating ethylene and H_2O_2. *Frontiers in Plant Science*, *13*, 967672.

Rodriguez, R. J., & Roossinck, M. (2012). Viruses, fungi and plants: Cross-Kingdom communication and mutualism. In: *Biocommunication of fungi* (pp. 219–227). Springer.

Romeis, T., & Herde, M. (2014). From local to global: CDPKs in systemic defense signaling upon microbial and herbivore attack. *Current Opinion in Plant Biology*, *20*, 1–10.

Rouphael, Y., & Colla, G. (2020). *Biostimulants in agriculture* (Vol. 11, pp. 40). Frontiers Media SA.

Saijo, Y., Loo, E. P. i., & Yasuda, S. (2018). Pattern recognition receptors and signaling in plant–microbe interactions. *The Plant Journal*, *93*(4), 592–613.

Sewelam, N., Kazan, K., & Schenk, P. M. (2016). Global plant stress signaling: Reactive oxygen species at the cross-road. *Frontiers in Plant Science*, *7*, 187.

Shaffique, S., Farooq, M., Adhikari, A., Khan, M. A., Kang, S. M., & Lee, I. J. (2025). Genomic and functional insights into *Bacillus cereus* SH-10 for mitigating lead stress in soybean. *Plant, Cell & Environment*, *48*.

Shaffique, S., Farooq, M., Kang, S.-M., & Lee, I.-J. (2024). Recent advances in biochemical reprogramming network under drought stress in soybean. *Journal of Soil Science and Plant Nutrition*, *24*(2), 1692–1703.

Shaffique, S., Imran, M., Wani, S. H., Pande, A., Rahim, W., Aaqil khan, M., Kang, S.-M., & Lee, I.-J. (2023). Role of plant growth-promoting rhizobacteria mitigating drought stress in maize. In: *Maize improvement: Current advances in yield, quality, and stress tolerance under changing climatic scenarios* (pp. 323–333). Springer.

Shaffique, S., Khan, M. A., Wani, S. H., Pande, A., Imran, M., Kang, S.-M., Rahim, W., Khan, S. A., Bhatta, D., & Kwon, E.-H. (2022). A review on the role of endophytes and plant growth promoting rhizobacteria in mitigating heat stress in plants. *Microorganisms*, *10*(7), 1286.

Singh, A. K., Kumar, S., & Sinha, T. (2021). Antioxidants in plant–microbe interaction. In: *Antioxidants in plant-microbe interaction* (pp. 3–20). Springer.

Singh, R. P., Shelke, G. M., Kumar, A., & Jha, P. N. (2015). Biochemistry and genetics of ACC deaminase: A weapon to "stress ethylene" produced in plants. *Frontiers in Microbiology*, *6*, 937.

Sun, X., Xu, Z., Xie, J., Hesselberg-Thomsen, V., Tan, T., Zheng, D., Strube, M. L., Dragoš, A., Shen, Q., & Zhang, R. (2022). *Bacillus velezensis* stimulates resident rhizosphere *Pseudomonas stutzeri* for plant health through metabolic interactions. *The ISME Journal*, *16*(3), 774–787.

Van Gerrewey, T., & Chung, H. S. (2024). MAPK cascades in plant microbiota structure and functioning. *Journal of Microbiology*, *62*(3), 231–248.

Wheeler, H., & Luke, H. (1963). Microbial toxins in plant disease. *Annual Reviews in Microbiology*, *17*(1), 223–242.

Yadav, R., & Tarafdar, J. (2003). Phytase and phosphatase producing fungi in arid and semi-arid soils and their efficiency in hydrolyzing different organic P compounds. *Soil Biology and Biochemistry*, *35*(6), 745–751.

Yi, Y., Li, Z., Song, C., & Kuipers, O. P. (2018). Exploring plant-microbe interactions of the rhizobacteria *Bacillus subtilis* and *Bacillus mycoides* by use of the CRISPR-Cas9 system. *Environmental Microbiology*, *20*(12), 4245–4260.

Zainurin, N., Imran, M., Shaffique, S., Khan, M. A., Kang, S.-M., Injamum-UL-Hoque, M., Das, A. K., Yun, B.-W., & Lee, I.-J. (2025). New insights into the synergistic interaction between *Pseudomonas qingdaonensis* NZ 1 and silicon to mitigate drought stress in rice. *Microorganisms*, *13*(5), 1046.

Zipfel, C. (2014). Plant pattern-recognition receptors. *Trends in Immunology*, *35*(7), 345–351.

11 Enzymatic Mechanisms of Plant Resistance to Pathogens and Herbivores

P. Saranraj, P. Sivasakthivelan, Alexander Machado Cardoso, B. Lokeshwari, and K. Gayathri

11.1 INTRODUCTION

Plants have evolved a sophisticated system of enzymatic defense against herbivores and pathogens and have the inducible types of resistance into enzyme defense mechanisms. These enzymes are important in identifying threats by the use of pattern recognition receptors (PRRs), transformation of the defense signal, and employment of direct and indirect countermeasures to minimize the damage and normalize homeostasis (Kurnik et al. 2015). The production and control of reactive oxygen species (ROS) is a significant function of these enzymes. This oxidative burst has a direct inhibitory effect on pathogens, but it also has an enhancing effect on physical barriers, cross-linking cell wall components and acts as a signaling molecule in the activation of other defense genes. This is linked with phytohormonal signaling, including salicylic acid (SA), jasmonic acid (JA), and ethylene (ET), which modulate the growth-defenses and form the systemic acquired resistance (SAR). Hydrolytic enzymes provide an exert direct anti-pathogen activity (García Bossi et al. 2020).

The chitinases break down chitin in fungal cell walls, as well as insect exoskeletons, interfering with the structure and releasing oligosaccharides, eliciting amplified immune responses. β-1,3-glucanases are applied to fungal glucans, and their synergistic effect with chitinases results in antifungal defense, which is often used by pathogen-derived effector proteins that serve as enzyme inhibitors. Protease inhibitors (PIs) are essential against the herbivores; they attach to digestive proteins in the insect gut, which inhibit protein digestion and cause developmental defects and larval death (Shao et al. 2013). Polyphenol oxidases (PPOs) and lipoxygenases (LOXs) are activated and produce reactive quinones and oxylipins, which are used to inhibit insect feeding and as interplant communication molecules. Additionally, the JA pathway serves as a worldwide controller of the anti-herbivore defense mechanism. Biosynthetic enzymes are essential to produce JA-isoleucine, which activates defense genes and produces toxic or antinutritive metabolites. The Brassicales' glucosinolate–myrosinase system is a pre-formed enzymatic defense (Bacete et al. 2017).

Myrosinase releases toxic isothiocyanates to deter herbivores by breaking down glucosinolates in tissue. Thioredoxins (TRXs) and glutaredoxins (GRXs) function as redox-regulating enzymes that regulate defense signals. They mitigate cysteine oxidative changes, which influence transcription factors and balance SA- and JA-mediated signals, optimizing defense responses to specific threats. Enzymes like benzoic acid 2-hydroxylase (BA2H), which produces SA, and JA-carboxyl methyltransferase (JMT), which volatilizes JA, help control resource allocation between defenses against biotrophic and necrotrophic pathogens and herbivores. Enzymatic responses represent a finely regulated circuit that recognizes pathogens/herbivores, transduces signals, and expresses hydrolytic/antimicrobial defenses (Aziz et al. 2023). These novel enzymes enable crop engineering for sustainability.

11.2 PLANT DEFENSE MECHANISMS

Plant defense systems are bipartite, consisting of both constitutive and inducible components. These include both inducible responses and the enzymatic basis for physical barriers, with the inducible response initiated by threats recognized by plant PRRs. This activates intracellular signaling pathways, including mitogen-activated protein kinases (MAPKs) and calcium-dependent protein kinases (CDPKs), leading to defense gene transcription (Walters et al. 2013). NADPH oxidases produce ROS as antimicrobials, and the synthesis of defense phytohormones such as SA, JA, and ET is regulated, enhancing responses to biotic threats. This complex system optimizes resource allocation and improves defense effectiveness.

11.2.1 Structural and Biochemical Defenses

The cell wall is the primary structural defense, a dynamic structure modulated by plant enzymes. It is triggered by the interactions with pathogen-associated molecular patterns (PAMPs) or tissue injury, and the enzymes of this type, i.e., peroxidases (PODs) and laccases, are activated. These enzymes promote oxidative cross-linking of phenolic compounds and polymerization of monolignols, thereby improving the mechanical integrity of the cell wall. The end product of this process is a lignified, hydrophobic, and antimicrobial matrix that gives the plant a strong resistance to invasion and degradation by pathogens, thus enabling

DOI: 10.1201/9781003669920-11

survival under stresses. Callose is deposited immediately at invasion sites by CalS (callose synthase), which enhances the cell wall strength and closes the plasmodesmata that suppress the formation of fungi as well as microorganisms. Thus, enzymatic processes form protective walls at the tissue level (Coomber et al. 2024). Suberin is a complex (heteropolymer) produced by enzymes in the bark and wounds, thereby covering the damaged tissues against pathogens as well as retaining reserves of water. These enzymatic activities are regulated to ensure that the native immunities of the plant are efficient as desired (Srivastava et al. 2024).

11.2.2 Role of Secondary Metabolites

Phenolics, terpenoids, and nitrogen-containing compounds are some of the secondary metabolites that are primarily used by plants as a defense against environmental hazards. These compounds are produced with the help of enzymes such as phenylalanine ammonia-lyase (PAL), which converts phenylalanine into cinnamic acid, a building block of flavonoids that are used as antioxidants (Espinoza et al. 2017). PAL isoforms and enzymes such as chalcone synthase increase resistance to pathogens such as bacteria and fungi. Terpene synthases (TPS) synthesize different terpenoids that may poison herbivores or may attract herbivore predators, such as (E)-b-caryophyllene in injured maize roots. Moreover, there are complex synthesis processes of nitrogen compounds, such as alkaloids and glucosinolates, where myrosinase in the Brassicaceae decomposes glucosinolates into toxic derivatives when attacked by herbivores, which strengthen the defense mechanism of the plant (Gale et al. 2019).

11.2.3 Enzyme-Mediated Responses in Plant Defense

Enzymes are essential in the plant defense against biotic factors, with chitinases and β-1,3-glucanases being some of the major enzymes expressed in the pathogenesis-related (PR) protein family. Chitinases are used in degrading the cell walls of fungi and the insect exoskeleton, and β-1,3-glucanases are used to degrade fungal glucan. This enzymatic activity leads to the discharge of the oligosaccharides, which boost the immune system of the plant. PIs interfere with the digestion of herbivores as they block the major digestive enzymes, which causes nutritional deficiency and possible developmental problems among insects (Kumari et al. 2024). Regulation of PIs by use of jasmonate signaling is triggered by herbivory, and transgenic approaches to improve certain inhibitors have increased pest resistance. Furthermore, oxidative stress leads to the quick generation of ROS, which are caused by respiratory burst oxidase homolog NADPH oxidases based on the superoxide radicals. ROS facilitates lignification, cross-linkage of cell wall proteins, and signal defense gene expression of the multifaceted contribution of enzymes to plant immunity (Xie et al. 2020).

11.3 PLANT–PATHOGEN INTERACTIONS

11.3.1 Mechanisms of Pathogen Invasion and Colonization

Cellulases and polygalacturonases are target enzymes that microbial pathogens use in breaking down the plant structural barriers and avoiding immune mechanisms called cell wall-degrading enzymes (CWDEs). CWDE is also used by plant-parasitic nematodes, which is synthesized by the gut symbiotic bacteria to mediate plant root invasion. The effectors play a critical role in the immunological defense mechanism of plants because they are delivered into the host cells by the pathogens. These include a great variety of proteins and phosphatases in the evolutionary process between the plant hosts and the produced enzymes of the pathogens. It is a complex process of interaction between plants and their diversity during evolution and responses to the biological strategies (Mishra et al. 2023; Zheng et al. 2023).

11.3.2 Plant Recognition of Pathogen-Associated Molecular Patterns (PAMPs)

The pathogens avoid pattern-triggered immunity (PTI) to cause effector-triggered immunity (ETI), which is a powerful defense mechanism in plants, with the help of effector proteins. It is composed of intracellular NLR receptors that sense the presence of pathogen effectors, resulting in the activation of a hypersensitive response (HR) that is characterized by programmed cell death (PCD) to trap biotrophic pathogens (Razmjou et al. 2019). Plants have a distinct set of cysteine proteins that are called the metacaspases, which also take part in the processing of important protein substrates. PAMPs are also an augmenter and maintain the presence of PTI enzymatic defenses, induce the expression of antimicrobial secondary metabolites, and systemic PR protein expression (chitinases and glucanases). The basis of this immunity is that the signaling hormone SA increases the local resistance, producing SAR, such that the plant could react to a defense response. ETI is a local, cell-guided, cell-bound, enzyme-dependent mechanism, which involves localized cell death and systemic priming to adapt to complex pathogen virulence.

11.3.3 Effector-Triggered Immunity (ETI)

The pathogens use effector proteins to circumvent PTI and induce ETI, an effective defense mechanism in plants. It consists of intracellular NLR receptors that sensitize pathogen effectors, leading to the activation of a HR, which is characterized by PCD to entrap biotrophic pathogens (Razmjou et al. 2019). Plants possess a separate group of cysteine proteins known as metacaspases, which are also involved in the processing of significant protein substrates. ETI also augments and sustains PTI enzymatic defenses, induces antimicrobial secondary metabolite expression, and systemic PR protein expression (chitinases and glucanases). This immunity is anchored on the fact that signaling hormone SA augments the local resistance and generates SAR, so that the plant may respond to a defense mechanism. ETI is a local, cell-bound, enzyme-dependent process that utilizes localized cell death, as well as systemic priming to cope with complex pathogen virulence.

11.4 PLANT–HERBIVORE INTERACTIONS

Plants and herbivores are engaged in a complex co-evolutionary relationship that extends beyond simple consumption. Plants have sophisticated inducible defense systems that activate precisely when attacked (Furuichi et al. 2012). This response relies on the plant's ability to recognize herbivore- and damage-associated molecular patterns (HAMPs) and reprogram them. Complex enzymatic cascades control signal perception, hormone biosynthesis, and defensive compound production. Figure 11.1 illustrates that higher nutrient levels increase plant toxins and biomass, boosting uninfected herbivores but lowering the parasite virulence with no net change in herbivore population.

11.4.1 Biochemical Signals in Plant–Herbivore Interactions

Plant defense mechanisms against herbivores depend on advanced perception systems that differentiate between herbivore attacks and non-biological damage, achieved through the detection of herbivore-associated molecular patterns (HAMPs) and damage-associated molecular patterns (DAMPs). Oral secretions from herbivores introduce chemical markers for herbivory when present in plants,

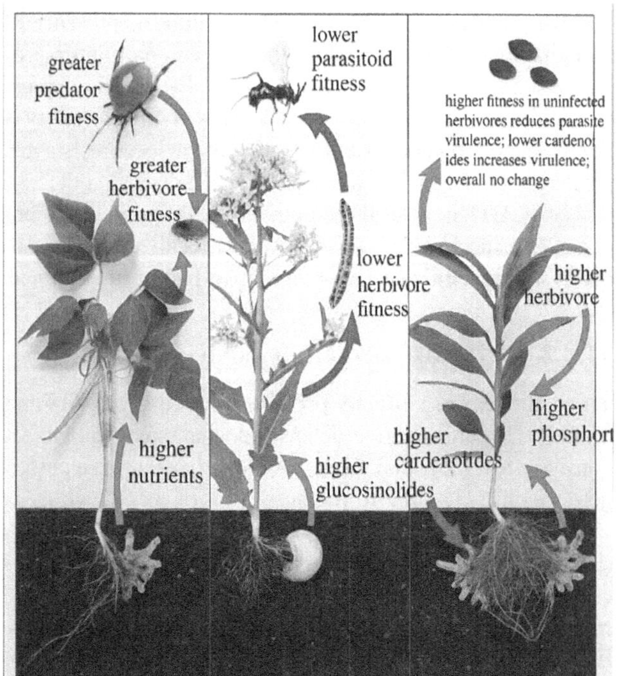

FIGURE 11.1 Effects of nutrient levels on Plant–Herbivore interactions.

while DAMPs signal from cellular damage. These patterns activate membrane-localized receptors, triggering a signaling cascade that involves NADPH oxidase and ROS, as well as increases in cytosolic calcium levels (Pusztahelyi 2018). The defense response is primarily regulated by JA, which interacts with SA pathways to the response, facilitating extensive activation of defensive enzymes and strategies through transcriptional strategies.

11.4.2 Induction of Defense Enzymes by Herbivores

The jasmonate (JA) signaling pathway is the primary regulatory module for the induction of defense enzymes by herbivory, triggering a cascade of transcription that upregulates defensive proteins (Ali et al. 2020). PIs, which lower insect digestive efficacy, and enzymes such as chitinases that target herbivores' gut mucosa. Herbivore attack turns on the biosynthesis of metabolites, such as lipoxygenase (LOX), which allows JA biosynthesis, generating a defense signaling feedback. Biosynthetic processes for secondary compounds such as glucosinolates and non-protein amino acids involve transcriptional activation, with defense depending on herbivore type and damaged plant tissue to enable resource allocation optimization in defense reactions.

11.4.3 Pathogen- and Herbivore-Induced Responses

Plant immunity involves complex interactions between defense signaling pathways, notably the antagonistic effects of SA and JA. JA triggers defenses against necrotrophic pathogens and herbivores, while SA targets biotrophic pathogens (Remick et al. 2023). This hormonal interplay influences the enzymatic response to threats; oxidative bursts against pathogens employ varied NADPH oxidases and peroxidases, and herbivory elevates enzymes like β-1,3-glucanases, which facilitate antifungal defense. Such dual responses can also help deter opportunistic pathogens exploiting plant injury (Geisler et al. 2000).

11.5 ENZYMATIC DEFENSE MECHANISMS

The plant's enzymatic defense system, which consists of proteins categorized by mechanism, is a direct and potent defense against biotic invaders. Oxidative enzymes create a toxic oxidative environment or strengthen structural barriers at the attack site. Peroxidases and PPOs produce ROS and cross-link cell wall polymers, respectively, weakening the invader and strengthening the plant's tissues. Another major category is hydrolytic enzymes, which directly degrade attacker structural and functional molecules. Chitinases target chitin in fungal cell walls and insect exoskeletons, proteases disrupt herbivore gut peritrophic matrix or microbial proteins, and β-1,3-glucanases hydrolyze glucan polymers in fungal cell walls (Zhao et al. 2024). The coordinated induction and deployment of these oxidative and hydrolytic enzymes form a biochemical front line that directly threatens pathogens and herbivores.

11.5.1 Enzymatic Roles in Plant Defense

Plant defense enzymes perform direct antibiosis, signal potentiation, and structural reinforcement. Enzymes that produce toxins against invaders cause direct antibiosis. Oxidase and peroxidases produce ROS with direct antimicrobial activity, and lytic enzymes like chitinases and glucanases degrade fungal cell walls and insect digestive systems' peritrophic matrix. Signal potentiation requires enzymes that synthesize defense phytohormones like JA or process endogenous danger signals. Pectin methylesterase (PME) demethylesterifies homogalacturonan to release oligogalacturonide (OGA) fragments, which act as DAMPs to boost systemic defense signaling (Oliveira et al. 2016). Structural reinforcement uses enzymes to cross-link cell wall proteins and phenolics, strengthening the apoplast, or to synthesize and deposit callose papillae at pathogen entry sites, creating a mechanical barrier. Enzymatic systems coordinate spatially and temporally to promote a complex, multi-layered plant immune response (Rashid and Chung 2017).

11.5.2 Oxidative Enzymes

The "oxidative burst," or controlled, localized production of ROS, is essential to the plant immune system. While ROS accumulation causes oxidative cellular damage, its spatially and temporally restricted generation is essential for ETI (Galindo-Trigo et al. 2020). ROS production directly damages invading pathogen cells by oxidizing cellular macromolecules, strengthens the plant cell wall by cross-linking structural components, and acts as a key secondary messenger to activate defense-related genes and downstream defense signaling. Figure 11.2 illustrates the classification of the oxidative enzymes.

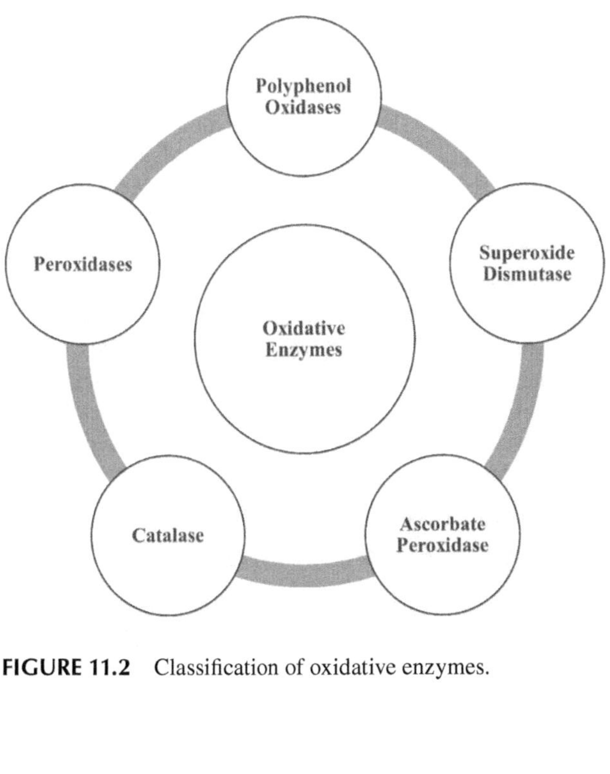

FIGURE 11.2 Classification of oxidative enzymes.

11.5.2.1 Peroxidases (POX)

Class III peroxidases (PRXs) are essential enzymes for plant defense, oxidizing substrates with hydrogen peroxide (H_2O_2). These enzymes aid in the oxidative cross-linking of cell wall polymers like lignins, extensins, and suberin monomers. The apoplastic boundary becomes stronger, preventing pathogen invasion and microbial enzymatic degradation (Engelsdorf et al. 2017). PRXs are essential to plant tissue oxidative defense and structural reinforcement. They biosynthesize antimicrobial phenolic compounds and maintain ROS homeostasis. This role prevents cellular damage and aids defensive signaling, improving the plant's pathogen-fighting ability.

11.5.2.2 Polyphenol Oxidases (PPOs)

PPOs assist plants in protecting themselves through the oxygenation of monophenols and o-diphenols to o-quinones. These compounds can react non-enzymatically with nucleophilic residues in order to alkylate important proteins and amino acids in the digestive systems of herbivores, depleting the nutritional value of plant tissues and leading to oxidative stress. This biochemical response offers a strong defense mechanism against insect herbivores, which is antinutritive and toxic. O-quinones polymerization forms melanin-like compounds at injury sites, and they can be used to seal wounds and keep off herbivores (Jomova et al. 2024). PPO responds quickly and, in response to damage caused to cell structures by herbivores, enables the enzyme, which is normally present in chloroplasts, to react with its vacuolar-based phenolic substrates. Such a dynamic interaction is a good example of complex and adaptive herbivore defenses exhibited by plants.

11.5.2.3 Superoxide Dismutase (SOD)

In plants, superoxide dismutase (SOD) plays a vital role in defending against ROS by dismutating superoxide radical ($O_2^{\cdot -}$) into molecular oxygen (O_2) and hydrogen peroxide (H_2O_2). Although this reaction contributes to an oxidative environment, the plant's immune response depends on converting the superoxide radical into hydrogen peroxide. Hydrogen peroxide activates systemic defense genes and is a substrate for apoplastic peroxidases (PRXs), which cross-link cell wall polymers (Blaschek and Pesquet 2021). Thus, SOD reduces the cytotoxic effects of elevated superoxide levels during oxidative bursts and promotes hydrogen peroxide-dependent signaling pathways and biochemical processes for structural defense. The localized distribution of SOD metallo-isoforms – Cu/Zn-SODs in the cytosol, chloroplasts, and apoplast; Fe-SODs in chloroplasts; and Mn-SODs in mitochondria enhances ROS regulation to meet each organelle's metabolic and defensive needs.

11.5.2.4 Catalase (CAT)

Catalase (CAT) is an essential enzyme in the ROS scavenging network, dismutating hydrogen peroxide (H_2O_2) into water and molecular oxygen. Rather than eliminating H_2O_2, precise regulation is crucial for plant immunity. Although

H_2O_2 is essential for defense signaling, excessive accumulation can cause PCD and oxidative damage, which pathogens like necrotrophs use to induce tissue necrosis by inhibiting the plant's oxidative burst. Suppressing CAT activity at the infection site leads to H_2O_2 accumulation for defense, while maintaining CAT activity in healthy tissues prevents oxidative damage and its propagation. This shows that ROS signaling requires spatial and temporal control. The characteristics of ROS signals, duration, and spatial distribution for the CAT are also affected by a dynamic interaction between enzymes like SOD and peroxidase (Shaat and Galal 2004). This integrated system creates a ROS-generating and scavenging circuit for plant stress and pathogen defense.

11.5.2.5 Ascorbate Peroxidase (APX)

Ascorbate peroxidase (APX) is a key enzyme in the AsA-GSH cycle that scavenges hydrogen peroxide (H_2O_2). APX is essential for fine-scale H_2O_2 concentration regulation in sensitive compartments like chloroplasts, cytosols, and peroxisomes due to its high affinity for H_2O_2 compared to CAT. APX activity is essential for preventing oxidative damage during biotic stress while allowing for limited H_2O_2 fluxes for intracellular signaling. Specific APX isoforms can be a host strategy to enhance the HR, a localized PCD that confines biotrophic pathogens. A recent rice study found that a fungal effector directly inhibited a chloroplastic APX isoform, illustrating this complex interaction. The pathogen suppressed this defense component, but the plant used the resulting H_2O_2 accumulation and chloroplast damage to initiate a powerful HR, turning virulence into defense (Choudhary et al. 2020). This shows the complexity and dynamic nature of the redox battle, where pathogen virulence strategies and plant immune countermeasures depend on APX activity regulation.

11.5.3 Hydrolytic Enzymes

Plants use hydrolytic enzymes as molecular scissors to degrade pathogen and herbivore polymers. Targeting microbial components like chitin and β-1,3-glucan in fungal cell walls or peritrophic matrix and digestive proteins in insect guts, these enzymes have direct antimicrobial and antiherbivory effects (Saldo and Szpyrka 2009). PR proteins, such as chitinases (PR-3, PR-4, PR-8, PR-11) and β-1,3-glucanases (PR-2), are key defense effectors that compromise invader integrity and release endogenous elicitors to boost immune responses.

11.5.3.1 β-1,3-Glucanases and Chitinases

Chitinases and β-1,3-glucanases function together to boost plant antifungal defense. Chitinases (EC 3.2.1.14) break down chitin, a β-1,4-linked N-acetylglucosamine polymer found in fungal cell walls and insect exoskeletons. At the same time, β-1,3-glucanases (EC 3.2.1.39) degrade fungal cell walls. Their coordinated action osmotically lyses fungal hyphal tips, severely degrading the pathogen. Enzymatic activity releases chito- and gluco-oligosaccharides, which boost defense signaling and immune responses, in addition to direct antibiosis. Transgenic hydrolytic enzyme expression has effectively protected crops from fungal diseases (Ren et al. 2021). Crop protection applications have become more dependent on the identification and characterization of novel, highly active isoforms from resistant wild crop relatives, which frequently have unique substrate specificities or better catalytic properties than controlled varieties.

11.5.3.2 Lysozymes

Lysozymes (Muramidases) are hydrolytic enzymes that break β-1,4-glycosidic bonds between N-acetylmuramic acid and N-acetyl-D-glucosamine residues in bacterial peptidoglycan. Historically important in animal innate immunity, orthologous enzymes in plants directly fight bacteria. SA signaling pathways upregulate plant lysozyme genes in response to bacterial infections (van Dam et al. 2004). These enzymes hydrolyze the essential peptidoglycan scaffold of the bacterial cell wall, compromising the cellular envelope's structural integrity and causing osmotic lysis, a powerful defense against phytopathogenic bacteria.

11.5.3.3 Proteases

Proteases in plants have many defense functions beyond protein degradation. They directly repel herbivores by hydrolyzing essential proteins in the insect digestive tract, preventing nutrient absorption. Furthermore, specific proteases are needed to mature and activate defense signaling molecules like systemin precursors and other endogenous peptide elicitors (Shi et al. 2018). They also help immune signaling networks post-translationally degrade central regulatory proteins to stop defense responses, preventing autoimmunity and allowing resource reallocation after threat reduction. These enzymes initiate biotrophic pathogen-limiting PCD by precisely cleaving cellular targets.

11.5.3.4 Lipoxygenases (LOX)

Lipoxygenases (LOXs) dioxygenate polyunsaturated fatty acids to form fatty acid hydroperoxides, the oxylipin pathway's substrates. LOXs are regulatory nodes that start defense signaling cascades, especially those mediated by JA in response to herbivores and necrotrophic pathogens (Ton et al. 2009). Some LOX-derived oxylipins have antibiotic or antinutritive properties in addition to signaling, providing immediate defense. LOX isoforms' activities affect the defense response's functional specificity. The channel of 13-LOXs substrates toward JA biosynthesis, which is necessary for chewing insect resistance, while 9-LOX pathways generate alternative oxylipin signatures that can paradoxically promote susceptibility or resistance, depending on the pathogen.

11.6 ENZYMES INVOLVED IN CELL WALL MODIFICATIONS

The plant cell wall is a dynamic and responsive interface, actively remodeled during plant–pathogen interactions. This enzymatic reinforcement strategy enhances the wall's physical integrity as a hurdle to penetration and its chemical resistance to microbial enzymes, forming a critical component of the plant's innate immune responses.

11.6.1 Callose Synthase and Callose Deposition

Callose, a polymer of β-1,3-glucans, is deposited in papillae at sites of pathogen attack or mechanical injury to prevent invasion. The presence of signaling cascades between defense-specific signaling and cytosolic influx and ROS production via defense receptors triggers plasma membrane localization of CalS complexes that contribute to the synthesis of callose. In this deposition, fungi, bacteria, and insects are effectively prevented. CalS activity and localization are regulated by complex interactions of plasma membrane proteins. Arabidopsis thaliana requires CalS8/PEN1 isoform to develop papillae against powdery mildew fungi and accumulates at the potential penetration sites before invasion to identify non-host resistance (Siddique and Garnevska 2018). This indicates the ability of the spatiotemporal control and localization of a single enzymatic complex to influence plant–pathogen interactions.

11.6.2 Pectin Methylesterase (PME)

Plant defense relies on PME to remodel cell walls by demethylesterifying homogalacturonan. This process changes the pectin matrix's physicochemical properties, affecting wall rigidity and porosity, and exposes carboxyl groups that can cross-link with calcium ions to strengthen the immunity-enhancing apoplastic barrier. Demethylesterified pectin also protects against pathogen-derived CWDEs by serving as a substrate for polygalacturonase-inhibiting proteins (PGIPs) and is selectively cleaved by plant pectinases to release oligogalacturonides (OGs), which act as DAMPs to boost immune signaling. PME function is tightly controlled by endogenous PME inhibitor (PMEI) proteins. In tomatoes, silencing a specific PME isoform made them more susceptible to Botrytis cinerea, highlighting the importance of PME defense. PME's pro-defense function was highlighted by a decreased ability to fortify the cell wall and produce the OGs needed for a strong immune response, resulting in increased susceptibility.

11.7 GENETIC REGULATION AND MOLECULAR BASIS OF ENZYMATIC RESISTANCE

Plant defense mechanisms are under complex regulation involving the interplay of transcription factors and signaling pathways, and are regulated by SA, JA, and ET. Transcription factors such as WRKY, MYC, ERF, and bZIP interact with gene promoters to initiate enzymatic defenses. The SA pathway induces NPR1, which increases the binding of TGA transcription factors, resulting in the elevated expression of pathogen-resistant PR genes producing enzymes such as chitinases and β-1,3-glucanases (Gale et al. 2019). On the other hand, MYC2 and homologs trigger proteinase inhibitor and volatile organic compound-related enzyme genes via the JA pathway. Two kinds of transcriptional systemic resistance found are SAR and induced systemic resistance (ISR). Local infections cause an increase in SA levels and PR gene expression, whereas positive rhizobacteria use JA/ET signaling to condition plants in anticipation of future stress without the initial expression of PR genes (Franzener et al. 2018). Genetic engineering and CRISPR-Cas9 technology developments are targeting increased transcription factor expression and defense gene promoter element modification to create crop types with sustainable enzymatic resistance.

11.8 ROLE OF MICROBIAL AND ENDOPHYTIC ENZYMES IN PLANT DEFENSE

Holobiont plants depend on diverse microorganisms, particularly endophytes, for immune system priming through enzymes like chitinases and glucanases. PGPR activates systemic resistance via JA/ET pathways, while rhizobacteria enhance immunity by disrupting fungal pathogens (Remick et al. 2023). AM fungi facilitate symbiosis and bolster plant defenses through enzyme production. Microbial biocontrol agents such as *Trichoderma harzianum* and *Bacillus thuringiensis* leverage these microbial interactions to reduce pesticides and promote sustainable agriculture by enhancing plant defenses and activating ISR.

11.8.1 Endophyte-Mediated Induction of Host Defense Enzymes

Endophytes have the capacity to enhance biotic stress tolerance in plants by the rapid activation of defense enzymes. Their release of elicitors such as MAMPs, secondary metabolites, and volatile organic compounds triggers defense signaling pathways that are regulated by JA and SA. As an example, *T. harzianum* fungus boosts the PAL activity, and the amplified production of the phenolic compounds, lignin, and SA strengthens the defenses of plants (Gachomo et al. 2014). Endophytes also promote the PR proteins and the ROS-generating enzymes to increase a defensive oxidative response. Moreover, they regulate antioxidant enzymes that regulate the levels of ROS, which coordinate defense signaling. Sustainable agriculture is based on endophyte bioinoculants to enhance crop defense to reduce the level of pesticide dependence and enhance agricultural resilience.

11.8.2 Rhizobacteria-Induced Systemic Resistance (ISR)

The beneficial bacteria, including *Pseudomonas* and *Bacillus*, promote the resistance of plants to the damage caused by Rhizobacteria by improving the plant JA and ET pathways, which are known as the systemic resistance (ISR). ISR allows faster responses to pathogens as opposed to SAR, which depends on SA. It involves the stimulation of enzymes associated with defense and the build-up of antimicrobial agents, especially the high levels of PAL. The ISR also increases the production of lipoxygenase (LOX) to produce JA to regulate defense genes against herbivores and pathogens (Hasanuzzaman et al. 2017). Complex signaling cascades are involved in the initiation of ISR triggered by elicitors such as flagellin and lipopolysaccharides. It also enhances PR protein activation, increasing resistance to fungal pathogens, including *Botrytis cinerea*, and the ISR-inducing microbes are relevant in sustainable crop protection approaches (Ren et al. 2021).

11.8.3 Mycorrhizal Fungi and Enzyme Activation in Host Plants

AM fungi engage in a significant symbiotic relationship with plants, enhancing nutrient acquisition and strengthening plant defenses. The presence of arbuscules triggers "mycorrhiza-induced resistance" (MIR), which bolsters the plant's immune system against pathogens, particularly soil-borne ones. Fungi such as *Rhizophagus irregularis* stimulate the phenylpropanoid pathway, increasing the expression and activity of enzymes like PAL and CAD, which fortify cell walls. AM fungi also elevate the activity of defensive enzymes like peroxidases, assisting in the controlled oxidative burst necessary for antimicrobial responses. While mycorrhizal colonization may initially suppress certain PR proteins, the importance of strigolactones in this process is highlighted. They have the potential benefits of AM fungal inoculants in agriculture, promoting plant health and creating resilient cropping systems (Furuichi et al. 2012).

11.9 ENVIRONMENTAL FACTORS INFLUENCING ENZYMATIC DEFENSE

The efficacy and dynamics of plant enzymatic defense mechanisms are not static traits but are profoundly modulated by a range of abiotic and biotic environmental factors. Table 11.1 summarizes key factors modulating plant defense enzymes. Biotic and abiotic stresses, nutrient levels, light, and beneficial microbes regulate enzymes like PAL and POD via signaling molecules such as JA, SA, and ROS, priming or inducing defense responses.

11.10 ABIOTIC STRESS ON ENZYME ACTIVITY

Drought, salinity, temperature extremes, and heavy metal contamination impair plant metabolism and defense enzymes by inducing oxidative stress via excessive ROS. Severe drought reduces PAL levels, hindering defensive phenolics and lignin synthesis, increasing pathogen susceptibility. Moderate stress may enhance plant defense through priming. The heat shock response produces heat shock proteins (HSPs) that stabilize enzymes under stress (Gachomo et al. 2014). Plant defense mechanisms depend on stress severity and the upregulation of antioxidant enzymes like SOD and APX to mitigate oxidative damage. The interplay of various stressors, including drought and pathogen attacks, threatens the plant's regulatory capabilities amid climate change.

11.10.1 Nutrient Availability and Defense Enzyme Modulation

The macronutrients like the potassium (K), phosphorus (P), and nitrogen (N) determine the nutritional status of plants, affecting their defense mechanisms. Some of the amino acids and enzymes that nitrogen produces in the growth-differentiation balance hypothesis include PAL and peroxidases (POD). The quantity of nitrogen can result in the excessive growth of vegetation, and this makes the vegetation more susceptible to being attacked by herbivores and pathogens since the defense mechanisms have been diverted to the growth. Furthermore, the defense mechanism and cell walls can be strengthened with silicon (Si) to fight pathogens (Lakshmi et al. 2018). The nutrient management, therefore, plays a role in increasing the crop production and resistance to diseases of sustainable agriculture (Simić et al. 2004; Zechmann 2020).

TABLE 11.1
Factors Modulating Plant Defense Enzyme Induction and Their Signaling Networks

S. No	Factors	Signaling Molecules	Effect on Defense Enzymes	References
1	Biotic stress (pathogen/herbivore attack)	Jasmonic acid (JA), salicylic acid (SA), reactive oxygen species (ROS)	Rapidly induces the synthesis of PAL, LOX, chitinases, glucanases, PPO, and POD.	Lakshmi et al. (2018)
2	Abiotic stress (drought, salinity, heavy metals)	Abscisic acid (ABA), ROS	Can prime or induce defense enzymes like POD, SOD, and CAT to manage oxidative stress, often leading to cross-tolerance against biotic threats.	Simić et al. (2004)
3	Nutrient availability (e.g., N, P, K)	Hormonal crosstalk (e.g., JA, ABA)	Deficiency can enhance phenolic and PI production as part of a resource allocation shift toward defense.	Zechmann (2020)
4	Quality	Phytochrome signaling, circadian clock	PAL and JA-responsive genes, making defenses more responsive to attack.	Feng et al. (2018)
5	Plant beneficial microbes (PGPR)	JA/SA Pathways	Priming activates the defense pathways, leading to a faster and stronger enzyme induction.	Liu et al. (2020)

11.10.2 Role of Elicitors and Signaling Molecules

Plants possess defense enzymes triggered by the signaling molecules referred to as elicitors. They can be exogenous (PAMPs, effectors, microbial signals) and endogenous (DAMPs of broken cells). These elicitors trigger the immune receptor signaling networks governed by phytohormones such as SA, JA, and ET (Feng et al. 2018). The SA pathway is expressed to generate the chitinases and glucanases to resist biotrophic pathogens, while the JA pathway and ET trigger LOX and PPOs to resist necrotrophic pathogens and herbivores (Liu et al. 2020). Plant growth regulators and chemical elicitors like acibenzolar-S-methyl and jasmonate analogues have been found to enhance the plant defenses by increasing the activity of defense enzymes.

11.10.3 Impact of Pesticides and Fertilizers on Enzyme-Based Resistance

Systemic fungicides play a pivotal role in enhancing plant defense mechanisms by not only reducing infections but also activating specific defense genes. These fungicides can lead to detrimental effects on beneficial microbiomes, lowering the plant's natural defenses. Additionally, fertilizer can significantly affect plant health, and the synthetic fertilizers may inadvertently raise disease susceptibility, whereas organic amendments tend to foster a supportive environment for beneficial microbiomes, facilitating the induction of defense enzymes. The integrated management approach is essential. Such systems must evaluate the impacts of pesticides on plant immunity, aiming to enhance sustainability while bolstering resistance to pests and diseases (Quaglia et al. 2011).

11.11 CHALLENGES AND LIMITATIONS

Despite advancements in enzymatic defense mechanisms for crop protection, significant challenges persist. Variability in defense efficacy is influenced by plant genotype, pathogen and herbivore specificity, developmental stage, and environmental conditions, complicating broad-spectrum solutions (Hasanuzzaman et al. 2017). The defense enzymes can improve resistance, but often the best option is due to resource allocation trade-offs. Functional redundancy among enzyme isoforms complicates identifying specific defensive roles. Multi-omics approaches (transcriptomics, proteomics, metabolomics) are essential for addressing this complexity. An integrated paradigm combining molecular biology, biochemistry, ecology, and precision breeding is crucial for developing resilient crop varieties with effective enzymatic defense systems (Kaur et al. 2022).

11.12 CONCLUSION

Enzymatic mechanisms are a central part of the complex plant immunity system. This defense relies on the interaction of oxidative and hydrolytic enzymes. Oxidative enzymes such as peroxidases (POX), PPOs, SOD, CAT, and APX create an environment that does not favor invaders but increases the protection of structures. Hydrolytic enzymes, which include chitinases, β-1,3-glucanases, and proteases, contribute by breaking down parts of the fungal cell walls and insect exoskeletons. These enzymatic activities are regulated by signaling networks that are highly mediated by phytohormones such as SA, JA, and ET. This hormonal coordination is part of the systemic resistance, particularly SAR and ISR, which are frequently supplemented by beneficial microorganisms. The environmental conditions are also very important in determining the efficiency of these enzymatic defenses. Although some progress has been made in these mechanisms, some of the challenges, such as the resource allocation of defense, redundancy on enzymes, and variable resistance, remain. Genetic engineering and sustainable farming are opportunities to strengthen the plant's enzymatic barrier, thus increasing crop resistance, reducing the use of chemical pesticides, and contributing to integrated pest and disease control. These enhancements are critical in securing food with the complicated issues in world agriculture.

REFERENCES

Ali, M., Li, Q.H., Zou, T., Wei, A.M., Gombojab, G. and Lu, G. 2020. The chitinase gene positively regulates hypersensitive and defense responses of pepper to *Colletotrichum acutatum* infection. Int. J. Mol. Sci. 21(18): 6624.

Aziz, M.A., Sabeem, M., Kutty, M.S., Rahman, S., Alneyadi, M.K. and Alkaabi, A.B. 2023. Enzyme stabilisation and thermotolerance function of the intrinsically disordered LEA2 proteins from date palm. Sci. Rep. 13(1): 11878.

Bacete, L., Mélida, H., Pattathil, S., Hahn, M.G., Molina, A. and Miedes, E. 2017. Characterisation of plant cell wall damage-associated molecular patterns regulating immune responses. In L. Shan and P. He (Eds.), Plant pattern recognition receptors. Springer, pp. 13–23.

Blaschek, L. and Pesquet, E. 2021. Phenoloxidases in plants—How structural diversity enables functional specificity. Front. Plant Sci. 12: 754601.

Choudhary, A., Kumar, A. and Kaur, N. 2020. ROS and oxidative burst: Roots in plant development. Plant Divers 42(1): 33–43.

Coomber, A., Saville, A. and Ristaino, J.B. 2024. Evolution of *Phytophthora infestans* on its potato host since the Irish potato famine. Nat. Commun. 15(1): 6488.

Engelsdorf, T., Will, C., Hofmann, J., Schmitt, C., Merritt, B.B., Rieger, L. et al. 2017. Cell wall composition and penetration resistance against the fungal pathogen *Colletotrichum higginsianum* are affected by impaired starch turnover in *Arabidopsis mutants*. J. Exp. Bot. 68(3): 701–713.

Espinoza, C., Liang, Y. and Stacey, G. 2017. Chitin receptor CERK1 links salt stress and chitin-triggered innate immunity in Arabidopsis. Plant J. 89(5): 984–995.

Feng, W., Kita, D., Peaucelle, A., Cartwright, H.N., Doan, V., Duan, Q. et al. 2018. The FERONIA receptor kinase maintains cell-wall integrity during salt stress through Ca2+ signalling. Curr. Biol. 28(5): 666–675.e5.

Franzener, G., Schwan-Estrada, K.R.F., Moura, G. S., Kuhn, O. J. and Stangarlin, J. R. 2018. Induction of defense enzymes and control of anthracnose in cucumber by *Corymbia citriodora* aqueous extract. Summa Phytopathol. 44(1): 10–16.

Furuichi, T., Iida, H., Sokabe, M. and Tatsumi, H. 2012. Expression of *Arabidopsis* MCA1 enhanced mechanosensitive channel activity in the *Xenopus laevis* oocyte plasma membrane. Plant Signal. Behav. 7(8): 1022–1026.

Gachomo, E.W., Jno Baptiste, L., Kefela, T., Saidel, W.M. and Kotchoni, S.O. 2014. The Arabidopsis CURVY1 (CVY1) gene encoding a novel receptor-like protein kinase regulates cell morphogenesis, flowering time and seed production. BMC Plant Biol. 14(1): 221.

Gale, U., Kemat, A. and Krens, F. 2019. Effect of calcium (Ca^{2+}) on pectin and pectin methyl esterases (PME) in normal and hyperhydric leaves of *Arabidopsis* cultured *in vitro*. J. Plant Physiol. 243: 153048.

Galindo-Trigo, S., Blanco-Touriñán, N., DeFalco, T.A., Wells, E.S., Gray, J.E., Zipfel, C. et al. 2020. CrRLK1L receptor-like kinases HERK1 and ANJEA are female determinants of pollen tube reception. EMBO Rep. 21(5): e48466.

García Bossi, J., Kumar, K., Barberini, M.L., Domínguez, G.D., Rondón Guerrero, Y.D.C. and Marino-Buslje, C. 2020. The role of P-type IIA and P-type IIB Ca_2^+-ATPases in plant development and growth. J. Exp. Bot. 71(4): 1239–1248.

Geisler, M., Axelsen, K.B., Harper, J.F. and Palmgren, M.G. 2000. Molecular aspects of higher plant P-type Ca_2^+-ATPases. Biochim. Biophys. Acta 1465(1–2): 52–78.

Hasanuzzaman, M., Nahar, K., Hossain, M.S., Mahmud, J.A., Rahman, A. and Inafuku, M. 2017. Coordinated actions of glyoxalase and antioxidant defense systems in conferring abiotic stress tolerance in plants. Int. J. Mol. Sci. 18(1): 200.

Jomova, K., Alomar, S.Y., Alwasel, S.H., Nepovimova, E., Kuca, K. and Valko, M. 2024. Several lines of antioxidant defense against oxidative stress: Antioxidant enzymes, nanomaterials with multiple enzyme-mimicking activities, and low-molecular-weight antioxidants. Arch. Toxicol. 98(5): 1323–1367.

Kaur, S., Samota, M.K., Choudhary, M., Choudhary, M., Pandey, A.K. and Sharma, A. 2022. How do plants defend themselves against pathogens-biochemical mechanisms and genetic interventions. Physiol. Mol. Biol. Plants. 28(2): 485–504.

Kumari, K., Gusain, S. and Joshi, R. 2024. Engineering cold resilience: Implementing gene editing tools for plant cold stress tolerance. Planta. 261(2): 2.

Kurnik, K., Treder, K., Skorupa-Kłaput, M., Tretyn, A. and Tyburski, J. 2015. Removal of phenol from synthetic and industrial wastewater by potato pulp peroxidases. Water Air Soil Pollut. 226(8): 254.

Lakshmi, S., Shashidhara, G.M., Madhu, G.M., Muthappa, R., Vivek, H.K. and Nagendra Prasad, M.N. 2018. Characterisation of peroxidase enzyme and detoxification of phenols using peroxidase enzyme obtained from *Zea mays* L. waste. Appl. Water Sci. 8(7): 207.

Liu, X., Yu, Y., Liu, Q., Deng, S., Jin, X. and Yin, Y. 2020. A Na_2CO_3-responsive chitinase gene from *Leymus chinensis* improves pathogen resistance and saline-alkali stress tolerance in transgenic tobacco and maize. Front. Plant Sci. 11: 1–12.

Mishra, N., Jiang, C., Chen, L., Paul, A., Chatterjee, A. and Shen, G. 2023. Achieving abiotic stress tolerance in plants through antioxidative defense mechanisms. Front. Plant Sci. 14: 1110622.

Oliveira, M.D.M., Varanda, C.M.R. and Félix, M.R.F. 2016. Induced resistance during the interaction between pathogen x plant and the use of resistance inducers. Phytochem. Lett. 15: 152–158.

Quaglia, M., Ederli, L., Pasqualini, S. and Zazzerini, A. 2011. Biological control agents and chemical inducers of resistance for postharvest control of *Penicillium expansum* Link. On an apple fruit. Postharvest Biol. Technol. 59(3): 307–315.

Rashid, M.H. and Chung, Y.R. 2017. Induction of systemic resistance against insect herbivores in plants by beneficial soil microbes. Front. Plant Sci. 8: 1816.

Razmjou, J., Jafary, M. and Borzoui, E. 2019. Host plant preference and life table of *Brevicoryne brassicae*. J. Crop Prot. 8(2): 201–214.

Remick, B.C., Gaidt, M.M. and Vance, R.E. 2023. Effector-triggered immunity. Annu. Rev. Immunol. 41: 453–481.

Ren, H., Wu, Y., Ahmed, T., Qi, X. and Li, B. 2021. Response of resistant and susceptible bayberry cultivars to infection of the twig blight pathogen by histological observation and gibberellin-related gene expression. Pathogens. 10(4): 402.

Saldo, S. and Szpyrka, E. 2009. Ecotoxicological view of protection of apple orchards against insect pests in Poland. Pestycydy 2009(1–4): 15–26.

Shaat, M.N.M. and Galal, A. 2004. Response of citrus fruits to preharvest antioxidant spraying and infection with *Alternaria* fruit rot and green mould. Ann. Agric. Sci. 49(2): 747–758.

Shao, X., Wang, H., Xu, F. and Cheng, S. 2013. Effects and possible mechanisms of tea tree oil vapour treatment on the main disease in postharvest strawberry fruit. Postharvest Biol. Technol. 77: 94–101.

Shi, Z., Wang, F., Lu, Y. and Deng, J. 2018. A combination of chitosan and salicylic acid to control postharvest green mould caused by *Penicillium digitatum* in grapefruit. Sci. Hortic. 233: 54–60.

Siddique, M.I. and Garnevska, E. 2018. Citrus value chain (s): A survey of the Pakistan citrus industry. In The agricultural value chain. Intech Open.

Simić, A., Soković, M.D., Ristić, M., Grujić-Jovanović, S., Vukojević, J. and Marin, P.D. 2004. The chemical composition of some Lauraceae essential oils and their antifungal activities. Phytother. Res. 18(9): 713–717.

Srivastava, V., De Guzman, C. and Fernandes, S.B. 2024. Beat the heat: Breeding, genomics, and gene editing for high nighttime temperature tolerance in rice. Curr. Opin. Plant Biol. 82: 102659.

Ton, J., van der Ent, S., van Hulten, M., Pozo, M., van Oosten, V. and van Loon, L.C. 2009. Priming as a mechanism behind induced resistance against pathogens, insects and abiotic stress. IOBC/WPRS Bull. 44: 3–13.

van Dam, N.M., Witjes, L. and Svatos, A. 2004. Interactions between aboveground and belowground induction of glucosinolates in two wild *Brassica* species. New Phytol. 161(3): 801–810.

Walters, D.R., Ratsep, J. and Havis, N.D. 2013. Controlling crop diseases using induced resistance: Challenges for the future. J. Exp. Bot. 64(5): 1263–1280.

Xie, H., Shi, J., Shi, H., He, K. and Wang, Z. 2020. Aphid fecundity and defenses in wheat exposed to a combination of heat and drought stress. J. Exp. Bot. 71(9): 2713–2722.

Zechmann, B. 2020. Subcellular roles of glutathione in mediating plant defense during biotic stress. Plants 9(9): 1067.

Zhao, S., Li, M., Ren, X., Wang, C., Sun, X. and Sun, M. 2024. Enhancement of broad-spectrum disease resistance in wheat through key genes involved in systemic acquired resistance. Front. Plant Sci. 15: 1355178.

Zheng, M., Liu, Y., Zhang, G., Yang, Z., Xu, W. and Chen, Q. 2023. The applications and mechanisms of superoxide dismutase in medicine, food, and cosmetics. Antioxidants 12(9): 1675.

12 Regulation of Enzyme Activity in Response to Environmental Pollutants

M. Manigandan, Yuvalakshmi Logesan, K. Gayathri, B. Lokeshwari, and P. Saranraj

12.1 INTRODUCTION

Biosphere includes a dynamic interaction of paramount significance in the interplay of biota and their chemical environment. The basic building blocks of this interaction are enzymatic systems, and, hence, it is understandable why they are considered the ideal catalysts in the triggering of biochemical reactions that sustain life. These macromolecules are not only central to the core of the basic metabolic processes but also form the major defensive mechanism against chemical attack. The widespread infiltrations of anthropogenic contaminants, or xenobiotics, comprise a collection of biochemical stressors such as organophosphate esters, pharmaceuticals, personal care products (PPCPs), heavy metals, and micro/nanoplastics (Joseph et al., 2025; Xie et al., 2025). These substances often include highly reactive functional groups like esters, amides, and glycosidic bonds, which disrupt cellular homeostasis by interfering with the currently vital enzyme processes, causing oxidative stress, and impairing physiological processes in both fauna and flora (Liao et al., 2025; Zheng et al., 2025).

The ability of organisms to survive and adapt to polluted environments is inherently connected with the process of controlling their enzymatic activity. Exposure to xenobiotics elicits a multi-tiered response comprising transcriptional upregulation of already existing enzyme pools (e.g., cytochrome P450 monooxygenases and glutathione S-transferases [GSTs]), allosteric alterations of pre-existing enzyme pools, and post-translational modifications (Joseph et al., 2025). Moreover, synergistic relationships between plants and the microbiomes that they host, in particular, in the rhizosphere, indicate a prevalent ecological aspect of this regulation system (Hang et al., 2025; Liu et al., 2025). Endophytic and rhizospheric microbes also increase the resilience of hosts by releasing extracellular polymers and degradative enzymes, and indirectly regulating the intrinsic defenses of plants, through modulatory effects on the enzyme system (Liu et al., 2025). Therefore, elucidating the regulatory mechanisms is not only a biochemical pursuit but also an ecological necessity.

The general objective of the review is, thus, to map out, in minute details the complex regulation of the enzymatic activity under the influence of environmental pollution. It aims to integrate the existing information on the catalytic processes and adaptability of natural enzymes and novel biomimetics, including the hydrolytic nanozymes (Liao et al., 2025). Critical review of activity regulation strategies, including the regulation of Lewis acid–base equilibrium, coordination microenvironment, and substrate selectivity is discussed critically. It is hoped that this review will offer a holistic idea of enzyme adaptation to pollution, therefore guiding the design of improved bioremediation and biosensing system methods, accurate biomonitoring systems, and methods to improve food quality and ecosystems.

12.2 BASIC PROCESSES OF ENZYME CONTROL

The enzymatic control of pharmaceutical pollutants follows a multi-stage detoxification system that is often referred to as a green liver system (Sandermann Jr., 1994). This system involves a transcriptional and post-translational regulation of the most important enzymes across three metabolic phases. During phase I, enzymes, including cytochrome P450 monooxygenases, peroxidases, and laccases, are upregulated (Pilon-Smits, 2005; Morsi et al. 2020). These enzymes add reactive functional groups through oxidation, hydroxylation, and hydrolysis and thus activate the pollutants to be conjugated (Dietz and Schnoor, 2001). The important enzymatic controls and adaptive responses to pharmaceutical pollutants were illustrated in Table 12.1.

The conjugation reactions of phase II of xenobiotic metabolism are catalyzed by a group of enzymes such as GSTs, uridine diphosphate glycosyltransferases (UGTs), and sulfotransferases (STs) (Bartha et al., 2014). These enzymes are involved in the processing of the pollutants activated by the covalent binding to the endogenous molecules, e.g., glutathione or sugars. Such enzymes become more expressed in the presence of xenobiotics and, consequently, improve solubility and decrease toxicity (Sun et al., 2019). The conjugated metabolites formed during phase III are transported into vacuoles or apoplast by ATP-binding cassette (ABC) transporters, and cellular integrity is maintained (LeFevre et al., 2015; Zheng and Guo, 2021). This enzymatic process is commonly initiated by oxidative stress caused by contaminants, which results in the activation of signaling pathways that cause the expression of the detoxification genes and overcome the phytotoxic impact of pharmaceutical contaminants.

12.2.1 Laws of Adaptation of Metabolism to Chemical Stress

Plants detoxify the pollutants by increasing major detoxifying enzymes, thus orchestrating degradation or sequestration

DOI: 10.1201/9781003669920-12

TABLE 12.1
Enzymatic Control and Adaptive Responses to Pharmaceutical Pollutants in Plants

Process/Level	Mechanisms of Enzymes	Functional Role	Key References
		Detoxification Phases	
Phase I: Activation	Cytochrome P450s, peroxidases, laccases	Introduces reactive groups via oxidation/hydroxylation, activating pollutants for conjugation.	Pilon-Smits (2005); Morsi et al. (2020)
Phase II: Conjugation	GSTs, UGTs, sulfotransferases	Conjugates activated pollutants with sugars/glutathione to increase solubility and reduce toxicity.	Bartha et al. (2014); Sun et al. (2019)
Phase III: Sequestration	ABC transporters	Isolates conjugated metabolites in vacuoles or apoplast to protect cellular integrity.	LeFevre et al. (2015); Zheng and Guo (2021)
		Adaptation Principles	
Quantum effects	Quantum tunneling; radical pair mechanism (RPM)	Enhances reaction rates and regulates ROS signaling pathways via quantum phenomena.	Schreiner (2020); Ivanov et al. (2017)
Metabolic restructuring	Upregulation of P450s, GSTs; flavonoid synthesis	Coordinates biotransformation of pollutants with synthesis of protective metabolites.	Mosharaf et al. (2024); Hurmat et al. (2025)
		Hierarchical Control	
Molecular level	Enzyme activation; miRNA regulation	Rapid breakdown of pollutants and fine-tuning of stress-responsive gene expression.	Chen et al. (2025); Mosharaf et al. (2024); Li et al. (2025)
Cellular/physiological level	Altered root development; ROS biomarkers	Manifests molecular changes as physiological adaptations and oxidative stress responses.	Zheng et al. (2025); Garcíaet al. (2025); Hurmat et al. (2025)
Ecosystem level	Plant–microbial symbiosis	Enhances large-scale phytoremediation through coordinated community interactions.	Hang et al. (2025); Liu et al. (2025)

pathways (Saravanan et al., 2021). One of the important concepts of critical detoxification reaction is the rate acceleration of reactions through quantum tunneling, i.e., the transmission of protons or electrons through energy barriers that cannot be approached in classical mechanics (Schreiner, 2020). This is of particular importance in electron-transfer processes during oxidative degradation. Furthermore, the radical pair mechanism (RPM) may regulate metabolic processes in response to chemical stress by altering the reactive oxygen species (ROS) signaling and scavenging signaling pathways (Ivanov et al., 2017). The spin-state dynamics of radical intermediates potentially sensitive to magnetism dictate the effectiveness of such redox reactions and thereby fine-tune metabolic processes. One major method toward modeling these enzymatic processes is by computational methods, including quantum mechanical/molecular mechanical (QM/MM) simulations, which provide an atomistic understanding of how enzyme structure maximizes efficiency and enables adaptation to chemical stressors (van der Kamp and Mulholland, 2013). The quantum-biochemical models support the metabolic restructuring that is complex and allows plants to survive under polluted conditions.

Plants adapt to chemical stress by increasing ROS-scavenging enzymes to overcome high oxidative stress caused by pollutants. This protective mechanism often involves radical reactions and RPM – a quantum effect of spin correlations – which control the rate and selectivity of radical scavenging processes (Ivanov et al., 2017). Also, quantum tunneling can significantly boost enzymatic detoxification by increasing the rate of electron and proton transfer between substrates, thus improving the kinetics of ROS-scavenging and avoiding oxidative stress on plant cells (Schreiner, 2020).

When organisms sense the presence of pollutants as alien stressors, the metabolic change to the chemical stress occurs by a series of molecular reactions. An important mechanism is the transcriptional and translational activation of enzymes that are important in detoxification, such as cytochrome P450s in phase I metabolism and GSTs in phase II conjugation (Mosharaf et al., 2024). This is a biotransformation process under the control of enzymes that convert more toxic compounds into excretable end products in the metabolic process. At the same time, the synthesis of protective secondary metabolites, such as flavonoids, is stimulated to reduce the effects of oxidative damage caused by pollutants (Hurmat et al., 2025). They play an essential role in cellular homeostasis and survival in contaminated environments, as these are structural biochemical adaptations.

12.2.2 Biological Control Hierarchical Levels

The complex multi-layered hierarchical control mechanism is a way for biological systems to oscillate their response to environmental pollutants. This regulatory system starts at the molecular level, where the activity or inactivity of prevalent enzymes and transcription factors takes place in a fast manner. As an example, organic contaminants can be broken down by special enzymes, like laccases (Chen et al., 2025), whereas inorganic pollutants can activate the expression of the detoxification enzymes cytochrome P450s and GSTs (Mosharaf et al., 2024). MicroRNAs are normally

TABLE 12.2
Transcriptional Regulation of Important Detoxification Enzyme Systems

Enzyme System/Process	Core Molecular Reaction	Regulatory Trigger	Key References
General transcriptional regulation	Transcription factors (e.g., AhR, Nrf2) bind to promoter elements (XRE, ARE), initiating gene transcription.	Chemical stress sensed via cellular signaling pathways.	Li et al. (2024); Joseph et al. (2025)
Gene induction by pollutants	Signal transduction cascades activate specific transcription factors, leading to gene expression.	Pollutant binding to sensors (e.g., AhR) or induction of oxidative/electrophilic stress.	Li et al. (2024); Mosharaf et al. (2024)
Cytochrome P450 monooxygenases (CYPs)	Oxidation (e.g., hydroxylation) of lipophilic pollutants, increasing hydrophilicity.	Transcriptional upregulation via pollutant exposure (e.g., atrazine, PFOA).	Mokhosoev et al. (2024); Hu et al. (2025)
Glutathione S-transferases (GSTs)	Conjugation of electrophilic compounds/metabolites with glutathione (GSH).	Co-induced with CYPs under chemical stress.	Alnasser (2025); Zhou et al. (2025)
Oxidoreductases	Reduction of quinones and scavenging of ROS (e.g., $O_2{\bullet}^-$, H_2O_2) to prevent cellular damage.	Induction by oxidative stress from pollutant exposure.	García-Muñoz et al. (2025); Qin et al. (2025)

used to coordinate such processes and post-transcriptional regulators that refine the expression of genes by targeting mRNA stress responses (Li et al., 2025).

These molecular changes occur both cellular and physiological as adaptive metabolic changes. Plants, e.g., can respond to triclosan-induced changes in root development by disrupting polar auxin transport (Zheng et al., 2025) or can synthesize defensive secondary metabolites, e.g., flavonoid, to mitigate oxidative damage (Hurmat et al., 2025). Animal cells also develop oxidative stress with exposure to pollutants, which is demonstrated by the presence of such biomarkers as catalase and malondialdehyde (MDA) (García et al., 2025). Eventually, these coordinated reactions are part of the community and ecosystem scales, whereby rhizosphere and wetland symbiotic plantmicrobial relations are allied to overall resilience, where phytoremediation occurs in large-scale and phytotransformative processes of phytotransformation, and release of functional enzymes (Hang et al., 2025; Liu et al., 2025).

12.3 DETOXIFICATION ENZYMES TRANSCRIPTIONAL REGULATION

Transcriptional regulation of the detoxification enzymes is one of the primary reactions to environmental pollutants that enable an organism to accelerate the creation of the proteins needed to overcome the toxic chemicals. This regulation is organized through complex cross-cellular signaling pathways thatsignal chemical stress and stimulate certain transcription factors. These later attach to promoter sequences of important enzyme-coding genes, leading to their joint expression (Li et al. 2024; Joseph et al., 2025). As an example, cytochrome P450 monooxygenases, GSTs, and other oxidoreductases can be expressed due to exposure to heavy metals and organic xenobiotics, which can form a critical defense mechanism and counter cellular damage and maintain metabolic homeostasis (Cao et al., 2025; Krówczyńska and Pietrowska-Borek, 2025). The enzymatic mechanisms involved in pollutant detoxification are summarized in Table 12.2.

12.3.1 Induction of Gene by Exposure to Pollutants

The detection of pollutants by cellular sensors triggers gene induction pathways, initiating signal transduction cascades. Some of the relevant pathways include transcription factors like planar aromatic hydrocarbon receptor (AhR) that binds planar aromatic hydrocarbons and the Nrf2-Keapl system that reacts to oxidative and electrophilic stress. When activated, these factors are translocated to the nucleus and bind to a particular set of DNA sequences, including the xenobiotic-responsible element (XRE) or the antioxidant-responsible element (ARE), and thus trigger the expression of the detoxification genes (Li et al., 2024; Mosharaf et al., 2024). This rapid transcriptional response is conserved across species, for example, in zebrafish, antidepressants can change the expression of antioxidant genes, whereas in insects such as *Cydia pomonella*, plant allelochemicals can cause a global transcriptomic change, with many detoxification pathways.

12.3.2 Enzyme Systems Cytochrome P450 Monooxygenases

Cytochrome P450 monooxygenases (CYPs) are a heme-thiolate superfamily of enzymes that catalyze the phase I metabolism of an enormous range of lipophilic contaminants, comprising pharmaceuticals, pesticides, and persistent organic contaminants. Their main mechanism of action includes the addition of a hydroxyl group, which elevates the hydrophilicity of the compound to allow further conjugation (Mokhosoev et al., 2024). They are greatly inducible in their expression, e.g., the rice transcriptome demonstrates that many CYP genes are upregulated in response to the herbicide atrazine, facilitating its metabolic detoxification (Tan et al., 2015). Likewise, in microalgae, certain CYP isoforms are highly induced in the process of biodegradation of contaminants including octocrylene and perfluorooctanoic acid (PFOA), indicating their evolutionarily conserved role in xenobiotic detoxification (Hu et al., 2025; Li et al., 2025).

12.3.3 Enzyme Systems – Glutathione S-transferases and Oxidoreductases

Phase II enzymes such as GSTs play a central role in conjugating electrophilic contaminants (or their phase I metabolites) with reduced glutathione (GSH) so that they can be excreted. This system plays an important role in the detoxification of pesticides, heavy metals, and organic hydrocarbons (Alnasser, 2025). It is common to induce them in parallel with CYPs; transcriptomic analysis of the fish *Trachinotus ovatus* and the insect *C.pomonella* demonstrates that GST and CYP genes are simultaneously induced in response to chemical stress (Cao et al., 2025; Zhou et al., 2025). This system is supplemented by oxidoreductases, such as NADPH-quinone oxidoreductase (NQO1) and other peroxidases that help mitigate oxidative stress commonly linked with the presence of pollutants, lowering levels of reactive quinones and scavenging ROS to prevent cell lesions (García-Muñoz et al., 2025; Qin et al., 2025).

12.4 POST-TRANSLATIONAL MODIFICATIONS AND DIRECT MODULATION

The enzyme activity can be easily and rapidly modified by environmental pollutants via post-translational modulations (PTMs); no change in gene expression is required. One important pathway is nitrosative stress, whereby the pollutants activate the formation of reactive nitrogen species (RNS). It results in the protein S-nitrosylation, that is, covalent binding of a nitric oxide (NO) group to thiols of cysteine, and tyrosine nitration, which can severely alter the structure, activity, and stability of the proteins (Jindal & Seth, 2023). An example is that these NO-mediated PTMs play a crucial role in the control of the plant antioxidant defense system and acclimation to abiotic stress conditions like exposure to heavy metals and salinity. S-nitrosylation is reversible and thus can be used as a dynamic redox-sensing pathway, allowing metabolic pathways to respond to oxidative and nitrosative stresses caused by pollutants.

12.4.1 Pollutants Allosteric Inhibition and Activation

Allosteric sites are different sites that can be occupied by natural compounds and pollutants that regulate enzyme activity with high accuracy, as opposed to the active site of the enzyme. One of the best illustrations of allosteric inhibition is the effect of immunosuppressant mycophenolate mofetil (MMF). Its active metabolite, mycophenolic acid (MPA) is an allosteric, non-competitive inhibitor of inosine-5-monophosphate dehydrogenase (IMPDH), which plays a role in the production of the guanine nucleotides. The binding triggers a conformational alteration that inhibits the activity of IMPDH, which is the basic mechanism of its therapeutic and environmental toxicity. On the other hand, the protection of allosteric activity can be provoked by natural compounds. The chlorogenic acid polyphenol (CGA) has been reported to interact with Estrogen Receptor 1 (ESR1) with high affinity, and the interaction has been demonstrated by surface plasmon resonance (Wu et al., 2025). This binding is hypothesized to be an allosteric one, and it affects the conformation of the receptor and consequent transcriptional activity to trigger neuroprotective mechanisms to respond to MMF-induced toxicity. It describes a two-fold paradigm, whereby the pollutants can be allosteric inhibitors, and dietary phytochemicals can be allosteric activators of defense mechanisms.

Figure 12.1 illustrates the molecular mechanism of protein dysregulation by environmental pollutants. In the

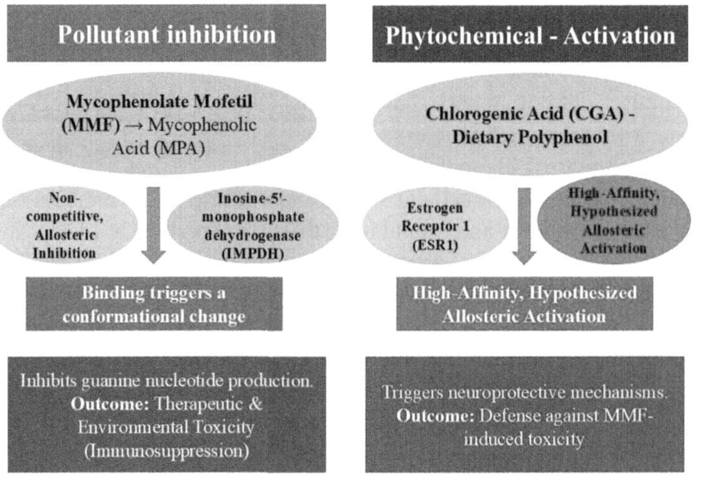

FIGURE 12.1 Molecular mechanisms of protein dysregulation by environmental pollutants.

image panel (a) shows the indirect modulation via oxidative/nitrosative stress leading to PTMs such as S-nitrosylation. The (b) panel of the image illustrates direct modulation by pollutant binding to allosteric sites causing inhibition, NO – nitric oxide; MDA – reaction electrophile; PTM – post-translational modification.

12.4.2 Covalent Changes Brought About by the Xenobiotic Exposure

One of the common effects caused by xenobiotic exposure is oxidative stress, which results in permanent covalent changes on cellular macromolecules, one of the major forms of enzyme inactivation and cytotoxicity. One of the important effects is lipid peroxidation, in which polyunsaturated fatty acids in cell membranes are the targets of the ROS. Such a reaction is a chain reaction that produces very reactive electrophilic aldehydes, including MDA. MDA can be used to create stable covalent adducts with proteins, such as enzymes, by reacting with Michael addition with lysine, cysteine, and histidine residues. As one example, the activation of the MMF led to a 4.6-fold rise in the MDA levels in PC12 cells, which revealed the wide presence of lipid peroxidation and, consequently, the cross-linking of a protein and its inactivation (Wu et al., 2025). The resultant covalent protein modification can cause enzyme inactivation, aggregation, and eventually disrupt the normal functions of the cells, such as mitochondrial respiration and intracellular signaling. The oxidative damages caused by xenobiotics are a long-term fingerprint of covalent adduct deposition.

12.5 OXIDATIVE STRESS INDIRECT REGULATION

Another significant indirect way through which the activity of enzymes is affected by environmental pollutants is the activation of oxidative stress. Many pollutants disrupt the metabolism of the cells, thereby causing excessive production of ROS. This triggers a pro-oxidant atmosphere that overburdens the antioxidant defense mechanism, leading to a chain of deleterious events. The resultant oxidative stress indirectly controls enzyme activity by changing the cell redox state, inducing oxidative stress on proteins, and activating particular redox-sensitive signaling pathways that regulate enzyme activity (Wu et al., 2025; Zhang et al., 2025).

12.5.1 Generation of Reactive Oxygen Species by the Pollutants

The surplus production of ROS, such as superoxide anion ($O_2\bullet^-$), hydrogen peroxide (H_2O_2), and hydroxyl radical (•OH), is directly catalyzed by environmental pollutants (Figure 12.2). The addition of the immunosuppressant MMF in PC12 cells led to a 2.8-fold rise in the level of ROS, which proved the strong pro-oxidant effect (Wu et al., 2025). On a similar note, the polycyclic aromatic hydrocarbon (PAH) pyrene is a significant stimulant of ROS generation in the rhizosphere of the plant, where the levels of the H_2O_2, OH, and O_2 were found to be 1.5–7.8 times the control samples (Zhang et al., 2025). Moreover, organic pollutant bisphenol A (BPA) causes a severe disturbance of redox homeostasis in *Chlamydomonas reinhardtii*, causing the intensification of ROS and the initiation of oxidative damage processes, including ferroptosis (Carbó et al., 2023). These results prove the fact that various pollutants, including pharmaceuticals and industrial chemicals, have in common the possibility to trigger the ROS formation, and, therefore, the onset of the state of oxidative stress.

12.5.2 Modulation of Cell Redox State and Effect on Redox-Sensitive Enzymes

The increase in the levels of the pollutant triggers a significant shift in the cellular redox state, which swings the balance toward the oxidation state and affects the activity of the redox-sensitive enzymes directly. One of the major effects is the deactivation of antioxidant defense enzymes themselves. An example is that MMF treatment led to a reduction of the activity of superoxide dismutase (SOD) by 40, which is an important enzyme that neutralizes superoxide radicals (Wu et al., 2025). This forms a vicious cycle of increasing oxidative damage. In addition to direct

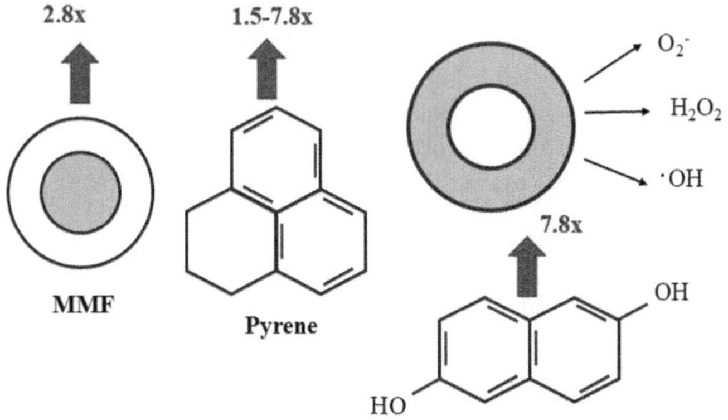

FIGURE 12.2 Generation of ROS by the pollutants.

inhibition, the disturbed redox condition is a signal. The protective role of chlorogenic acid (CGA) against MMF is related to its direct association with the estrogen receptor 1 (ESR1), which can be regulated by the cellular redox condition (Pal et al., 2024). RS such as hydroxyl radicals (·OH) directly catalyze the destruction of pollutants that include pyrene in microbial systems, and the modified redox state contributes to the enhanced functional capacity of the degrading microbial community (Ajao et al., 2025). The redox state is a master controller therefore, it inactivates certain enzymes and activates others as part of the response to stress and detoxification processes.

12.6 ADAPTIVE SIGNIFICANCE AND FUNCTIONAL CONSEQUENCES

Enzymatic reaction to the environmental pollutants is not simply a response but a complex adaptive mechanism. Organisms are able to modulate enzyme activity to transform hazardous substances into less toxic products, store them to reduce metabolic disturbance, and begin to be excreted, which guarantees their survival in a polluted environment and allows the ecosystem to continue functioning (Pal et al., 2024; Liao et al., 2025).

12.6.1 Biotransformation and Detoxification Pathways

Biotransformation catalyzed by specific enzymatic systems is the primary method of detoxification. The cytochrome P450 monooxygenases (CYPs) are the most important enzymatic family in the primary oxidative change in a vast range of organic pollutants, including pharmaceuticals, pesticides, and polycyclic aromatic hydrocarbons (PAHs). These enzymes add functional moieties that make the xenobiotics susceptible to conjugation (Li et al., 2023; Tan et al., 2015). An example is microalgal systems, where CYP450s are involved in the crucial first decarboxylation of PFOA and hence initiate its biodegradation to less persistent and shorter-chain compounds (Li et al., 2025). After the phase I oxidation, phase II conjugation enzymes, especially GSTs, catalyze the covalent conjugation of xenobiotic molecules or their metabolites with glutathione and increase aqueous solubility and fecal excretion (Alnasser, 2025; Zhou et al., 2025). Furthermore, because hydrolytic enzymes are used to hydrolyze ester, amide, and glycosidic bonds in organophosphate pesticides and other similar pollutants, their toxicity is directly nullified by the hydrolytic enzymes themselves, or the nanozymes that mimic these enzymes (Liao et al., 2025).

12.6.2 Sequestration and Discharge Cycles

In cases where complete biochemical detoxification is impossible, organisms employ sequestration and excretion mechanisms to limit the distribution of the pollutants in critical cellular organelles. These mechanisms involve enzyme processes. In both plant and microbial taxa, exposure to heavy metals leads to the biosynthesis of chelators like phytochelatins and metallothioneins, the synthesis of which is strictly controlled by the enzymatic pathways. These chelators trap metals to be deposited or used in the cell wall matrices (Jobby et al., 2024; Zhang & Li, 2025) to reduce the cytosolic level of reactive metal ions. On organic contaminants, active efflux forms one of the major excretion forms. The energy-dependent transporter proteins (serving effectively as an enzymatic pump) in the bacterial cells use the ATP hydrolysis or proton motive force to get rid of the toxic substances (Priyadarshanee et al., 2022). Also, extracellular ligninolytic enzymes, laccases, and peroxidases that are produced by fungi and bacteria, in the context of bioremediation, polymerize or immobilize the pollutants on the soil organic matter, and thus effectively trap pollutants in the environment, making them less available (Shah et al., 2024; Qin et al., 2025).

12.7 APPLICATIONS AND IMPLICATIONS

The understanding of how enzymes are regulated by environmental pollutants has far-reaching practical applications that can be used to develop advanced biomarkers, optimize bioremediation systems, and enhance ecological and human health risk evaluation (Jiang et al., 2025; Thanigaivel et al., 2025).

12.7.1 Advancement of Complex Biomarkers to Monitor the Environment

The basis of sophisticated biomarker systems is enzymatic and stress-response systems. Sensitive, sub-lethal signals of pollutant exposure have been reported as the induction of the specific enzymes, such as CYP450s, GSTs, or antioxidant systems (e.g., SOD, catalase) in bioindicator species (Patel et al., 2024; García-Muñoz et al., 2025). The use of this principle in advanced biosensing technologies is achieved. High-throughput toxicity screening is currently being developed using genetically engineered whole-cell bacterial bioreporters that emit a detectable signal (e.g., bioluminescence) in response to stress caused by pollutants (Lautan et al., 2025; Sulivan et al., 2025). In addition, the existence of a positive relationship between the contaminant levels in keratinized tissues (e.g., hair, spines) and internal oxidative-stress biomarkers in wildlife (e.g., hedgehogs) supports the use of these non-invasive samples as a surrogate of internal enzymatic disturbance and contaminant burden (García-Muñoz et al., 2025).

12.7.2 Bioremediation Strategies: Improving by Understanding Enzymes

Bioremediation strategies are transformed by a deeper understanding of enzymes and their functions. The modern approaches focus on bioaugmentation using microbial strains that contain strong degradative enzymes (e.g., laccases, CYP450s, dehalogenases) and biostimulation methods aimed at enhancing the activity of indigenous enzymes (Martínez-Burgos et al., 2024; Abu-Tahon et al., 2025). The immobilization of the enzyme to the nano-biochar or other carriers improves the stability and re-use of the catalyst, especially with ex-situ treatment of wastewater (Ibrahim et al., 2025). Synthetic materials that replicate the action of natural enzymes (the so-called nanozymes) offer robust and economical alternatives to pollutant degradation and pollutant detection and can be made using engineered composites, specifically SiO2@APTCuMnO2, which has proven to be highly efficient and reusable (Chen et al., 2025; Liao et al., 2025). Moreover, genetic engineering via CRISPR also offers a precise recombination of microbial enzymatic pathways, which spawn so-called super-degraders, specialized for certain contaminants (Ibrahim et al., 2025).

12.7.3 The Risk Assessment of Ecological and Human Health

One of the parts of proper risk assessment is enzyme-mediated processes. The detailed knowledge of biotransformation pathways is required since specific enzyme reactions could trigger pro-carcinogenic intermediates or produce metabolites more toxic and mobile than the original compound (Kim et al., 2025; Wang et al., 2025). The adverse outcome pathway (AOP) framework merges molecular initiating events (e.g., inhibition of a certain enzyme (e.g., acetylcholinesterase through pesticides) or molecular) with downstream physiological and population consequences (Kuppan et al., 2024; Wu et al., 2025). Comprehensive and efficient substitutes to animal testing alternatives are provided by high-throughput in-vitro assays, which measure enzymatic inhibition and cellular stress response, and computational ToxPi analyses (Li et al., 2025; Thiruvengadam et al., 2025).

12.8 CONCLUSION AND FUTURE PERSPECTIVES

Control of enzymatic activity has been one of the foundations of the biological response to environmental pollution, and it is a complicated web of processes that define the destiny and impacts of contaminants. The major regulatory processes create a multi-level defense system. Genetically, there is transcriptional upregulation of those genes involved in detoxification enzymes (e.g., CYP, GST, mer operon) in response to pollutant exposure via particular transcription factors and signaling cascades of ROS, hormones, and second messengers. Post-translational changes include S-nitrosylation and nitration of tyrosine and provide quick reversible regulation of enzymatic activity in response to nitrosative stress. In addition, long-term adaptive responses are fine-tuned by epigenetic changes of a form of DNA methylation, and histone acetylation, which may also result in transgenerational tolerance. The interaction between microbiomes and plants represents a greater degree of control, where the rhizosphere attracts and maintains microbial communities whose combined enzyme repertoires will improve host resilience and their ability to degrade. New frontiers will see ecotoxicology be revolutionized, multi-omics (metagenomics, transcriptomics, proteomics, metabolomics) combined with improved bioinformatics and machine learning will facilitate system predictions of enzymatic behaviors and pollutant fate in multi-level biological contexts. The necessity to consider the ecotoxicology of complex mixtures of pollutants (e.g., nanoplastics and heavy metals or pharmaceuticals) is increasing, and its regulation by co-exposure is a specific and highly fascinating topic that requires deeper research. The creation of in-situ, real-time biosensing networks with the help of nano-enabled sensors and IoT technology will give dynamic information about enzymatic biomarkers and pollutant fluxes. Lastly, synthetic biology presents unparalleled possibilities for the design and engineering of novel enzymatic pathways and consortia toward targeted bioremediation of intractable pollutants like PFAS and microplastics.

REFERENCES

Abu-Tahon, MA, Housseiny, MM, Aboelmagd, HI, Daifalla, N, Khalili, M, Isichei, AC, Ramadan, A, Abu El-Saad, AM, Seddek, NH, Ebrahim, D, Ali, YH, Saeed, IK, Rikabi, HA, Eltaib, L. 2025. A holistic perspective on the efficiency of microbial enzymes in bioremediation process: mechanism and challenges: a review. Int J Biol Macromol. 308(Part 2):142278.

Ajao, AT, Awe, S, Aborisade, WT, Jimoh, FA, Ajao, FJ. 2025. Bioremediation of soil contaminated with cassava mill effluents a structural, functional, and metagenomics approach. Next Res 2(3):100439.

Alnasser, SM. 2025. The role of glutathione S-transferases in human disease pathogenesis and their current inhibitors. Genes Dis. 12(4):101482.

Bartha, B, Huber, C, Schröder, P. 2014. Uptake and metabolism of diclofenac in *Typha latifolia* – how plants cope with human pharmaceutical pollution. Plant Sci. 227:12–20.

Cao, X, Li, G, Ye, X, Sattar, A. 2025. Transcriptome and detoxification enzyme–related gene analysis of the *Cydia pomonella* response to walnut husks. Ecotoxicol Environ Saf. 300:118430.

Carbó, M, Chaturvedi, P, Álvarez, A, Pineda-Cevallos, D, Ghatak, A, González, PR, Cañal, MJ, Weckwerth, W, Valledor, L. 2023. Ferroptosis is the key cellular process mediating Bisphenol A responses in Chlamydomonas and A promising target for enhancing microalgae-based bioremediation. J Hazard Mater. 448:130997.

Chen, X, Tao, W, Xu, R, Mo, T, Yang, X, Cavaco-Paulo, A, Su, J, Wang, H. 2025. Construction of a stable manganese/copper core-shell laccase-like enzyme for recyclable dye decolorization and phenolic pollutant detection. J Hazard Mater. 495:139131.

Dietz, AC, Schnoor, JL. 2001. Advances in phytoremediation. Environ Health Perspect. 109:163–168.

García-Muñoz, J, Rodríguez Somoza, E, Nuñez Zafra, I, Fernández-Casado, D, Portillo-Moreno, Á, Míguez-Santiyán, MP, Soler Rodríguez, F, López-Beceiro, A, Fidalgo, LE, Martínez-Morcillo, S, Pérez-López, M. 2025. Oxidative stress biomarkers, metal(loid)s and persistent organic pollutants in

European hedgehogs: an integrated approach to environmental contamination. Environ Res. 280:121957.

Hang, S, Wu, C, Peng, G, Li, F, Ge, F, Gan, J, Li, L, Li, J. 2025. Rhizosphere endophytes as allies in plant defense against heavy metals and organic pollutants in soil: advances and applications. Appl Soil Ecol. 211:106119.

Hu, JJ, Pan, CG, Xu, XR, Peng, FJ, Liu, S, Pei, J, Liu, ZZ, Liang, H, Yu, K. 2025. Cytochrome P450-mediated biotransformation of octocrylene in *Tetradesmus obliquus*: integrative insights from transcriptomics, proteomics, molecular docking, and density functional theory calculation. J Hazard Mater. 496:139512.

Hurmat, Shri, R, Sood, P, Kaur, G. 2025. Exploring different environmental impacts: how abiotic factors shape flavonoid synthesis in plants. Biocatal Agric Biotechnol. 68:103726.

Ibrahim, A, Oginga, B, Zhang, Y, Ling, W, Tang, L, Elatafi, E, Abady, M, Gao, Y. 2025. Bioremediation of soils with emerging organic contaminants using immobilized microorganisms. Environ Technol Innov. 40:104345.

Ivanov, KL, Wagenpfahl, A, Deibel, C, Matysik, J. 2017. Spin-chemistry concepts for spintronics scientists. Beilstein J Nanotechnol. 8:1427–1445.

Jiang, L, Shi, Y, Bai, L, Zhao, Q, Peng, H, Yang, X, Liu, Y, Qu, G, Zhou, Q, Sun, Y, Fu, J, Liu, J, Pei, Z, Lv, J, Yuan, R, Zhao, L, Li, Y, Li, J, Mao, L, et al. 2025. Breakthroughs in new pollutant research: analytical methods, environmental processes, toxicological mechanisms, and control strategies. J Environ Sci. [advance online publication].

Jindal, A, Seth, CS. 2023. Nitric oxide mediated post-translational modifications and its significance in plants under abiotic stress. In: Khan MIR, Iqbal N, Poór P, Ferrante A, Singh VP, Tripathi DK, Fotopoulos V, editors. Nitric oxide in developing plant stress resilience. Academic Press. p. 233–250.

Jobby, R, Sarkar, M, Bose, R, Srivastava, S, Suprasanna, P. 2024. Chromiomics: chromium detoxification and approaches for engineering tolerance in plants. Environ Pollut. 350:123991.

Joseph, R, Apollon, W, De Oliveira, AC. 2025. The stressing point: how plants respond to environmental stimuli. Biotechnol Adv. 84:108691.

Kim, JW, Tung, HC, Yang, B, Pant, R, Guan, X, Feng, Y, Xie, W. 2025. Heme-thiolate monooxygenase cytochrome P450 1B1, an old dog with many new tricks. Pharmacol Rev. 77(3):100045.

Krówczyńska, N, Pietrowska-Borek, M. 2025. From signal perception to adaptive responses: a comprehensive review of plant mechanisms under cadmium, lead, and aluminum stress. Environ Exp Bot. 238:106240.

Kuppan, N, Padman, M, Mahadeva, M, Srinivasan, S, Devarajan, R. 2024. A comprehensive review of sustainable bioremediation techniques: eco friendly solutions for waste and pollution management. Waste Manag Bull. 2(3):154–171.

Lautan, O, Chen, SN, Lu, CC, Huang, YH, Chen, HL, Chih, PS, Chen, RJ. 2025. Effect-based high-throughput screening of environmental water samples: focus on liver toxicity pathways and endocrine disruption. J Hazard Mater. 496:139432.

LeFevre, GH, Müller, CE, Li, RJ, Luthy, RG, Sattely, ES. 2015. Rapid phytotransformation of benzotriazole generates synthetic tryptophan and auxin analogs in Arabidopsis. Environ Sci Technol. 49:10959–10968.

Li, R, Zhao, Y, Dai, J, Zhang, N, Yuan, J, Liu, D, He, Z, Yang, J, Gu, S. 2025. Emerging roles of RNA methylation modifications in environmental metalloid or metal pollutants toxicity. Ecotoxicol Environ Saf. 299:118384.

Li, S, Chu, Y, Ren, N, Ho, S.H. 2023. Cytochrome P450 enzyme-based biotransformation of pharmaceuticals and personal care products (PPCPs) by microalgae in the aquatic environment. Chem Eng J. 476:146557.

Li, Y, Chen, B, Yang, S, Jiao, Z, Zhang, M, Yang, Y, Gao, Y. 2025. Advances in environmental pollutant detection techniques: enhancing public health monitoring and risk assessment. Environ Int. 197:109365.

Li, Z, Yang, J, Cai, X, Zeng, X, Zou, J. J, & Xing W. 2024. A systematic review on the role of miRNAs in plant response to stresses under the changing climatic conditions. Plant Stress. 14:100674.

Liao, X, Wang, X, Li, B, Wang, L, Chen, Y. 2025. Hydrolytic nanozymes: classification, activity regulation, and applications in sensing and degradation of environmental pollutants. Coordin Chem Rev. 541:216807.

Liu, L, Fan, X, Han, Y, Wang, H. 2025. Enhanced removal of heavy metals by wetland plant-microbiome symbiont: prospect of potential strategies and mechanisms for environmental heavy metal regulation. Pedosphere. 35(1):116–136.

Martínez-Burgos, WJ, Vandenberghe, LPS, de Mello, AFM, de Carvalho, JC, Valladares-Diestra, KK, Manzoki, MC, Scapini, T, Pozzan, R, Liew, RK, Thomaz-Soccol, V, Soccol, CR. 2024. Bioremediation strategies against pesticides: an overview of current knowledge and innovations. Chemosphere. 364:142867.

Mokhosoev, IM, Astakhov, DV, Terentiev, AA, Moldogazieva, NT. 2024. Cytochrome P450 monooxygenase systems: diversity and plasticity for adaptive stress response. Prog Biophys Mol Biol. 193:19–34.

Morsi, R, Bilal, M, Iqbal, HM, Ashraf, SS. 2020. Laccases and peroxidases: the smart, greener and futuristic biocatalytic tools to mitigate recalcitrant emerging pollutants. Sci Total Environ. 714:136572.

Mosharaf, MK, Gomes, RL, Cook, S, Alam, MS, Rasmusssen, A. 2024. Wastewater reuse and pharmaceutical pollution in agriculture: uptake, transport, accumulation and metabolism of pharmaceutical pollutants within plants. Chemosphere. 364:143055.

Pal, P, Pramanik, K, Ghosh, SK, Mondal, S, Mondal, T, Soren, T, Maiti, TK. 2024. Molecular and eco-physiological responses of soil-borne lead (Pb^{2+})-resistant bacteria for bioremediation and plant growth promotion under lead stress. Microbiol Res. 287:127831.

Patel, A, Tiwari, S, Khandelwal, A, Singh, C, Pandey, N, Tiwari, A, Yadav, S, Prasad, SM. 2024. Plants as biomarkers for monitoring environmental pollution. In: Mishra R, Madhav S, Dhaka RK, Garg P, editors. Biomarkers in environmental and human health biomonitoring. Academic Press. p. 169–184.

Pilon-Smits, E. 2005. Phytoremediation. Annu Rev Plant Biol. 56:15–39.

Priyadarshanee, M, Chatterjee, S, Rath, S, Dash, HR, Das, S. 2022. Cellular and genetic mechanism of bacterial mercury resistance and their role in biogeochemistry and bioremediation. J Hazard Mater. 423(Part A):126985.

Qin, W, Lin, Y, Xu, L, Li, Q, Tang, A, Li, Q, Liang, H, Liu, Y. 2025. NADH flavin oxidoreductase and catalase-induced reactive oxygen species and key enzymes synergistically drive oxidative degradation of xanthene dye Rose Bengal in Aspergillus flavus A5P1. Int J Biol Macromol. 314:144136.

Sandermann, H Jr. 1994. Higher plant metabolism of xenobiotics: the 'green liver' concept Pharmacogenetics. 4:225–241.

Saravanan, A, Kumar, PS, Vo, DVN, Jeevanantham, S, Karishma, S, Yaashikaa, PR. 2021. A review on catalytic-enzyme degradation of toxic environmental pollutants: microbial enzymes. J Hazard Mater. 419:126451.

Schreiner, PR. 2020. Quantum mechanical tunneling is essential to understanding chemical reactivity. Trends Chem. 2:980–989.

Shah, K, Ray, S, Bose, H, Pandey, V, Wohlschlegel, JA, Mahendra, S. 2024. Proteomics insights into the fungal-mediated bioremediation of environmental contaminants. Curr Opin Biotechnol. 90:103213.

Sulivan, J, Thomas, L, Gérald, T. 2025. A bioelectronic tongue to estimate the toxicological intensity of pollutants in wastewater treatment plant. Water Res. 279:123470.

Sun, C, Dudley, S, McGinnis, M, Trumble, J, Gan, J. 2019. Acetaminophen detoxification in cucumber plants via induction of glutathione S-transferases. Sci Total Environ. 649:431–439.

Tan, LR, Lu, YC, Zhang, JJ, Luo, F, Yang, H. 2015. A collection of cytochrome P450 monooxygenase genes involved in modification and detoxification of herbicide atrazine in rice (*Oryza sativa*) plants. Ecotoxicol Environ Saf. 119:25–34.

Thanigaivel, S, Shaji, A, Kamalesh, R, Thamarai, P, Saravanan, A, Anbarasu, K, Vickram, AS. 2025. Advances in smart biosensors for real-time pollutant monitoring and detoxification: mechanisms, applications, and challenges. TrAC Trend Anal Chem. 193:118483.

Thiruvengadam, M, Chi, HY, Choi, HJ, Jung, BS, Lee, SB, Park, Y, Jeon, D, Ciftci, F, Shariati, MA, Kim, SH. 2025. Sustainable and smart nano-biosensors: integrated solutions for healthcare, environmental monitoring, agriculture, and food safety. Ind Crop Prod. 233:121337.

van der Kamp, MW, Mulholland, AJ. 2013. Combined quantum mechanics/molecular mechanics (QM/MM) methods in computational enzymology. Biochemistry. 52:2708–2728.

Wang, J, Chen, Y, Zhang, Z, Yan, J, Ling, W, Chen, X, Gao, Y. 2025. Occurrence, fate and impact of steroidal estrogens and xenoestrogens in plants: a review. J Environ Manage. 389:126194.

Wu, J, Guo, JJ, Tong, FC, Zhang, SL, Zhang, JL, Chen, MY, Yu, HC. 2025. Chlorogenic acid counteracts mycophenolate mofetil-induced neurotoxicity in zebrafish and PC12 cells: a potential bioremediation strategy for aquatic pharmaceuticals. Environ Int. 203:109771.

Xie, Y., Lyu, Y, Irshad, S, Liu, X, Jiang, Y, Sun, Y, Zhang, Z, Wei, H, He, H, An, C, Zhang, P. 2025. Combined interactions and ecotoxicological effects of micro/nanoplastics and organic pollutants in soil–plant systems: a critical overview. Environ Sci Adv. 4(8):1166–1180.

Zhang, H, Li, G. 2025. Microbial strategies for enhancing nickel nanoparticle detoxification in plants to mitigate heavy metal stress. Phyton-Int J Exp Bot. 94(5):1367–1399.

Zheng, L, You, L, Wu, S, He, S, Ning, K, Lei, J, Wang, C, Wang, G, Han, H, Guo, H, Wang, J, Zhang, Y, Liu, J, Li, C. 2025. The avoidance response in plant root to the environmental pollutant triclosan. J Hazard Mater. 496:139216.

Zheng, S, Guo, H. 2021. The ABCs of membrane transport in plants. In: Merchant SS, editor. Annual plant reviews online. John Wiley & Sons, Ltd. p. 255–290.

Zhou, JM, Zhu, TF, Guo, HY, Xian, L, Liu, BS, Zhang, N, Zhang, TY, Zhu, KC, Zhang, DC. 2025. Identification and functional analysis of glutathione S-transferase gene family in *Trachinotus ovatus*: transcriptomic response to environmental and pathogenic stress. Dev Comp Immunol. 167:105371.

13 Role of Enzyme Systems in Plant Growth and Development

K. Gayathri, P. Saranraj, M. Manigandan, Glaucio Dire Feliciano, and B. Lokeshwari

13.1 INTRODUCTION

Plants are the base autotrophs, meaning these organisms produce biomass via hot chemical processes that capture energy from the sun to convert inorganic precursors into organic matter. The production of organic matter via photosynthesis, and the subsequent growth, development, and survival of a complex multicellular organism from a single fertilized egg, is due to the activity of a large and exceptionally diverse group of molecular machines: enzymes (Rahman et al., 2025). Enzymes are the forgotten heroes of plant biology, the critical agents that carry out the genetic blueprint with fidelity and efficiency. Without enzymes, the chemistry that constitutes life would take place far too slowly for life to be sustained. Enzymes are the kinetic force that convert the static information in DNA, into the dynamic processes of growth, metabolism, or adaptation. From the vast canopy of a rainforest to a single blade of grass in a picked lawn, every act of plant form or function is a reflection of the regulated activity of enzymes (Sathya et al., 2025).

Enzymes are typically proteins though catalytic RNA molecules also exist that accelerate biochemical reactions by several orders of magnitude by lowering the activation energy required. They achieve this feat without being consumed in the process, operating through a specific region known as the active site. The classic "lock and key" model, proposed by Emil Fischer, has been refined into the "induced fit" hypothesis, where both the enzyme and substrate undergo conformational changes to achieve optimal catalytic alignment (Wang et al., 2025). This specificity is paramount; it ensures that the thousands of reactions within a plant cell occur in a controlled and directed manner, preventing metabolic chaos. Furthermore, enzyme activity is not a simple on/off switch but is exquisitely regulated by the cell through a multitude of mechanisms, including allosteric modulation by effector molecules, covalent modifications like phosphorylation, transcriptional control of gene expression, and compartmentalization within organelles. This sophisticated regulation allows plants to respond with remarkable plasticity to their environment, adjusting their metabolic flux in real-time to cues such as light, temperature, nutrient availability, and pathogen attack (Huang and Jin, 2025).

The centrality of enzymes becomes immediately apparent when examining the primary engines of plant life: photosynthesis and respiration. The process of photosynthesis, which converts light energy into chemical energy, is a cascade of enzymatic reactions. In the light-dependent reactions, the enzyme adenosine triphosphate (ATP) synthase acts as a rotary motor, using a proton gradient to synthesize ATP, the universal energy currency of the cell. Simultaneously, ferredoxin-NADP+ reductase (FNR) catalyzes the production of NADPH, a reductant (Sulaman et al., 2025). In the Calvin cycle, the enzyme ribulose-1,5-bisphosphate carboxylase/oxygenase (Rubisco) holds the singular distinction of being responsible for the majority of carbon entry into the biosphere. However, its dual nature, also catalyzing a wasteful oxygenation reaction in photorespiration, highlights how the properties of a single enzyme can have global implications for agricultural productivity. Similarly, cellular respiration, which liberates the energy stored in carbohydrates, is entirely enzyme-dependent. The glycolytic pathway, the Krebs cycle in the mitochondria, and the electron transport chain (ETC) are each a sequence of reactions catalyzed by specific enzymes, with pyruvate kinase and the various dehydrogenases serving as critical control points to meet the energy demands of the cell (Pandey et al., 2025).

Beyond energy metabolism, enzymes are the architects of the plant's physical structure. The plant cell wall, a complex and dynamic extracellular matrix, is constantly being synthesized, remodeled, and degraded by enzymatic activities. Cellulose synthase (CesA) complexes in the plasma membrane spin the crystalline microfibrils of cellulose, the primary load-bearing component. Concurrently, a variety of hydrolytic and transglycosylation enzymes, including xyloglucan endotransglucosylase/hydrolase (XTH) and expansins, facilitate selective loosening of the wall matrix in order for cells to expand (Sodhi et al., 2025). In addition, the enzymes polygalacturonase (PG) and pectin methylesterase execute pectin breakdown in a systematic manner during ripening of fruit or abscission of organs, ultimately resulting in softening of tissue and separation of tissue. Finally, the uptake of necessary nutrients such as nitrogen or sulfur is dependent on enzyme systems. Nitrate/nitrite reductase provides sequential reduction of inorganic nitrate to ammonium, which is converted to amino acids through the glutamine synthetase (GS)/glutamate synthase (GS/GOGAT) cycle, and is an important integrated cycling process between carbon and nitrogen metabolism (Abas et al., 2025). Perhaps one of the most advanced functions of enzymes is in the biosynthesis and deactivation

DOI: 10.1201/9781003669920-13

of plant hormones, the chemical messengers that regulate growth and development. The specific spatial and temporal concentrations of hormones, such as auxin, gibberellin, and ethylene, determine phenotypic outcomes (Riseh et al., 2024). Enzymes such as the YUCCA flavin monooxygenases for auxin biosynthesis and ACC synthase for ethylene biosynthesis are tightly coordinated, ensuring that hormones are synthesized when and where they are needed. On the flip side, enzymes such as cytokinin oxidase/dehydrogenase (CKX) break down hormones to end their signal (Han et al., 2025). Enzymatic control also comes into play for defense and stress responses. When plants experience a biotic or abiotic stressor, it activates its enzymatic defense mechanism. For example, NADPH oxidases purposefully produce reactive oxygen species (ROS) as a signaling molecule, and there is a complementary antioxidant system (consisting of superoxide dismutase, catalase, and ascorbate peroxidase) to scavenge excessive ROS and prevent cellular damage. Similarly, the phenylpropanoid pathway, initiated by the enzyme phenylalanine ammonia-lyase (PAL), produces an enormous array of secondary metabolites for defense, including lignin, flavonoids, and phytoalexins (Chattaraj et al., 2025).

13.2 TYPES OF ENZYMES IN PLANTS

13.2.1 Oxidoreductases

Oxidoreductase enzymes carry out oxidation–reduction reactions, which include the transfer of electrons among molecules. If a molecule loses electrons, it is said to be oxidized. When a molecule gains electrons, it is reduced. These enzymes are essential in processes that need to convert energy or in redox signaling. In plants, they are essential in both photosynthesis and respiration, as they are responsible for loading and unloading electrons through ETCs in the chloroplast and the mitochondria, respectively. Important examples include the enzyme Rubisco that catalyzes carbon fixation by carboxylating RuBP. Furthermore, cytochrome c oxidase in the mitochondrion catalyzes the reduction of oxygen to water as the final electron acceptor during oxidative phosphorylation. There are also oxidoreductases involved in stress responses, such as ascorbate peroxidase and superoxide dismutase, that scavenge ROS, and nitrate reductase (NR) that is involved in nitrogen assimilation (Kim et al., 2025).

13.2.2 Transferases

Transferases are enzymes that catalyze the transfer of a functional group, such as a phosphate, methyl, or glycosyl group, from a donor molecule to an acceptor molecule. This property permits transferases to partake in metabolic regulation, signal transduction, and the biosynthesis of cellular building blocks. Concerning plant metabolism, perhaps the most prominent subclass of transferases are kinases and enzymes that phosphorylate proteins and thereafter regulate innumerable processes. For instance, sucrose synthase, a glycosyltransferase, transfers glucose to fructose and is a major aspect of sucrose metabolism. Another relevant example is glutathione S-transferase (GST) enzyme that conjugates glutathione to herbicides and other toxins to tag them for detoxification and sequestration (Ugalde et al., 2025).

13.2.3 Hydrolases

Hydrolases catalyze the cleavage of chemical bonds through the addition of water—an example of hydrolysis. This is one of the most common forms of enzymatic reaction and it is critical for macromolecule degradation, stored reserve mobilization, and the remodeling of cellular structures. In plants, hydrolases are vital to seed germination where amylases will degrade starch and proteases will degrade stored proteins to produce energy and building blocks for the growing seedling. They are also fundamental to cell wall dynamics where enzymes such as PG are responsible for softening fruit and pectin methylesterase modify cell wall structures. Moreover, ATPase hydrolases, such as the plasma membrane H^+-ATPases, utilize the energy from ATP hydrolysis to pump protons and to form electrochemical gradients for the active transport of solutes (Chang et al., 2025).

13.2.4 Lyases

Lyases catalyze the cleavage of C–C, C–O, C–N, and other bonds by means other than hydrolysis or oxidation. These reactions often result in the formation of a double bond or the addition of a group to a double bond. A common reaction is decarboxylation, the removal of a carboxyl group releasing CO_2. Lyases are enzymes in central metabolism and the biosynthesis of important compounds. The most abundant enzyme on Earth, Rubisco, is technically a lyase as it catalyzes the carboxylation (addition to a double bond) of ribulose-1,5-bisphosphate. Another critically important lyase is 1-aminocyclopropane-1-carboxylic acid (ACC) synthase, which is the rate-limiting enzyme in the biosynthesis of the plant hormone ethylene, a regulator of ripening and senescence (Ye et al., 2025).

13.2.5 Isomerases

Isomerases catalyze the intramolecular rearrangement of atoms within a molecule, converting a substrate from one isomer to another. These reactions include racemization (changing between D- and L-forms), cis–trans isomerization, and intramolecular oxidoreductions (tautomerization). By interconverting isomers, these enzymes ensure that metabolic intermediates are in the correct form for subsequent reactions in a pathway. A classic example from glycolysis is triose phosphate isomerase, which rapidly interconverts dihydroxyacetone phosphate and glyceraldehyde-3-phosphate, allowing both products of the aldolase reaction to be used for energy production. Isomerases are also involved in the photoprotective xanthophyll cycle in chloroplasts, where they convert violaxanthin to zeaxanthin via antheraxanthin (Van Gelder et al., 2023).

13.2.6 Ligases

Ligases, commonly referred to as synthetases, function as catalysts that facilitate the joining of two molecules through a reaction that relies on the hydrolysis of a high-energy diphosphate bond within a nucleoside triphosphate – e.g., ATP. The dependency of this ligation on energy is necessary for the assembly of increasingly complex molecules from simpler precursors. Ligases are required for activity both during the biosynthesis of new molecules and during substrate activation. GS is a ligase necessary for assimilating toxic ammonium as the amino acid glutamine during nitrogen metabolism. Similarly, acetyl-CoA carboxylase is the first committed step of fatty acid biosynthesis. Acetyl-CoA carboxylase ligates a carboxyl group to acetyl-CoA ultimately yielding malonyl-CoA. Finally, aminoacyl-tRNA synthetases are ligases that link specific amino acids to their cognate tRNA and are critical for maintaining fidelity during protein synthesis (Ye et al., 2025).

13.2.7 Translocases

Translocases are a newly recognized group of enzymes that catalyze the movement of ions or molecule across membranes or move ions or molecules within membranes. They are essential for cellular energetics, compartmentalization, and transport. What is arguably the most well-known example of a translocase is ATP synthase, an incredible molecular machine that acts as a rotary motor. Located in the thylakoid membrane of chloroplasts and inner mitochondrial membrane, ATP synthase generates ATP from ADP and inorganic phosphate using energy provided by a proton gradient (proton translocation) in a process known as oxidative phosphorylation or photophosphorylation. Other translocases also mediate the exchange of metabolites across organellar membranes; for instance, the chloroplast phosphate translocator imports phosphate and exports triose phosphates (Jung et al., 2025).

Table 13.1 illustrates how oxidoreductases form the core of energy transduction and redox management, with Rubisco driving global carbon fixation, ATP synthase generating cellular energy, and enzymes like NR and the antioxidant trio (SOD, CAT, APX) linking metabolism with nutrient assimilation and stress defense. Transferases emerge as the primary architects of biosynthesis and signal relay, catalyzing critical steps from the initiation of glycolysis by hexokinase and the assimilation of nitrogen by GS to the construction of the cell wall by CesA and the activation of defense via GST. The hydrolases are the master remodelers and mobilizers, responsible for breaking down stored reserves like starch (amylase) and proteins (proteases) to fuel growth, while simultaneously modifying the cell wall through the coordinated action of PG and pectin methylesterase to enable expansion, abscission, and fruit ripening. Lyases are shown to be pivotal in creating structural and signaling molecules, such as the ethylene precursor from ACC synthase and the phenylpropanoid precursors from PAL, while also playing a specialized role in carbon concentration via PEPC in certain plant types. The isomerases, represented by triosephosphate isomerase, ensure metabolic efficiency by optimizing substrate pools, and the ligases perform the energy-intensive task of building essential molecules, from amino acids (GS) to fatty acids (ACCase) and regulating protein turnover via ubiquitin ligases. Finally, the translocases, exemplified by the dual-function ATP synthase, highlight the intimate link between bioenergetics and transport, using ion gradients to power the synthesis of ATP.

13.3 ENZYMES IN CENTRAL CARBON METABOLISM

Central carbon metabolism is the fundamental biochemical network that converts energy from sunlight or food into usable chemical energy for the cell, while also generating the essential building blocks for life (Ehinmitan et al., 2024). This network is driven by a series of enzymes biological catalysts that accelerate and regulate these vital chemical reactions. The two primary pillars of this system are photosynthesis, which captures and stores energy, and respiration, which releases that stored energy to power cellular activities (Figure 13.1).

13.3.1 Photosynthesis

Photosynthesis is the process by which plants, algae, and some bacteria convert light energy into chemical energy, stored in the bonds of carbohydrates like glucose. This process is divided into two interconnected sets of reactions: the light-dependent reactions and the light-independent Calvin cycle (Noor et al., 2025).

13.3.1.1 The Light-Dependent Reactions

The light-dependent reactions occur in the thylakoid membranes of chloroplasts, where chlorophyll and other pigments absorb sunlight. This absorbed light energy is used to split water molecules (a process called photolysis), releasing oxygen as a byproduct, and generating two crucial energy-carrier molecules: ATP and NADPH. Two enzymes are central to this process. FNR is the final enzyme in the photosynthetic ETC. It catalyzes the transfer of electrons from the protein ferredoxin to the molecule NADP+, reducing it to NADPH. This NADPH acts as a potent reducing agent, carrying high-energy electrons to the next stage of photosynthesis. The other critical enzyme, ATP synthase, is a remarkable molecular machine that harnesses the flow of protons (H+) down their concentration gradient from the thylakoid lumen back into the stroma. This proton motive force, generated by the ETC, drives the rotation of part of

TABLE 13.1
Enzymes and Their Roles in Plant Physiology

Enzyme Class	Specific Enzyme	Role in Plant Growth and Development	References
Oxidoreductases	Rubisco (ribulose-1,5-bisphosphate carboxylase/oxygenase)	Catalyzes the initial carbon fixation step in the Calvin cycle of photosynthesis. It is the most abundant enzyme on Earth and central to primary productivity.	Kim et al. (2025)
	ATP synthase	Uses the proton gradient across membranes (in chloroplasts and mitochondria) to synthesize ATP from ADP and inorganic phosphate, the primary energy currency of the cell.	Kim et al. (2025)
	Nitrate reductase (NR)	Performs the first and rate-limiting step in nitrogen assimilation, reducing nitrate (NO_3^-) to nitrite (NO_2^-).	Kim et al. (2025)
	Superoxide dismutase (SOD), catalase (CAT), ascorbate peroxidase (APX)	The antioxidant defense system that scavenges reactive oxygen species (ROS) to protect cells from oxidative damage during stress.	Kim et al. (2025)
	ACC oxidase (ACO)	The final enzyme in the ethylene biosynthesis pathway, producing the hormone ethylene which regulates fruit ripening, senescence, and abscission.	Kim et al. (2025)
Transferases	Hexokinase	The first enzyme in glycolysis, it phosphorylates glucose, trapping it in the cell and initiating its breakdown for energy.	Ugalde et al. (2025)
	Glutamine synthetase (GS)	Assimilates ammonium (NH_4^+) into the amino acid glutamine, a central step in nitrogen metabolism and the synthesis of other amino acids and nucleotides.	Ugalde et al. (2025)
	Cellulose synthase (CesA)	The central enzyme complex that synthesizes cellulose, the primary structural component of the plant cell wall, from UDP-glucose.	Ugalde et al. (2025)
	Receptor-like kinases (RLKs)	Phosphorylate target proteins in signal transduction pathways, enabling the cell to perceive and respond to hormones, pathogens, and environmental signals.	Ugalde et al. (2025)
	Glutathione S-transferase (GST)	Conjugates glutathione to herbicides, toxins, and secondary metabolites, facilitating their detoxification and compartmentalization.	Ugalde et al. (2025)
Hydrolases	α-amylase	Breaks down starch into sugars during seed germination, mobilizing stored energy reserves to fuel the growth of the embryo.	Chang et al. (2025)
	Polygalacturonase (PG)	Degrades pectin in the middle lamella of the cell wall, a process in fruit softening, organ abscission (e.g., leaf fall), and pollen tube growth.	Chang et al. (2025)
	Proteases	Hydrolyze proteins, involved in the mobilization of seed storage proteins, programmed cell death, and the regulation of specific signaling pathways.	Chang et al. (2025)
	Pectin methylesterase (PME)	Removes methyl groups from pectin, altering cell wall properties, porosity, and susceptibility to degradation by other enzymes.	Chang et al. (2025)
	ATPase (e.g., H^+-ATPase)	Hydrolyzes ATP to pump protons (H^+) across the plasma membrane, creating an electrochemical gradient that drives the transport of other nutrients.	Chang et al. (2025)
Lyases	ACC synthase (ACS)	Regulatory enzyme in ethylene biosynthesis, catalyzing the formation of ACC from SAM (S-adenosylmethionine).	Ye et al. (2025)
	Phenylalanine ammonia-lyase (PAL)	The first and committed enzyme in the phenylpropanoid pathway, producing precursors for lignin, flavonoids, and other defense compounds.	Ye et al. (2025)
	Phosphoenolpyruvate carboxylase (PEPC)	Fixes CO_2 into oxaloacetate in C4 and CAM plants, concentrating CO_2 for Rubisco and improving photosynthetic efficiency under hot, dry conditions.	Ye et al. (2025)
Isomerases	Triosephosphate isomerase (TPI)	Rapidly interconverts dihydroxyacetone phosphate and glyceraldehyde-3-phosphate in glycolysis, ensuring efficient energy extraction from sugars.	Van Gelder et al. (2023)
	VieB protein (Xanthophyll cycle)	Catalyzes the isomerization of violaxanthin to antheraxanthin and zeaxanthin, a photoprotective mechanism to dissipate excess light energy.	Van Gelder et al. (2023)
Ligases	Glutamine synthetase (GS)	A ligase that catalyzes the ATP-dependent condensation of glutamate and ammonium to form glutamine. (Note: GS is both a ligase and a transferase in function.)	Ye et al. (2025)
	Acetyl-CoA carboxylase (ACCase)	Catalyzes the first committed step in fatty acid biosynthesis, carboxylating acetyl-CoA to form malonyl-CoA.	Ye et al. (2025)
	Ubiquitin ligase (E3)	Recognizes specific target proteins and facilitates their tagging with ubiquitin, marking them for degradation by the proteasome in regulatory processes.	Ye et al. (2025)
Translocases	ATP synthase	Functions as a translocase by facilitating the movement of protons (H^+) across a membrane, using the energy of this flow to power ATP synthesis.	Jung et al. (2025)

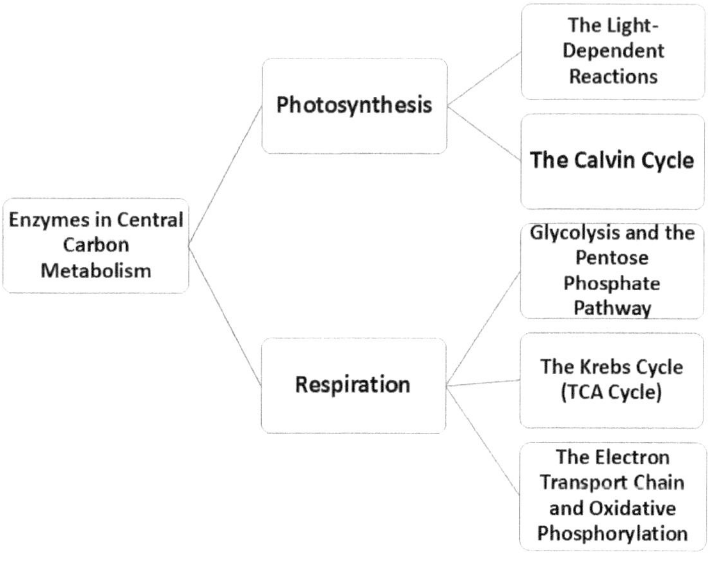

FIGURE 13.1 Enzymes in central carbon metabolism.

the ATP synthase complex, which in turn catalyzes the synthesis of ATP from ADP and inorganic phosphate, a process known as photophosphorylation (Rashkov et al., 2025).

13.3.1.2 The Calvin Cycle

The Calvin cycle, which takes place in the stroma of the chloroplast, uses the ATP and NADPH produced in the light reactions to fix atmospheric carbon dioxide into organic sugar molecules. The most pivotal enzyme in this cycle, and arguably on Earth, is Rubisco (ribulose-1,5-bisphosphate carboxylase/oxygenase). Rubisco catalyzes the first major step of carbon fixation: the attachment of a CO_2 molecule to a five-carbon sugar called ribulose-1,5-bisphosphate (RuBP). This reaction initiates a series of steps that ultimately produce glyceraldehyde-3-phosphate (G3P), which is used to regenerate RuBP and to build glucose and other carbohydrates. Rubisco's abundance is a direct result of its notoriously slow catalytic rate; cells must produce massive amounts of it to sustain adequate rates of photosynthesis, making it the most prevalent enzyme on the planet (Chen et al., 2025).

13.3.2 Respiration

Cellular respiration is an organism's method of breaking down certain organic molecules, such as glucose, in order to unlock the stored energy within those molecules, which can then be used to create ATP. This takes place in multiple steps: glycolysis, the Krebs cycle, and oxidative phosphorylation (Evans et al., 2024).

13.3.2.1 Glycolysis and the Pentose Phosphate Pathway

Glycolysis is a universal metabolic pathway that occurs in the cytoplasm, where one molecule of glucose is converted to two molecules of pyruvate. This ten-step process, with each step being catalyzed by a specific enzyme, produces a small net gain of ATP and NADH. Glycolysis does not require oxygen, and it is often the major source of energy for many anaerobic organisms. Running in parallel to glycolysis is the pentose phosphate pathway (PPP). The PPP does not produce energy, but it serves two important functions: the production of NADPH for biosynthetic reactions and antioxidant defense, and the production of ribose-5-phosphate, which is a precursor for nucleotides (DNA and RNA) (Evans et al., 2024).

13.3.2.2 The Krebs Cycle (TCA Cycle)

If pyruvate is subsequently transported into the mitochondrial matrix, where oxygen is available, glycolysis will continue with this step, which converts pyruvate to acetyl-CoA. This acetyl-CoA is then used as a substrate for the Krebs cycle (or tricarboxylic acid or TCA cycle). The Krebs cycle is a hub of metabolism within cells because it fully oxidizes acetyl-CoA, producing carbon dioxide and generating a large pool of high-energy electron carriers (NADH and $FADH_2$) in the process. The cycle, in addition to pyruvate, produces important metabolic intermediates, or carbon skeletons, that can be used in the synthesis of amino acids, nucleotides, and cellular components. This central pathway of metabolism is tightly regulated by important enzymes such as citrate synthase, isocitrate dehydrogenase, and alpha-ketoglutarate dehydrogenase (Bathe et al., 2021).

13.3.2.3 The Electron Transport Chain and Oxidative Phosphorylation

The ETC and oxidative phosphorylation take place on the inner mitochondrial membrane. The NADH and $FADH_2$ molecules generated during glycolysis and the Krebs cycle will donate their high-energy electrons to the series of protein complexes (I–IV) in the ETC. As the electrons are passed from one complex to the next, energy is released and used to pump protons (H+) from the mitochondrial matrix into the intermembrane space. This movement generates a steep electrochemical gradient, a form of potential energy. The enzyme ATP synthase, like its counterpart in chloroplasts, then allows protons to flow back down their gradient into the mitochondrial matrix. As protons flow back down their chemiosmotic gradient, ATP synthase rotates and catalyzes the phosphorylation of ADP into ATP. This is called oxidative phosphorylation, as this process uses energy from the oxidation of electron carriers ($NADH/FADH_2$) to subsequently act to phosphorylate ADP. It is decidedly the most efficient ATP producer in the cell and it is where the majority of ATP is produced during aerobic respiration (Rix and Liu, 2021).

13.4 PLANT METABOLIC PATHWAYS

13.4.1 Nitrogen Assimilation: From Inert Gas to Organic Molecules

Nitrogen is a fundamental component of amino acids, proteins, nucleic acids, and chlorophyll, making its assimilation from the environment a cornerstone of plant growth. The process transforms inert inorganic nitrogen, primarily in the form of nitrate (NO_3^-) from the soil or ammonium (NH_4^+), into vital organic compounds. This transformation is driven by a series of critical enzymes. The first step often involves NR, a highly regulated cytosolic enzyme that performs the rate-limiting reduction of nitrate to nitrite. Nitrite is then further reduced to ammonium in chloroplasts by nitrite reductase (NiR). The potentially toxic ammonium is immediately and efficiently channeled into amino acids via the GS/glutamate synthase (GS–GOGAT) cycle. In this cycle, GS catalyzes the ATP-dependent incorporation of ammonium into glutamate to form glutamine. Subsequently, glutamate synthase (GOGAT) transfers the amide group from glutamine to 2-oxoglutarate, yielding two molecules of glutamate. This cycle not only detoxifies ammonium but also serves as the primary conduit for generating the amino groups for nearly all nitrogenous compounds in the plant, effectively linking carbon skeleton metabolism with nitrogen organization (Qaderi et al., 2025).

13.4.2 Sulfur Assimilation: ATP Sulfurylase and APS Reductase

The amino acids cysteine and methionine, coenzymes like CoA, and defense compounds are all important macronutrients that sulfur plays a key role in. The incorporation of sulfur from soil sulfate (SO_4^{2-}) into organic compounds generally requires energy and is highly regulated. The first process steps of sulfate assimilation are sulfate activation by ATP sulfurylase, which uses ATP to activate sulfate into adenosine 5′-phosphosulfate (APS), a high-energy intermediate subsequently used as the activated sulfur compound in reactions. The next committed step for sulfate assimilation occurs when APS reductase catalyzes the reduction of APS to sulfite (SO_3^{2-}), using glutathione as an electron donor. Sulfite is then further reduced down to sulfide (S^{2-}) by sulfite reductase. Eventually, O-acetylserine (thiol) lyase incorporates sulfide into the amino acid skeleton of O-acetylserine to produce cysteine. Cysteine is the principal sulfur donor for all other organic sulfur compounds in cells. Regulation of APS reductase particularly and other enzymes ensures that sulfate assimilation is appropriately coordinated with nitrogen and carbon metabolism and that sulfate assimilation is responsive to stress condition when sulfur-containing defense compounds must be produced (Xu et al., 2025).

13.4.3 Lipid Biosynthesis: Acetyl-CoA Carboxylase and Fatty Acid Synthase

Lipids are crucial for plant life as the primary components of cellular membranes, as energy-rich storage reserves in seeds, and as protective surface coatings. The *de novo* biosynthesis of fatty acids, the building blocks of lipids, occurs in the plastids and is initiated by two enzyme complexes. The first committed and highly regulated step is catalyzed by acetyl-CoA carboxylase (ACCase), which carboxylates acetyl-CoA to form malonyl-CoA. This reaction consumes ATP and biotin as a cofactor, and its product, malonyl-CoA, provides the two-carbon units for chain elongation. The subsequent sequential addition of these two-carbon units to a growing fatty acid chain is performed by the fatty acid synthase (FAS) complex. The plant FAS is a dissociated type II system, in which each catalytic step is carried out by an independent enzyme. The primary reactions encompass the condensation of malonyl-ACP with the growing acyl-ACP chain by β-ketoacyl-ACP synthase (KAS), followed by further reduction, dehydration, and a second reduction. This cycle is repeated, generally until 16-carbon palmitic acid is produced, which will ultimately be a precursor for elongation and desaturation to generate the diversity of fatty acids in plant membranes and storage oils (Huercano et al., 2025).

13.4.4 Amino Acid and Nucleotide Biosynthesis

The biosynthesis of amino acids and nucleotides is the final step in the integration of carbon, nitrogen, and sulfur assimilation, providing the direct monomers that build proteins and genetic material. These routes and pathways are complex and involve a complex network of enzymes that construct the biosystems of the products of core metabolism. Amino acid biosynthesis comes from a few starting precursors from the glycolytic pathway, the PPP, and the Krebs cycle. The family of amino acids derived from glutamate (for example, glutamate, glutamine, proline, and arginine) comes from the precursor 2-oxoglutarate; likewise, the aspartate family of amino acids (for example, aspartate, asparagine, lysine, threonine, methionine, isoleucine) is derived from oxaloacetate. Transaminases (aminotransferases) are a particular group of enzymes involved in amino acid and metabolism, whose role is to transfer amino groups between amino acids and keto acids to produce new amino acids. Similarly, nucleotide biosynthesis is divided into *de novo* and salvage pathways. The *de novo* synthesis of purines and pyrimidines intricately assembles the rings from precursors, including amino acids (aspartate, glutamine, glycine), CO_2, and one-carbon units from the folate pool. Enzymes like glutamine phosphoribosyl pyrophosphate amidotransferase for purines and aspartate transcarbamoylase for pyrimidines catalyze committed steps in these pathways. The coordinated activity of these biosynthetic enzymes ensures a steady supply of building blocks to support rapid growth, cell division, and gene expression (Heidarzadeh, 2025).

13.5 ENZYMES IN THE CELL WALL

13.5.1 Cell Wall Synthesis: Cellulose Synthase (CesA)

The synthesis of the primary cell wall is a precisely orchestrated process, and the enzyme complex CesA plays the starring role. Residing within the plasma membrane, CesA complexes are nanomachines that catalyze the polymerization of glucose molecules from the substrate UDP-glucose into β-1,4-glucan chains, which crystallize outside the cell to form cellulose microfibrils. These microfibrils constitute the primary load-bearing network of the cell wall, akin to steel rebar in concrete (Shang et al., 2025). The organization of these microfibrils, directed by the cortical microtubule cytoskeleton, dictates the direction of cell expansion; cells expand perpendicular to the orientation of the microfibrils. The CesA complex itself is a rosette-shaped structure, and its activity and trafficking are highly regulated, ensuring that cellulose is deposited in the correct quantity and pattern during cell growth and differentiation. Mutations in CesA genes lead to severe defects in plant morphology, highlighting their non-redundant and critical function in determining cellular architecture and mechanical strength (Cheng et al., 2025).

13.5.2 Cell Wall Loosening and Expansion: Expansins, XTH, and Pectinases

The orderly loosening of the rigid cell wall is essential for cell expansion in a plant cell, allowing confined turgor pressure to promote cell growth. This loosening process is achieved by a plethora of proteins and enzymes that function as molecular engineers. Expansins are unique wall-loosening proteins that are believed to break the noncovalent hydrogen bonds between bundles of cellulose microfibrils and matrix polysaccharides such as xyloglucan. This apparent "wall creep" facilitates movement of the polymers while maintaining the covalent bonds equipping irreversible cell wall extension. The enzyme class xyloglucan endotransglucosylase/hydrolases (XTH) and other classes complete a "cut-and-sew" operation. Some XTHs cleave the existing xyloglucan chains, which hold bundles of cellulose microfibrils, and re-attach the newly cut xyloglucan chain to another xyloglucan chain, thereby reorganizing the xyloglucan–cellulose matrix system and inserting new confines of the xyloglucan polymer bulk, while maintaining net strength (Shang et al., 2025). Other enzymes, such as pectin methylesterases (PMEs), fine tune the pectin matrix by removing methyl groups from pectin homogalacturonan and adjusting wall electrostatic properties. Changing demethylated pectin creates a space to interact with calcium ions and forms stiffer gel networks, while the pectin is also an attractive substrate for pectin-degrading enzymes. The controlled action of PMEs, therefore, regulates wall porosity, hydration, and stiffness, making the wall more amenable to expansion or, conversely, more rigid (Zhang et al., 2025).

13.5.3 Cell Wall Degradation: Polygalacturonases, Pectin Lyases, and Cellulases

Wall synthesis and loosening are crucial for growth; the controlled degradation of wall components is equally vital for specific developmental and physiological processes. This targeted dismantling is executed by hydrolytic and lyase enzymes. PGs and pectin lyases are primary actors in pectin breakdown. PGs hydrolyze the α-1,4-glycosidic bonds in the backbone of de-esterified pectin, while pectin lyases cleave via a β-elimination mechanism. The synergistic action of these enzymes is essential for the dramatic softening of fruits during ripening, as they dissolve the pectin-rich middle lamella that glues cells together (Munzert and Engelsdorf, 2025). This same enzymatic activity is crucial for organ abscission, facilitating the clean separation of cells in the abscission zone to allow a leaf or fruit to fall. Furthermore, pollen tubes and root hairs secrete these enzymes to penetrate and navigate through maternal tissues or soil, respectively. In more specialized contexts, cellulases (or β-1,4-endoglucanases) are deployed to break down cellulose itself. This is not a widespread process, as cellulose is the main structural scaffold, but it is critical in processes like the formation of aerenchyma (air channels in waterlogged roots) or in the creation of hydathodes (water-secreting pores), where specific cells are selectively degraded to form open channels (Anderson and Pelloux, 2025).

13.6 ENZYMES IN HORMONE METABOLISM

Plant hormones, or phytohormones, are not simply synthesized and then become active; their levels are dynamically and precisely controlled by a complex network of biosynthetic and deactivating enzymes. This precise regulation allows plants to fine-tune their growth, development, and stress responses in a rapidly changing environment Figure 13.2.

13.6.1 Auxin Biosynthesis (TAA/YUC) and Conjugation

The primary auxin, indole-3-acetic acid (IAA), is predominantly synthesized from the amino acid tryptophan via a

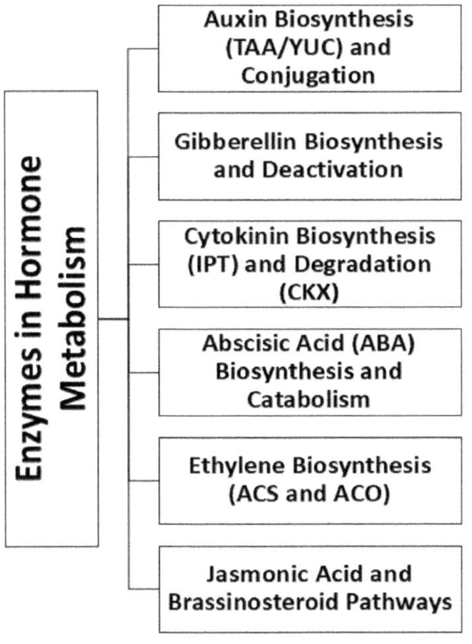

FIGURE 13.2 Enzymes in hormone metabolism.

two-step pathway that is spatially separated within the cell. The first step is catalyzed by enzymes from the tryptophan aminotransferase of *Arabidopsis* (TAA) family, which convert tryptophan to indole-3-pyruvic acid (IPyA). This intermediate is then rapidly converted to IAA by enzymes of the YUCCA (YUC) family, which are flavin monooxygenases. The YUC enzymes are frequently regarded as the rate-limiting factor and are necessary for optimal plant development. Plants utilize conjugation enzymes to avoid auxin accumulation that can reach toxic levels and form reversible storage forms. IAA can often be conjugated to sugars or amino acids by a class of specific synthetases (e.g., GH3 enzymes) and result in biological inactivity. The conjugated auxins can quickly be hydrolyzed back to active IAA by specific hydrolases when and where the plant needs it, providing a rapid and local method for controlling auxin levels (Vijayanathan et al. 2025).

13.6.2 Gibberellin Biosynthesis and Deactivation

Biosynthesis of gibberellin (GA) is a complex process that occurs in plastids, the endoplasmic reticulum, and the cytosol. Gibberellin biosynthesis comprises multiple steps of oxidation catalyzed by cytochrome P450 monooxygenases and 2-oxoglutarate-dependent dioxygenases. There are two classes of dioxygenases that are the principle regulatory controls in this pathway. The GA 20-oxidases catalyze the terminal biosynthetic steps to form bioactive GAs, and the GA 3-oxidases catalyze the final activation step. While GA synthesis is important, plants must also regulate these levels of active GAs with deactivating enzymes. In fact, GA 2-oxidases are the primary enzymes that catalyze the irreversible conversion of bioactive GAs (or their immediate precursors) to inactive forms. The opposing processes of 3-oxidation (activation) and 2-oxidation (deactivation) represent the primary mechanism for control of processes like stem elongation, seed germination, and floral transition (Ishida et al., 2025).

13.6.3 Cytokinin Biosynthesis (IPT) and Degradation (CKX)

Cytokinin biosynthesis begins with the rate-limiting step catalyzed by isopentenyl transferase (IPT) enzymes. IPT transfers an isopentenyl group from dimethylallyl diphosphate to an AMP/ADP/ATP molecule, creating the core isopentenyladenine-type cytokinin backbone. These precursors are then modified into the various active forms of cytokinins. Conversely, cytokinin levels are tightly controlled by degradation, primarily through the action of cytokinin oxidase/dehydrogenase (CKX) enzymes. CKX enzymes irreversibly degrade active cytokinins by cleaving the N^6-isopentenyl side chain, and they are major determinants of cytokinin abundance in tissues. The relative activity of IPT (biosynthesis) versus CKX (degradation) fundamentally regulates processes such as cell division, shoot meristem maintenance, and leaf senescence (Chen et al., 2025).

13.6.4 Abscisic Acid (ABA) Biosynthesis and Catabolism

Abscisic acid (ABA), the stress hormone, is synthesized de novo in response to drought and other stresses via the carotenoid pathway in plastids. The critical regulatory step is the cleavage of the carotenoid precursor, 9′-cis-neoxanthin, by 9-cis-epoxycarotenoid dioxygenase (NCED), which is considered the rate-limiting enzyme in ABA biosynthesis. ABA levels are rapidly downregulated through catabolism to ensure the hormone signal is transient. The primary catabolic pathway is initiated by cytochrome P450 enzymes, specifically CYP707A family members, which catalyze the 8′-hydroxylation of ABA, leading to its irreversible inactivation. The dynamic balance between NCED-driven biosynthesis and CYP707A-mediated catabolism allows for the rapid accumulation and subsequent removal of ABA, enabling precise control over stomatal aperture and the expression of stress-responsive genes (Li et al., 2025).

13.6.5 Ethylene Biosynthesis (ACS and ACO)

The gaseous hormone ethylene is synthesized from the amino acid methionine in a well-defined two-step pathway, the Yang cycle. The first committed and primary regulatory step is the conversion of S-adenosylmethionine (SAM) to 1-aminocyclopropane-1-carboxylic acid (ACC), catalyzed by ACC synthase (ACS). ACS is a highly regulated enzyme, with its gene expression and protein stability being controlled by a wide range of developmental and environmental signals. The second step is the oxidation of ACC to ethylene, a reaction catalyzed by ACC oxidase (ACO). While both are important, ACS activity is generally the control point for ethylene production, making it central to processes like fruit ripening, leaf senescence, and the triple response in seedlings (Khan et al., 2024).

13.6.6 Jasmonic Acid and Brassinosteroid Pathways

Jasmonic acid (JA) and brassinosteroids (BRs) are both lipid-derived hormones with complex biosynthesis pathways. JA is synthesized from α-linolenic acid released from chloroplast membranes. An enzyme early in the pathway is lipoxygenase (LOX), which adds molecular oxygen to the fatty acid. Later, allene oxide synthase (AOS) and allene oxide cyclase (AOC) create the characteristic cyclopentanone ring, and further β-oxidation steps complete the synthesis of the active hormone, jasmonoyl-isoleucine (JA-Ile). The BR pathway, in contrast, is a triterpenoid pathway that shares enzymes with gibberellin biosynthesis. A critical regulatory point is the reaction catalyzed by DWARF4, a cytochrome P450 enzyme that performs multiple C-22 hydroxylation steps. This enzyme is essential for producing the bioactive brassinolide, and its activity is a major determinant of BR levels, which in turn control cell elongation, division, and photomorphogenesis (Zhang et al., 2024).

13.7 CONCLUSION

In conclusion, enzymes are biochemical factors in plant development and physiology. They catalyze the most basic processes of life, including the fixation of atmospheric carbon during photosynthesis, the release of energy from respiration, and the conversion of inorganic nutrients into organic macromolecules. Enzymes also direct morphological development through the careful synthesis and degradation of phytohormones as well as dynamic synthesis and remodeling of the cell wall. This precise catalysis implies that genetic instructions are connected to environmental signals to facilitate phenotypic plasticity and resilience. Therefore, enzymes move beyond being mere catalysts – they are the executors responsible for converting the static genetic code into dynamic processes of growth and development and environmental interaction. A thorough understanding of this enzymatic governance will provide an essential basis to inform novel biotechnological efforts to use catalysis precision to enhance crop productivity, stress resilience, and sustainable crop production practices.

REFERENCES

Abas, S.A., Abdulhamed, Z.A. and Abood, N.M. 2025. Impact of climate change on the growth, yield, and sunflower cultivar performance under the influence of plant density. *Iraqi Journal of Agricultural Sciences 56*(Special): 10–19.

Anderson, C.T. and Pelloux, J. 2025. The dynamics, degradation, and afterlives of pectins: Influences on cell wall assembly and structure, plant development and physiology, agronomy, and biotechnology. *Annual Review of Plant Biology 76.*

Bathe, U., Leong, B.J., McCarty, D.R., Henry, C.S., Abraham, P.E. and Wilson, M.A. 2021. The moderately (d)efficient enzyme: Catalysis-related damage in vivo and its repair. *Biochemistry 60*: 3555–3565.

Chang, T.S., Wu, J.Y., Ding, H.Y. and Wang, T.Y. 2025. Enzymatic glycosylation of ganoderma terpenoid via bacterial glycosyltransferases and glycoside hydrolases. *Biomolecules 15*(5): 655.

Chattaraj, S., Samantaray, A., Ganguly, A. and Thatoi, H. 2025. Employing plant growth-promoting rhizobacteria for abiotic stress mitigation in plants: With a focus on drought stress. *Discover Applied Sciences 7*(1): 68.

Chen, C., Yan, Y., Li, D., Dong, W., Zhang, Y. and Tao, P. 2025. Identification and expression profiling of the cytokinin synthesis gene family IPT in maize. *Genes 16*(4): 415.

Cheng, Z., Wang, C., Tang, F., Zhou, Y., Zhu, C. and Ding, Y. 2025. The cell wall functions in plant heavy metal response. *Ecotoxicology and Environmental Safety 299*: 118326.

Ehinmitan, E., Losenge, T., Mamati, E., Ngumi, V., Juma, P. and Siamalube, B. 2024. BioSolutions for green agriculture: Unveiling the diverse roles of plant growth-promoting rhizobacteria. *International Journal of Microbiology 2024*(1): 6181491.

Evans, S.E., Franks, A.E., Bergman, M.E., Sethna, N.S., Currie, M.A. and Phillips, M.A. 2024. Plastid ancestors lacked a complete Entner-Doudoroff pathway, limiting plants to glycolysis and the pentose phosphate pathway. *Nature Communications 15*(1): 1102.

Han, X., Shen, Y., Sun, L., Shen, J., Mao, Y., Fan, K. and Wang, Y. 2025. Phyllospheric application of *Bacillus mucilaginosus* mediates the recovery of tea plants exposed to low-temperature stress by alteration of the leaf endophytic community and plant physiology. *BMC Microbiology 25*(1): 177.

Heidarzadeh, A. 2025. Role of amino acids in plant growth, development, and stress responses: A comprehensive review. *Discover Plants 2*(1): 1–31.

Huang, S. and Jin, S. 2025. Enhancing drought tolerance in horticultural plants through plant hormones: A strategic coping mechanism. *Frontiers in Plant Science 15*: 1502438.

Huercano, C., Moya-Barrientos, M., Cuevas, O., Sanchez-Vera, V. and Ruiz-Lopez, N. 2025. ER–plastid contact sites as molecular crossroads for plastid lipid biosynthesis. *BMC Biology 23*(1): 139.

Ishida, T., Zhang, Y., Zhu, H., Fudano, S., Peng, Y., Seto, Y. and Yamaguchi, S. 2025. Stepwise deactivation of gibberellins during rice internode elongation. *Proceedings of the National Academy of Sciences 122*(23): e2415835122.

Jung, K., Jo, I.H., Choi, B.Y. and Kim, J. 2025. Structure-based pipeline for plant enzymes: Pilot study identifying novel ginsenoside biosynthetic UGTs. *BioTech 14*(3): 73.

Khan, S., Alvi, A.F., Saify, S., Iqbal, N. and Khan, N.A. 2024. The ethylene biosynthetic enzymes, 1-aminocyclopropane-1-carboxylate (ACC) synthase (ACS) and ACC oxidase (ACO): The less explored players in abiotic stress tolerance. *Biomolecules 14*(1): 90.

Kim, J., Xu, Y., Riu, K.Z., Kim, J.H., Hyun, C.G. and Boo, K.H. 2025. Identification of oxidoreductase genes for phytoecdysteroids biosynthesis in Spinach through comparative transcriptomics. *Plant Physiology and Biochemistry 110216.*

Li, Y., Jin, L., Li, W., Wang, K., Su, H., Mao, H. and Yan, W. 2025. Modulation of an evolutionarily conserved epigenetic regulon controlling abscisic acid catabolism enhances drought tolerance in wheat. *New Phytologist 228.*

Munzert, K.S. and Engelsdorf, T. 2025. Plant cell wall structure and dynamics in plant–pathogen interactions and pathogen defence. *Journal of Experimental Botany 76*(2): 228–242.

Noor, M., Basharat, M.A.R.H.N., Arif, M.S., Qadir, G., Khalid, S., Hussein, S., Asif, A., Rehman, F.U., Shaukat, M.F. and Mehmood A. 2025. Extraction techniques and purification methods: Precision tools for pure innovation. *Pure and Applied Biology 14*(2): 224–248.

Pandey, R., Singh, C., Mishra, S., Abdulraheem, M.I. and Vyas, D. 2025. Silicon uptake and transport mechanisms in plants: Processes, applications and challenges in sustainable plant management. *Biologia Futura 76*(1): 19–31.

Qaderi, M.M., Evans, C.C. and Spicer, M.D. 2025. Plant nitrogen assimilation: A climate change perspective. *Plants 14*(7): 1025.

Rahman, M.A., Lee, S.H., Park, H.S., Min, C.W., Woo, J.H., Choi, B.R. and Lee, K.W. 2025. Light quality plays a crucial role in regulating germination, photosynthetic efficiency, plant development, reactive oxygen species production, antioxidant enzyme activity, and nutrient acquisition in alfalfa. *International Journal of Molecular Sciences 26*(1): 360.

Rashkov, G.D., Stefanov, M.A., Misra, A.N. and Apostolova, E.L. 2025. The role of light-harvesting complex II organization in the efficiency of light-dependent reactions in the photosynthetic apparatus of *Pisum sativum* L. *Plants 14*(12): 1846.

Riseh, R.S., Vatankhah, M., Hassanisaadi, M. and Barka, E.A. 2024. Unveiling the role of hydrolytic enzymes from soil biocontrol bacteria in sustainable phytopathogen management. *Frontiers in Bioscience-Landmark 29*(3): 105.

Rix, G. and Liu, C.C. 2021. Systems for in vivo hypermutation: A quest for scale and depth in directed evolution. *Current Opinion in Chemical Biology 64*: 20–26.

Sathya, T.A., Alarjani, K.M., Elshikh, M.S., Flanetraj, S.R. and Ponnuswamy, V. 2025. Co-composting of green leaves and kitchen waste: Characterization of organic amendments, microbial activity and analysis of defence enzymes in plants. *Biomass Conversion and Biorefinery 15*(17): 24331–24342.

Shang, E., Tu, Q., Yu, Z. and Ding, Z. 2025. Cell wall dynamic changes and signaling during plant lateral root development. *Journal of Integrative Plant Biology 67*(3): 632–648.

Sodhi, G.K., Wijesekara, T., Kumawat, K.C., Adhikari, P., Joshi, K., Singh, S., Farda, B., Ghorbani, M., Hosseini, A., Zulfiqar, S.M., Sardar, S., Javed, S., Zubair, M., Ren, J., Haider, M., Seleiman, M.F. and Riseh, R.S. 2025. Nanomaterials–plants–microbes interaction: Plant growth promotion and stress mitigation. *Frontiers in Microbiology 15*: 1516794.

Sulaman, S., Nadeem, M., Shabaan, M., Orman, S., Anwar-ul-Haq, M. and Zulfiqar, U. 2025. Exogenous application of nitrogen (N) and potassium (K) improves drought tolerance in plants: A review. *Journal of Soil Science and Plant Nutrition*: 1–16.

Trudeau, D.L., Edlich-Muth, C., Zarzycki, J., Scheffen, M., Goldsmith, M., Khersonsky, O., Bruner, S.D., Alcalde, R., Höhne, M.G., Friedrich, T., Steffler, F., Leibiger, B.M., Buckel, W., Miller, B.R., Erb, S.J., Tawfik, D.S. and Bar-Even, A. 2018. Design and *in vitro* realization of carbon-conserving photorespiration. *Proceedings of the National Academy of Sciences 115*: E11455–E11464.

Van Gelder, K., Oliveira-Filho, E.R., García-García, J.D., Hu, Y., Bruner, S.D. and Hanson, A.D. 2023. Directed evolution of aerotolerance in sulfide-dependent thiazole synthases. *ACS Synthetic Biology 12*: 963–970.

Vijayanathan, M., Faryad, A., Abeywickrama, T.D., Christensen, J.M. and Neilson, E.H.J. 2025. The auxin gatekeepers: Evolution and diversification of the YUCCA family. *The Plant Journal 124*.

Wang, Y., Huang, W., Li, X. and Zhang, Y. 2025. The deubiquitinating enzyme AMSH1 contributes to plant immunity through regulating the stability of BDA1. *Plants 14*(3): 429.

Xu, Z., Liu, D., Zhu, J., Zhao, J., Shen, S., Wang, Y. and Yu, P. 2025. Catalysts for sulfur: Understanding the intricacies of enzymes orchestrating plant sulfur anabolism. *Planta 261*(1): 16.

Ye, Y., Zhang, Y., Zhou, Y. and Gao, Y. 2025. Molecular engineering of alginate lyases and the potential agricultural applications of their enzymatic products. *Journal of Agricultural and Food Chemistry 73*(10): 5666–5684.

Zhang, G., Zhai, N., Zhu, M., Zheng, K., Sang, Y., Li, X. and Xu, L. 2025. Cell wall remodeling during plant regeneration. *Journal of Integrative Plant Biology 67*(4): 1060–1076.

Zhang, Z., Zhang, T., Lu, L., Qiu, S., Huang, Z., Wang, Y. and Zeng, R. 2024. Synergistic interaction between brassinosteroid and jasmonate pathways in rice response to cadmium toxicity. *Science of The Total Environment 954*: 176369.

14 Enzyme Systems in Plant Seed Germination and Early Development

Salar Farhangi-Abriz, Seyyed Amirreza Mousavi, Morteza Alami-Milani, and Parisa Aghaei-Gharachorlou

14.1 INTRODUCTION

Following dormancy, seed germination, a process critical for seedling growth and pivotal to biodiversity and agriculture because of its effect on yield, uniformity, and stress resilience, begins (Rajjou et al. 2012). It begins with imbibition, where water uptake reactivates metabolism, ruptures the seed coat, and allows radicle emergence. Stored macromolecules are mobilized to fuel embryonic growth, with enzymes driving reserve breakdown and respiration. Inefficient enzymatic activity can delay germination, reduce vigor, and increase susceptibility to stresses such as drought and salinity (Alhammad et al. 2023). Hydrolytic enzymes drive germination by degrading stored reserves into soluble nutrients for the embryo. Amylases convert starch to sugars for glycolysis and ATP, proteases release amino acids for protein synthesis and osmotic balance, and lipases mobilize lipids for gluconeogenesis via the glyoxylate cycle. These enzymes are synthesized or activated after imbibition, largely regulated by gibberellins, which induce enzyme production, and abscisic acid (ABA), which maintains dormancy (Ali and Elozeiri 2017)

As germination progresses into early seedling development, enzyme systems extend their influence to support root and shoot elongation, photosynthetic establishment, and defense mechanisms. To scavenge reactive oxygen species (ROS) produced during metabolic resurgence, antioxidant enzymes, including superoxide dismutase (SOD), catalase (CAT), and peroxidase (POD), become increasingly active, thus averting cellular damage and ensuring viability (Han and Yang 2015). In parallel, enzymes involved in hormone biosynthesis and signaling, like those in the GA and auxin pathways, regulate cell division and differentiation, facilitating the transition to autotrophic growth. For instance, in dicotyledonous plants, beta-oxidation enzymes in peroxisomes convert stored lipids into carbohydrates, sustaining the seedling until chlorophyll synthesis begins (Rajjou et al. 2012; Carrera-Castaño et al. 2020). As depicted in Figure 14.1, the schematic overview illustrates the interconnected enzyme systems, including hydrolases for reserve mobilization, metabolic pathways for energy production, and antioxidants for ROS mitigation, alongside their hormonal and epigenetic regulation during seed germination.

This chapter delves into the intricate enzyme systems underpinning plant seed germination and early development, synthesizing current knowledge on their biochemical mechanisms, regulatory pathways, and responses to abiotic and biotic factors. By examining case studies across diverse plant species, from cereals to legumes, we aim to elucidate how these systems contribute to germination success and seedling establishment.

14.2 MAJOR ENZYME SYSTEMS ACTIVATED DURING SEED IMBIBITION AND GERMINATION

Upon imbibition, the initial phase of water uptake, a rapid cascade of hydrolytic enzymes is activated to break down the abundant storage macromolecules within the seed. This enzymatic activity is paramount for supplying the necessary energy and foundational building blocks that fuel the developing embryo and facilitate radicle protrusion (Anumala et al. 2025; Turk et al. 2025).

14.2.1 Hydrolases for Reserve Mobilization

Hydrolases constitute a primary class of enzymes, indispensable for converting complex stored compounds into simpler, metabolically usable forms. Their actions initiate the vital process of reserve mobilization, providing essential nutrients for the growing embryo (Turk et al. 2025).

14.2.1.1 Amylases

These enzymes, particularly α-amylase, are central to starch degradation, a predominant carbohydrate reserve in many seeds. They catalyze the hydrolysis of α-1,4-glycosidic linkages in starch polymers, transforming them into soluble sugars such as glucose and maltose, which are then used as primary energy sources (Bozdar et al. 2025). The activity of α-amylase significantly increases during germination, and specific treatments like seed priming can enhance this activity, leading to faster and more uniform germination. For instance, waxy maize inbreds, rich in amylopectin, exhibit higher α-amylase activity, correlating with improved germination and seedling vigor (Talukder et al. 2025). In *Cucurbita moschata* (pumpkin) seeds, α-amylase has been purified and characterized for its efficiency in sugar production, highlighting its key role (Posoongnoen et al. 2025).

14.2.1.2 Proteases

Stored proteins serve as crucial sources of amino acids required for synthesizing new structural proteins and enzymes during cell division and elongation in the germinating seedling. Proteases catalyze the breakdown of these

DOI: 10.1201/9781003669920-14

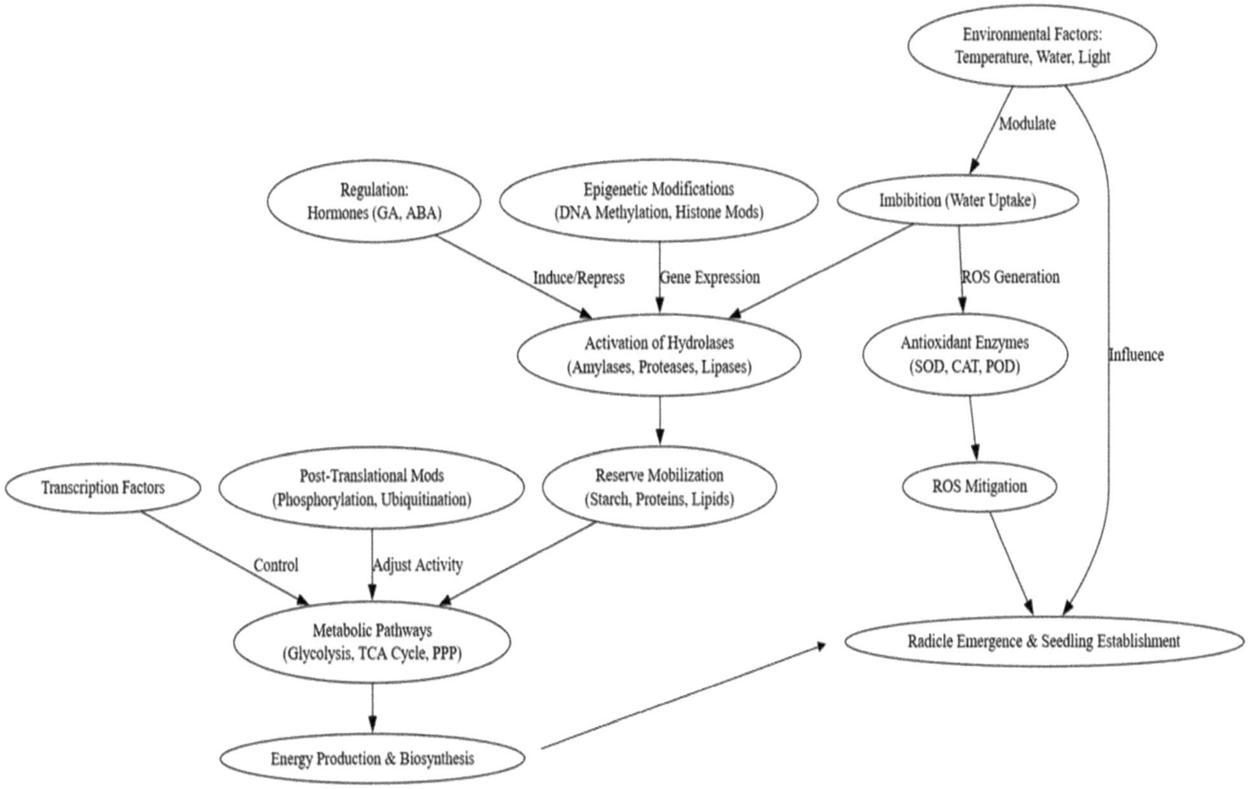

FIGURE 14.1 Schematic diagram summarizing enzyme systems, regulatory mechanisms, and environmental influences in plant seed germination.

large storage proteins into their constituent amino acids, ensuring an adequate nitrogen supply for early development. Aspartic proteases are broadly distributed in plants and play diverse roles in development, including seed germination and responses to stress (Yu and Feng 2025). By inhibiting cell wall remodeling enzymes, including proteases, in the micropylar endosperm, the tomato transcription factor (TF) SlLBD40 regulates germination speed, implying a role in endosperm weakening (Zhang et al. 2025).

14.2.1.3 Lipases

For seeds rich in lipids, such as oats or maize, lipases are essential for hydrolyzing triacylglycerols (TAGs) stored in lipid droplets into free fatty acids and glycerol. These fatty acids are subsequently channeled into the β-oxidation pathway in peroxisomes and the glyoxylate cycle to generate acetyl-CoA, which then feeds into gluconeogenesis to produce sugars, providing both energy and precursors for membrane synthesis (do Nascimento et al. 2025; Turk et al. 2025). Oleosin, a structural protein of lipid droplets, is vital for their stabilization and subsequent mobilization during seed germination and seedling growth. Exogenous carnitine treatments in maize have been shown to increase the activities of lipase, isocitrate lyase (ICL), and malate synthase (MS), thereby accelerating lipid catabolism and mitochondrial respiration, which promotes germination (Kaur et al. 2025; Turk et al. 2025).

14.2.2 Metabolic Enzymes for Energy and Biosynthesis

Integrating these mobilized nutrients into central metabolic pathways that produce ATP and intermediate metabolites requires other metabolic enzymes, and this is crucial beyond direct reserve mobilization.

14.2.2.1 Glycolysis and TCA Cycle Enzymes

As central pathways of carbon metabolism, glycolysis and the tricarboxylic acid (TCA) cycle facilitate energy transformation and supply building blocks for biosynthetic processes. Key enzymes such as *Hexokinase* (HXK1) initiate glucose metabolism, while pyruvate kinase (PKM2) is a pivotal regulator of glycolysis, converting phosphoenolpyruvate to pyruvate (Bozdar et al. 2025; Kim et al. 2025a; Upadhyay et al. 2025). In maize, citrate synthase (CS) activity, crucial for the TCA cycle, increases significantly with carnitine treatments, supporting lipid-derived respiration (Turk et al. 2025). Isocitrate dehydrogenase is upregulated in *Quercus acutissima* seeds during deep desiccation stages, indicating active energy metabolism (Chen and Shen 2025). These pathways ensure that sugars and other carbon skeletons from reserves are efficiently converted into ATP and precursors for synthesizing new cellular components (Bozdar et al. 2025; Zheng et al. 2025).

14.2.2.2 Pentose Phosphate Pathway (PPP) Enzymes

The pentose phosphate pathway (PPP) operates to optimize energy utilization, produce precursors for nucleic acid synthesis, and generate NADPH, which is essential for reductive biosynthesis and antioxidant production. The rate-limiting step of the oxidative phase is catalyzed by enzymes like glucose-6-phosphate dehydrogenase (G6PDH), and they generate NADPH (Aburto et al. 2025; Bozdar et al. 2025). This pathway becomes particularly important under oxidative stress, rerouting metabolic flux from glycolysis to the PPP to maintain NADPH levels. In *Fraxinus hupehensis* seeds, the PPP, alongside glycolysis and the TCA cycle, provides energy during germination (Wei et al. 2025).

14.3 BIOCHEMICAL PATHWAYS AND MECHANISMS OF RESERVE MOBILIZATION

The mobilization of stored reserves is a precisely orchestrated process involving several interconnected biochemical pathways, ensuring a continuous and regulated supply of energy and building blocks for seedling growth. Starch degradation primarily occurs in the endosperm or cotyledons, initiated by the concerted action of α-amylase and other starch-hydrolyzing enzymes. The resulting glucose and maltose are actively transported to the embryo, where they are consumed for respiration, which provides energy, and biosynthesis, which creates new cellular materials (Bozdar et al. 2025; Turk et al. 2025). A complex regulatory mechanism involving the target of rapamycin (TOR) signaling pathway was highlighted by Komiya et al. (2025), which regulates starch degradation via the phosphorylation of α-glucan water dikinase (CmGWD), a key enzyme in the process. In rice, OsNUDX23 also regulates early seed germination by modulating starch metabolism, indicating the complexity of this pathway (Dai et al. 2025).

Storage proteins are broken down by various proteases into free amino acids and small peptides (Matsui and Matsubayashi 2025). These amino acids are then efficiently transported and utilized for the synthesis of new structural proteins, enzymes, and other nitrogen-containing compounds vital for cell proliferation and expansion during the initial phases of seedling establishment (Farhangi-Abriz and Ghassemi-Golezani 2016).

In seeds, lipids, primarily triacylglycerols, are broken down by lipases into fatty acids and glycerol (do Nascimento et al. 2025). The fatty acids undergo β-oxidation in peroxisomes, converting them into acetyl-CoA. Acetyl-CoA subsequently enters the glyoxylate cycle, which bypasses the decarboxylation steps of the TCA cycle to produce four-carbon compounds like succinate, crucial for gluconeogenesis. This gluconeogenic pathway synthesizes sugars, thereby supplementing carbohydrate supplies essential for early seedling development before photosynthesis becomes fully active. Glycerol is also metabolized into glycolytic intermediates, further contributing to energy production (Turk et al. 2025).

14.4 REGULATION OF ENZYME ACTIVITY AND GENE EXPRESSION

Enzyme activity during seed germination is under sophisticated multi-level control, involving intricate hormonal signals, precise genetic regulation, and dynamic post-translational modifications (PTMs) (Özden et al. 2025). These mechanisms ensure that enzyme functions are finely tuned to internal developmental cues and external environmental signals (Zheng et al. 2025).

14.4.1 Hormonal Control

Phytohormones are central to orchestrating germination by modulating the expression and activity of key enzymes. GAs are crucial germination-promoting hormones that counteract the inhibitory effects of ABA to break dormancy and induce the synthesis of hydrolytic enzymes, particularly α-amylase (Berqdar et al. 2025). GA treatment significantly improves germination rates and seed vitality, as observed in sainfoin, by altering hormone and secondary metabolite levels (Luo et al. 2025). Exogenous 2,4-epibrassionolide (EBR) priming in rice increases GA levels and induces the expression of GA biosynthesis genes, promoting germination under heat stress (Qian et al. 2025).

ABA is a primary dormancy-maintaining hormone that represses the expression of genes encoding hydrolytic enzymes and inhibits germination (Bozdar et al. 2025; Kępczyński et al. 2025). It primarily exerts its effects by activating TFs like *ABSCISIC ACID INSENSITIVE3* (ABI3) and ABI5, which bind to promoters of germination-related enzyme genes, thereby suppressing their transcription (Du and Zhang 2025; Huang et al. 2025). Through the action of NIN-LIKE PROTEIN 8 (NLP8), nitrate, independently of the nitric oxide pathway, can notably attenuate ABA signaling, where NLP8 physically interacts with ABI3 and ABI5 to inhibit their transcriptional functions and ABA-induced accumulation, consequently facilitating dormancy release without affecting ABA content (Huang et al. 2025). In Luffa, a decrease in LcNCED2 levels, an enzyme involved in ABA biosynthesis, coincides with an increase in α-amylase activity during germination, underscoring ABA's repressive role (Fang et al. 2025). According to Chen et al. (2025), applying melatonin can greatly reduce ABA levels, thus balancing GAs to promote maize seed germination when it's hot.

Auxins (IAA), cytokinins (CTK), and ethylene also finely regulate germination. Cytokinins stimulate *Striga hermonthica* germination, while high concentrations trigger embryo-like structures (Brun et al. 2025). In contrast, free bases of cytokinin are potent inhibitors of broomrape germination. Auxins, along with GAs and ABA, play a significant role in determining the seed's chemistry and physiology (Chattha et al. 2025). Melatonin improves seed germination and promotes lateral root formation and flowering (Gamalero and Glick 2025). Also, it encourages the expression of genes related to photosynthesis enzymes, which shield plants from phytopathogens, according to Gamalero and Glick (2025). A dynamic balance between basal growth and stress responses is ensured by the complex crosstalk among these phytohormones, including brassinosteroids, strigolactones, salicylic acid, nitric oxide, melatonin, polyamines, and hydrogen sulfide (Chauhan et al. 2025).

14.4.2 Epigenetic Mechanisms and RNA Modifications

During seed dormancy, germination, and early development, gene expression is dynamically regulated by epigenetic modifications, like DNA methylation, histone modifications, and RNA methylation, all without changing the DNA sequence.

14.4.2.1 DNA Methylation

The addition of methyl groups to cytosine residues, which this modification entails, plays a crucial role in regulating fruit ripening and seed development (Ge et al. 2025). Hyper-methylation, a result of the high expression of DNA methyltransferase PnCMT2 in stubborn *Panax notoginseng* seeds, rises during after-ripening and contributes to dormancy by inhibiting genes related to embryo development. To regulate dormancy release, these changes affect gene transcription levels (Ge et al. 2025,).

14.4.2.2 Histone Modifications

Acetylation of histone proteins, primarily occurring in the nucleus, regulates chromatin structure, thereby controlling gene transcription on a large scale (Li et al. 2025a). For the exact control of plant organ formation and development, this process is critical. Histone acetylation and deacetylation enzymes dynamically modulate gene accessibility, impacting the expression of enzyme-encoding genes during germination (Li et al. 2025a).

14.4.2.3 RNA Modifications

RNA N6-methyladenosine (m6A) modification is a post-transcriptional regulatory mechanism that influences gene expression during germination by affecting mRNA stability and translation efficiency (Cui et al. 2025; Li et al. 2025b). In tomato, the m6A modification level significantly increases during germination, and the expression of the SlYTP9 reader gene markedly affects seed germination (Cui et al. 2025). In rice, the RNA methyltransferase gene OsMTA1 participates in seed germination and salt stress responses, further highlighting the role of epitranscriptome profiles (Li et al. 2025b).

14.4.3 Transcription Factors and Signaling Pathways

As master regulators, TFs directly control the expression of genes that code for enzymes in various processes. TFs such as NIN-LIKE PROTEIN 8 (NLP8), ABSCISIC ACID INSENSITIVE3 (ABI3), and ABI5 play critical roles in hormonal crosstalk and enzyme regulation (Huang et al. 2025). By targeting genes encoding cell wall remodeling enzymes like expansin, xyloglucan endotransglucosylase/hydrolase, and endo-β-mannanase, SlLBD40 in tomato directly regulates endosperm weakening and germination speed. Wang et al. (2025) showed that NAC-type TFs like NAC90, NAC36, and NAC61 negatively regulate NHP and SA biosynthesis, which affects plant immunity. MYC2 contributes to plant growth from seed germination to development (Kim et al. 2025b).

Plant hormone signal transduction is a highly enriched pathway with many differentially expressed genes during seed dormancy release and germination (Luo et al. 2025; Quan et al. 2025). In addition to integrating stress responses, mitogen-activated protein kinase (MAPK) signaling pathways are crucial for positively regulating cold tolerance in rice (Ding et al. 2025; Wu et al. 2025). ROS, in addition to being harmful, act as signaling molecules, influencing gene expression and hormone crosstalk to regulate antioxidant enzyme genes (Kaya et al. 2025; Wang et al. 2025).

14.5 POST-TRANSLATIONAL MODIFICATIONS (PTMs)

Enzymes can be rapidly and reversibly regulated through PTMs such as phosphorylation, ubiquitination, and acetylation. These modifications enable plants to rapidly adjust their metabolic flux landscape to match cellular demands and environmental fluctuations.

14.5.1 Phosphorylation

This modification rapidly and reversibly alters enzyme organization and function, enabling plants to adjust metabolic fluxes. Enzymes involved in glycolysis and the TCA cycle are common targets of phosphorylation, which can induce functional changes (Zheng et al. 2025). Kinases and phosphatases, whose opposing enzymatic activities are critical regulatory factors, play essential roles in ABA signaling and other signal transduction pathways during germination (Wu et al. 2025b).

14.5.2 Ubiquitination

This PTM's precise orchestration of plant hormone signaling is achieved by modulating protein activity, localization, assembly, and interaction (Du and Zhang 2025). While mono-ubiquitination can enhance the localization of ABA receptors to the plasma membrane and improve signal perception, E3 ubiquitin ligases, for example, promote their degradation through polyubiquitination (Du and Zhang 2025). SnRK1 subunits undergo various PTMs, including phosphorylation, SUMOylation, ubiquitination, and acetylation, integrating energy homeostasis, stress adaptation, and hormonal crosstalk (Xu et al. 2025).

14.5.3 Acetylation

Acetyltransferases and deacetylases' collaborative efforts led to this widespread modification, which then regulates crucial cellular processes in plants. By influencing enzyme stability, subcellular distribution, and activity, non-histone protein acetylation can finely regulate almost all key cellular functions and processes (Li et al. 2025a).

14.6 ROLE OF ANTIOXIDANT ENZYME SYSTEMS IN STRESS PROTECTION

Seed germination is a metabolically intense process, inherently accompanied by the generation of ROS, which, if not properly managed, can lead to severe oxidative damage. Antioxidant enzyme systems are thus vital for detoxifying ROS and maintaining cellular redox homeostasis, thereby protecting germinating seeds and enhancing seedling vigor and subsequent plant growth.

As the first line of defense against oxidative stress, SOD works by catalyzing the dismutation of highly reactive superoxide radicals ($O_2^{\bullet-}$) into less harmful hydrogen peroxide (H_2O_2) (Chaachouay et al. 2025; Fatemeh et al. 2025; Srivastava et al. 2025). Enhanced SOD activity is frequently observed in seeds treated with priming agents or phytohormones, improving stress tolerance (Turk et al. 2025; Yılmaz and Çiftçi 2025). For example, exogenous melatonin significantly increased SOD activity in maize under salt and heat stress, reducing oxidative damage (Chen et al. 2025; Li et al. 2025; Zhang et al. 2025). CAT plays a crucial role in removing H_2O_2 by converting it into water and oxygen. Increased CAT activity is a common response to various seed priming treatments and stress mitigation strategies (Rudiyanto et al. 2025; Zhang et al. 2025). In maize, a resistant cultivar exhibited a greater increase in α-amylase and antioxidant content after combined cold and high moisture stress, facilitating starch decomposition and rapid removal of O_2^- and H_2O_2 (Meng et al. 2025). Peroxidases (POD, APX), including guaiacol peroxidase (POD) and ascorbate peroxidase (APX), also detoxify H_2O_2, often using different reducing substrates (Ferreyra et al. 2025; Rudiyanto et al. 2025). The ascorbate–glutathione cycle, with APX as a key component, is essential for maintaining cellular redox balance (Havii et al. 2025; Singh et al. 2025). Spermidine application to aged sorghum seeds boosted the activity of various antioxidant enzymes, including peroxidases and glutathione transferases, curbing ROS and enhancing germination vigor (Xing et al. 2025). Glutathione S-transferases (GSTs) exhibit dual enzymatic activities as both transferases and peroxidases, involved in detoxification and stress responses (Khan et al. 2025; Lin et al. 2025).

14.7 REDOX SIGNALING AND STRESS RESILIENCE

Beyond direct detoxification, ROS, particularly H_2O_2, can act as important signaling molecules during germination, influencing gene expression and hormone crosstalk. A finely tuned balance of ROS levels, maintained by the concerted action of antioxidant enzymes, is crucial for seed activation and stress resilience (Dake et al. 2025; Wang et al. 2025). For instance, TIP3 aquaporins in soybean are found to transport H_2O_2, highlighting its signaling role in seed germination (Sudhakaran et al. 2025). By managing abiotic stresses and maintaining redox balance in plants, melatonin, a key antioxidant, plays a crucial role (Turk et al. 2025).

14.8 ENVIRONMENTAL MODULATION OF ENZYME ACTIVITY

Environmental factors are critical modulators of enzyme activity during germination, dictating the timing and overall success of seedling establishment. Temperature profoundly affects enzyme kinetics, metabolic rates, and the overall speed of germination. Each species has an optimal temperature range for germination where enzyme activities are maximized (Koutsovoulou et al. 2025; Susko 2025). For instance, *Leonurus cardiaca* seeds germinate optimally between 25/15°C and 30/20°C (Susko 2025). Extreme temperatures, both low and high, can inhibit or even denature enzymes, leading to delayed or failed germination (Harmancı et al. 2025; Zhou et al. 2025). Cold stress significantly impacts rice growth by reducing photosynthetic efficiency and altering metabolic pathways, whereas heat stress causes morphophysiological changes, biochemical disturbances, and affects enzymatic activities (Khanzada et al. 2025; Zhou et al. 2025). Seed priming can mitigate temperature-induced inhibitory effects by enhancing enzyme activities (Anumala et al. 2025; Hua et al. 2025).

Imbibition initiates metabolic reactivation and enzyme synthesis in germination, as water is vital for enzyme hydration and reserve mobilization. Drought or osmotic stress (e.g., PEG) reduces germination and seedling growth by restricting water flow and enzymatic activity (Gülöksüz et al. 2025). Seeds from warmer, drier *Dianthus langeanus* populations show greater tolerance, reflecting adaptive enzymatic responses. Seed priming improves germination under water stress by enhancing water uptake and activating stress-response mechanisms (Almakas et al. 2025; Nowicki et al. 2025). Oxygen is vital for aerobic respiration, which generates the ATP needed for numerous enzymatic reactions and biosynthetic processes during germination. Low oxygen levels can inhibit metabolic enzymes and impair overall germination success. Light acts as a critical environmental signal, influencing seed germination by modulating hormone levels and enzyme expression (Chen et al. 2025). Some seeds are photoblastic, requiring light for germination, while others are inhibited by it. By affecting ROS intensity and promoting germination, different light qualities, like red and blue light, can significantly regulate antioxidant enzymes (e.g., SOD, CAT, APX, GR) and the expression of genes linked to the ascorbate–glutathione pathway (Halimeh 2025; Rahman et al. 2025). Phytochrome is required for light-inhibited germination in *Aethionema arabicum*, demonstrating its role in regulating photoblastic responses (Mérai et al. 2025). As detailed in Table 14.1, the enzymatic systems crucial for plant seed germination, encompassing hydrolases for reserve mobilization, metabolic enzymes for energy production, and antioxidant enzymes for ROS mitigation, are intricately regulated by hormonal signals, epigenetic modifications,

TABLE 14.1
Enzymatic Systems, Regulation, and Environmental Influences in Plant Seed Germination

Aspect	Details
Key enzyme systems	*Hydrolases*: Amylases (α-amylase degrades starch to glucose/maltose), proteases (break down proteins to amino acids), lipases (hydrolyze lipids to fatty acids/glycerol). *Metabolic Enzymes*: Glycolysis (hexokinase, pyruvate kinase), TCA cycle (citrate synthase), pentose phosphate pathway (glucose-6-phosphate dehydrogenase). *Antioxidant Enzymes*: Superoxide Dismutase, Catalase, Peroxidase mitigate ROS during metabolic reactivation.
Role in germination	Hydrolases mobilize stored reserves in the endosperm/cotyledons to provide energy and building blocks for radicle protrusion. Metabolic enzymes fuel ATP production and biosynthesis during imbibition and early respiration. Antioxidant enzymes protect against oxidative stress from ROS generated during metabolic resurgence, ensuring germination success.
Regulation mechanisms	*Hormonal Control*: Gibberellins induce hydrolytic enzyme synthesis (e.g., α-amylase in aleurone layer); abscisic acid represses it to maintain dormancy. Cytokinins, auxins, ethylene, and melatonin modulate germination speed and vigor. *Epigenetic Modifications*: DNA methylation (e.g., in *Panax notoginseng*), histone acetylation, and RNA m6A modification (e.g., in tomato) regulate enzyme gene expression. *Transcription Factors*: NLP8, ABI3, ABI5, SlLBD40 control enzyme genes, influencing endosperm weakening and germination timing. *Post-Translational Modifications*: Phosphorylation, ubiquitination, acetylation adjust enzyme activity rapidly.
Environmental modulation	*Temperature*: Optimal ranges enhance enzyme kinetics; extremes delay or inhibit germination. *Water*: Imbibition triggers enzyme activation; drought/osmotic stress impairs activity. *Oxygen*: Essential for aerobic respiration to support enzymatic processes. *Light*: Modulates hormone levels and enzyme expression (e.g., phytochrome in *Aethionema arabicum* for photoblastic seeds).

and environmental factors such as temperature and water availability.

14.9 IMPACT OF SEED PRIMING ON ENZYMATIC ACTIVITIES

Seed priming and biotechnological approaches enhance germination, seedling vigor, and stress tolerance by pre-activating enzyme systems. Priming, achieved through methods such as hydropriming, osmopriming, chemical priming (e.g., KNO_3, ascorbic acid, melatonin), and biopriming with plant-growth-promoting rhizobacteria (PGPR), initiates metabolic processes without radicle emergence. These treatments stimulate hydrolytic enzymes (e.g., α-amylase) and antioxidants (SOD, CAT, POD), improving reserve mobilization and stress defense (Farhangi-Abriz and Ghassemi-Golezani 2019; Chen et al. 2025; Yılmaz and Çiftçi 2025; Zhang et al. 2025). For instance, melatonin priming boosted enzyme activity and antioxidant capacity in maize under heat stress (Chen et al. 2025), while ascorbic acid enhanced mitochondrial AsA-GSH cycle activity in aged oat seeds (Mi et al. 2025).

Seed priming significantly enhances tolerance to abiotic stresses like drought, salinity, and heat by boosting antioxidant enzyme activities, modulating hormone signaling pathways, and activating stress-responsive genes (Farhangi-Abriz and Torabian 2018; Zhang et al. 2025; Zhuang et al. 2025). Silica nanoparticle priming, for instance, accelerated the mobilization of seed-stored reserves in cucumber, leading to increased germination rates and enhanced drought tolerance (Zhuang et al. 2025). Stress memory can be induced by priming through the upregulation of genes involved in antioxidant defense and polyamine metabolism, which in turn boosts tolerance to salinity and drought (Anumala et al. 2025).

Nanopriming, utilizing nanoparticles (e.g., zinc oxide, silica) to treat seeds, and cold plasma technology, a non-thermal method using partially ionized gas, promote seed metabolism, activate antioxidant enzymes, and improve germination under stress (Ahmed et al. 2025; Chandrasekaran et al. 2025). Plasma-activated water (PAW) has been shown to increase bell pepper germination percentage by up to 26% and promote plant growth by regulating antioxidant enzymes (Ferreyra et al. 2025). Non-thermal plasma, by inducing epigenetic changes such as DNA methylation that regulate stress-related genes, further supports seed performance (Ahmed et al. 2025).

14.10 FUTURE TRENDS AND CHALLENGES

Future research aims to integrate these advanced techniques to develop more resilient crops and sustainable agricultural practices. The convergence of artificial intelligence (AI) with omics technologies, genome editing, and protein design promises rapid advances in crop improvement, allowing for the engineering of enzymes with enhanced efficiency and specificity. AI-driven breeding is already showing potential to predict and enhance stress tolerance in high-elevation extremophytes (Kaushik 2025). Harnessing beneficial microbial interactions within the seed microbiome through engineering approaches offers a promising avenue to enhance germination success, plant nutrition, and stress tolerance. This includes the use of PGPR as biofertilizers and their genomic analysis (Guardado-Fierros et al. 2025; Liu et al. 2025; Novello et al. 2025). Developing crop varieties that can withstand multiple abiotic stresses, such as cold, drought, salinity, and heat, is an urgent need for global food security. Manipulating enzyme systems through biotechnological approaches will be crucial for achieving these complex traits (Khanzada et al. 2025; Zhou et al. 2025). Ensuring the safe and responsible application of biotechnological innovations, especially with genome-edited products and enzyme modifications, is crucial. The overarching goal is to optimize enzyme systems through low-cost, eco-friendly approaches to improve seed quality, enzyme system activation, and robust seedling establishment, contributing to resilient and sustainable agriculture globally (Fico et al. 2025; Magar et al. 2025; Rhaman 2025). Addressing challenges such as upscaling limitations, nanotoxicity concerns, and environmental risks is crucial for viable implementation.

14.11 CONCLUSION

In conclusion, seed germination and early seedling development stand as pivotal milestones in the plant life cycle, governed by a symphony of enzyme systems that facilitate reserve mobilization, energy metabolism, and oxidative stress management. These biochemical actors facilitate the smooth transition from dormancy to vigorous growth, involving hydrolases breaking down starch, proteins, and lipids, and metabolic enzymes powering glycolysis, the TCA cycle, and PPPs. Antioxidant enzymes like SOD, CAT, and POD further protect emerging seedlings by neutralizing ROS, while intricate regulatory networks spanning phytohormones (for example, GA and ABA), epigenetic modifications, TFs, and PTMs adapt these processes to environmental variables such as temperature, water, light, and species-specific traits in monocots, dicots, orthodox, and recalcitrant seeds. Biotechnological advancements, including seed priming, nanopriming, and genetic tools like CRISPR Cas, integrated with omics and AI, promise to amplify enzyme efficiency, bolstering germination rates, stress resilience, and crop yields. Nonetheless, addressing regulatory, scalability, and ecological challenges is essential for sustainable application. Ultimately, harnessing these enzyme insights will fortify agricultural productivity and global food security against escalating climatic pressures.

REFERENCES

Aburto, C., Parada-Goddard, V. and San Martín, A. 2025. The activation of pentose phosphate pathway flux by hydrogen peroxide is not regulated by NADPH-mediated feedback inhibition. bioRxiv. 2025: 03.

Ahmed, N., Yong, L.X., Yang, J.H.C. and Siow, K.S. 2025. Review of non-thermal plasma technology and its potential impact on food crop seed types in plasma agriculture. Plasma Chem. Plasma Process. 45: 421–462.

Alhammad, B.A., Ahmad, A., Seleiman, M.F. and Tola, E. 2023. Seed priming with nanoparticles and 24-epibrassinolide improved seed germination and enzymatic performance of *Zea mays* L. in salt-stressed soil. Plants. 12: 690.

Ali, A.S. and Elozeiri, A.A. 2017. Metabolic processes during seed germination. Adv. Seed Biol. 2017: 141–166.

Almakas, A., Elrys, A.S., Desoky, E.S.M., Al-Shuraym, L.A., Alhag, S.K., Alshaharni, M.O., Alnadari, F., NanNan, Z., Farooq, Z., El-Tarabily, K.A. and Zhao, T. 2025. Enhancing soybean germination and vigor under water stress: The efficacy of bio-priming with sodium carboxymethyl cellulose and gum arabic. Front. Plant Sci. 15: 1475148.

Anumala, V., Rime, J., Sarma, P., Mashine, L., Chaupoo, A.S. and Singh, Y.D. 2025. Deciphering of multifaceted seed priming technologies through physiological and molecular approaches: An updated overview. Plant Mol. Biol. Rep. 2025: 1–33.

Berqdar, L., Salem, M.A., Wang, J.Y., Alrudayan, A., Jamil, M. and Al-Babili, S. 2025. Development of a rapid and efficient protocol for seed germination and seedling establishment of *Oryza coarctata*. Bio-Protocol. 15: e5203.

Bozdar, B., Ahmed, N., Tu, P. and Li, Z.H. 2025. Beyond energy: How small-molecule sugars fuel seed life and shape next-generation crop technologies. J. Agron. Crop Sci. 211: e70050.

Brun, G., Billard, E., Hýlová, A., Montiel, G., Plačková, L., Doležal, K., Puech-Pagès, V., Simier, P., Wicke, S., Spíchal, L. and Delavault, P. 2025. Divergent control of seed germination by cytokinins in weedy broomrapes and witchweeds. Plant Cell Physiol. 2025: pcaf032.

Carrera-Castaño, G., Calleja-Cabrera, J., Pernas, M., Gómez, L. and Oñate-Sánchez, L. 2020. An updated overview on the regulation of seed germination. Plants. 9: 703.

Chaachouay, N., Ansari, M.K.A., Houssni, M., Zidane, L. and Husen, A. 2025. Enzymatic antioxidant defense systems in plants. In Role of antioxidants in mitigating plant stress. Academic Press. 2025: 37–50.

Chandrasekaran, U., Hong, W.J. and Kim, H. 2025. Subtilases: A major prospect to the genome editing in horticultural crops. Front. Plant Sci. 15: 1532074.

Chattha, F.A., Kousar, S., Naeem, F. and Ahmad, S. 2025. Connotation of plant growth regulators on seed germination. J. Exp. Agric. Int. 47: 343–356.

Chauhan, R., Yadav, P., Sahu, K. and Majumdar, S.K. 2025. Light as a key abiotic driver in enhancing germination and phytochemicals of chickpea cultivars. Environ. Ecol. 43: 169–174.

Chen, C., Li, D., Yan, Y., Yin, C., Shi, Z., Zhang, Y. and Tao, P. 2025. Facilitating maize seed germination under heat stress via exogenous melatonin. Int. J. Mol. Sci. 26: 1608.

Chen, H. and Shen, Y. 2025. Comparative proteomic analysis of desiccation responses in recalcitrant *Quercus acutissima* seeds. Plant Mol. Biol. 115: 68.

Cui, J., Zhao, J., Zhang, H., Li, J., Jiang, L. and Wang, N. 2025. Dynamic changes in tomato (*Solanum lycopersicum*) RNA m6A modification during seed germination and under microgravity conditions. Horticulturae. 11: 282.

Dai, X., Kong, W., Wu, F., Pan, Z., Gao, L., Mo, B., Yu, Y. and Luo, W. 2025. OsNUDX23 regulates early seed germination by modulating ROS balance and starch metabolism in rice. Front. Plant Sci. 16: 1581800.

Dake, D., Supriya, L., Kumar, A. and Gudipalli, P. 2025. Melatonin-mediated methylglyoxal homeostasis and regulation of autophagy during seed germination under PEG-induced drought stress in upland cotton. bioRxiv. 2025: 02.

Ding, G., Li, Z., Iqbal, Z., Zhao, M., Cui, Z., Cao, L., Zhou, J., Lei, L., Luo, Y., Bai, L. and Yang, G. 2025. Identifications of genes involved in ABA and MAPK signaling pathways positively regulating cold tolerance in rice. Plants. 14: 498.

do Nascimento, N.N., Cansian, A.B.M., de Sousa, J.S., Negrão, F.N., Tardioli, P.W. and Vieira, A.M.S. 2025. Plants lipases: Challenges, recent advances, and future prospects - A review. Bioprocess Biosyst. Eng. 2025: 1–19.

Du, C. and Zhang, Z. 2025. Precision tuning of ABA signaling by ubiquitination of ABA receptors: Modulating protein activity and localization. J. Exp. Bot. 76: 2610–2613.

Farhangi-Abriz, S. and Ghassemi-Golezani, K. 2016. Improving amino acid composition of soybean under salt stress by salicylic acid and jasmonic acid. J. Appl. Bot. Food Qual. 89.

Farhangi-Abriz, S. and Ghassemi-Golezani, K. 2019. Jasmonates: Mechanisms and functions in abiotic stress tolerance of plants. Biocatal. Agric. Biotechnol. 20: 101210.

Farhangi-Abriz, S. and Torabian, S. 2018. Nano-silicon alters antioxidant activities of soybean seedlings under salt toxicity. Protoplasma. 255: 953–962.

Fatemeh, G., Saideh, K. and Laleh, A. (2025). Effect of exogenous melatonin application on the expression of Catalase, Superoxide dismutase and Menthol Biosynthesis genes in *Mentha pulegium* L. Plant Sci. Today. 12(2): 2348–1900.

Ferreyra, M.G., Caffaro, M.M., Santamaría, B., Zilli, C., Hernández, A., Fina, B.L., Vélez, A.S., Balestrasse, K.B. and Prevosto, L. 2025. Plasma-activated water produced by a moderately high energy-efficient 1-liter reactor: Effects on germination and growth of tomato and bell pepper plants. Plants. 14: 722.

Fico, B., Andrada, H.E., Alves, F.B., da Silva, E.E., Reinaldi, J.S., Tavares, D.C., Squarisi, I.S., Nuevo, L.G., Montanha, G.S., de Carvalho, H.W. and Vaca Chávez, F. 2025. Amine-polyether-epoxide nanoplatform-driven seed germination, plant growth, and nutrient uptake for sustainable agriculture. ACS Omega. 10: 12453–12466.

Gamalero, E. and Glick, B.R. 2025. How melatonin affects plant growth and the associated microbiota. Biology. 14: 371.

Ge, N., Jia, J.S., Wang, Q.Y., Li, C.L., Huang, M. and Chen, J.W. 2025. A genome-wide survey of DNA methylation reveals that hypermethylation regulates after-ripening and dormancy of recalcitrant *Panax notoginseng* seeds. J. Exp. Bot. 76(12): 3359–3377.

Guardado-Fierros, B.G., Lorenzo-Santiago, M.A., Gumiere, T., Aid, L., Rodriguez-Campos, J. and Contreras-Ramos, S.M. 2025. Glyphosate biodegradation by airborne plant growth-promoting bacteria: Influence on soil microbiome dynamics. Agriculture. 15: 362.

Gülöksüz, T., Eker, A.H., Ertürk, N., Kadıoglu, N. and Demir, İ. 2025. Electrical conductivity relates seed germination and seedling emergence in *Tagetes* seed lots. Bahçe. 54: 26–30.

Halimeh, H. 2025. Red light induced seed germination and seedling growth by modulating antioxidant defense system, Rubisco, and NADPH oxidase activities in *Capsicum frutescens*. BMC Plant Biol. 25: 519.

Han, C. and Yang, P. 2015. Studies on the molecular mechanisms of seed germination. Proteomics. 15: 1671–1679.

Harmancı, P., Ergin, N., Kulan, E.G., Yaman, E. and Kaya, M.D. 2025. Mitigation of adverse effects of drought and low temperature stresses by seed priming in sunflower during germination. Selcuk J. Agric. Food Sci. 39: 20–30.

Havii, V., Palyvoda, Y., Kuchmenko, O., Stamirowska-Krzaczek, E., Tomaszewska, M. and Kocira, A. 2025. Biochemical mechanisms of drought resistance in soft wheat under modeling of water deficiency and effects of seed treatment with metabolically active substances. Agric. Eng. 29.

Hua, Y., Dong, L., Sun, S., Wang, K., Zou, Y., Gao, Y., Gong, T., Hu, G. and Qin, L. 2025. Metabonomics and physiology revealed the critical function of 5-phosphoribosylamine and antioxidant enzymes in enhancing aged oat seed germination. BMC Plant Biol. 25: 28.

Huang, Z., Han, X., He, K., Ye, J., Yu, C., Xu, T., Zhang, J., Du, J., Fu, Q. and Hu, Y. 2025. Nitrate attenuates abscisic acid signaling via NIN-LIKE PROTEIN8 in Arabidopsis seed germination. Plant Cell. 37: koaf046.

Kaur, M., Sinha, K., Eastmond, P.J. and Bhunia, R.K. 2025. Exploiting lipid droplet metabolic pathway to foster lipid production: Oleosin in focus. Plant Cell Rep. 44: 12.

Kaushik, P. 2025. AI-driven breeding enhances stress tolerance in high-elevation extremophytes: A proof-of-concept study with cross-component validation. bioRxiv. 2025: 02.

Kaya, C., Ashraf, M., Ahmad, P. and Corpas, F.J. 2025. Deciphering the interplay between redox switches and signaling networks in plant stress responses. Crit. Rev. Plant Sci. 2025: 1–30.

Kępczyński, J., Wójcik, A. and Dziurka, M. 2025. After-ripening is associated with changes in the sensitivity of *Avena fatua* L. Caryopses to abscisic acid, as well as changes in the abscisic acid and bioactive gibberellins contents in embryos. Plants. 14: 463.

Khan, A., Kanwal, F., Ullah, S., Fahad, M., Tariq, L., Altaf, M.T., Riaz, A. and Zhang, G. 2025. Plant secondary metabolites—Central regulators against abiotic and biotic stresses. Metabolites. 15: 276.

Khanzada, A., Yan, K., Hu, W., Malko, M., Khan, K.A., Bao, Y., Elboughdiri, N. and Li, Y. 2025. Heat stress response mechanisms and resilience strategies in wheat. J. Agron. Crop Sci. 211: e70023.

Kim, H., Takegahara, N. and Choi, Y. 2025a. The impact of IgSF11-PKM2 pathway on gene expression in osteoclasts. microPublication Biol. 2025: 10–17912.

Kim, Y.H., Park, C.W., Lee, K.M., Hong, C.O., Son, H.J., Kim, K.K., Park, H.C. and Kim, Y.J. 2025b. The roles of MYC2 transcription factor in JA-signaling pathway in plants. J. Plant Biol. *68*(2) 113–131.

Komiya, S., Pancha, I., Shima, H., Igarashi, K., Tanaka, K. and Imamura, S. 2025. Target of rapamycin signaling regulates starch degradation via α-glucan water dikinase in a unicellular red alga. Plant Physiol. 197: kiaf106.

Koutsovoulou, K., Thanos, C.A. and Daskalakou, E.N. 2025. Ecophysiology of seed germination in riparian trees and implications for restoration and conservation. Plant Species Biol. 40: 233–244.

Li, R., Li, X., He, Y., Chen, X., Li, J. and Zhuang, C. 2025a. Progress of acetylation modification in plants. Agronomy. 15: 1910.

Li, S., Xu, L., Li, Y., Waqar, A., Hu, Z., Yang, M., Zhao, Y., Qi, Z., Chen, Q. and Hu, L. 2025b. Advances in salinity tolerance of soybean: Molecular mechanism and breeding strategy. Food Energy Secur. 14: e70073.

Lin, C., Zhang, Z., Zhang, Z., Long, Y., Shen, X., Zhang, J. and Wang, Y. 2025. The role of glutathione S-transferase in the regulation of plant growth, and responses to environmental stresses. Phyton. 94.

Liu, K., Deng, F., Zeng, F., Chen, Z.H., Qin, Y. and Chen, G. 2025. Plant growth-promoting rhizobacteria improve drought tolerance of crops: A review. Plant Growth Regul. 2025: 1–15.

Luo, Y., Wang, K., Cheng, J. and Nan, L. 2025. Transcriptome analysis of onobrychis viciifolia during seed germination reveals GA3-inducible genes associated with phenylpropanoid and hormone pathways. Int. J. Mol. Sci. 26: 2335.

Magar, B.N., Chapagaee, P. and Bohora, A. 2025. Evaluating the efficacy of organic and inorganic seed priming methods in promoting cucumber germination and growth. Turk. J. Agric. Food Sci. Technol. 13: 406–412.

Matsui, S. and Matsubayashi, Y. 2025. An activity-based proteomics with two-dimensional polyacrylamide gel electrophoresis (2D-PAGE) for identifying target proteases in Arabidopsis apoplastic fluid. Bio-Protocol. 15: e5226.

Meng, X., Cao, Y., Lv, Y., Wang, L. and Wang, Y. 2025. Integrating physiological, metabolome and transcriptome revealed the response of maize seeds to combined cold and high soil moisture stresses. Physiol. Plant. 177: e70096.

Mérai, Z., Xu, F., Hajdu, A., Kozma-Bognár, L. and Dolan, L. 2025. Phytochrome A is required for light-inhibited germination of *Aethionema arabicum* seed. New Phytol. 2025.

Mi, C., Hong, L., Sun, S., Zhao, S., Dou, L. and Mao, P. 2025. Ascorbic acid priming restores the seed vigor by enhancing the mitochondrial AsA-GSH cycle and related gene expression in the aged oat seeds. Physiol. Plant. 177: e70190.

Novello, B.D., Barroso, A.A., da Silva, D.R., Bacha, A.L., Roncatto, E. and Albrecht, A.J. 2025. Auxinic herbicides and glyphosate inhibit horseweed seed production and germination. Adv. Weed Sci. 43: e020250057.

Nowicki, M., Nowakowska, M., Nowak, K., Szczechura, W. and Kaminski, P. 2025. Seed priming and abiotic stress tolerance in carrot: Unraveling the mechanisms of improved germination. PloS One. 20: e0318753.

Özden, E., Ermiş, S., Yildirim, E. and Demir, İ. 2025. Perspective chapter: Insights into seed germination–physiological and environmental mechanisms. In *Physiology*. IntechOpen. https://doi.org/10.5772/intechopen.1010636

Posoongnoen, S., Thummavongsa, T., Junthip, J., Jandaruang, J. and Preecharram, S. 2025. Potential application of α-amylase from germinating Cucurbita moschata duchesne seeds in eco-friendly bioethanol production. Sci. Rep. 15: 25499.

Qian, J., Mo, X., Wang, Y. and Li, Q. 2025. Seed priming with 2,4-epibrassionolide enhances seed germination and heat tolerance in rice by regulating the antioxidant system and plant hormone signaling pathways. Antioxidants. 14: 242.

Quan, C.Q., Tang, D.F., Jiang, J.P. and Zhu, Y.X. 2025. Transcriptome sequencing reveals molecular mechanism of seed dormancy release of *Zanthoxylum nitidum*. Zhongguo Zhong Yao Za Zhi. 50: 102–110.

Rahman, M.A., Lee, S.H., Park, H.S., Min, C.W., Woo, J.H., Choi, B.R., Rahman, M.M. and Lee, K.W. 2025. Light quality plays a crucial role in regulating germination, photosynthetic efficiency, plant development, reactive oxygen species production, antioxidant enzyme activity, and nutrient acquisition in alfalfa. Int. J. Mol. Sci. 26: 360.

Rajjou, L., Duval, M., Gallardo, K., Catusse, J., Bally, J., Job, C. and Job, D. 2012. Seed germination and vigor. Annu. Rev. Plant Biol. 63: 507–533.

Rhaman, M.S. 2025. Seed priming before the sprout: Revisiting an established technique for stress-resilient germination. Seeds. 4: 29.

Rudiyanto, Y.F., Alfiana, O.H., Putri, A.L., Rizkiantoro, R. and Siswoyo, T.A. 2025. Antioxidant responses of black glutinous rice to drought and salinity stresses at different growth stages. Open Agric. 10: 20250417.

Singh, K., Asghar, M.A., Jobbágy, K., Kulman, K., Szalai, G., Hamow, K., Soltész, A., Polgári, D., Gulyás, Z. and Kocsy, G. 2025. Different modulation of the redox homeostasis and hormone levels by ascorbate, hydrogen peroxide and hydrogen sulfide in maize. Physiol. Plant. 177: e70215.

Srivastava, S.P., Kopasz-Gemmen, O., Thurman, A., Rajendran, B.K., Selvam, M.M., Kumar, S., Srivastava, R., Suresh, M.X., Kumari, R., Goodwin, J.E. and Inoki, K. 2025. The molecular determinants regulating redox signaling in diabetic endothelial cells. Front. Pharmacol. 16: 1563047.

Sudhakaran, S., Thakral, V., Mandlik, R., Mahakalkar, B., Sharma, Y., Kumar, V., Deshmukh, R., Sharma, T.R. and Sonah, H. 2025. Potential role of TIP3 aquaporins in the transport of H_2O_2 and boric acid during seed development and germination in soybean (*Glycine max* L.). Plant Cell Environ. 2025.

Susko, D.J. 2025. Seed germination ecology of the medicinal plant motherwort (*Leonurus cardiaca*). Seed Sci. Res. 2025: 1–8.

Talukder, Z.A., Chhabra, R., Basu, S., Gain, N., Mishra, S.J., Kumar, A., Zunjare, R.U., Muthusamy, V. and Hossain, F. 2025. High amylopectin in waxy maize synergistically affects seed germination and seedling vigour over traditional maize genotypes. J. Appl. Genet. 66: 267–278.

Turk, H., Genisel, M. and Dumlupinar, R. 2025. The regulatory role of exogenous carnitine applications in lipid metabolism, mitochondrial respiration, and germination in maize seeds (*Zea mays* L.). Life. 15: 631.

Upadhyay, S., Bhardwaj, M., Kumar, S.P., Khan, S., Kumar, A. and Hassan, M.I. 2025. Impact of cancer-associated PKM2 mutations on enzyme activity and allosteric regulation: Structural and functional insights into metabolic reprogramming. Biochemistry. 64: 1463–1475.

Wang, Y., Liu, X., Sun, X., Mao, X., Wang, Z., Peng, J., Yang, Z., Ali, F., Wang, Z. and Li, F. 2025. The promotive and repressive effects of exogenous H2O2 on Arabidopsis seed germination and seedling establishment depend on application dose. Physiol. Plant. 177: e70098.

Wei, Y., Deng, J., Tang, H., Xu, J., Zhou, M. and Ye, J. 2025. Integrated transcriptomic and metabolomic analyses reveal regulatory networks governing hub metabolic pathways in *Fraxinus hupehensis* seeds during germination. Tree Physiol. 45: tpaf032.

Wu, B., Liang, H., Lv, J., Liu, R. and Ye, N. 2025b. Key regulators of seed germination: Kinases and phosphatases. Seeds. 4: 30.

Wu, J., Cao, X., Sun, X., Chen, Y., Zhang, P., Li, Y., Ma, C., Wu, L., Liang, X., Fu, Q. and Li, L. 2025a. OsEL2 regulates rice cold tolerance by MAPK signaling pathway and ethylene signaling pathway. Int. J. Mol. Sci. 26: 1633.

Xing, Y., Zhang, H., Liu, C., Liu, C. and Zhou, Y. 2025. Spermidine revives aged sorghum seed germination by boosting antioxidant defense. Antioxidants. 14: 349.

Xu, Q., Kong, F. and Yang, W. 2025. SnRK1 as the core node integrating energy homoeostasis, stress adaptation and hormonal crosstalk in plants. Plant Cell Environ. 2025.

Yılmaz, H. and Çiftçi, V. 2025. Enhancing cadmium tolerance in common bean plants by seed priming with putrescine. Black Sea J. Agric. 8: 9–10.

Yu, X. and Feng, T. 2025. The multifaceted roles of plant aspartic proteases. J. Exp. Bot. 2025: eraf147.

Zhang, J., Liu, L., Dong, D., Xu, J., Li, H., Deng, Q., Zhang, Y., Huang, W., Zhang, H. and Guo, Y.D. 2025. The transcription factor SlLBD40 regulates seed germination by inhibiting cell wall remodeling enzymes during endosperm weakening. Plant Physiol. 197: kiaf022.

Zheng, K., Martinez, M.D.P., Bouzid, M., Balparda, M., Schwarzländer, M. and Maurino, V.G. 2025. Regulation of plant glycolysis and the tricarboxylic acid cycle by post-translational modifications. Plant J. 122: e70142.

Zhou, L., Ullah, F., Zou, J. and Zeng, X. 2025. Molecular and physiological responses of plants that enhance cold tolerance. Int. J. Mol. Sci. 26: 1157.

Zhuang, D., Li, H.B., Wang, Y., Zhou, D. and Zhao, L. 2025. Nanoparticle-elicited eustress intensifies cucumber plant adaptation to water deficit. Environ. Sci. Technol. 59: 3613–3623.

15 Metabolic Engineering of Secondary Metabolites for Improved Crop Protection and Yield

Weria Weisany

15.1 INTRODUCTION

Plants produce an extraordinary array of secondary metabolites—specialized compounds that, while not essential for basic cellular functions or primary growth, play pivotal roles in plant survival, ecological interactions, and adaptation to environmental challenges. Unlike primary metabolites such as sugars, amino acids, and nucleotides that are conserved across all organisms, secondary metabolites are structurally diverse and often species-specific (Otles and Özyurt 2021). They function as chemical defense agents against herbivores, pathogens, and competing plants, act as signaling molecules in plant–microbe interactions, and contribute to the sensory and nutritional quality of plant-derived foods (Lone et al. 2024). Their ecological importance is mirrored by their extensive utility in human industries, including pharmaceuticals, nutraceuticals, cosmetics, and food additives.

In the context of agriculture, secondary metabolites are increasingly recognized as integral to plant resilience and productivity. Compounds such as phytoalexins, glucosinolates, flavonoids, and alkaloids can enhance tolerance to biotic and abiotic stressors, thereby reducing dependence on synthetic agrochemicals (Jan et al. 2021). Moreover, many of these metabolites influence plant development, reproductive success, and postharvest characteristics, linking them indirectly to yield and quality traits. However, the natural levels and expression patterns of these compounds are often suboptimal or unstable under field conditions, limiting their practical benefits (Reshi et al. 2023).

With the advancement of metabolic engineering, scientists now have the ability to modify plant metabolic pathways with increasing precision. This multidisciplinary approach integrates molecular biology, genetic engineering, systems biology, synthetic biology, and computational modeling to reprogram metabolic networks and redirect biosynthetic fluxes toward desired end-products. Key strategies include overexpression of biosynthetic genes, silencing of competing pathways, introduction of heterologous genes, modulation of regulatory elements (e.g., transcription factors, promoters), and use of genome editing tools such as CRISPR/Cas9. When applied to secondary metabolism, these interventions can dramatically increase the production of beneficial compounds, minimize anti-nutritional factors, or generate novel metabolites with enhanced bioactivity.

In recent years, the convergence of omics technologies (e.g., transcriptomics, metabolomics, proteomics) with genome-editing platforms has further expanded the toolkit for rational metabolic design (Zhang et al. 2025). Researchers can now identify bottlenecks, regulatory nodes, and gene–metabolite associations with unprecedented resolution, enabling a shift from empirical to predictive engineering. These advances hold great promise for developing crops with enhanced immunity, better nutritional profiles, improved flavor and aroma, and greater adaptability to environmental stresses (Mmbando 2024).

The primary objective of this chapter is to provide a comprehensive overview of the metabolic engineering of secondary metabolites, with a focus on its applications in enhancing crop protection and yield. The chapter will begin by introducing the major classes of secondary metabolites and their roles in plant physiology and ecology. It will then explore the biosynthetic pathways, regulatory networks, and current strategies for metabolic reprogramming, supported by recent case studies in model and crop species. Challenges associated with metabolic flux balancing, pleiotropy, and regulatory hurdles will be discussed, along with emerging trends such as synthetic circuits, machine learning (ML) in pathway prediction, and climate-resilient metabolic design. Ultimately, this chapter aims to highlight the transformative potential of metabolic engineering in creating next-generation crops tailored for the demands of sustainable agriculture and global food security.

15.2 CLASSIFICATION AND FUNCTIONAL ROLES OF PLANT SECONDARY METABOLITES

Plants synthesize an impressive diversity of secondary metabolites—specialized compounds that are structurally varied and typically restricted to specific plant lineages. These compounds are categorized based on their biosynthetic origin, chemical structure, and functional roles. The major classes include alkaloids, phenolic compounds, terpenoids, and glucosinolates, each contributing uniquely to the plant's physiology, ecological interactions, and adaptive responses.

15.2.1 Alkaloids

Alkaloids are nitrogen-containing secondary metabolites derived mainly from amino acid precursors. They are often

DOI: 10.1201/9781003669920-15

biologically active and serve as potent defensive agents against herbivores and microbial pathogens due to their bitter taste and toxic effects (Bhatla and Lal 2023). Well-known examples include nicotine (from tobacco), morphine (from opium poppy), and quinine (from Cinchona species). Beyond defense, alkaloids may also influence plant growth and intercellular signaling. In agriculture, alkaloid profiles have been associated with pest resistance, though their concentration must be carefully managed due to potential toxicity to humans and livestock.

15.2.2 Phenolic Compounds

Phenolics, including flavonoids, tannins, lignins, and phenolic acids, are among the most abundant and widely distributed plant secondary metabolites. They play crucial roles in UV protection, structural integrity (via lignification), pathogen defense, and allelopathic interactions (Lin et al. 2016). Flavonoids, in particular, are multifunctional—acting as antioxidants, colorants, and signal molecules in symbiotic relationships such as legume–rhizobia interactions. In crops, high phenolic content is often correlated with improved stress tolerance and enhanced nutritional and antioxidant properties, making them attractive targets for biofortification (Kołton et al. 2022).

15.2.3 Terpenoids (Isoprenoids)

Terpenoids, synthesized from isopentenyl diphosphate (IPP) units, represent the largest and most structurally diverse class of secondary metabolites. They include essential oils, carotenoids, sterols, and plant hormones (gibberellins, abscisic acid) (Bergman et al. 2024). Terpenoids contribute to plant defense by acting as repellents or toxins to pests and pathogens, while also functioning in attracting pollinators through volatile emissions. Certain terpenoids, such as artemisinin (from *Artemisia annua*), have notable pharmaceutical applications (Li et al. 2023). In crops, manipulation of terpenoid biosynthesis can improve both pest resistance and postharvest shelf-life.

15.2.4 Glucosinolates

Glucosinolates are sulfur- and nitrogen-containing compounds predominantly found in Brassicaceae (broccoli, mustard, canola). Upon tissue disruption, they are hydrolyzed by the enzyme myrosinase to produce bioactive compounds such as isothiocyanates and nitriles, which have strong antimicrobial and anti-insect properties (Cheng et al. 2024). These metabolites serve as chemical defenses and also contribute to the flavor, aroma, and nutritional quality of cruciferous vegetables (Connolly et al. 2021). Breeding or engineering crops with optimized glucosinolate profiles offers potential for integrated pest management and improved dietary health benefits (Figure 15.1).

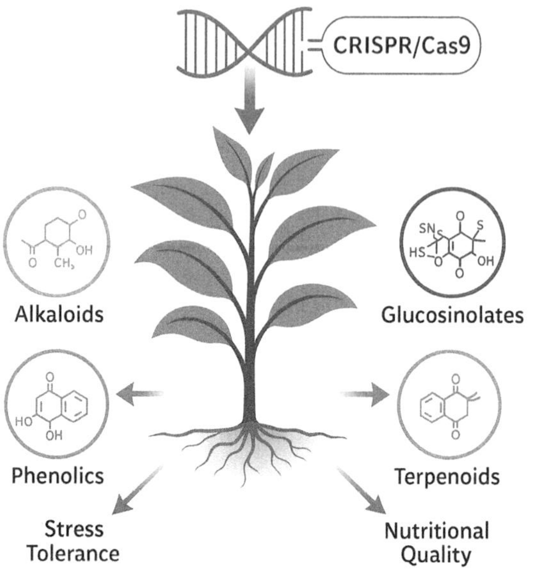

FIGURE 15.1 Metabolic engineering approaches used to enhance secondary metabolite production for improved crop protection, stress resilience, and yield.

15.2.5 Ecological and Physiological Significance

Collectively, secondary metabolites form a chemical language through which plants interact with their environment. They regulate responses to biotic stressors (insects, fungi, bacteria) and abiotic challenges (UV radiation, salinity, drought), often acting as signaling molecules in complex defense cascades (Aqaei et al. 2020; Manvelian et al. 2021). These metabolites also mediate plant–plant and plant–microbe interactions, influence pollination and seed dispersal, and modulate plant architecture and reproductive timing through hormonal cross-talk (Ketehouli et al. 2025). From a physiological perspective, secondary metabolites intersect with primary metabolism, often competing for precursors and cofactors. This metabolic integration underscores the importance of maintaining a delicate balance between growth and defense—a balance that can be fine-tuned through metabolic engineering (Figure 15.1).

15.2.6 Relevance to Crop Resilience and Nutritional Quality

The strategic manipulation of secondary metabolites offers immense potential to enhance crop resilience against stress, reduce reliance on chemical inputs, and improve the nutritional and functional quality of plant-based foods. For instance, elevating flavonoid levels can improve antioxidant capacity in fruits; increasing glucosinolate content can enhance pest resistance in brassicas; and optimizing terpenoid emission profiles can reduce herbivore attacks in cereals and legumes (Trejo-Téllez et al. 2019). As consumer demand for natural, functional, and sustainable food systems rises, secondary metabolites are becoming central to crop breeding and engineering programs. Their dual role in plant protection and human health makes them ideal targets for metabolic enhancement in next-generation crops.

15.3 BIOSYNTHETIC PATHWAYS AND REGULATORY NETWORKS

The biosynthesis of secondary metabolites in plants is a complex and highly regulated process involving diverse enzymes, specialized intermediates, distinct subcellular compartments, and tightly controlled gene expression networks. Understanding these components is essential for the rational design of metabolic engineering strategies that can enhance metabolite accumulation without disrupting essential physiological processes.

15.3.1 Key Enzymes, Intermediates, and Cellular Compartmentalization

Secondary metabolites arise when central metabolic precursors chiefly acetyl-CoA, shikimate-derived intermediates, and mevalonate (MVA) or 2-C-methyl-D-erythritol 4-phosphate (MEP)-derived isoprenoid units are diverted into specialized biosynthetic branches coordinated by enzyme families exhibiting high substrate and regio-/stereospecificity. Alkaloids typically originate from amino acids and are assembled and diversified by decarboxylases, methyltransferases, and cytochrome P450 monooxygenases, among others (Guirimand et al. 2010). Phenolic compounds stem largely from the phenylpropanoid network, initiated by phenylalanine ammonia-lyase (PAL) and elaborated by polyketide-type enzymes such as chalcone synthase (CHS) and tailoring enzymes like flavonoid 3′-hydroxylase (F3′H) (Hanhineva 2008).

Subcellular compartmentation is a defining feature of specialized-metabolite biosynthesis: plastids house carotenoid formation and much of the MEP-derived isoprenoid pathway, providing the prenyl precursors isopentenyl diphosphate (IPP) and dimethylallyl diphosphate (DMAPP), together with plastidial enzymes required for terpenoid scaffold biosynthesis. (Wanke et al. 2001). Many oxidative tailoring steps are catalyzed by cytochrome P450s that are embedded in the endoplasmic reticulum (ER), positioning these enzymes for efficient access to lipophilic substrates and for assembly of multi-enzyme complexes on ER membranes (Guengerich et al. 2011). Vacuoles function as major sequestration sites for diverse secondary metabolites (e.g., alkaloids, phenolics, glucosinolates), mitigating autotoxicity and enabling regulated storage and mobilization via dedicated tonoplast transporters (de Brito Francisco and Martinoia 2018). Such spatial organization promotes metabolic channeling—physical or functional clustering of pathway enzymes that minimizes diffusion losses, protects unstable intermediates, and enhances flux control—thereby increasing pathway efficiency and regulatory flexibility (de Brito Francisco and Martinoia 2018) (Figure 15.2).

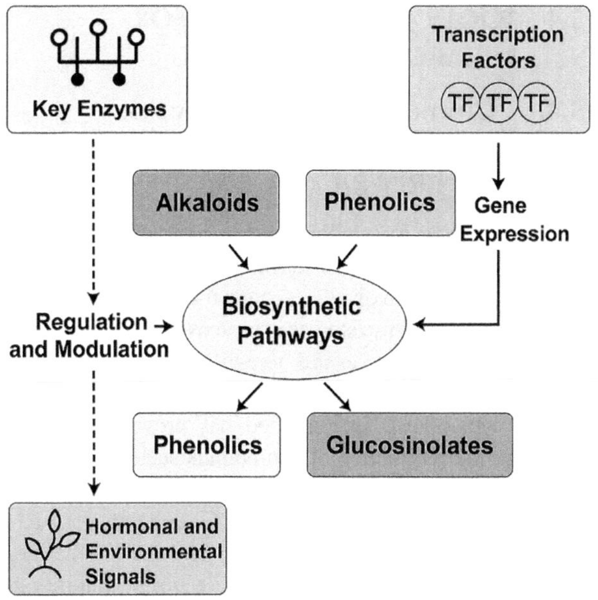

FIGURE 15.2 Graphical abstract illustrating the biosynthetic pathways and regulatory networks of plant secondary metabolites. Key metabolite classes—alkaloids, phenolics, terpenoids, and glucosinolates—are synthesized via specific enzymatic routes, regulated by transcription factors, and modulated by hormonal and environmental signals. These interconnected systems control metabolite accumulation, influencing plant defense, signaling, and adaptation.

15.3.2 Hormonal and Environmental Modulation

Plant secondary metabolism is tightly wired to hormone signaling and environmental inputs. Jasmonates (via COI1–JAZ derepression) rapidly activate alkaloid and terpenoid pathways during herbivory, often in synergy with ethylene through EIN3/EIL1 to potentiate defense gene expression (Alharbi et al. 2025). Salicylic acid, through NPR1-dependent transcriptional reprogramming, promotes phenolic accumulation during systemic acquired resistance and frequently antagonizes JA-driven routes to tailor defenses against biotrophs versus herbivores/necrotrophs (Chen et al. 2021). Abscisic acid integrates abiotic stress cues (drought/salinity) via SnRK2–AREB/ABF modules and can enhance flavonoid/anthocyanin biosynthesis to bolster oxidative stress protection, with the magnitude and direction of cross-talk depending on context (Kumar et al. 2019). Overall, dynamic, sometimes synergistic interactions between jasmonic acid (JA) and ethylene (ET) signaling pathways, and frequently antagonistic crosstalk between jasmonic acid (JA) and salicylic acid (SA) pathways, integrate hormonal networks with environmental cues to ensure that specialized metabolites are produced with appropriate composition, timing, and tissue specificity in accordance with developmental and ecological demands.

15.4 TOOLS AND STRATEGIES FOR METABOLIC ENGINEERING

15.4.1 Genetic Engineering Approaches: Overexpression, Gene Silencing, and CRISPR/Cas Editing

The foundation of metabolic engineering in plants lies in the application of genetic tools that modulate the expression of biosynthetic genes involved in secondary metabolism. One of the most widely used strategies is gene overexpression, where key biosynthetic enzymes are upregulated using strong constitutive or inducible promoters to increase the flux toward desired metabolites. This method has proven effective in enhancing the production of compounds such as flavonoids, alkaloids, and terpenoids. Conversely, gene silencing technologies such as RNA interference (RNAi) and antisense RNA are employed to suppress undesirable branches in the metabolic network, thereby redirecting precursor availability and improving product specificity (Leung and Whittaker 2005).

The introduction of CRISPR/Cas-based genome editing has significantly advanced the field by allowing precise, targeted modifications in plant genomes. CRISPR/Cas9 and its derivatives—such as base editors and CRISPR activation (CRISPRa) or interference (CRISPRi) systems—enable knockout, knock-in, or fine-tuning of genes and regulatory sequences (Sun et al. 2024). This technology allows for transgene-free editing and has been successfully applied to modulate metabolite pathways in various crop species, offering both functional validation and trait improvement (Figure 15.3).

15.4.2 Synthetic Biology and Modular Pathway Reconstruction

Synthetic biology offers a more predictive and modular approach to engineering metabolic pathways. One of its core principles is pathway reconstruction, in which biosynthetic genes are assembled into standardized, synthetic cassettes for targeted expression in plant hosts or heterologous systems like yeast or *Nicotiana benthamiana* (Lee et al. 2024). These modular constructs enable precise control over gene expression and metabolic flux, allowing the production of complex metabolites outside of their native context.

Synthetic biology also leverages orthogonal regulatory systems, including artificial promoters, repressors, riboswitches, and feedback loops to dynamically regulate gene expression in response to developmental or environmental cues (Chappell et al. 2015). Furthermore, pathway refactoring—which involves the rational redesign of gene order, codon usage, and regulatory elements—can alleviate pathway bottlenecks and enhance flux efficiency. The development of synthetic metabolons, or enzyme scaffolds that co-localize multiple biosynthetic enzymes, further enhances substrate channeling and minimizes intermediate loss, thereby improving metabolite yield and pathway performance.

15.4.3 Promoter Selection, Multigene Stacking, and Regulatory Fine-Tuning

Precise and context-specific gene expression is essential for the success of metabolic engineering in plants. Promoter

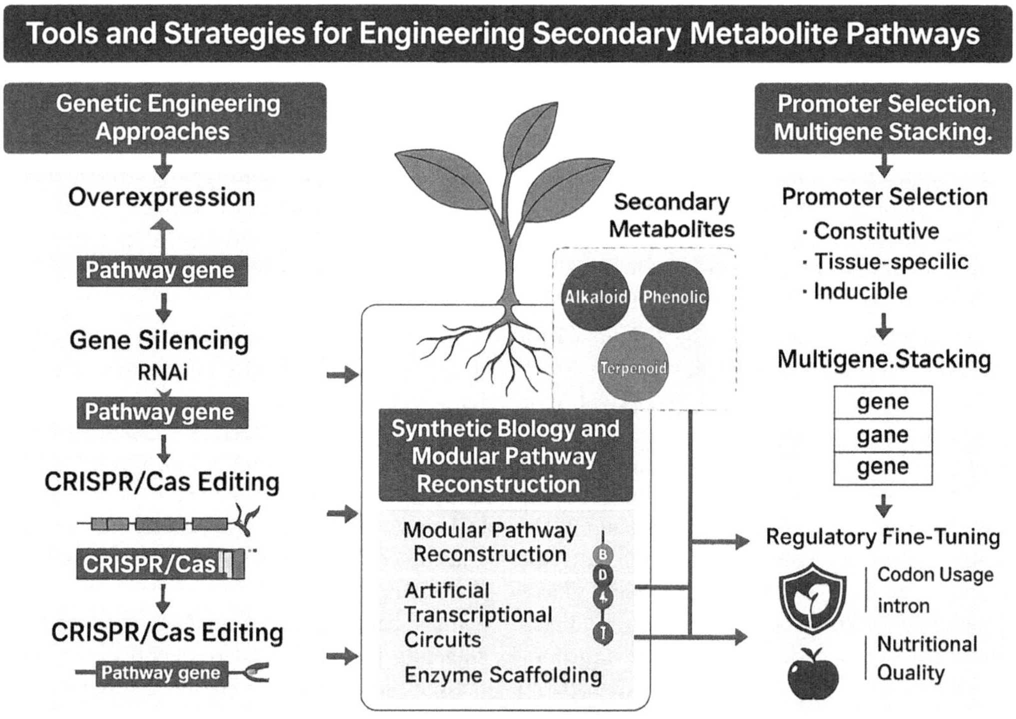

FIGURE 15.3 Integrated strategies for engineering plant secondary metabolite pathways, combining genetic engineering, synthetic biology, and regulatory optimization to enhance metabolite production for improved crop defense, stress tolerance, and nutritional value.

selection plays a critical role in this regard, with options ranging from strong constitutive promoters (CaMV 35S, Ubiquitin) to tissue-specific or inducible promoters that restrict expression to certain organs or conditions. For example, trichome-specific promoters have been used to enhance the production of volatile terpenoids in glandular structures without affecting other tissues (Holtorf et al. 1995). Complex metabolic pathways often require the simultaneous expression of multiple genes, which can be achieved through multigene stacking (Zhu et al. 2020). Techniques such as polycistronic vectors, **Golden Gate cloning**, and site-specific recombination systems including **Cre/lox** (*Causes recombination/loxP locus of X-over P1*) and **FLP/FRT** (*Flippase/Flippase Recognition Target*) systems facilitate the stable genomic integration and coordinated expression of multiple transgenes. (Snaith et al. 1995).

To further enhance pathway control, regulatory fine-tuning strategies are employed. These include optimizing untranslated regions (UTRs) to improve mRNA stability and translation efficiency, inserting introns to boost transcriptional activity, and adapting codon usage to match the host expression system. Together, these regulatory modifications help balance pathway activity, reduce metabolic burden, and ensure stable metabolite production over time.

15.5 ENGINEERING SECONDARY METABOLITES FOR CROP PROTECTION

One of the most impactful applications of metabolic engineering is in the enhancement of plant defense systems through the strategic manipulation of secondary metabolite pathways (Weisany and Khosropour 2023). Plants rely heavily on a repertoire of these metabolites—including alkaloids, flavonoids, terpenoids, and glucosinolates—to defend against a broad spectrum of biotic and abiotic stresses (Weisany et al. 2011, 2024a). By engineering the biosynthesis or regulation of these compounds, crops can be made more resistant to pathogens, herbivores, and environmental stressors, reducing reliance on chemical pesticides and improving overall sustainability.

Secondary metabolites serve as natural deterrents and antimicrobial agents. For example, the enhancement of phenolic content in crops like tomatoes and rice has been shown to improve resistance to fungal pathogens such as *Colletotrichum nymphaeae* (Karimi et al. 2016). Likewise, the overproduction of terpenoid volatiles in maize and cotton can repel insect pests or attract their natural predators (von Mérey et al. 2013). In Arabidopsis, glucosinolate profiles have been engineered to provide protection against aphids and root-feeding nematodes (Smith et al. 2016). In addition to direct toxicity or deterrence, some secondary metabolites act as signaling molecules, priming systemic acquired resistance or mediating defense gene expression.

Across horticultural systems, nano-enabled delivery of plant essential oils has demonstrably strengthened disease control and postharvest protection. In strawberries, silver-nanoparticle encapsulation of *Thymus daenensis* and *Anethum graveolens* essential oils markedly enhanced antifungal activity against *C. nymphaeae*—the causal agent of anthracnose—compared with free oils, highlighting synergy between EO chemistries (thymol, carvacrol, carvone) and nanoscale carriers that improve stability and bioavailability (Weisany et al. 2019). Extending this approach, mesoporous silica (SBA-15) nano-encapsulation of *Mentha aquatica* essential oil significantly increased growth inhibition of *Botrytis cinerea* and *C. nymphaeae* over 5–15 days, demonstrating sustained release and prolonged antifungal efficacy relevant to postharvest management (Weisany et al. 2022). More recently, copper-oxide nanoparticle encapsulation of fenugreek and coriander essential oils further enhanced suppression of *C. nymphaeae*, underscoring the versatility of EO–nanocarrier combinations to counter emerging resistance to synthetic fungicides and to reduce postharvest losses in horticultural value chains (Weisany et al. 2024b).

An important group of defense-related metabolites are phytoalexins—low-molecular-weight antimicrobial compounds synthesized de novo in response to pathogen attack. Engineering phytoalexin biosynthesis, such as resveratrol in grape or momilactone in rice, has been shown to boost resistance against a range of microbial invaders. On the other hand, anti-nutritional compounds like cyanogenic glycosides or certain alkaloids, while potentially harmful to human health in excess, play a crucial role in herbivore deterrence and need to be carefully managed when engineering crops for defense. Ultimately, enhancing plant defense through metabolic engineering of secondary metabolites offers a sustainable, environmentally friendly alternative to agrochemicals, contributing to both crop resilience and food safety. However, these modifications must be balanced to avoid trade-offs in growth or nutritional quality, underscoring the need for precise regulatory control and a systems-level understanding of plant metabolism.

15.6 ENGINEERING SECONDARY METABOLITES TO IMPROVE YIELD AND QUALITY

While secondary metabolites are traditionally viewed through the lens of defense and ecological interactions, their engineering has far-reaching consequences that extend into primary metabolism, crop quality traits, and developmental regulation. Understanding these broader implications is essential for ensuring that metabolic interventions deliver both agronomic and consumer-oriented benefits without compromising plant fitness.

One of the primary concerns in engineering secondary metabolism is its impact on primary metabolic pathways and source–sink relationships. Secondary metabolites are often derived from central metabolic intermediates such as acetyl-CoA, pyruvate, shikimate, and amino acids, which are also precursors for primary metabolic processes like respiration, protein synthesis, and hormone biosynthesis (Thirumurugan et al. 2018). Excessive redirection of these intermediates toward secondary pathways can cause metabolic imbalances, growth retardation, or yield penalties. For example, overproduction of lignin or flavonoids can lead to altered carbon partitioning and reduced biomass accumulation (Hancock et al. 2007). To mitigate these effects, it is crucial to employ strategies that balance metabolite flux—such as temporal or tissue-specific expression, inducible promoters, or feedback-regulated synthetic circuits.

Beyond plant physiology, metabolic engineering of secondary metabolites plays a critical role in enhancing organoleptic and postharvest qualities of crops. Many metabolites are responsible for flavor, aroma, and color, which are key determinants of consumer acceptance. For instance, increasing anthocyanin or betalain content can improve the visual appeal and antioxidant potential of fruits and vegetables (Stintzing and Carle 2004). Engineering volatile organic compounds (VOCs), such as terpenoids and esters, has been successful in restoring or enhancing aroma in modern strawberries (Tiwari et al. 2020). Additionally, secondary metabolites such as phenolics and tannins can enhance postharvest stability by reducing microbial spoilage, oxidative damage, and enzymatic browning, thereby extending shelf life and reducing food loss along the supply chain (Karnwal et al. 2025).

Furthermore, several secondary metabolites exert regulatory effects on plant growth and development, directly or indirectly influencing physiological and morphological traits (Reshi et al. 2023). For example, flavonoids modulate auxin transport, affecting root and shoot architecture; terpenoids serve as precursors to phytohormones like gibberellins and abscisic acid, influencing seed germination, flowering, and stress responses. Phenolics have also been shown to play roles in stomatal movement, cell wall modification, and senescence (Ali et al. 2025). By manipulating these compounds, researchers can potentially modulate growth patterns, developmental timing, and stress adaptation, opening avenues for crop ideotype design tailored to specific environments (Figure 15.4).

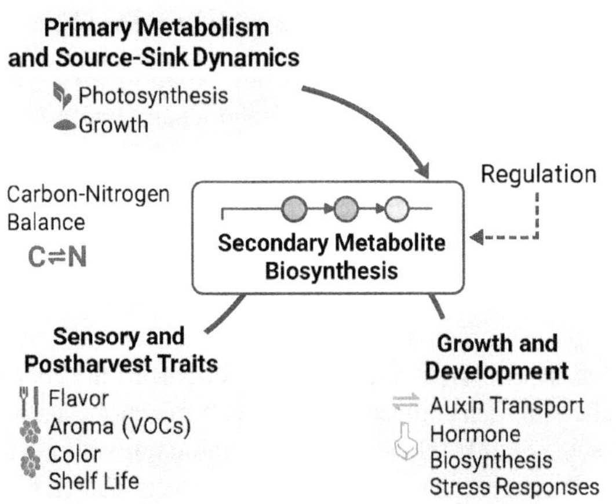

FIGURE 15.4 Summary of how engineering secondary metabolites affects primary metabolism, crop quality, and plant development, enhancing stress tolerance, flavor, color, shelf life, and growth through integrated metabolic and regulatory pathways.

15.7 SYSTEMS BIOLOGY AND OMICS INTEGRATION

The successful engineering of secondary metabolite pathways increasingly depends on comprehensive systems-level insights provided by omics technologies. By integrating genomics, transcriptomics, proteomics, and metabolomics, researchers can unravel the complex regulatory networks that govern secondary metabolism, identify rate-limiting steps, and prioritize gene targets for intervention. These high-throughput platforms form the foundation for predictive metabolic engineering and rational pathway optimization.

At the core, genomic tools such as whole-genome sequencing and gene annotation enable the identification of biosynthetic gene clusters (BGCs) and the mapping of metabolic gene families across species (Dason et al. 2025). These data are essential for comparative pathway analysis, especially in non-model crops with poorly characterized secondary metabolism. Transcriptomic profiling (e.g., RNA-Seq) further refines this understanding by revealing tissue-specific, stress-inducible, or developmentally regulated expression patterns of biosynthetic genes, transcription factors, and regulatory RNAs (Liu et al. 2025). Proteomics complements transcriptomics by confirming the presence, abundance, and post-translational modifications of key enzymes and regulatory proteins involved in metabolite biosynthesis (Sun et al. 2021). Finally, metabolomics—using techniques such as LC–MS, GC–MS, and NMR—quantifies and characterizes the end products and intermediates of these pathways, helping to link gene function with metabolic phenotype (Khanijou et al. 2022).

The integration of these omics layers is greatly enhanced by computational modeling and flux balance analysis (FBA). These systems biology tools allow researchers to simulate the behavior of complex secondary metabolite networks under varying conditions and genetic modifications. By constructing genome-scale metabolic models (GEMs) or using constraint-based approaches, it is possible to predict bottlenecks, competing pathways, and the impact of specific genetic interventions on overall metabolite yield (Durot et al. 2009). Fluxomics, in particular, provides dynamic insight into the flow of metabolic intermediates, helping to identify reactions that limit throughput and offering guidance on how to redistribute flux toward target metabolites (Emwas et al. 2022).

In parallel, advances in data-driven design and predictive validation are streamlining the engineering process. ML algorithms and AI-based modeling tools are now employed to analyze large-scale omics datasets and predict optimal gene targets, enzyme variants, and regulatory circuits for desired outcomes (Wang and Dwivedi 2025). These predictive models are then experimentally validated using CRISPR/Cas systems, synthetic promoters, or inducible constructs in model organisms or crop species. Feedback loops between model predictions and experimental results create an iterative cycle that accelerates the rational design of metabolite pathways.

Together, omics integration and systems biology approaches provide a powerful framework for moving beyond trial-and-error engineering toward a more predictive, scalable, and efficient strategy for tailoring plant secondary metabolism. These technologies are central to unlocking the full potential of metabolic engineering in both fundamental research and applied agricultural biotechnology.

15.8 FUTURE DIRECTIONS AND EMERGING TRENDS

As the complexity of plant metabolic networks becomes increasingly apparent, emerging tools such as genome-wide association studies (GWAS), ML, and synthetic regulatory elements are opening new frontiers in secondary metabolite engineering. These advances not only improve our understanding of natural metabolic diversity but also provide precise and predictive tools for designing climate-resilient, high-value crops.

GWAS offer a powerful strategy to connect genetic variation with metabolite phenotypes across diverse germplasm. By correlating natural allelic differences with the accumulation of specific secondary metabolites, GWAS enables the discovery of novel biosynthetic genes, regulatory loci, and quantitative trait loci (QTLs) involved in metabolite production. This approach has been successfully used to identify flavonoid and glucosinolate pathway regulators in Arabidopsis, alkaloid variation in poppy, and aroma volatiles in tomato. When integrated with ML algorithms, GWAS outputs can be further analyzed to identify complex gene–metabolite relationships, prioritize causal variants, and predict engineering targets. ML models, trained on large omics datasets, also assist in optimizing pathway designs, predicting off-target effects, and identifying synergistic gene combinations that maximize metabolite yield while minimizing growth penalties.

Alongside computational prediction, advances in precision gene regulation are enabling more refined control of metabolite biosynthesis. Synthetic promoters, designed to respond to specific spatial, temporal, or environmental cues, allow the targeted expression of pathway genes in defined tissues or in response to stress. These can be built using cis-regulatory element libraries or de novo sequence design and can be fine-tuned for expression strength and inducibility. In addition, RNA-based regulatory tools, including riboswitches, synthetic RNA scaffolds, and CRISPR-Cas-guided RNA effectors (e.g., Cas13 or dCas13), are being developed to modulate transcript stability, translation, or localization in response to endogenous or exogenous signals. These tools offer reversible and non-transgenic options for dynamic metabolic control.

One of the most promising directions is the development of climate-smart crops through stress-inducible metabolic reprogramming. Under abiotic stress conditions—such as drought, salinity, or heat—plants activate specific secondary metabolite pathways that enhance stress tolerance. Engineering crops to boost or mimic these pathways in a stress-responsive manner allows for metabolite accumulation only when needed, conserving resources during normal growth. For example, engineering drought-inducible expression of osmoprotectant-producing pathways (e.g., proline, flavonoids) has shown promise in improving water-use efficiency and oxidative stress resistance. Similarly, salt-inducible terpenoid biosynthesis can enhance membrane stability and protect against ion toxicity.

15.9 CONCLUSION

Over the past decade, the field of plant metabolic engineering has undergone a remarkable transformation, driven by advances in molecular biology, omics technologies, synthetic biology, and genome editing. These innovations have not only expanded our understanding of complex biosynthetic pathways but have also enabled the successful reprogramming of plant secondary metabolism for improved crop protection, enhanced nutritional quality, and extended postharvest shelf life. Notable examples include the engineering of anthocyanin-enriched tomatoes for antioxidant benefits, glucosinolate-enhanced Brassica crops for pest resistance, and flavonoid-rich cereals for improved stress resilience. Such breakthroughs showcase the potential of targeted metabolite enhancement in both model and non-model crop species.

The promise of metabolic engineering lies in its ability to contribute to sustainable agriculture, especially under the looming challenges of climate change, resource scarcity, and global food security. By enabling crops to better withstand biotic and abiotic stresses through intrinsic biochemical defenses, metabolic engineering reduces the need for external chemical inputs such as pesticides and synthetic fertilizers. Moreover, the tailoring of metabolic pathways for health-promoting phytochemicals and natural preservatives can meet growing consumer demand for functional foods and clean-label products, further aligning agricultural innovation with public health and environmental stewardship.

Looking ahead, the key challenge lies in bridging the gap between laboratory innovations and real-world agricultural deployment. While controlled-environment experiments have demonstrated proof-of-concept success, field-level translation requires robust expression systems, stable transgene integration, minimal fitness trade-offs, and adaptability to diverse agroecological zones. Regulatory considerations, public acceptance, and intellectual property rights will also shape the trajectory of metabolically engineered crops. To address these hurdles, future research must emphasize context-specific trait optimization, high-throughput field phenotyping, and participatory breeding approaches that integrate farmers and stakeholders into the technology adoption process. In conclusion, plant metabolic engineering holds transformative potential for building resilient, high-value crop systems tailored for the 21st century. By harnessing the synergy of biological insight, technological innovation, and systems-level integration, researchers and breeders are well-positioned to design the next generation of climate-smart, nutrient-rich, and environmentally sustainable crops.

REFERENCES

Alharbi K, Alnusaire TS, Alayafi AA, Abdulmajeed AM, ALrashidi AA, Alghanem S, Al-Balawi SM, Anazi HK, Soliman MH (2025) Synergistic effects of biochar and methyl jasmonate nanoparticles in enhancing maize resilience to heat stress. Journal of Plant Growth Regulation 45 (1): 1–21.

Ali J, Chen R, Mukarram M, Tonğa A, Khan KA, Ghramh HA, Jamnická G, Li Q, Kurjak D (2025) Multifaceted roles of ethylene in plants: From traditional functions to modern insights into tritrophic interactions for sustainable agriculture. Plant Stress 17:100925. https://doi.org/10.1016/j.stress.2025.100925

Aqaei P, Weria W, Marjan D, Javad R, Struik PC (2020) Response of maize (Zea mays L.) to potassium nano-silica application under drought stress. Journal of Plant Nutrition 43 (9):1205–1216. https://doi.org/10.1080/01904167.2020.1727508

Bergman ME, Kortbeek RWJ, Gutensohn M, Dudareva N (2024) Plant terpenoid biosynthetic network and its multiple layers of regulation. Progress in Lipid Research 95:101287. https://doi.org/10.1016/j.plipres.2024.101287

Bhatla SC, Lal MA (2023) Secondary metabolites. In: *Plant physiology, development and metabolism* (pp. 765–808). Springer.

Chappell J, Watters KE, Takahashi MK, Lucks JB (2015) A renaissance in RNA synthetic biology: New mechanisms, applications and tools for the future. Current Opinion in Chemical Biology 28:47–56. https://doi.org/10.1016/j.cbpa.2015.05.018

Chen H, Li M, Qi G, Zhao M, Liu L, Zhang J, Chen G, Wang D, Liu F, Fu ZQ (2021) Two interacting transcriptional coactivators cooperatively control plant immune responses. Science Advances 7 (45):eabl7173. https://doi.org/10.1126/sciadv.abl7173

Cheng B, Ran R, Qu Y, Verkerk R, Henry R, Dekker M, He H (2024) Advancements in balancing glucosinolate production in plants to deliver effective defense and promote human health. Agriculture Communications 2 (2):100040. https://doi.org/10.1016/j.agrcom.2024.100040

Connolly EL, Sim M, Travica N, Marx W, Beasy G, Lynch GS, Bondonno CP, Lewis JR, Hodgson JM, Blekkenhorst LC (2021) Glucosinolates from cruciferous vegetables and their potential role in chronic disease: Investigating the preclinical and clinical evidence. Frontiers in Pharmacology 12:767975. https://doi.org/10.3389/fphar.2021.767975

Dason MS, Corà D, Re A (2025) Sequence modeling tools to decode the biosynthetic diversity of the human microbiome. mSystems 10 (7):e0033325. https://doi.org/10.1128/msystems.00333-25

de Brito Francisco R, Martinoia E (2018) The vacuolar transportome of plant specialized metabolites. Plant and Cell Physiology 59 (7):1326–1336.

Durot M, Bourguignon PY, Schachter V (2009) Genome-scale models of bacterial metabolism: Reconstruction and applications. FEMS Microbiology Reviews 33 (1):164–190. https://doi.org/10.1111/j.1574-6976.2008.00146.x

Emwas A.-H, Szczepski K, Al-Younis I, Lachowicz JI, Jaremko M (2022) Fluxomics – New metabolomics approaches to monitor metabolic pathways. Frontiers in Pharmacology 13. https://doi.org/10.3389/fphar.2022.805782

Guengerich FP, Sohl CD, Chowdhury G (2011) Multi-step oxidations catalyzed by cytochrome P450 enzymes: Processive vs. distributive kinetics and the issue of carbonyl oxidation in chemical mechanisms. Archives of Biochemistry and Biophysics 507 (1):126–134. https://doi.org/10.1016/j.abb.2010.08.017

Guirimand G, Courdavault V, St-Pierre B, Burlat V (2010) Biosynthesis and regulation of alkaloids. In: Plant developmental biology-biotechnological perspectives (Vol. 2, pp 139–160). Springer.

Hancock JE, Loya WM, Giardina CP, Li L, Chiang VL, Pregitzer KS (2007) Plant growth, biomass partitioning and soil carbon formation in response to altered lignin biosynthesis in *Populus tremuloides*. New Phytologist 173 (4):732–742.

Hanhineva K (2008) Metabolic engineering of phenolic biosynthesis pathway and metabolite profiling of strawberry (Fragaria × Ananassa). Kuopion yliopiston julkaisuja. C, Luonnontieteet ja ympäristötieteet 231:1–89.

Holtorf S, Apel K, Bohlmann H (1995) Comparison of different constitutive and inducible promoters for the overexpression of transgenes in *Arabidopsis thaliana*. Plant Molecular Biology 29 (4):637–646. https://doi.org/10.1007/bf00041155

Jan R, Asaf S, Numan M, Lubna & Kim K.-M (2021) Plant secondary metabolite biosynthesis and transcriptional regulation in response to biotic and abiotic stress conditions. Agronomy 11 (5):968.

Karimi K, Ahari AB, Weisany W, Pertot I, Vrhovsek U, Arzanlou M (2016) *Funneliformis mosseae* root colonization affects *Anethum graveolens* essential oil composition and its efficacy against *Colletotrichum nymphaeae*. Industrial Crops and Products 90:126–134. https://doi.org/10.1016/j.indcrop.2016.06.024

Karnwal A, Kumar G, Singh R, Selvaraj M, Malik T, Al Tawaha ARM (2025) Natural biopolymers in edible coatings: Applications in food preservation. Food Chemistry: X 25:102171. https://doi.org/10.1016/j.fochx.2025.102171

Ketehouli T, Sossah FL, Panwala R, Suazo Tejada AK, Goss EM, Garcia FHS, Vallad GE, Martins SJ (2025) Secondary metabolites in plant-microbe interactions. Journal of Applied Microbiology 136 (6). https://doi.org/10.1093/jambio/lxaf124

Khanijou JK, Kulyk H, Bergès C, Khoo LW, Ng P, Yeo HC, Helmy M, Bellvert F, Chew W, Selvarajoo K (2022) Metabolomics and modelling approaches for systems metabolic engineering. Metabolic Engineering Communications 15:e00209. https://doi.org/10.1016/j.mec.2022.e00209

Kołton A, Długosz-Grochowska O, Wojciechowska R, Czaja M (2022) Biosynthesis regulation of folates and phenols in plants. Scientia Horticulturae 291:110561. https://doi.org/10.1016/j.scienta.2021.110561

Kumar M, Kesawat MS, Ali A, Lee S.-C, Gill SS, Kim HU (2019) Integration of abscisic acid signaling with other signaling pathways in plant stress responses and development. Plants 8 (12):592.

Lee SB, Lee S.-e, Lee H, Kim J.-S, Choi H, Lee S, Kim B.-G (2024) Engineering *Nicotiana benthamiana* for chrysoeriol production using synthetic biology approaches. Frontiers in Plant Science 15. https://doi.org/10.3389/fpls.2024.1458916

Leung RK, Whittaker PA (2005) RNA interference: From gene silencing to gene-specific therapeutics. Pharmacology & Therapeutics 107 (2):222–239. https://doi.org/10.1016/j.pharmthera.2005.03.004

Li C, Zha W, Li W, Wang J, You A (2023) Advances in the biosynthesis of terpenoids and their ecological functions in plant resistance. International Journal of Molecular Sciences 24 (14):11561.

Lin D, Xiao M, Zhao J, Li Z, Xing B, Li X, Kong M, Li L, Zhang Q, Liu Y, Chen H, Qin W, Wu H, Chen S (2016) An overview of plant phenolic compounds and their importance in human nutrition and management of type 2 diabetes. Molecules 21 (10). https://doi.org/10.3390/molecules21101374

Liu Z, Zhang Y, Zhao Q, Liu H, Zhou Y, Qin A, Li C, Yan L, Li M, Gao P, Song X, Xie Y, Guo E, Kong L, Guan L, Sun AGX (2025) Single-cell RNA sequencing reveals developmental trajectories and environmental regulation of callus formation in Arabidopsis. Stress Biology 5 (1):57. https://doi.org/10.1007/s44154-025-00255-4

Lone R, Bhat A, Nazim N, Malla NA, Rohella GK, Mohamed HI (2024) Role of phenolics in Plant–Microbe interaction: A review. Plant Phenolics in Biotic Stress Management: 1–33.

Manvelian J, Weisany W, Tahir NA-r, Jabbari H, Diyanat M (2021) Physiological and biochemical response of safflower (*Carthamus tinctorius* L.) cultivars to zinc application under drought stress. Industrial Crops and Products 172:114069. https://doi.org/10.1016/j.indcrop.2021.114069

Mmbando GS (2024) Omics: A new, promising technologies for boosting crop yield and stress resilience in African agriculture. Plant Stress 11:100366. https://doi.org/10.1016/j.stress.2024.100366

Otles S, Özyurt VH (2021) Chapter 11 – Biotransformation in the production of secondary metabolites. In: Atta ur R (ed) Studies in natural products chemistry (Vol. 68, pp 435–457). Elsevier. https://doi.org/10.1016/B978-0-12-819485-0.00007-4

Reshi ZA, Ahmad W, Lukatkin AS, Javed SB (2023) From nature to lab: A review of secondary metabolite biosynthetic pathways, environmental influences, and in vitro approaches. Metabolites 13 (8). https://doi.org/10.3390/metabo13080895

Smith JD, Woldemariam MG, Mescher MC, Jander G, De Moraes CM (2016) Glucosinolates from host plants influence growth of the parasitic plant *Cuscuta gronovii* and its susceptibility to aphid feeding. Plant Physiology 172 (1):181–197. https://doi.org/10.1104/pp.16.00613

Snaith MR, Murray JA, Boulter CA (1995) Multiple cloning sites carrying loxP and FRT recognition sites for the Cre and Flp site-specific recombinases. Gene 166 (1):173–174. https://doi.org/10.1016/0378-1119(95)00579-8

Stintzing FC, Carle R (2004) Functional properties of anthocyanins and betalains in plants, food, and in human nutrition. Trends in Food Science & Technology 15 (1):19–38. https://doi.org/10.1016/j.tifs.2003.07.004

Sun X, Zhang H, Jia Y, Li J, Jia M (2024) CRISPR-Cas9-based genome-editing technologies in engineering bacteria for the production of plant-derived terpenoids. Engineering Microbiology 4 (3):100154. https://doi.org/10.1016/j.engmic.2024.100154

Sun Y, Wang X, Chen Z, Qin L, Li B, Ouyang L, Peng X, He H (2021) Quantitative proteomics and transcriptomics reveals differences in proteins during anthers development in *Oryza longistaminata*. Frontiers in Plant Science 12. https://doi.org/10.3389/fpls.2021.744792

Thirumurugan D, Cholarajan A, Raja S, Vijayakumar R (2018) An introductory chapter: Secondary metabolites. Secondary Metabolites-Sources and Applications 1:13.

Tiwari S, Kate A, Mohapatra D, Tripathi MK, Ray H, Akuli A, Ghosh A, Modhera B (2020) Volatile organic compounds (VOCs): Biomarkers for quality management of horticultural commodities during storage through e-sensing. Trends in Food Science & Technology 106:417–433. https://doi.org/10.1016/j.tifs.2020.10.039

Trejo-Téllez LI, Estrada-Ortiz E, Gómez-Merino FC, Becker C, Krumbein A, Schwarz D (2019) Flavonoid, nitrate and glucosinolate concentrations in *Brassica* species are differentially affected by photosynthetically active radiation, phosphate and phosphite. Frontiers in Plant Science 10. https://doi.org/10.3389/fpls.2019.00371

von Mérey GE, Veyrat N, D'Alessandro M, Turlings T (2013) Herbivore-induced maize leaf volatiles affect attraction and feeding behavior of *Spodoptera littoralis* caterpillars. Frontiers in Plant Science 4. https://doi.org/10.3389/fpls.2013.00209

Wang Q, Dwivedi Y (2025) Recent developments in omics studies and artificial intelligence in depression and suicide. Translational Psychiatry 15 (1):275. https://doi.org/10.1038/s41398-025-03497-y

Wanke M, Skorupinska-Tudek K, Swiezewska E (2001) Isoprenoid biosynthesis via 1-deoxy-D-xylulose 5-phosphate/2-C-methyl-D-erythritol 4-phosphate (DOXP/MEP) pathway. Acta Biochimica Polonica 48 (3):663–672.

Weisany W, Amini J, Samadi S, Hossaini S, Yousefi S, Struik PC (2019) Nano silver-encapsulation of *Thymus daenensis* and *Anethum graveolens* essential oils enhances antifungal potential against strawberry anthracnose. Industrial Crops and Products 141:111808. https://doi.org/10.1016/j.indcrop.2019.111808

Weisany W, Khosropour E (2023) Chapter 8 – Engineered nanomaterials in crop plants drought stress management. In: Husen A (ed) Engineered nanomaterials for sustainable agricultural production, soil improvement and stress management (pp 183–204). Academic Press. https://doi.org/10.1016/B978-0-323-91933-3.00005-2

Weisany W, Razmi J, Eshaghabadi AH, Pashang D (2024a) Silicon nanoparticles (SiNP): A novel and sustainable strategy for mitigating environmental stresses in plants. Journal of Soil Science and Plant Nutrition 24 (2):2167–2191. https://doi.org/10.1007/s42729-024-01790-1

Weisany W, Samadi S, Tahir N.A.-r, Amini J, Hossaini S (2022) Nano-encapsulated with mesoporous silica enhanced the antifungal activity of essential oil against *Botrytis cinerea* (Helotiales; Sclerotiniaceae) and *Colletotrichum nymphaeae* (Glomerellales; Glomerellaceae). Physiological and Molecular Plant Pathology 122:101902.

Weisany W, Sohrabi Y, Heidari G, Siosemardeh A, Ghassemi-Golezani K (2011) Physiological responses of soybean (*Glycine max* L.) to zinc application under salinity stress. Australian Journal of Crop Science 5 (11):1441–1447.

Weisany W, Soufiania SP, Razmi J, Eshaghadadi AH (2024b). Nano-encapsulation of fenugreek and coriander essential oils using copper oxide NPs: Novel approach for augmenting their effectiveness against *Colletotrichum nymphaeae*. Industrial Crops and Products 219:119051. https://doi.org/10.1016/j.indcrop.2024.119051

Zhang C, Singh RP, Yadav P, Kumar I, Kaushik A, Roychowdhury R, Mubeen M, Singh SK, Kumar A, Wang J (2025) Recent advances in biotechnology and bioengineering for efficient microalgal biofuel production. Fuel Processing Technology 270:108199. https://doi.org/10.1016/j.fuproc.2025.108199

Zhu Q, Wang B, Tan J, Liu T, Li L, Liu Y-G (2020) Plant synthetic metabolic engineering for enhancing crop nutritional quality. Plant Communications 1 (1):100017. https://doi.org/10.1016/j.xplc.2019.100017

16 Genetic Modifications to Enhance Enzyme Systems for Crop Improvement

Md. Marufur Rahman, Md. Ashraful Alam, and Akbar Hossain

16.1 INTRODUCTION

Enzymes are the primary regulators of plant physiology. They act as catalysts for biochemical reactions during metabolism and the development of the plant. Enzymes are crucial for crop growth and development. Central metabolic pathways are related to energy production, carbon partitioning, and the biosynthesis of structural and storage compounds associated with enzymes. Malic enzymes, hexokinase, and sucrose-cleaving enzymes are the key enzymes directly involved in these pathways (Ruan, 2014; Sturm & Tang, 1999; Sun, Han, Meng, Lin, & Sui, 2019; Vanderwall & Gendron, 2023). Lipoxygenases and aldehyde dehydrogenases play vital roles in plant defenses. They can detoxify harmful compounds such as heavy metals. They also synthesize signaling molecules involved in stress responses. Overall, they improve the plant resilience system to stress (Tola, Jaballi, Germain, & Missihoun, 2020; Viswanath et al., 2020).

Enzyme assembly forms complexes that facilitate various metabolic pathways in the cell. This can enhance the efficiency and regulation of metabolic pathways. Furthermore, it helps plants adapt to unfavorable environments and increases their productivity (Winkel, 2004). Enzyme cofactors are important for metabolism and stress responses. These are required for catalytic activity (da Fonseca-Pereira, Monteiro-Batista, Araújo, & Nunes-Nesi, 2023). Genomic advances have revealed the prevalence of metabolic–gene clusters. It can easily determine the evolutionary adaptation and specialization of the enzyme systems in plants. This technology helps produce a diverse range of metabolites (Lou, Pichersky, & Last, 2022; Schläpfer et al., 2017).

The objectives of crop improvement programs are yield enhancement, stress tolerance, and increased nutrient efficiency. It also focuses on sustainable agricultural production. Historically, crop improvement has been achieved through breeding programs that target specific traits. Traits such as robust root systems and high-stress tolerance are targeted for crop improvement. For example, high-yield maize is characterized by enhanced stress tolerance and high plant density (Ahmad, Ali, Hafeez, & Malik, 2021).

The concept of enhanced NUE is the yield per unit of fertilizer input. It minimizes input costs and adverse environmental effects, while maintaining productivity. Root architecture optimization is a vital strategy for improving nutrient efficiency. Nutrient uptake and translocation are major components of nutrient use efficiency. Nutrient-signaling-related genes can be modified through breeding and genome-editing technologies to improve NUE (Chen & Liao, 2017; Sathee et al., 2022). An integrated approach using biofertilizers, exogenous hormone treatments, and biostimulants can enhance the stress tolerance of crops. It enhances the antioxidant activity, nutrient uptake, and defense systems of plants (Liu et al., 2024; Singhal et al., 2023; Sun, Shahrajabian, Kuang, & Wang, 2024; Swain et al., 2023). To overcome the problems of adverse climate, a multidimensional approach, such as the use of biotechnological tools, conventional breeding, and agronomic practices, is required (Ahmad et al., 2021; González Guzmán, Cellini, Fotopoulos, Balestrini, & Arbona, 2022).

Enhanced enzyme systems are the most important for yield improvement, stress tolerance, and industrial applications of crops. Genetic modification may play a vital role in yield improvement. Modification of photosynthesis-related enzymes can enhance photosynthesis, which ultimately affects crop yield. Plants overexpressing phosphoenolpyruvate carboxylase (PEPC)-related genes exhibited increased CO_2 levels around Rubisco. It can also enhance plant metabolism and tolerance against biotic and abiotic stress (Nazari et al., 2024). Stress response-regulating enzyme uridine diphosphate (UDP)-glycosyltransferases (UGTs) modification helps to adapt the plant under stress (Gharabli, Della Gala, & Welner, 2023).

In the case of industrial applications of crops, efficient biofuel production is the most important aspect of the industry. Cell wall enzymes such as CesA, GH9, and GT61 can enhance biomass production and enzymatic digestibility (Rocha Meneses, Ferreira, Karimi, Orupõld, & Kikas, 2019; Wang et al., 2016). Cutting-edge tools, such as CRISPR/Cas, can be used to precisely modify enzyme systems. After successful genetic modification, photosynthetic efficiency, nutrient uptake, and stress resilience were enhanced in staple crops (Chen et al., 2024).

16.2 ENZYME SYSTEMS CRITICAL FOR CROP PERFORMANCE

Enzyme systems play vital roles in crop performance. It regulates various physiological processes and events. Enzyme systems significantly affect carbohydrate metabolism, nutrient cycling, and plant defense. In this section, we discuss some critical enzyme systems for crop performance, especially photosynthetic, nitrogen-assimilation-related, cell wall biosynthesis, stress-responsive, and hormone biosynthesis and signaling enzymes (Table 16.1).

DOI: 10.1201/9781003669920-16

TABLE 16.1
Key Enzymes Related to Crop Performance

Categories	Enzyme	Function	Key References
Photosynthetic enzymes	Rubisco (ribulose-1,5-bisphosphate carboxylase/oxygenase)	• Carbon fixation • Limiting enzyme rate in photosynthesis	Orr et al. (2016), Perdomo, Capó-Bauçà, Carmo-Silva, and Galmés (2017), von Caemmerer (2020)
	Rubisco activase	• Regulates rubisco activity	von Caemmerer (2020)
	Phosphoenolpyruvate carboxylase (PEPC) Pyruvate orthophosphate dikinase (PPDK) NADP-malic enzyme (NADP-ME)	• Photosynthesis	Li, Wang, Lan, and Yu (2024), Nazari et al. (2024), Perdomo et al. (2017), Singh, Garai, Das, Thakur, and Tripathy (2022), von Caemmerer and Furbank (2016)
	Carbonic anhydrase (CA)	• Carbon fixation	Nazari et al. (2024), Singh et al. (2022), von Caemmerer and Furbank (2016)
Nitrogen assimilation	Nitrate reductase (NR) Nitrite reductase (NiR) Glutamine synthetase (GS)	• Inorganic nitrogen (nitrate and nitrite) converts into available form • Influencing nitrogen use efficiency (NUE), crop yield • Leaf senescence	Andrews, Lea, Raven, and Lindsey (2004), Fortunato et al. (2023), Kishorekumar, Bulle, Wany, and Gupta (2020), Lea and Miflin (2011), Li, Hu, and Chu (2017), Liu, Hu, and Chu (2022), Miflin and Habash (2002)
	Glutamine-2-oxoglutarate aminotransferase (GOGAT)	• Grain formation and grain filling • Efficient nitrogen remobilization	
	Asparagine synthetase (ASN) Glutamate dehydrogenase (GDH) Carbamoyl phosphate synthetase (CPSase) Alanine aminotransferase	• Nitrogen assimilation	Kishorekumar et al. (2020), Lea and Miflin (2011), Liu et al. (2022), Miflin and Habash (2002)
Cell wall biosynthesis	Cellulose synthases (CESAs)	• Cellulose production • Stress tolerance	Yin et al. (2024)
	Cellulose synthase-like proteins (OsCSLD4)	• Cell wall polysaccharide synthesis • Salt stress tolerance, • Abscisic acid biosynthesis	Zhao et al. (2022)
	Glycosyltransferases (GTs)	• Pectins, xylans, xyloglucans, mannans, and arabinogalactans synthesis • Influencing plant growth, biomass quality, and stress responses	Amos and Mohnen (2019), Zhong, Cui, and Ye (2019)
	Polygalacturonases (PGs)	• Modify pectin structure • Affecting drought tolerance	Zhang, Hou, et al. (2021)
	Laccases and peroxidases	• Lignin biosynthesis • Secondary cell wall formation, contributing to stress adaptation	Meents, Watanabe, and Samuels (2018), Zhong et al. (2019)
Stress responsive	Cytochrome P450s	• Secondary metabolites biosynthesis, • Contribute to plant defense and resilience against stress	Chakraborty, Biswas, Dey, Bhattacharjee, and Chakrabarty (2023), Pandian, Sathishraj, Djanaguiraman, Prasad, and Jugulam (2020)
	Class III peroxidases Ascorbate oxidases	• Managing reactive oxygen species (ROS), act as antioxidants, maintain cellular homeostasis under stress	Batth, Singh, Kumari, and Mustafiz (2017), Kidwai, Ahmad, and Chakrabarty (2020)
	NADP-malic enzyme	• Stabilizes cytoplasmic pH, • Controls stomatal aperture, • Modulate ethylene levels to promote growth under stress	Chen, Wang, Ding, Zhang, and Li (2019), Shahid et al. (2023)
	Carbohydrate metabolic enzymes	• Linked to yield stability under salt stress	Stasnik, Vollmann, Großkinsky, and Jonak (2024)
	ER-associated ubiquitin-conjugating enzymes (ZmGA20ox3)	• Phytohormone biosynthesis • Regulate stress tolerance	Liu, Chen, et al. (2023), Yu and Xie (2024)
Hormone biosynthesis	Lipoxygenases Allene oxide synthase Allene oxide cyclase 12-Oxophytodienoate reductase	• Jasmonates biosynthesis	Shi et al. (2023b), Wasternack and Song (2017)

(Continued)

TABLE 16.1 *(Continued)*
Key Enzymes Related to Crop Performance

Categories	Enzyme	Function	Key References
	Ent-copalyl diphosphate synthase Ent-kaurene synthase Ent-kaurene oxidase GA 20-oxidase	• Gibberellin biosynthesis	Gao and Chu (2020)
	ACC synthase and ACC oxidase	• Ethylene biosynthesis	Fatma et al. (2022), Zhou et al. (2020)
	DWF4, CPD, and BR6ox	• Brassinosteroids biosynthesis	Vukašinović et al. (2021)
	SABATH methyltransferases GH3 acyl acid-amido synthetases Peptidyl hydrolases Methylesterases E3 ubiquitin ligases and 14-3-3 proteins	• Regulate hormone activation, inactivation, and homeostasis • Hormone signaling • Mediating protein stability; Nutrient use efficiency	Westfall, Muehler, and Jez (2013) Camoni, Visconti, Aducci, and Marra (2018), Kelley (2018)

16.2.1 Regulatory Roles of Photosynthetic Enzymes in Crop

Photosynthesis is an essential biochemical process in crop growth. Enzymes are related to different photosynthetic cycles, such as C_4 and C_3. It plays a vital role in energy transfer and in carbon fixation. The overall photosynthetic efficiency also depends on the availability of enzymes. In the photosynthetic process, enzymes are classified into two groups: C_3 and C_4 pathway enzymes (Miyao, 2003). Atmospheric inorganic CO_2 is converted into organic carbon molecules via photosynthetic CO_2 fixation or carboxylation. Plants use this molecule for energy and growth. These enzymes are related to the event named the C_3 pathway. Ribulose bisphosphate carboxylase/oxygenase (RuBisCO) is an enzyme directly involved in CO_2 fixation, and its oxygenase activity produces glycolate as a byproduct. The glycolate oxidase (GOX) enzyme converts glycolate into glyoxylate and H_2O_2 during photorespiration. Another photosynthetic enzyme, PEPC, is involved in the carboxylation process. Generally, it is inactive in the C_3 pathways but may have some intermediate functions (Rintamäki & Aro, 1985; Singh et al., 2022). Pyruvate orthophosphate dikinase (PPDK), NADP-malate dehydrogenase (MDH), NADP-malic enzyme (NADP-ME), and PEPC are key enzymes in the C_4 pathway that enhance the rate of photosynthesis and yield, and enhance water-use efficiency in plants (Miyao, 2003; Singh et al., 2022; Taniguchi et al., 2008; Taylor, Slack, & McPherson, 1974).

The photosynthesis rate in C_3 plants is lower than that in C_4 plants. Photosynthetic activity in C_3 plants can be enhanced by incorporating C_4 photosynthetic enzymes. PEPC overexpression in C_4 plants in *Arabidopsis* resulted in enhanced photosynthetic activity. It also helps with stomatal closure and drought tolerance in plants. Conversely, it decreases the availability of reactive oxygen species (ROS) (Rintamäki & Aro, 1985; Singh et al., 2022). Photosynthesis may not be significantly enhanced by the overproduction of a single enzyme. The integration of different photosynthetic enzymes may produce high levels of photosynthetic products (Li et al., 2024; Miyao, 2003; Taniguchi et al., 2008). Environmental cues such as light and temperature regulate photosynthetic enzyme activity. NADP-MDH and PPDK activities may decrease under low-temperature and high-light conditions (Taylor et al., 1974). Photosynthetic enzymes play vital roles in carbon fixation and plant growth. Understanding enzyme regulation and using cutting-edge technologies paves the way for crop improvement.

16.2.2 Nitrogen Use Efficiency in Plants Regulates the Assimilation of Enzymes Such as NR and GS by Environmental Cues

Nitrogen assimilation is essential for crop growth. In this process, atmospheric inorganic nitrogen is converted to organic nitrogen. Plants can uptake and incorporate organic nitrogen sources into their tissues. This process is highly regulated by enzymes. Nitrate reductase (NR), nitrite reductase (NiR), and glutamine synthetase (GS) are key regulatory enzymes involved in nitrogen assimilation (Heath-Pagliuso, Huffaker, & Allard, 1984). These enzymes play a significant role in the transformation of inorganic nitrogen into organic nitrogen compounds. Nitrate NiR and NR are present as isoenzymes and depend on regulatory control. Nitrate availability mainly induces NR and NiR activity (Cabello, de la Haba, González-Fontes, & Maldonado, 1998; Heath-Pagliuso et al., 1984). During the nitrogen starvation period in plants, NiR, GS, and GOGAT activities are restored by ammonium and potassium nitrate. In contrast, GS activity increases with ammonium chloride (Balotf, Kavoosi, & Kholdebarin, 2016; Stancheva & Dinev, 1995). Amino acids, such as asparagine and glutamine, may also reduce NiR induction (Sivasankar & Oaks, 1996).

Environmental factors such as light, nitrogen sources, and metabolites influence nitrogen assimilation. Light is an essential factor for the high-level synthesis of NiR, NR, and chloroplastic GS. NR activity is controlled by nitrate, and its full induction depends on light (Balotf et al., 2016; Cabello et al., 1998; Heath-Pagliuso et al., 1984; Seith, Setzer, Flaig, & Mohr, 1994; Sivasankar & Oaks, 1996; Stancheva & Dinev, 1995). Nitrate and ammonium induce GS. Light and nitrogen sources specifically regulate GS activity (Balotf et al., 2016; Cabello et al., 1998; Seith et al., 1994; Stancheva & Dinev, 1995). Enzyme induction may be greatly reduced in the absence of light (Seith et al., 1994; Sivasankar & Oaks, 1996; Stancheva & Dinev, 1995). GS and NR exhibited tissue-specific expression patterns and were highly regulated by environmental factors. These findings provide information on adaptation to diverse environments and ensure efficient nitrogen use.

16.2.3 Complementary Roles of Expansins and Cellulose Synthase in Plant Cell Wall Biosynthesis and Remodeling

Enzymes play vital roles in plant cell wall construction and remodeling processes. Cellulose synthase and expansions are key players in this process. Cellulose synthase polymerizes glucose units and synthesizes cellulose microfibrils (CMFs). CMFs are key components of the cell walls. CMFs and polysaccharides contain non-covalent bonds. Cellulases are endo- and exo-glucanases that break down cellulose. These activities enhance cell wall extensibility and accessibility for other enzymes (Bunterngsook, Eurwilaichitr, Thamchaipenet, & Champreda, 2015; Ding et al., 2022; Kim, Lee, Bang, Choi, & Kim, 2009; Lee, Lee, Ko, Kim, & Choi, 2010).

Expansions are major non-catalytic proteins that can lose the cell wall without hydrolyzing cellulose. It can increase the efficiency of cellulose by up to 5–7.6 times (Bunterngsook, Mhuantong, Champreda, Thamchaipenet, & Eurwilaichitr, 2014; Kim et al., 2009; Lee et al., 2010; Zhang, Cui, et al., 2021; Zhang, Su, et al., 2021). Expansin-pretreated cellulose shows higher hydrolysis and enzyme binding (Ding et al., 2022; Zhang, Cui, et al., 2021). Expansins typically bind to specific domains and amino acid residues. It disrupts hydrogen bonds and increases substrate accessibility to enzymes (Ding et al., 2022; Lee et al., 2010; Orłowski et al., 2018). Expansins genes have huge industrial importance for biofuel production. Expansin-overexpressing plants show higher biomass digestibility and cell size increment (Mira et al., 2024). In the cotton fiber cell elongation stage, the expression levels of expansin and endo-1,4-beta-glucanase are high, indicating cell wall loosening and growth. In contrast, low expression levels during cell wall synthesis indicate cell wall remodeling (Shimizu et al., 1997). Therefore, expansins and cellulose play complementary roles in cell wall remodeling and biosynthesis.

16.2.4 The Coordinated Action of Antioxidant and Detoxification Enzymes in Plant Stress Tolerance

Plants experience oxidative damage due to environmental stressors. Stress-responsive enzymes detoxify harmful compounds and help organisms adapt to stress. These enzymes can be classified into two groups: antioxidant and detoxification enzymes. The major enzymes such as superoxide dismutase (SOD) and catalase (CAT) are antioxidant enzymes, whereas glutathione S-transferase (GST) is a detoxification enzyme (Anjum et al., 2016; Rajput et al., 2021). These enzymes help maintain cellular homeostasis and growth in unfavorable environments.

SOD works as a stress signal in plants. Under stress conditions, SOD can be converted to hydrogen peroxide (H_2O_2). This reduces oxidative damage to the plant cells. In addition, SOD regulates different antioxidant activities (Lee, Kim, & Lee, 2001; Rajput et al., 2021; Sarker & Oba, 2018). SOD produces H_2O_2, which can accumulate and become toxic to the cells. CAT converts H_2O_2 into water and oxygen to prevent toxicity. Ascorbate peroxidase (APX) is an enzyme with a higher affinity for H_2O_2 than that of CAT. This enzyme also helps reduce H_2O_2 levels and functions in the water of different cellular compartments. It plays a significant role in the ascorbate–glutathione cycle (Anjum et al., 2016; Lee et al., 2001; Mishra et al., 2023; Rajput et al., 2021; Sarker & Oba, 2018).

Glutathione- or thioredoxin-based systems also reduce organic hydroperoxides and H_2O_2 in presence of glutathione peroxidase (GPX). It may control redox homeostasis and stress responses in plants (Bela et al., 2015; Rajput et al., 2021). GST is the main detoxification enzyme. It catalyzes the binding of glutathione to toxic substances. It also helps remove and detoxify toxic compounds. In saline environments, GST gene families have been found to be upregulated in different studies (Gaafar et al., 2022; Kiremit & Arslan, 2024; Rajput et al., 2021). For example, Zeta, Tau, and Lambda were differentially expressed in response to the stress. Detoxification capacity is also improved by the expression of these genes (Gaafar et al., 2022). GST plays a significant role in enhancing tolerance to abiotic stress and herbicide resistance in plants. Sometimes it also worked with different antioxidant enzymes (Gaafar et al., 2022; Kiremit & Arslan, 2024).

These enzymatic activities can serve as biochemical markers to identify stress response. Stress-tolerant plants can be evaluated based on their stress levels under different conditions (Kiremit & Arslan, 2024; Sahoo & Tiwari, 2022). Enzyme-related gene expression can be modified using biostimulants, such as cyanobacterial extracts, to improve plant resilience. Gene expression patterns can be modified through a plant-breeding approach (Gaafar et al., 2022; Mishra et al., 2023; Rajput et al., 2021). Their coordinated action is vital for plant survival under an unfavorable environment, and their activity patterns can be used as indicators of stress response and as targets for enhancing stress tolerance. Stress-responsive enzymes, particularly antioxidants and detoxification enzymes, are crucial for managing oxidative stress. The combined action of different enzymes is important for plant adaptation under stress conditions, and the pattern of enzyme activity may give ideas for improving stress tolerance.

16.2.5 Building Resilient Crops: Engineering Hormone Pathways for Yield and Stress Tolerance

Enzymes are related to hormone biosynthesis and signaling. They play significant roles in plant growth, development, and stress response. Major hormone (auxin, gibberellin, and ethylene)-related enzyme-targeted modifications offer promising strategies for improving crop quality. SABATH methyltransferases, methylesterases, GH_3 acyl-amido synthetases, and hormone peptidyl hydrolases regulate the biosynthesis of bioactive hormone forms. These hormones affect plant growth (Westfall et al., 2013). Ethylene biosynthesis is directly related to a multistep enzymatic process. Ethylene is responsive to specific developmental stages and environmental stress (Fatma et al., 2022; Shibuya & Clark, 2006). The modification of ethylene and abscisic acid biosynthesis and signaling enhances plant adaptation to stress conditions and nutrient-deficient soils (Fatma et al., 2022; Hedden & Phillips, 2000; Huang, Ji, Zhang, & Li, 2024). In the present era, advances in the understanding of hormonal pathways offer enzyme-coding gene identification and cloning for crop improvement (Hedden & Phillips, 2000).

A complex network is associated with hormonal signaling. It also depends on the activation or repression of transcription factors by proteins in response to hormones. For example, plant defense mechanisms depend on jasmonate signaling. This hormone signaling is fine-tuned by activators and repressors (Shi et al., 2023a; Wasternack & Song, 2017). Crosstalk among hormones can also regulate gene expression. The interaction of melatonin with auxin, cytokinin, gibberellin, ethylene, and abscisic acid regulates stress responses (Altaf et al., 2023).

Modification of hormone biosynthetic genes can alter hormone levels. Using genetic engineering, gene modification, and hormone levels, desired traits such as delayed senescence can be conferred (Cardoso, Gori, Da-Silva, & Brunetti, 2020; Shibuya & Clark, 2006). Inhibiting cytokinin oxidase/dehydrogenase may enhance cytokinin biosynthesis. Increased cytokinin levels can enhance grain yield, although this may vary among species (Arora & Sen, 2022). The integration of plant breeding and biotechnology with hormone pathway engineering may be a key approach for improving yield and developing resilient crops (Hedden & Phillips, 2000; Huang et al., 2024). Species-specific responses and the complexity of hormonal networks highlight the need for continued research and development of integrated strategies.

16.3 GENETIC MODIFICATION TECHNIQUES FOR ENZYME ENHANCEMENT

Enzyme enhancement in crops through genetic modification targets crop productivity, quality, and stress tolerance. This technique also focuses on multiple enzyme and metabolite pathways. The main objective of this process is enzyme optimization. Genetic modification of the cytochrome b6f complex, such as the introduction of genes such as PEPC and NADP-malic enzymes, can enhance carbon fixation and reduce photorespiration. As a result, photosynthetic efficiency and crop yield can be successfully enhanced (Nazari et al., 2024). Cell wall biosynthesis gene modification can enhance enzyme digestibility and biomass yields. This is important for bioenergy-producing crops (Wang et al., 2016). Genome editing tools, such as clustered regularly interspaced palindromic repeats (CRISPR/Cas9), transcription activator-like effector nucleases (TALENs), and zinc finger nucleases, can precisely edit the genome. Past studies on genome editing have focused on traits such as nutrient content, stress tolerance, and yield improvement (Arora & Narula, 2017; Gharabli et al., 2023; Sedeek, Mahas, & Mahfouz, 2019). Therefore, genetic modification holds promise for enhancing enzyme activity in crops.

16.3.1 Transgenic Enhancement of Enzyme Systems: A Promising Approach for Crop Improvement

The most used technique for crop improvement is the transgenic approach, which enhances enzyme systems. Recent investigations on enzyme genes, such as endogenous, synthetic, or heterologous genes, and their overexpression using a transgenic approach, have shown significant promise for crop improvement. Table 16.2 shows a list of enzyme-related genes modified through the transgenic approach.

In rice and potato, photosynthetic efficiency can be significantly improved by the overexpression of C_4 cycle enzyme genes. GS OsGS1;1 and OsGS2 overexpressed transgenic rice showed high stress tolerance capacity. Late-embryogenesis-abundant (LEA) proteins overexpressed in plants show abiotic stress tolerance (Anwar & Kim, 2020; Bajaj, Targolli, Liu, Ho, & Wu, 1999; Habben et al., 2014; James et al., 2018; Yu et al., 2016). Although some problems arise in metabolic balance, it has a significant impact on the enzyme system for crop improvement.

Some crops lack resistance genes, especially disease- and pest-resistance genes. The introgression of foreign genes into a species results in new traits through a transgenic approach, and commercially released transgenic plants are a great example of this (Yu, Yang, Fu, & Li, 2022). Compared to overexpressed gene expression, transgene expression is more effective because it is regulated by complex and redundant pathways. Proper gene selection and field trials are important for transgenic approaches (Fazal et al., 2024). Therefore, a transgenic approach is effective for improving the enzyme systems of crops.

TABLE 16.2
List of Enzyme-Related Genes Modified Through Transgenic Approach

Crops	Enzyme	Function	Key References
Rice	Glutamine synthetase genes OsGS1;1 and OsGS2	Drought and salinity tolerance Higher chlorophyll content and water retention	James et al. (2018)
	AtGolS2	Drought tolerance and increases grain yield	Selvaraj et al. (2017)
	Sucrose synthase (SUS) genes	Improves hull size and grain weight	Fan et al. (2019)
	C4-cycle enzyme genes (such as phosphoenolpyruvate carboxylase, PEPC)	Photosynthetic efficiency enhancement	Häusler, Hirsch, Kreuzaler, and Peterhänsel (2002)
Wheat	AtbZIP69	Enhances drought and low nitrogen stress tolerance	Yan et al. (2025)
	V-type H+ pyrophosphatase gene EdVP1	Higher K uptake capacity in roots	Zhou et al. (2020)
Maize	ZmSKL1 and ZmSKL2	Improves drought stress tolerance, reduced reactive oxygen species accumulation	Liu, Li, Liang, Zhang, and Wu (2023)
Sugarcane	SoSPS1 gene	Increases sucrose content	Anur, Mufithah, Sawitri, Sakakibara, and Sugiharto (2020)
Sweet potato	9-Cis-epoxycarotenoid dioxygenase gene, IbNCED1	Reduce plant height	Zhou et al. (2023)
	Sucrose non-fermenting-1-related protein kinase-1 gene, IbSnRK1	Starch content enhancement	Ren, He, Zhao, Zhai, and Liu (2019)
Tomato	Xyloglucan xyloglucosyltransferase/ endohydrolase	Maintain the cell wall integrity during fruit ripening	Miedes, Herbers, Sonnewald, and Lorences (2010)
	Saussurea involucrata dehydrin gene (SiDHN)	Cell membrane damage reduction Cold and drought tolerance	Guo et al. (2019)
	Expansin	Improves the viscosity and consistency of juice and paste Enhancing product rheological properties	Kalamaki et al. (2003)
Tobacco	Tomato carotenoid ε-hydroxylase gene (SlLUT1)	Drought stress tolerance	Zhou, Deng, Kong, Li, and Meng (2013)
	Alfalfa glutathione S-transferase gene	Improves resistance to saline–alkali stresses	Du et al. (2019)
	C4-cycle enzyme genes (such as phosphoenolpyruvate carboxylase, PEPC)	Photosynthetic efficiency enhancement	Häusler et al. (2002)

16.3.2 Enhancing Crop Enzyme Systems through Precision Genome Editing

Gene-editing technologies are powerful tools for the modification of enzyme-coding genes. Enzyme efficiency can be enhanced or suppressed by gene modification. The most used genome-modifying tool is CRISPR/Cas9. Gene knockout, base editing, and gene expression regulation can be performed using this technique (Ahmar et al., 2020; Arora & Narula, 2017; Wang, Zhang, Zhang, & Zhu, 2021). TALENs and ZFNs are also useful for genome editing. CRISPR/Cas is the simplest method among all editing methods. These methods can be used to enhance the enzyme system in crops for crop improvement (Abdallah, Prakash, & McHughen, 2015; Ahmar et al., 2020; Wang et al., 2021). Table 16.3 represents a list of enzyme-related genes modified through different editing techniques.

Enzymes are directly related to traits such as yield, nutritional quality, disease resistance, and stress tolerance, and genome editing goals are mainly trait improvement. Therefore, enhancing the enzyme system through genome editing can be a potential part of crop improvement (Arora & Narula, 2017; Xu, Hua, & Lang, 2019; Zhang, Massel, Godwin, & Gao, 2018). Crop growth, development, and metabolism can be modified by the activity of enzyme-coding genes. Target gene modification can increase the efficiency of enzymes (Gao, 2021). Desired traits, such as plant architecture, can be modified through quantitative trait engineering. A wide range of variations can be created through the editing of gene regulatory regions (Chen et al., 2023; Rodríguez-Leal, Lemmon, Man, Bartlett, & Lippman, 2017).

Genome editing can be categorized into three types: type I, type II, and type III. Modification type I includes the alteration of several nucleotides. Enzyme activity can be fine-tuned through type I modifications (Yang, Xu, Shen, & Yan, 2021). Replacement of alleles or new gene insertion is classified as type II and type III, respectively. The replaced allele may be used as a substitute for a less-efficient enzyme. However, new gene insertion can introduce novel functions (Abdallah et al., 2015). Therefore, precise genome editing may pave the way for enzyme enhancement and crop improvement.

16.3.3 Promoter Engineering: A Precision Tool for Tailoring Gene Expression and Enhancing Crop Traits

In plant biotechnology, promoter modification is an effective technique for controlling gene expression. The promoter region may be a target for modification. Synthetic promoters

TABLE 16.3
List of Enzyme-Related Genes Modified through Different Editing Techniques

Editing Techniques	Genes	Enzymes	Crops	Functions	Key References
CRISPR/Cas	4-Coumarate ligase (4CL) gene	4-Coumarate ligase	Chickpea	Lignin accumulation, drought tolerance	Badhan, Ball, and Mantri (2021)
	MaACO1	Aminocyclopropane-1-carboxylate oxidase 1	Banana	Extended shelf life of fruit, delayed ripening	Shekhawat, Ram, and Soni (2024)
	StPPO2 (StPPO)	Polyphenol oxidase	Potato	Reduced tuber enzymatic browning	Shekhawat et al. (2024)
	GBSS/GBSSI	Granule-bound starch synthase	Cassava	Amylose synthesis	Shekhawat et al. (2024)
	OsBADH2	Betaine aldehyde dehydrogenase 2	Rice	Biosynthesis of aroma	Shekhawat et al. (2024)
	OsALS	Acetolactate synthase (ALS)	Rice	Herbicide resistance	Prajapati and Tyagi (2024)
	OsNRAMP5	Natural resistance-associated macrophage protein 5	Rice	Mn/Cd uptake	Kaur et al. (2025)
	Wx/GBSS (waxy)	Granule bound starch synthase	Maize	Amylose synthesis	Shekhawat et al. (2024)
	TaSBEIIa	Starch-branching enzyme IIa	Wheat	High amylose	Kaur et al. (2025)
TALENs	Acetolactate synthase (ALS) gene	Acetolactate synthase	Tobacco	Amino acid biosynthesis	Zhang et al. (2013)
	ZmPDS (phytoene desaturase), ZmIPK1A and ZmIPK (inositol phosphate kinases), and ZmMRP4 (multidrug resistance-associated protein 4)		Maize	Encode enzymes involved in different metabolic pathways	Liang, Zhang, Chen, and Gao (2014)
	Mitochondrial genes orf79 and orf125		Rice and rapeseed	Cytoplasmic male sterility	Liang et al. (2014)
ZFNs	Acetohydroxyacid synthase (AHAS)	Acetohydroxyacid synthase	Wheat	Resistance to imidazolinone herbicides	Kazama et al. (2019)
	IPK1 gene		Wheat	Plant metabolism	Townsend et al. (2009)
	Acetolactate synthase (ALS, also known as SuRA and SuRB)	Acetolactate synthase	Tobacco	Herbicide-resistance	Ran et al. (2018)

and modified cis elements can affect tissue-specific expression. The insertion or deletion of cis elements is also part of vivo promoter modification. Promoter components, such as cis elements, can regulate gene expression. Modification of cis-regulatory elements can enhance enzyme systems (Ali & Kim, 2019; Gondalia et al., 2025; Pandiarajan & Grover, 2018; Rodríguez-Leal et al., 2017; Shi et al., 2023). It can enhance the enzyme systems of crops and help improve their traits. A list of modified promoters and their function in crops is given in Table 16.4.

CRISPR/Cas9 and CRISPR–Cas12a are usually used to edit promoters to create variations in the genome. Grain size and starch content have been modified by promoter editing in rice. Tomato plant architecture has been modified using CRISPR-based promoter editing (Bilichak et al., 2020; Rodríguez-Leal et al., 2017; Zhou et al., 2023). In vivo promoter modification has been used to develop disease-resistant plants (Gondalia et al., 2025). Modified promoters can regulate tissue-specific gene expression at specific growth stages. Some traits depend on specific stages of development. Promoter engineering is important for improving this trait without affecting other traits. A previous report showed that genes can be activated by inducible promoters in response to environmental factors (Kummari et al., 2020; Villao-Uzho, Chávez-Navarrete, Pacheco-Coello, Sánchez-Timm, & Santos-Ordóñez, 2023; Yasmeen, Wang, Riaz, Zhang, & Zuo, 2023). Quality traits are regulated by multiple genes via complex pathways. The spatial, temporal, or conditional expression of genes can be tailored using artificially designed promoter sequences. Recently, an artificial promoter has been used to improve nutritional quality (Ali & Kim, 2019; Kummari et al., 2020; Shi et al., 2023).

The identification and analysis of promoter activity are important for understanding gene expression patterns and regulation. Gene expression is directly related to enzymatic activity. Promoter engineering is a transgene-free crop improvement method; therefore, it is widely accepted and supports sustainable agricultural practices (Bilichak et al., 2020; Gondalia et al., 2025; Rodríguez-Leal et al., 2017; Zhou et al., 2023). Although this strategy has benefits, it still faces some obstacles to practical implementation. Achieving the predicted gene expression through promoter

TABLE 16.4
List of Modified Promoters and Their Function in Crops

Crop	Enzyme	Gene promoter	Function	Key References
Soybean	Ubiquitin-conjugating enzyme	GmUBC4 the pUceS8.3 promoter	Response to pest and nematode attack	Villao, Chávez-Navarrete, Pacheco-Coello, Sanchez Timm, and Santos (2023)
Maize	Subtilisin-like serine proteases	CLAVATA3/EMBRYO SURROUNDING REGION (CLE) gene promoters	Increased grain yield-related traits	Fragoso et al. (2022)
	Cytochrome P450 enzymes	P450s	Insecticide resistance	Liu et al. (2021)
Soybean	4-Hydroxyphenylpyruvate dioxygenase (HPPD)	HPPD	Herbicide resistance	Ye et al. (2022)
Rice	Proteases	GSX7R promoter Cry1Ab gene	Resistance to yellow stem borer	Siehl et al. (2014)
Tobacco		EaPHT1;2 promoter	Stress-tolerant	Lin et al. (2022)
Tomato	Ribonuclease	2087-bp SlPR10 promoter (pSlPR10)	Fruit quality and resistance to gray mold and water loss	Naveenarani et al. (2023)
Capsicum annuum	Sesquiterpene cyclase (EAS) and a hydroxylase (EAH)	pCaD	Disease-resistant	Ruan, Xiong, and Li (2024)
Rice		OrGSE gene (OrGSEp)	Pest-resistant crops	In, Lee, Woo, Park, and Choi (2020)
		OsGTL1 promoter	Stress response	Xue et al. (2018)

engineering is extremely difficult (Gondalia et al., 2025; Rodríguez-Leal et al., 2017). Despite these challenges, the use of integrated approaches, such as bioinformatics and high-throughput synthetic biology, is expected to enhance the efficacy of promoter engineering. The rapid expansion of new technologies has huge potential for improving crops.

16.4 CASE STUDIES IN ENZYME-BASED CROP IMPROVEMENT

Previous studies have shown the effectiveness of crop improvement based on enzymes in different fields of agriculture. Table 16.5 shows the list of enzyme-based crop improvement programs. The nutritional quality of crops is a target trait for crop improvement. Genome editing tools have been precisely employed to enhance nutritional quality. High-GABA tomatoes are now available in the market. Another example is high oleic acid-containing soybeans. Tomatoes and soybeans have been successfully improved using gene editing approaches (Aghdam, Palma, & Corpas, 2020; Duraiswamy et al., 2022; Nie, Yang, Zheng, & Hou, 2024). In addition, anti-nutritional factors, such as phytic acid and enzyme inhibitors, can also be reduced using these techniques (da Silva, 2019). The nutritional quality of maize, such as protein and starch synthesis, has also been successfully enhanced by modifying O2 and PBF genes. Specific gene manipulation may be promising for avoiding negative impacts on other traits (Ojiewo et al., 2020).

Post-harvest quality is also being improved using enzyme-based technology. Natural biocontrol against fungal pathogens in fruits and vegetables extends their shelf life without compromising nutritional quality (Ku & Ha, 2020). For example, gene editing has improved tomato firmness and shelf life (Zhang, Zheng, Yang, Messing, & Wu, 2016). Gene silencing has been shown to reduce aflatoxin contamination and allergen content in groundnuts (Sharma et al., 2023). Additionally, root enzymes can enhance nutrient uptake and overall crop performance.

16.5 EMERGING TRENDS AND FUTURE OF GENETIC MODIFICATION FOR ENHANCEMENT OF ENZYME SYSTEM

The integration of systems biology with genomics and breeding will be the key strategy for enhancing enzyme systems for crop improvement. Targeted enzyme engineering can

TABLE 16.5
List of Enzyme-Based Crop Improvement Programs

Crops	Enzymes/Gene Approach	Traits	Key References
Tomato	CRISPR/Cas9, TALENs	Enhanced firmness, nutrition, improved shelf life, stress tolerance	Zhang et al. (2016)
Groundnut	Gene silencing, allele pyramiding	Reduced aflatoxin, allergens, higher oleic acid, extended shelf life, marketability	Sharma et al. (2023)
Maize	Transcription factor modulation (O2, PBF)	Improved protein, starch balance, linked to yield, kernel quality	Ojiewo et al. (2020)
Fruits/vegetables	Enzyme biocontrol	Maintained nutritional content Reduced spoilage, longer shelf life	Ku & Ha (2020)

enhance productivity and resilience. In recent decades, we have observed the successful integration of metabolic modeling and systems biology. The identification of the photosynthetic enzyme RuBisCO and stress response enzyme classes III peroxidases and sucrose phosphate synthase are the most prominent discoveries (Li et al., 2024; Nagamine & Ezura, 2022; Oliveira-Filho, Voiniciuc, & Hanson, 2023). These techniques can be used to guide genetic manipulation. In addition, this will help identify enzymes that are suitable for crop improvement.

Currently, metabolic pathway reconstruction and enhancement using engineered or non-native enzymes have become common. In addition, structure-based recombination and continuous directed evolution are promising methods for enhancing enzyme systems (Jat et al., 2021; Kidwai et al., 2020). Future research should focus on thermostability, substrate specificity, and compatibility with the plant cellular environment. The identification of useful enzyme-related alleles is increasing owing to high-throughput phenotyping and advances in genomics. These techniques are useful for breeding programs (Oliveira-Filho et al., 2023). Soil enzyme activities can be increased through legume crop cultivation. Integrated nutrient management can also enhance soil enzyme activity, and soil enzyme systems and management practices are emerging topics (Borase et al., 2020; Mann, 1999; Miao et al., 2019).

16.6 CHALLENGES AND CONSIDERATIONS DURING GENETICALLY MODIFIED CROP DEVELOPMENT

Genetically modified plants face several challenges. Off-target gene activity is a major problem. The activities of metabolites may be altered. For example, C_4 pathway enzyme overexpression in C_3 plants alters metabolic fluxes. Metabolite changes affect plant growth and development (Han et al., 2019; Nazari et al., 2024). Environmental and food safety concerns are serious for genetically modified crops, which must undergo a rigorous regulatory process for safety. Containment and monitoring biosafety protocols are used to prevent the unintended spread or ecological impacts of modified crops (Nazari et al., 2024). Regulatory and biosafety concerns, public perception, acceptance, intellectual property, and access are also concerns for field applications. Consumer trust is the most important factor in the safety of GMO crops. Good communication regarding the usefulness, risks, and regulatory oversight of these new technologies is essential for their broader acceptance. Patents and intellectual property rights can limit access to new technologies, especially for farmers and in developing countries. Enhancing enzyme systems through genetic modification holds promise for improving crops. Addressing these challenges is important for effective agricultural use. The success and acceptance of this new innovation will depend on the proper management of these challenges.

16.7 CONCLUSION

Enzyme enhancement through genetic modification is a method for crop improvement. This approach can overcome the basic physiological limitations of crop growth and development. This chapter explains how transgenic and gene-editing approaches allow for the precise optimization of enzymes. The regulation of photosynthesis, nutrient use, and stress resilience by enzymes leads to improved yields and sustainability. Although genetic modification has made significant advancements, it may face some challenges. A systems biology approach can address these challenges. Enzyme modification has emerged as a fundamental element for increasing productivity. Crop improvement through enzyme modification is a promising approach to meet food security demands.

REFERENCES

Abdallah, N. A., Prakash, C. S., & McHughen, A. G. (2015). Genome editing for crop improvement: Challenges and opportunities. *GM Crops & Food*, *6*(4), 183–205. https://doi.org/10.1080/21645698.2015.1129937

Aghdam, M. S., Palma, J. M., & Corpas, F. J. (2020). NADPH as a quality footprinting in horticultural crops marketability. *Trends in Food Science & Technology*, *103*, 152–161. https://doi.org/10.1016/j.tifs.2020.07.002

Ahmad, M., Ali, Q., Hafeez, M., & Malik, A. (2021). Improvement for biotic and abiotic stress tolerance in crop plants. *Biological and Clinical Sciences Research Journal*, *2021*(1), e004.

Ahmar, S., Saeed, S., Khan, M. H. U., Ullah Khan, S., Mora-Poblete, F., Kamran, M., … Jung, K. H. (2020). A revolution toward Gene-editing technology and its application to crop improvement. *International Journal of Molecular Sciences*, *21*(16). https://doi.org/10.3390/ijms21165665

Ali, S., & Kim, W. C. (2019). A fruitful decade using synthetic promoters in the improvement of transgenic plants. *Frontiers in Plant Science*, *10*, 1433. https://doi.org/10.3389/fpls.2019.01433

Altaf, M. A., Sharma, N., Singh, J., Samota, M. K., Sankhyan, P., Singh, B., … Kumar, R. (2023). Mechanistic insights on melatonin-mediated plant growth regulation and hormonal cross-talk process in solanaceous vegetables. *Scientia Horticulturae*, *308*, 111570. https://doi.org/10.1016/j.scienta.2022.111570

Amos, R. A., & Mohnen, D. (2019). Critical review of plant cell wall matrix polysaccharide glycosyltransferase activities verified by heterologous protein expression. Frontiers in Plant Science, *10*. https://doi.org/10.3389/fpls.2019.00915

Andrews, M., Lea, P. J., Raven, J. A., & Lindsey, K. (2004). Can genetic manipulation of plant nitrogen assimilation enzymes result in increased crop yield and greater N-use efficiency? An assessment. *Annals of Applied Biology*, *145*(1), 25–40. https://doi.org/10.1111/j.1744-7348.2004.tb00356.x

Anjum, N. A., Sharma, P., Gill, S. S., Hasanuzzaman, M., Khan, E. A., Kachhap, K., … Tuteja, N. (2016). Catalase and ascorbate peroxidase-representative H2O2-detoxifying heme enzymes in plants. *Environmental Science and Pollution Research International*, *23*(19), 19002–19029. https://doi.org/10.1007/s11356-016-7309-6

Anur, R. M., Mufithah, N., Sawitri, W. D., Sakakibara, H., & Sugiharto, B. (2020). Overexpression of sucrose phosphate synthase enhanced sucrose content and biomass production in transgenic sugarcane. *Plants (Basel)*, *9*(2). https://doi.org/10.3390/plants9020200

Anwar, A., & Kim, J. K. (2020). Transgenic breeding approaches for improving abiotic stress tolerance: Recent progress and future perspectives. *International Journal of Molecular Sciences, 21*(8). https://doi.org/10.3390/ijms21082695

Arora, K., & Sen, S. (2022). Cytokinin oxygenase/dehydrogenase inhibitors: An emerging tool in stress biotechnology employed for crop improvement. *Frontiers in Genetics, 13*, 877510. https://doi.org/10.3389/fgene.2022.877510

Arora, L., & Narula, A. (2017). Gene editing and crop improvement using CRISPR-Cas9 system. *Frontiers in Plant Science, 8*, 1932. https://doi.org/10.3389/fpls.2017.01932

Badhan, S., Ball, A. S., & Mantri, N. (2021). First report of CRISPR/Cas9 mediated DNA-free editing of 4CL and RVE7 genes in chickpea protoplasts. *International Journal of Molecular Sciences, 22*(1). https://doi.org/10.3390/ijms22010396

Bajaj, S., Targolli, J., Liu, L.-F., Ho, T.-H. D., & Wu, R. (1999). Transgenic approaches to increase dehydration-stress tolerance in plants. *Molecular Breeding, 5*(6), 493–503. https://doi.org/10.1023/A:1009660413133

Balotf, S., Kavoosi, G., & Kholdebarin, B. (2016). Nitrate reductase, nitrite reductase, glutamine synthetase, and glutamate synthase expression and activity in response to different nitrogen sources in nitrogen-starved wheat seedlings. *Biotechnology and Applied Biochemistry, 63*(2), 220–229. https://doi.org/10.1002/bab.1362

Batth, R., Singh, K., Kumari, S., & Mustafiz, A. (2017). Transcript profiling reveals the presence of abiotic stress and developmental stage specific ascorbate oxidase genes in plants. *Frontiers in Plant Science, 8*, 198. https://doi.org/10.3389/fpls.2017.00198

Bela, K., Horváth, E., Gallé, Á, Szabados, L., Tari, I., & Csiszár, J. (2015). Plant glutathione peroxidases: Emerging role of the antioxidant enzymes in plant development and stress responses. *Journal of Plant Physiology, 176*, 192–201. https://doi.org/10.1016/j.jplph.2014.12.014

Bilichak, A., Sastry-Dent, L., Sriram, S., Simpson, M., Samuel, P., Webb, S., … Eudes, F. (2020). Genome editing in wheat microspores and haploid embryos mediated by delivery of ZFN proteins and cell-penetrating peptide complexes. *Plant Biotechnology Journal, 18*(5), 1307–1316. https://doi.org/10.1111/pbi.13296

Borase, D. N., Nath, C. P., Hazra, K. K., Senthilkumar, M., Singh, S. S., Praharaj, C. S., … Kumar, N. (2020). Long-term impact of diversified crop rotations and nutrient management practices on soil microbial functions and soil enzymes activity. *Ecological Indicators, 114*, 106322. https://doi.org/10.1016/j.ecolind.2020.106322

Bunterngsook, B., Eurwilaichitr, L., Thamchaipenet, A., & Champreda, V. (2015). Binding characteristics and synergistic effects of bacterial expansins on cellulosic and hemicellulosic substrates. *Bioresource Technology, 176*, 129–135. https://doi.org/10.1016/j.biortech.2014.11.042

Bunterngsook, B., Mhuantong, W., Champreda, V., Thamchaipenet, A., & Eurwilaichitr, L. (2014). Identification of novel bacterial expansins and their synergistic actions on cellulose degradation. *Bioresource Technology, 159*, 64–71. https://doi.org/10.1016/j.biortech.2014.02.004

Cabello, P., de la Haba, P., González-Fontes, A., & Maldonado, J. M. (1998). Induction of nitrate reductase, nitrite reductase, and glutamine synthetase isoforms in sunflower cotyledons as affected by nitrate, light, and plastid integrity. *Protoplasma, 201*(1), 1–7. https://doi.org/10.1007/BF01280705

Camoni, L., Visconti, S., Aducci, P., & Marra, M. (2018). 14-3-3 proteins in plant hormone signaling: Doing several things at once. *Frontiers in Plant Science, 9*, 297. https://doi.org/10.3389/fpls.2018.00297

Cardoso, A. A., Gori, A., Da-Silva, C. J., & Brunetti, C. (2020). Abscisic acid biosynthesis and signaling in plants: Key Targets to Improve Water Use Efficiency and Drought Tolerance. *Applied Sciences, 10*(18), 6322.

Chakraborty, P., Biswas, A., Dey, S., Bhattacharjee, T., & Chakrabarty, S. (2023). Cytochrome P450 gene families: Role in plant secondary metabolites production and plant defense. *Journal of Xenobiotics, 13*(3), 402–423. https://doi.org/10.3390/jox13030026

Chen, F., Chen, L., Yan, Z., Xu, J., Feng, L., He, N., … Liu, C. (2024). Recent advances of CRISPR-based genome editing for enhancing staple crops. *Frontiers in Plant Science, 15*, 1478398. https://doi.org/10.3389/fpls.2024.1478398

Chen, L., & Liao, H. (2017). Engineering crop nutrient efficiency for sustainable agriculture. *Journal of Integrative Plant Biology, 59*(10), 710–735. https://doi.org/10.1111/jipb.12559

Chen, Q., Wang, B., Ding, H., Zhang, J., & Li, S. (2019). Review: The role of NADP-malic enzyme in plants under stress. *Plant Science, 281*, 206–212. https://doi.org/10.1016/j.plantsci.2019.01.010

Chen, Y. H., Lu, J., Yang, X., Huang, L. C., Zhang, C. Q., Liu, Q. Q., & Li, Q. F. (2023). Gene editing of non-coding regulatory DNA and its application in crop improvement. *Journal of Experimental Botany, 74*(19), 6158–6175. https://doi.org/10.1093/jxb/erad313

da Fonseca-Pereira, P., Monteiro-Batista, R. C., Araújo, W. L., & Nunes-Nesi, A. (2023). Harnessing enzyme cofactors and plant metabolism. An essential partnership. *The Plant Journal, 114*(5), 1014–1036. https://doi.org/10.1111/tpj.16167

da Silva, R. R. (2019). Enzyme technology in food preservation: A promising and sustainable strategy for biocontrol of post-harvest fungal pathogens. *Food Chemistry, 277*, 531–532. https://doi.org/10.1016/j.foodchem.2018.11.022

Ding, S., Liu, X., Hakulinen, N., Taherzadeh, M. J., Wang, Y., Wang, Y., … Tu, T. (2022). Boosting enzymatic degradation of cellulose using a fungal expansin: Structural insight into the pretreatment mechanism. *Bioresource Technology, 358*, 127434. https://doi.org/10.1016/j.biortech.2022.127434

Du, B., Zhao, W., An, Y., Li, Y., Zhang, X., Song, L., & Guo, C. (2019). Overexpression of an alfalfa glutathione S-transferase gene improved the saline-alkali tolerance of transgenic tobacco. *Biology Open, 8*(9). https://doi.org/10.1242/bio.043505

Duraiswamy, A., Sneha, A. N., Jebakani, K. S., Selvaraj, S., Pramitha, J. L., Selvaraj, R., … Kumar, P. R. (2022). Genetic manipulation of anti-nutritional factors in major crops for a sustainable diet in future. *Frontiers in Plant Science, 13*, 1070398. https://doi.org/10.3389/fpls.2022.1070398

Fan, C., Wang, G., Wang, Y., Zhang, R., Wang, Y., Feng, S., … Peng, L. (2019). Sucrose synthase enhances hull size and grain weight by regulating cell division and starch accumulation in transgenic rice. *International Journal of Molecular Sciences, 20*(20). https://doi.org/10.3390/ijms20204971

Fatma, M., Asgher, M., Iqbal, N., Rasheed, F., Sehar, Z., Sofo, A., & Khan, N. A. (2022). Ethylene signaling under stressful environments: Analyzing collaborative knowledge. *Plants (Basel), 11*(17). https://doi.org/10.3390/plants11172211

Fazal, A., Yang, M., Han, H., Lu, G., Hao, C., Lai, X., … Yang, Y. (2024). Impact of dual Bt-transgenic maize (2A7) on soil microbial communities and enzyme activities: A comparative study with control variety Z58. *Environmental Research, 263*, 120006. https://doi.org/10.1016/j.envres.2024.120006

Fortunato, S., Nigro, D., Lasorella, C., Marcotuli, I., Gadaleta, A., & de Pinto, M. C. (2023). The role of glutamine synthetase

(GS) and glutamate synthase (GOGAT) in the improvement of nitrogen use efficiency in cereals. *Biomolecules*, *13*(12). https://doi.org/10.3390/biom13121771

Fragoso, R. R., Arraes, F. B. M., Lourenço-Tessutti, I. T., Miranda, V. J., Basso, M. F., Ferreira, A. V. J., … Grossi-de-Sa, M. F. (2022). Functional characterization of the pUceS8.3 promoter and its potential use for ectopic gene overexpression. *Planta*, *256*(4), 69. https://doi.org/10.1007/s00425-022-03980-6

Gaafar, R. M., Osman, M. E. H., Abo-Shady, A. M., Almohisen, I. A. A., Badawy, G. A., El-Nagar, M. M. F., & Ismail, G. A. (2022). Role of antioxidant enzymes and glutathione S-transferase in bromoxynil herbicide stress tolerance in wheat plants. *Plants (Basel)*, *11*(20). https://doi.org/10.3390/plants11202679

Gao, C. (2021). Genome engineering for crop improvement and future agriculture. *Cell*, *184*(6), 1621–1635. https://doi.org/10.1016/j.cell.2021.01.005

Gao, S., & Chu, C. (2020). Gibberellin metabolism and signaling: Targets for improving agronomic performance of crops. *Plant and Cell Physiology*, *61*(11), 1902–1911. https://doi.org/10.1093/pcp/pcaa104

Gharabli, H., Della Gala, V., & Welner, D. H. (2023). The function of UDP-glycosyltransferases in plants and their possible use in crop protection. *Biotechnology Advances*, *67*, 108182. https://doi.org/10.1016/j.biotechadv.2023.108182

Gondalia, N., Quiroz, L. F., Lai, L., Singh, A. K., Khan, M., Brychkova, G., … Spillane, C. (2025). Harnessing promoter elements to enhance gene editing in plants: Perspectives and advances. *Plant Biotechnology Journal*, *23*(5), 1375–1395. https://doi.org/10.1111/pbi.14533

González Guzmán, M., Cellini, F., Fotopoulos, V., Balestrini, R., & Arbona, V. (2022). New approaches to improve crop tolerance to biotic and abiotic stresses. *Physiologia Plantarum*, *174*(1), e13547. https://doi.org/10.1111/ppl.13547

Guo, X., Zhang, L., Wang, X., Zhang, M., Xi, Y., Wang, A., & Zhu, J. (2019). Overexpression of *Saussurea involucrata* dehydrin gene SiDHN promotes cold and drought tolerance in transgenic tomato plants. *PLoS One*, *14*(11), e0225090. https://doi.org/10.1371/journal.pone.0225090

Habben, J. E., Bao, X., Bate, N. J., DeBruin, J. L., Dolan, D., Hasegawa, D., … Schussler, J. R. (2014). Transgenic alteration of ethylene biosynthesis increases grain yield in maize under field drought-stress conditions. *Plant Biotechnology Journal*, *12*(6), 685–693. https://doi.org/10.1111/pbi.12172

Han, H., Ling, Z., Khan, A., Virk, A. K., Kulshrestha, S., & Li, X. (2019). Improvements of thermophilic enzymes: From genetic modifications to applications. *Bioresource Technology*, *279*, 350–361. https://doi.org/10.1016/j.biortech.2019.01.087

Häusler, R. E., Hirsch, H. J., Kreuzaler, F., & Peterhänsel, C. (2002). Overexpression of C(4)-cycle enzymes in transgenic C(3) plants: A biotechnological approach to improve C(3)-photosynthesis. *Journal of Experimental Botany*, *53*(369), 591–607. https://doi.org/10.1093/jexbot/53.369.591

Heath-Pagliuso, S., Huffaker, R. C., & Allard, R. W. (1984). Inheritance of nitrite reductase and regulation of nitrate reductase, nitrite reductase, and glutamine synthetase isozymes. *Plant Physiology*, *76*(2), 353–358. https://doi.org/10.1104/pp.76.2.353

Hedden, P., & Phillips, A. L. (2000). Manipulation of hormone biosynthetic genes in transgenic plants. *Current Opinion in Biotechnology*, *11*(2), 130–137. https://doi.org/10.1016/s0958-1669(00)00071-9

Huang, Y., Ji, Z., Zhang, S., & Li, S. (2024). Function of hormone signaling in regulating nitrogen-use efficiency in plants. *Journal of Plant Physiology*, *294*, 154191. https://doi.org/10.1016/j.jplph.2024.154191

In, S., Lee, H. A., Woo, J., Park, E., & Choi, D. (2020). Molecular characterization of a pathogen-inducible bidirectional promoter from hot pepper (*Capsicum annuum*). *Molecular Plant-Microbe Interactions*, *33*(11), 1330–1339. https://doi.org/10.1094/mpmi-07-20-0183-r

James, D., Borphukan, B., Fartyal, D., Ram, B., Singh, J., Manna, M., … Reddy, M. K. (2018). Concurrent overexpression of OsGS1;1 and OsGS2 genes in transgenic rice (*Oryza sativa* L.): Impact on tolerance to abiotic stresses. *Frontiers in Plant Science*, *9*, 786. https://doi.org/10.3389/fpls.2018.00786

Jat, H. S., Datta, A., Choudhary, M., Sharma, P. C., Dixit, B., & Jat, M. L. (2021). Soil enzymes activity: Effect of climate smart agriculture on rhizosphere and bulk soil under cereal based systems of north-west India. *European Journal of Soil Biology*, *103*, 103292. https://doi.org/10.1016/j.ejsobi.2021.103292

Kalamaki, M. S., Powell, A. L., Struijs, K., Labavitch, J. M., Reid, D. S., & Bennett, A. B. (2003). Transgenic overexpression of expansin influences particle size distribution and improves viscosity of tomato juice and paste. *Journal of Agricultural and Food Chemistry*, *51*(25), 7465–7471. https://doi.org/10.1021/jf0341666

Kaur, N., Qadir, M., Francis, D. V., Alok, A., Tiwari, S., & Ahmed, Z. F. R. (2025). CRISPR/Cas9: A sustainable technology to enhance climate resilience in major Staple Crops. *Frontiers in Genome Editing*, *7*, 1533197. https://doi.org/10.3389/fgeed.2025.1533197

Kazama, T., Okuno, M., Watari, Y., Yanase, S., Koizuka, C., Tsuruta, Y., … Arimura, S. I. (2019). Curing cytoplasmic male sterility via TALEN-mediated mitochondrial genome editing. *Nature Plants*, *5*(7), 722–730. https://doi.org/10.1038/s41477-019-0459-z

Kelley, D. R. (2018). E3 ubiquitin ligases: Key regulators of hormone signaling in plants. *Molecular & Cellular Proteomics*, *17*(6), 1047–1054. https://doi.org/10.1074/mcp.MR117.000476

Kidwai, M., Ahmad, I. Z., & Chakrabarty, D. (2020). Class III peroxidase: An indispensable enzyme for biotic/abiotic stress tolerance and a potent candidate for crop improvement. *Plant Cell Reports*, *39*(11), 1381–1393. https://doi.org/10.1007/s00299-020-02588-y

Kim, E. S., Lee, H. J., Bang, W. G., Choi, I. G., & Kim, K. H. (2009). Functional characterization of a bacterial expansin from Bacillus subtilis for enhanced enzymatic hydrolysis of cellulose. *Biotechnology and Bioengineering*, *102*(5), 1342–1353. https://doi.org/10.1002/bit.22193

Kiremit, M. S., & Arslan, H. (2024). Response of antioxidant enzymes of winter wheat (*Triticum aestivum* L.) to shallow and saline groundwater depths. *Journal of Agronomy and Crop Science*, *210*(5), e12752. https://doi.org/10.1111/jac.12752

Kishorekumar, R., Bulle, M., Wany, A., & Gupta, K. J. (2020). An overview of important enzymes involved in nitrogen assimilation of plants. *Methods in Molecular Biology*, *2057*, 1–13. https://doi.org/10.1007/978-1-4939-9790-9_1

Ku, H. K., & Ha, S. H. (2020). Improving nutritional and functional quality by genome editing of crops: Status and perspectives. *Frontiers in Plant Science*, *11*, 577313. https://doi.org/10.3389/fpls.2020.577313

Kummari, D., Palakolanu, S. R., Kishor, P. B. K., Bhatnagar-Mathur, P., Singam, P., Vadez, V., & Sharma, K. K. (2020). An update and perspectives on the use of promoters in plant genetic engineering. *Journal of Biosciences*, *45*, 119. https://doi.org/10.1007/s12038-020-00087-6

Lea, P. J., & Miflin, B. J. (2011). Nitrogen assimilation and its relevance to crop improvement. In C. H. Foyer & H. Zhang (Eds.), *Annual plant reviews online*(pp. 1–40), Blackwell Publishing Ltd. https://doi.org/10.1002/9781444328608.ch1

Lee, D. H., Kim, Y. S., & Lee, C. B. (2001). The inductive responses of the antioxidant enzymes by salt stress in the rice (*Oryza sativa* L.). *Journal of Plant Physiology*, *158*(6), 737–745. https://doi.org/10.1078/0176-1617-00174

Lee, H. J., Lee, S., Ko, H.-j, Kim, K. H., & Choi, I.-G. (2010). An expansin-like protein from *Hahella chejuensis* binds cellulose and enhances cellulase activity. *Molecules and Cells*, *29*(4), 379–385. https://doi.org/10.1007/s10059-010-0033-z

Li, A.-M., Liao, F., Qin, C.-X., Wang, M., Chen, Z.-L., Zhang, B.-Q., … Huang, D.-l. (2024). Sucrose phosphate synthase genes in plants: Its role and practice for crop improvement. *Journal of Agricultural and Food Chemistry*, *72*. https://doi.org/10.1021/acs.jafc.4c05068

Li, C., Wang, J., Lan, H., & Yu, Q. (2024). Enhanced drought tolerance and photosynthetic efficiency in *Arabidopsis* by overexpressing phosphoenolpyruvate carboxylase from a single-cell C4 halophyte *Suaeda aralocaspica*. *Frontiers in Plant Science*, *15*, 1443691. https://doi.org/10.3389/fpls.2024.1443691

Li, H., Hu, B., & Chu, C. (2017). Nitrogen use efficiency in crops: Lessons from *Arabidopsis* and rice. *Journal of Experimental Botany*, *68*(10), 2477–2488. https://doi.org/10.1093/jxb/erx101

Liang, Z., Zhang, K., Chen, K., & Gao, C. (2014). Targeted mutagenesis in *Zea mays* using TALENs and the CRISPR/Cas system. *Journal of Genetics and Genomics*, *41*(2), 63–68. https://doi.org/10.1016/j.jgg.2013.12.001

Lin, M., Yan, J., Ali, M. M., Wang, S., Tian, S., Chen, F., & Lin, Z. (2022). Isolation and functional characterization of a green-tissue promoter in Japonica rice (*Oryza sativa* subsp. Japonica). *Biology (Basel)*, *11*(8). https://doi.org/10.3390/biology11081092

Liu, L., Gallagher, J., Arevalo, E. D., Chen, R., Skopelitis, T., Wu, Q., … Jackson, D. (2021). Enhancing grain-yield-related traits by CRISPR-Cas9 promoter editing of maize CLE genes. *Nature Plants*, *7*(3), 287–294. https://doi.org/10.1038/s41477-021-00858-5

Liu, X., Hu, B., & Chu, C. (2022). Nitrogen assimilation in plants: Current status and future prospects. *Journal of Genetics and Genomics*, *49*(5), 394–404. https://doi.org/10.1016/j.jgg.2021.12.006

Liu, Y., Chen, Z., Zhang, C., Guo, J., Liu, Q., Yin, Y., … Liu, X. (2023). Gene editing of ZmGA20ox3 improves plant architecture and drought tolerance in maize. *Plant Cell Reports*, *43*(1), 18. https://doi.org/10.1007/s00299-023-03090-x

Liu, Y., Lan, X., Hou, H., Ji, J., Liu, X., & Lv, Z. (2024). Multifaceted ability of organic fertilizers to improve crop productivity and abiotic stress tolerance. *Review and Perspectives*, *14*(6), 1141.

Liu, Y., Li, A., Liang, M., Zhang, Q., & Wu, J. (2023). Overexpression of the maize genes ZmSKL1 and ZmSKL2 positively regulates drought stress tolerance in transgenic *Arabidopsis*. *Plant Cell Reports*, *42*(3), 521–533. https://doi.org/10.1007/s00299-022-02974-8

Lou, Y. R., Pichersky, E., & Last, R. L. (2022). Deep roots and many branches: Origins of plant-specialized metabolic enzymes in general metabolism. *Current Opinion in Plant Biology*, *66*, 102192. https://doi.org/10.1016/j.pbi.2022.102192

Mann, C. C. (1999). Genetic engineers aim to soup up crop photosynthesis. *Science*, *283* (5400), 314–316. https://doi.org/10.1126/science.283.5400.314

Meents, M. J., Watanabe, Y., & Samuels, A. L. (2018). The cell biology of secondary cell wall biosynthesis. *Annals of Botany*, *121*(6), 1107–1125.

Miao, F., Li, Y., Cui, S., Jagadamma, S., Yang, G., & Zhang, Q. (2019). Soil extracellular enzyme activities under long-term fertilization management in the croplands of China: A meta-analysis. *Nutrient Cycling in Agroecosystems*, *114*(2), 125–138. https://doi.org/10.1007/s10705-019-09991-2

Miedes, E., Herbers, K., Sonnewald, U., & Lorences, E. P. (2010). Overexpression of a cell wall enzyme reduces xyloglucan depolymerization and softening of transgenic tomato fruits. *Journal of Agricultural and Food Chemistry*, *58*(9), 5708–5713. https://doi.org/10.1021/jf100242z

Miflin, B. J., & Habash, D. Z. (2002). The role of glutamine synthetase and glutamate dehydrogenase in nitrogen assimilation and possibilities for improvement in the nitrogen utilization of crops. *Journal of Experimental Botany*, *53*(370), 979–987. https://doi.org/10.1093/jexbot/53.370.979

Mira, W., Heinz, O., Gonçalvez, A., Crema, L., Vicentini, R., Cardoso, S., … da Franca Silva, T. (2024). SacEXP32 sugarcane expansin gene expression increases cell size and improves biomass digestibility. *Journal of Plant Biochemistry and Biotechnology*, *33*(3), 313–325. https://doi.org/10.1007/s13562-024-00891-3

Mishra, N., Jiang, C., Chen, L., Paul, A., Chatterjee, A., & Shen, G. (2023). Achieving abiotic stress tolerance in plants through antioxidative defense mechanisms. *Frontiers in Plant Science*, *14*, 1110622. https://doi.org/10.3389/fpls.2023.1110622

Miyao, M. (2003). Molecular evolution and genetic engineering of C4 photosynthetic enzymes. *Journal of Experimental Botany*, *54*(381), 179–189. https://doi.org/10.1093/jxb/erg026

Nagamine, A., & Ezura, H. (2022). Genome editing for improving crop nutrition. *Frontiers in Genome Editing*, *4*, 850104. https://doi.org/10.3389/fgeed.2022.850104

Naveenarani, M., Swamy, H. K. M., Surya Krishna, S., Mahadevaiah, C., Valarmathi, R., Manickavasagam, M., … Appunu, C. (2023). Isolation and characterization of *Erianthus arundinaceus* phosphate transporter 1 (PHT1) gene promoter and 5' deletion analysis of transcriptional regulation regions under phosphate stress in transgenic tobacco. *Plants (Basel)*, *12*(21). https://doi.org/10.3390/plants12213760

Nazari, M., Kordrostami, M., Ghasemi-Soloklui, A. A., Eaton-Rye, J. J., Pashkovskiy, P., Kuznetsov, V., & Allakhverdiev, S. I. (2024). Enhancing photosynthesis and plant productivity through genetic modification. *Cells*, *13*(16). https://doi.org/10.3390/cells13161319

Nie, H., Yang, X., Zheng, S., & Hou, L. (2024). Gene-based developments in improving quality of tomato: Focus on firmness, shelf life, and pre- and post-harvest stress adaptations. *Horticulturae*, *10*, 641. https://doi.org/10.3390/horticulturae10060641

Ojiewo, C. O., Janila, P., Bhatnagar-Mathur, P., Pandey, M. K., Desmae, H., Okori, P., … Varshney, R. K. (2020). Advances in crop improvement and delivery research for nutritional quality and health benefits of groundnut (*Arachis hypogaea* L.). *Frontiers in Plant Science*, *11*, 29. https://doi.org/10.3389/fpls.2020.00029

Oliveira-Filho, E. R., Voiniciuc, C., & Hanson, A. D. (2023). Adapting enzymes to improve their functionality in plants: Why and how. *Biochemical Society Transactions*, *51*(5), 1957–1966. https://doi.org/10.1042/bst20230532

Orłowski, A., Artzi, L., Cazade, P.-A., Gunnoo, M., Bayer, E. A., & Thompson, D. (2018). On the distinct binding modes of expansin and carbohydrate-binding module proteins on crystalline and nanofibrous cellulose: Implications for cellulose degradation by designer cellulosomes. *Physical Chemistry Chemical Physics*, *20*(12), 8278–8293. https://doi.org/10.1039/C7CP07764E

Orr, D. J., Alcântara, A., Kapralov, M. V., Andralojc, P. J., Carmo-Silva, E., & Parry, M. A. (2016). Surveying rubisco diversity and temperature response to improve crop photosynthetic efficiency. *Plant Physiology*, *172*(2), 707–717. https://doi.org/10.1104/pp.16.00750

Pandian, B. A., Sathishraj, R., Djanaguiraman, M., Prasad, P. V. V., & Jugulam, M. (2020). Role of cytochrome P450 enzymes in plant stress response. *Antioxidants (Basel)*, *9*(5). https://doi.org/10.3390/antiox9050454

Pandiarajan, R., & Grover, A. (2018). In vivo promoter engineering in plants: Are we ready? *Plant Science*, *277*, 132–138. https://doi.org/10.1016/j.plantsci.2018.10.011

Perdomo, J. A., Capó-Bauçà, S., Carmo-Silva, E., & Galmés, J. (2017). Rubisco and rubisco activase play an important role in the biochemical limitations of photosynthesis in rice, wheat, and maize under high temperature and water deficit. *Frontiers in Plant Science*, *8*, 490. https://doi.org/10.3389/fpls.2017.00490

Prajapati, R., & Tyagi, K. (2024). CRISPR/Cas-mediated multiplex genome editing in plants and applications. In S. K. Upadhyay (Ed.), *Applications of genome engineering in plants* (pp. 20–39), John Wiley & Sons Ltd. https://doi.org/10.1002/9781394183913.ch2

Rajput, V. D., Harish, Singh, R. K., Verma, K. K., Sharma, L., Quiroz-Figueroa, F. R., … Mandzhieva, S. (2021). Recent developments in enzymatic antioxidant defence mechanism in plants with special reference to abiotic stress. *Biology (Basel)*, *10*(4). https://doi.org/10.3390/biology10040267

Ran, Y., Patron, N., Kay, P., Wong, D., Buchanan, M., Cao, Y.-Y., … Hayden, M. J. (2018). Zinc finger nuclease-mediated precision genome editing of an endogenous gene in hexaploid bread wheat (*Triticum aestivum*) using a DNA repair template. *Plant Biotechnology Journal*, *16*(12), 2088–2101. https://doi.org/10.1111/pbi.12941

Ren, Z., He, S., Zhao, N., Zhai, H., & Liu, Q. (2019). A sucrose non-fermenting-1-related protein kinase-1 gene, IbSnRK1, improves starch content, composition, granule size, degree of crystallinity and gelatinization in transgenic sweet potato. *Plant Biotechnology Journal*, *17*(1), 21–32. https://doi.org/10.1111/pbi.12944

Rintamäki, E., & Aro, E.-M. (1985). Photosynthetic and photorespiratory enzymes in widely divergent plant species with special reference to the Moss *Ceratodon purpureus*: Properties of ribulose bisphosphate carboxylase/oxygenase, phosphoenolpyruvate carboxylase and glycolate oxidase. *Journal of Experimental Botany*, *36*(11), 1677–1684.

Rocha Meneses, L., Ferreira, A., Karimi, J., Orupõld, S., & Kikas, K., T. (2019). *Genetic modification of cereal plants: A strategy to enhance bioethanol yields from agricultural waste. Industrial Crops and Products*, *150*, 112408. https://doi.org/10.1016/j.indcrop.2020.112408

Rodríguez-Leal, D., Lemmon, Z. H., Man, J., Bartlett, M. E., & Lippman, Z. B. (2017). Engineering quantitative trait variation for crop improvement by genome editing. *Cell*, *171*(2), 470–480.e478. https://doi.org/10.1016/j.cell.2017.08.030

Ruan, X. M., Xiong, X., & Li, J. F. (2024). Identification and application of an exocarp-preferential promoter for genetic engineering of tomato fruit. *Horticulture Research*, *11*(3), uhae035. https://doi.org/10.1093/hr/uhae035

Ruan, Y. L. (2014). Sucrose metabolism: Gateway to diverse carbon use and sugar signaling. *Annual Review of Plant Biology*, *65*, 33–67. https://doi.org/10.1146/annurev-arplant-050213-040251

Sahoo, A., & Tiwari, S. (2022). Antioxidants and antioxidative enzymes as potential biomarkers for assessing stress in plants. *International Journal of Plant and Environment*, *8*, 95–105. https://doi.org/10.18811/ijpen.v8i02.01

Sarker, U., & Oba, S. (2018). Catalase, superoxide dismutase and ascorbate-glutathione cycle enzymes confer drought tolerance of Amaranthus tricolor. *Scientific Reports*, *8*(1), 16496. https://doi.org/10.1038/s41598-018-34944-0

Sathee, L., Jagadhesan, B., Pandesha, P. H., Barman, D., Adavi, B. S., Nagar, S., … Chinnusamy, V. (2022). Genome editing targets for improving nutrient use efficiency and nutrient stress adaptation. *Frontiers in Genetics*, *13*, 900897. https://doi.org/10.3389/fgene.2022.900897

Schläpfer, P., Zhang, P., Wang, C., Kim, T., Banf, M., Chae, L., … Rhee, S. Y. (2017). Genome-wide prediction of metabolic enzymes, pathways, and gene clusters in plants. *Plant Physiology*, *173*(4), 2041–2059.

Sedeek, K. E. M., Mahas, A., & Mahfouz, M. (2019). Plant genome engineering for targeted improvement of crop traits. *Frontiers in Plant Science*, *10*, 114. https://doi.org/10.3389/fpls.2019.00114

Seith, B., Setzer, B., Flaig, H., & Mohr, H. (1994). Appearance of nitrate reductase, nitrite reductase and glutamine synthetase in different organs of the Scots pine (*Pinus sylvestris*) seedling as affected by light, nitrate and ammonium. *Physiologia Plantarum*, *91*(3), 419–426. https://doi.org/10.1111/j.1399-3054.1994.tb02969.x

Selvaraj, M. G., Ishizaki, T., Valencia, M., Ogawa, S., Dedicova, B., Ogata, T., … Ishitani, M. (2017). Overexpression of an *Arabidopsis* thaliana galactinol synthase gene improves drought tolerance in transgenic rice and increased grain yield in the field. *Plant Biotechnology Journal*, *15*(11), 1465–1477. https://doi.org/10.1111/pbi.12731

Shahid, M., Singh, U. B., Khan, M. S., Singh, P., Kumar, R., Singh, R. N., … Singh, H. V. (2023). Bacterial ACC deaminase: Insights into enzymology, biochemistry, genetics, and potential role in amelioration of environmental stress in crop plants. *Frontiers in Microbiology*, *14*, 1132770. https://doi.org/10.3389/fmicb.2023.1132770

Sharma, P., Pandey, A., Malviya, R., Dey, S., Karmakar, S., & Gayen, D. (2023). Genome editing for improving nutritional quality, post-harvest shelf life and stress tolerance of fruits, vegetables, and ornamentals. *Frontiers in Genome Editing*, *5*, 1094965. https://doi.org/10.3389/fgeed.2023.1094965

Shekhawat, P. K., Ram, H., & Soni, P. (2024). Genome editing for nutritional improvement of crops. In S. K. Upadhyay (Ed.), *Applications of genome engineering in plants* (pp. 122–152), John Wiley & Sons Ltd. https://doi.org/10.1002/9781394183913.ch7

Shi, L., Su, J., Cho, M. J., Song, H., Dong, X., Liang, Y., & Zhang, Z. (2023). Promoter editing for the genetic improvement of crops. *Journal of Experimental Botany*, *74*(15), 4349–4366. https://doi.org/10.1093/jxb/erad175

Shi, R., Yu, J., Chang, X., Qiao, L., Liu, X., & Lu, L. (2023). Recent advances in research into jasmonate biosynthesis and signaling pathways in agricultural crops and products. *Processes*, *11*, 736. https://doi.org/10.3390/pr11030736

Shibuya, K., & Clark, D. G. (2006). Ethylene. *Journal of Crop Improvement*, *18*(1–2), 391–412. https://doi.org/10.1300/J411v18n01_05

Shimizu, Y., Aotsuka, S., Hasegawa, O., Kawada, T., Sakuno, T., Sakai, F., & Hayashi, T. (1997). Changes in levels of mRNAs for cell wall-related enzymes in growing cotton fiber cells. *Plant and Cell Physiology*, *38*(3), 375–378. https://doi.org/10.1093/oxfordjournals.pcp.a029178

Siehl, D. L., Tao, Y., Albert, H., Dong, Y., Heckert, M., Madrigal, A., … Castle, L. A. (2014). Broad 4-hydroxyphenylpyruvate dioxygenase inhibitor herbicide tolerance in soybean with an optimized enzyme and expression cassette. *Plant Physiology*, *166*(3), 1162–1176. https://doi.org/10.1104/pp.114.247205

Singh, J., Garai, S., Das, S., Thakur, J. K., & Tripathy, B. C. (2022). Role of C_4 photosynthetic enzyme isoforms in C3 plants and their potential applications in improving agronomic traits in crops. *Photosynthesis Research*, *154*(3), 233–258. https://doi.org/10.1007/s11120-022-00978-9

Singhal, R. K., Fahad, S., Kumar, P., Choyal, P., Javed, T., Jinger, D., … Nawaz, T. (2023). Beneficial elements: New players in improving nutrient use efficiency and abiotic stress tolerance. *Plant Growth Regulation*, *100*(2), 237–265. https://doi.org/10.1007/s10725-022-00843-8

Sivasankar, S., & Oaks, A. (1996). Nitrate assimilation in higher plants: The effect of metabolites and light. *Plant Physiology and Biochemistry*, *34*, 609–620.

Stancheva, I., & Dinev, N. (1995). Response of wheat and maize to different nitrogen sources: II. Nitrate reductase and glutamine synthetase enzyme activities, and plastid pigment content. *Journal of Plant Nutrition*, *18*(6), 1281–1290. https://doi.org/10.1080/01904169509364979

Stasnik, P., Vollmann, J., Großkinsky, D. K., & Jonak, C. (2024). Leaf carbohydrate metabolic enzyme activities are associated with salt tolerance and yield stability in the climate-resilient crop Camelina sativa. *Plant Stress*, *14*, 100629. https://doi.org/10.1016/j.stress.2024.100629

Sturm, A., & Tang, G. Q. (1999). The sucrose-cleaving enzymes of plants are crucial for development, growth and carbon partitioning. *Trends in Plant Science*, *4*(10), 401–407. https://doi.org/10.1016/s1360-1385(99)01470-3

Sun, W., Shahrajabian, M. H., Kuang, Y., & Wang, N. (2024). Amino acids biostimulants and protein hydrolysates in agricultural sciences. *Plants*, *13*(2), 210. https://doi.org/10.3390/plants13020210

Sun, X., Han, G., Meng, Z., Lin, L., & Sui, N. (2019). Roles of malic enzymes in plant development and stress responses. *Plant Signaling & Behavior*, *14*(10), e1644596. https://doi.org/10.1080/15592324.2019.1644596

Swain, R., Sahoo, S., Behera, M., & Rout, G. R. (2023). Instigating prevalent abiotic stress resilience in crop by exogenous application of phytohormones and nutrient. *Frontiers in Plant Science*, 14, 1104874. https://doi.org/10.3389/fpls.2023.1104874

Taniguchi, Y., Ohkawa, H., Masumoto, C., Fukuda, T., Tamai, T., Lee, K., … Miyao, M. (2008). Overproduction of C4 photosynthetic enzymes in transgenic rice plants: An approach to introduce the C4-like photosynthetic pathway into rice. *Journal of Experimental Botany*, *59*(7), 1799–1809. https://doi.org/10.1093/jxb/ern016

Taylor, A. O., Slack, C. R., & McPherson, H. G. (1974). Plants under climatic stress: VI. Chilling and light effects on photosynthetic enzymes of sorghum and maize. *Plant Physiology*, *54*(5), 696–701. https://doi.org/10.1104/pp.54.5.696

Tola, A. J., Jaballi, A., Germain, H., & Missihoun, T. D. (2020). Recent development on plant aldehyde dehydrogenase enzymes and their functions in plant development and stress signaling. *Genes (Basel)*, *12*(1). https://doi.org/10.3390/genes12010051

Townsend, J. A., Wright, D. A., Winfrey, R. J., Fu, F., Maeder, M. L., Joung, J. K., & Voytas, D. F. (2009). High-frequency modification of plant genes using engineered zinc-finger nucleases. *Nature*, *459*(7245), 442–445. https://doi.org/10.1038/nature07845

Vanderwall, M., & Gendron, J. M. (2023). HEXOKINASE1 and glucose-6-phosphate fuel plant growth and development. *Development*, *150*(20). https://doi.org/10.1242/dev.202346

Villao, L., Chávez-Navarrete, T., Pacheco-Coello, R., Sanchez Timm, L., & Santos, E. (2023). Plant promoters: Their identification, characterization, and role in gene regulation. *Genes (Basel)*, *14*, 1226. https://doi.org/10.3390/genes14061226

Villao-Uzho, L., Chávez-Navarrete, T., Pacheco-Coello, R., Sánchez-Timm, E., & Santos-Ordóñez, E. (2023). Plant promoters: Their identification, characterization, and role in gene regulation. *Genes (Basel)*, *14*(6). https://doi.org/10.3390/genes14061226

Viswanath, K. K., Varakumar, P., Pamuru, R. R., Basha, S. J., Mehta, S., & Rao, A. D. (2020). Plant lipoxygenases and their role in plant physiology. *Journal of Plant Biology*, *63*(2), 83–95. https://doi.org/10.1007/s12374-020-09241-x

von Caemmerer, S. (2020). Rubisco carboxylase/oxygenase: From the enzyme to the globe: A gas exchange perspective. *Journal of Plant Physiology*, *252*, 153240. https://doi.org/10.1016/j.jplph.2020.153240

von Caemmerer, S., & Furbank, R. T. (2016). Strategies for improving C4 photosynthesis. *Current Opinion in Plant Biology*, *31*, 125–134. https://doi.org/10.1016/j.pbi.2016.04.003

Vukašinović, N., Wang, Y., Vanhoutte, I., Fendrych, M., Guo, B., Kvasnica, M., … Russinova, E. (2021). Local brassinosteroid biosynthesis enables optimal root growth. *Nature Plants*, *7*(5), 619–632. https://doi.org/10.1038/s41477-021-00917-x

Wang, T., Zhang, C., Zhang, H., & Zhu, H. (2021). CRISPR/Cas9-mediated gene editing revolutionizes the improvement of horticulture food crops. *Journal of Agricultural and Food Chemistry*, *69*(45), 13260–13269. https://doi.org/10.1021/acs.jafc.1c00104

Wang, Y., Fan, C., Hu, H., Li, Y., Sun, D., Wang, Y., & Peng, L. (2016). Genetic modification of plant cell walls to enhance biomass yield and biofuel production in bioenergy crops. *Biotechnology Advances*, *34*(5), 997–1017. https://doi.org/10.1016/j.biotechadv.2016.06.001

Wasternack, C., & Song, S. (2017). Jasmonates: Biosynthesis, metabolism, and signaling by proteins activating and repressing transcription. *Journal of Experimental Botany*, *68*(6), 1303–1321. https://doi.org/10.1093/jxb/erw443

Westfall, C. S., Muehler, A. M., & Jez, J. M. (2013). Enzyme action in the regulation of plant hormone responses. *Journal of Biological Chemistry*, *288*(27), 19304–19311. https://doi.org/10.1074/jbc.R113.475160

Winkel, B. S. (2004). Metabolic channeling in plants. *Annual Review of Plant Biology*, *55*, 85–107. https://doi.org/10.1146/annurev.arplant.55.031903.141714

Xu, J., Hua, K., & Lang, Z. (2019). Genome editing for horticultural crop improvement. *Horticulture Research*, *6*, 113. https://doi.org/10.1038/s41438-019-0196-5

Xue, M., Long, Y., Zhao, Z., Huang, G., Huang, K., Zhang, T., … Pei, X. (2018). Isolation and characterization of a green-tissue promoter from common wild rice (*Oryza rufipogon* Griff.). *International Journal of Molecular Sciences*, *19*(7). https://doi.org/10.3390/ijms19072009

Yan, J., Wang, D., He, Z., Li, X., Tang, W., Chen, K., … Chen, M. (2025). Overexpression of AtbZIP69 in transgenic wheat confers tolerance to nitrogen and drought stress. *Planta*, *261*(2), 25. https://doi.org/10.1007/s00425-024-04605-w

Yang, Y., Xu, C., Shen, Z., & Yan, C. (2021). Crop quality improvement through genome editing strategy. *Frontiers in Genome Editing*, *3*, 819687. https://doi.org/10.3389/fgeed.2021.819687

Yasmeen, E., Wang, J., Riaz, M., Zhang, L., & Zuo, K. (2023). Designing artificial synthetic promoters for accurate, smart, and versatile gene expression in plants. *Plant Communications*, *4*(4), 100558. https://doi.org/10.1016/j.xplc.2023.100558

Ye, M., Nayak, B., Xiong, L., Xie, C., Dong, Y., You, M., … You, S. (2022). The role of insect cytochrome P450s in mediating insecticide resistance. *Agriculture*, *12*(1), 53.

Yin, Q., Qin, W., Zhou, Z., Wu, A.-M., Deng, W., Li, Z., … Lu, W.-j. (2024). Banana MaNAC1 activates secondary cell wall cellulose biosynthesis to enhance chilling resistance in fruit. *Plant Biotechnology Journal*, *22*(2), 413–426. https://doi.org/10.1111/pbi.14195

Yu, F., & Xie, Q. (2024). ER-associated ubiquitin-conjugating enzyme: A key regulator of grain yield and stress resistance in crops. *Trends in Plant Science*, *29*(3), 286–289. https://doi.org/10.1016/j.tplants.2023.11.025

Yu, H., Yang, Q., Fu, F., & Li, W. (2022). Three strategies of transgenic manipulation for crop improvement. *Frontiers in Plant Science*, *13*, 948518. https://doi.org/10.3389/fpls.2022.948518

Yu, L. H., Wu, S. J., Peng, Y. S., Liu, R. N., Chen, X., Zhao, P., … Xiang, C. B. (2016). *Arabidopsis* EDT1/HDG11 improves drought and salt tolerance in cotton and poplar and increases cotton yield in the field. *Plant Biotechnology Journal*, *14*(1), 72–84. https://doi.org/10.1111/pbi.12358

Zhang, G., Hou, X., Wang, L., Xu, J., Chen, J., Fu, X., … Luan, S. (2021). PHOTO-SENSITIVE LEAF ROLLING 1 encodes a polygalacturonase that modifies cell wall structure and drought tolerance in rice. *New Phytologist*, *229*(2), 890–901. https://doi.org/10.1111/nph.16899

Zhang, P., Cui, M., Huang, R., Qi, W., Thielemans, W., He, Z., & Su, R. (2021). Enhanced enzymatic hydrolysis of cellulose by endoglucanase via expansin pretreatment and the addition of zinc ions. *Bioresource Technology*, *333*, 125139. https://doi.org/10.1016/j.biortech.2021.125139

Zhang, P., Su, R., Duan, Y., Cui, M., Huang, R., Qi, W., … Thielemans, W. (2021). Synergy between endo/exo-glucanases and expansin enhances enzyme adsorption and cellulose conversion. *Carbohydrate Polymers*, *253*, 117287. https://doi.org/10.1016/j.carbpol.2020.117287

Zhang, Y., Massel, K., Godwin, I. D., & Gao, C. (2018). Applications and potential of genome editing in crop improvement. *Genome Biology*, *19*(1), 210. https://doi.org/10.1186/s13059-018-1586-y

Zhang, Y., Zhang, F., Li, X., Baller, J. A., Qi, Y., Starker, C. G., … Voytas, D. F. (2013). Transcription activator-like effector nucleases enable efficient plant genome engineering. *Plant Physiology*, *161*(1), 20–27. https://doi.org/10.1104/pp.112.205179

Zhang, Z., Zheng, X., Yang, J., Messing, J., & Wu, Y. (2016). Maize endosperm-specific transcription factors O_2 and PBF network the regulation of protein and starch synthesis. *Proceedings of the National Academy of Sciences of the United States of America*, *113*(39), 10842–10847. https://doi.org/10.1073/pnas.1613721113

Zhao, H., Li, Z., Wang, Y., Wang, J., Xiao, M., Liu, H., … Zhang, Z. (2022). Cellulose synthase-like protein OsCSLD4 plays an important role in the response of rice to salt stress by mediating abscisic acid biosynthesis to regulate osmotic stress tolerance. *Plant Biotechnology Journal*, *20*(3), 468–484. https://doi.org/10.1111/pbi.13729

Zhong, R., Cui, D., & Ye, Z. H. (2019). Secondary cell wall biosynthesis. *New Phytologist*, *221*(4), 1703–1723. https://doi.org/10.1111/nph.15537

Zhou, B., Deng, Y. S., Kong, F. Y., Li, B., & Meng, Q. W. (2013). Overexpression of a tomato carotenoid ε-hydroxylase gene alleviates sensitivity to chilling stress in transgenic tobacco. *Plant Physiology and Biochemistry*, *70*, 235–245. https://doi.org/10.1016/j.plaphy.2013.05.035

Zhou, J., Liu, G., Zhao, Y., Zhang, R., Tang, X., Li, L., … Zhang, Y. (2023). An efficient CRISPR-Cas12a promoter editing system for crop improvement. *Nature Plants*, *9*(4), 588–604. https://doi.org/10.1038/s41477-023-01384-2

Zhou, Y., Li, Y., Qi, X., Liu, R., Dong, J., Jing, W., … Chen, M. (2020). Overexpression of V-type H(+) pyrophosphatase gene EdVP1 from *Elymus dahuricus* increases yield and potassium uptake of transgenic wheat under low potassium conditions. *Scientific Reports*, *10*(1), 5020. https://doi.org/10.1038/s41598-020-62052-5

Zhou, Y., Xiong, Q., Yin, C.-C., Ma, B., Chen, S.-Y., & Zhang, J.-S. (2020). Ethylene biosynthesis, signaling, and crosstalk with other hormones in rice. *Small Methods*, *4*(8), 1900278. https://doi.org/10.1002/smtd.201900278

Zhou, Y., Zhao, C., Du, T., Li, A., Qin, Z., Zhang, L., … Hou, F. (2023). Overexpression of 9-cis-epoxycarotenoid dioxygenase gene, IbNCED1, negatively regulates plant height in transgenic sweet potato. *International Journal of Molecular Sciences*, *24*(13). https://doi.org/10.3390/ijms241310421

17 Enzyme-based Biotechnological Approaches for Sustainable Agriculture

B. Lokeshwari, P. Saranraj, Abdel Rahman Mohammad Al-Tawaha, K. Gayathri, and S. Monisha

17.1 INTRODUCTION

Chemical-intensive technologies that degrade soil and pollute the environment are being replaced with enzyme-based biotechnology in sustainable agriculture. Eco-friendly agricultural applications that increase crop yield, soil fertility, and resource-use efficiency while reducing ecological footprints use enzymes as natural biocatalysts to accelerate biochemical reactions under moderate conditions (Alcalde et al. 2006). Modern agriculture relies on cellulases, amylases, proteases, lipases, phytases, ureases, dehydrogenases, and phosphatases. They improve soil health, nitrogen cycling, organic matter decomposition, and plant pathogen control. Enzyme biotechnology reduces synthetic fertilisers and pesticides, improves plant nutrient uptake, and turns organic wastes into biofertilisers and soil conditioners, promoting sustainable development. Enzyme innovations can help the global agricultural community overcome nutrient depletion, population growth, and changing climatic conditions (Carvalho et al. 2007).

Biofertilisers and biopesticides from enzyme-producing microorganisms solubilise nutrients, promote plant development, and control diseases. Nitrogenase and urease use nitrogen, while phosphatases mobilise insoluble phosphorus. Other enzymes turn agricultural leftovers into organic matter, improving soil fertility and biodiversity. Enzyme biotechnology uses agricultural and industrial waste to produce closed-loop systems that save resources and reduce greenhouse gas emissions (Traore et al. 2000). Recent advances in molecular biology, genetic engineering, and metagenomics have identified and optimised enzymes to improve stability and efficacy in different conditions (Paine et al. 2005). Recombinant DNA and protein engineering enable these enzymes to improve agricultural drought resilience and nutrient availability. Phytase and β-glucanase-expressing transgenic plants demonstrate enhanced nutrient absorption and stress tolerance, suggesting possibilities for enzyme-based crop enhancement tactics.

Enzyme-based biosensors are being developed for soil nutrient, pathogen, and pesticide monitoring in precision agriculture and resource optimisation. In addition to regulated nutrient release, nano-enzyme formulations improve seed treatments and enable enzyme-linked testing for quick environmental assessments (Fuentes-Zaragoza et al. 2010). These non-toxic, biodegradable, and energy-efficient solutions support the UN's Sustainable Development Goals (SDGs) for food security, responsible consumption, and climate action. Long-term agricultural productivity depends on the sustainable management of soil health, efficient nutrient cycling, and the adoption of environmentally friendly biotechnological approaches such as enzyme-based agricultural technologies. Enzymes compost organic matter, reducing waste and methane emissions and enhancing nutrient recovery in farming. These enzyme bioformulations for pest and nutrient management systems boost soil enzyme activity, microbial diversity, and plant efficiency. Enhancing germination and stress resistance with enzyme-assisted seed coatings and oxidation enzyme bioremediation for contaminated soils and rivers (Mahdieh and Rabbani 2013).

Immobilisation, formulation, and bioprocessing research are needed to address large-scale production, enzyme stability in active agricultural environments, and cost-effectiveness. Targeted and renewable enzyme-based methods create resilient agroecosystems that combine biological and technological systems to maintain agriculture. Due to advances in enzyme biotechnology, several innovative and sustainable approaches have been developed to enhance agricultural productivity, improve soil health, and reduce the dependence on synthetic chemical inputs. (Hefferon 2015). These enzymes are used to regulate nutrients, reduce environmental impacts, and increase crop and soil health for developing a smart, agricultural paradigm for the 21st century. The rise in global food demand has intensified the need for sustainable agricultural practices and innovative biotechnological solutions to enhance crop productivity while minimizing environmental impacts.

17.2 ENZYMES IN SOIL HEALTH AND FERTILITY MANAGEMENT

Soil enzymes that break down organic materials, cycle nutrients, and maintain soil structure are mostly microbial, but plant roots and soil fauna also release them. Their activities are affected by soil health and react rapidly to management changes, pollution, and organic amendments (Ezhilan et al. 2017). Enzymatic biochemical processes boost nutrient efficiency and resilience in agroecosystems.

17.2.1 Soil Enzyme Activities and Functions

The soil's biological community and plant-essential biogeochemical cycles don't involve enzymes. Carbon, nitrogen, phosphorus, and sulphur are cycled by two enzyme groups. Enzymes β-glucosidase, cellulase, and xylanase simplify

DOI: 10.1201/9781003669920-17

complex plant polysaccharides for microbial and plant metabolism. This process produces SOM and humus, which affect soil structure, water retention, and carbon sequestration (Jain et al. 2021). In soil, C-cycling enzymes measure organic carbon dynamics. In nitrogen cycling, enzymes convert organic N to plant-available inorganic N. Urease converts urea to ammonia and CO_2, while enzymes make peptides and amino acids. Chitin, found in fungal cell walls and insect exoskeletons, provides nitrogen in many soils. N-acetyl-β-D-glucosaminidase and chitinase help to mineralise nitrogen. This insoluble soil phosphorus requires phosphatase (acid and alkaline phosphomonoesterases). It makes plant-absorbed inorganic phosphate from inositol phosphates. In P-fixing soils, phosphatases reduce chemical P fertiliser use by using organic P (Pooja et al. 2023). The abundance and diversity of these genes in soil microbiomes suggest their importance to ecosystem function. Controlling these enzymes with cropping, organic amendments, and reduced tillage improves soil health and nutrient use. Biochemical functions of agricultural enzymes like proteases and cellulases are listed in Table 17.1.

17.2.2 Enzyme-Mediated Nutrient Management

Eutrophication, soil acidification, and greenhouse gas emissions result from synthetic fertilisers. The enzyme-based solutions adjust nutrient availability to plant needs using biological catalysis. Examples include urea fertilisers with urease inhibitors. N-(n-butyl) thiophosphoric triamide (NBPT) inhibits urease, slowing urea breakdown and reducing ammonia volatilisation, improving N-use efficiency and preventing N loss. Farmyard manure and compost stimulate the native microbial community to produce extracellular phosphatases, which mineralise both added and native soil organic P (Olivaeira et al. 2020). Biotechnology can create "enzyme-activated" fertilisers with enzymes or enzyme-producing microorganisms for controlled nutrient release. This can be solubilised by phytase, which hydrolyses the main organic matter. An enzymatically rich soil ecosystem allows farmers to feed soil instead of plants to increase fertility and reduce agriculture's environmental impact.

17.2.3 Microbial Enzymes for Biofertilisers

Biofertilisers and microorganism enzymes are interconnected. Living microorganisms in biofertilisers increase primary nutrient availability to seeds, plant surfaces, and soil, promoting growth. Enzymes from microbes make these products work. Biofertilisers like *Rhizobium* spp. for legumes and *Azotobacter* or *Azospirillum* for non-legumes convert atmospheric N_2 to ammonia using nitrogenase (Dutt and Meghwanshi 2024). Symbiotic and asymptotic fixation is enzyme-driven but not explicit. Extracellular phosphatases and phytases from *Pseudomonas*, *Bacillus*, *Penicillium*, and *Aspergillus* solubilise inorganic and mineralise organic phosphates and acidify the rhizosphere (Fernandes et al. 2024). Biofertilisers with phosphate-solubilizing microorganismss increase wheat, maize, and rice P uptake, reducing chemical P fertiliser use by 25–50% (Fasim et al. 2021). Under abiotic stresses or protectants, polymer-based coatings prolong these microbes' shelf-life and efficacy in field conditions.

17.3 ENZYME-BASED PLANT GROWTH PROMOTION

Enzymes play a critical role in regulating plant physiology and stress resilience to improve growth and yield. Beyond nutrition, these enzymes directly influence the

TABLE 17.1
Agricultural Enzymes and Their Functions

Classes of Enzymes	Specific Enzymes	Primary Functions in Agriculture	Applications	References
Phosphatases	Phytase	• Releases bound phosphorus (phytate) in animal feed, making it digestible. • Reduces phosphorus pollution in manure.	• Animal feed additive. • Soil amendment for organic phosphorus mobilisation.	Mulualem (2015)
Proteases	Subtilisin	• Breaks down proteins into amino acids. • Improves protein digestibility in feed.	• Animal feed additive. • Degradation of organic matter in soil/silage.	Sharma and Patel (2017)
Carbohydrases	Cellulase, hemicellulase, xylanase, amylase	• Breaks down complex carbohydrates into simple sugars. • Improves feed efficiency and nutrient availability.	• Animal feed additive. • Composting accelerator. • Soil conditioner to degrade crop residues.	Patel et al. (2017)
Nitrogen-metabolising Enzymes	Urease	• Catalyses the hydrolysis of urea to ammonia and CO_2. • Can lead to nitrogen volatilisation losses if not managed.	• Often inhibited from reducing nitrogen loss from urea fertilisers.	Fuentes-Zaragoza et al. (2010)
Oxidoreductases	Laccase, peroxidase	• Degrades lignin in plant cell walls. • Detoxifies soil pollutants and suppresses pathogens.	• Bioremediation of soils. • Lignin degradation for improved compost/feed.	Doan et al. (2021)
Chitinases	Chitinase	• Breaks down chitin, a key component of fungal cell walls and insect exoskeletons.	• Biocontrol agent against fungal pathogens and pests.	Nadeem et al. (2021)

plant's hormonal balance and developmental pathways. These enzymatic reactions enable biotechnologists to come up with novel approaches to programming the plants to perform better, e.g. strong root systems to generate profiles that are more balanced to grow well, and also to tolerate abiotic stresses (Mulualem 2015). These enzymatic interventions are a highly advanced approach to contemporary agriculture, which would minimise the use of growth regulators that are synthetic in nature and their communities.

17.3.1 Hormone-Modulating Enzymes

Hormones in plants are responsible for controlling growth, development, and defence. Complex biosynthetic and degradative enzymes regulate signalling molecules inside the cells. Auxin, cytokinin, gibberellin, and ethylene levels are regulated by plant growth-promoting rhizobacteria (PGPR) and fungi, affecting hormonal balance (Singhania et al. 2010). ACC deaminase is the most prevalent bacterial enzyme. Plants produce more ethylene in drought, salinity, and pathogen stress, slowing root growth and causing senescence. Enzymes from useful microorganisms in the rhizosphere convert ACC to A-ketobutyrate and ammonia, the plant's precursor to ethylene. Enzymatic hydrolysis reduces the plant's internal ethylene levels and growth-inhibiting effects by depleting ACC for ethylene synthesis, promoting root elongation and stress tolerance. Indole-3-acetic acid (IAA)-producing microbial enzymes help plants divide and elongate. IAA is made by bacteria using tryptophan monooxygenase and indole acetamide hydrolase (Pooja et al. 2023). Increasing lateral root and root hair density with rhizosphere IAA increases nutrient and water absorption surface area. Microbial strains with enhanced hormone-modulating enzyme expression or direct enzyme formulations can induce these beneficial physiological responses in seeds or soils. Figure 17.1 illustrates that the ACC deaminase decomposes ethylene precursor ACC to promote root growth and stress resistance for sustainable plant growth.

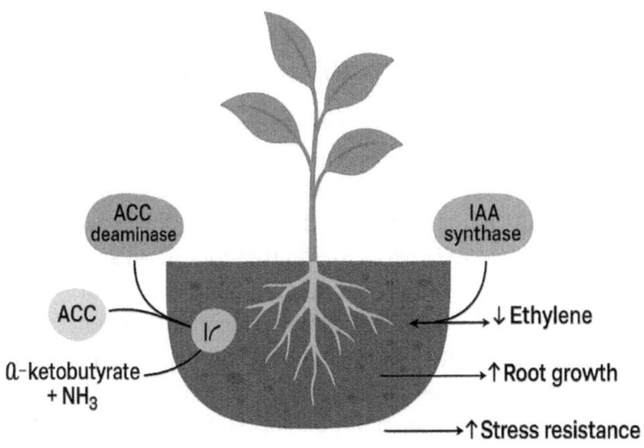

FIGURE 17.1 ACC deaminase enzyme pathway for sustainable plant growth promotion.

17.3.2 Enzymatic Stimulation of Root Development

The roots of plants are responsible for absorbing water and nutrients. Enzymes stimulate root growth by affecting cell walls and the environment. Cellulases, pectinases, and xylanases from rhizosphere microbes soften root tip cell walls, allowing expansion and elongation. Enzymatic degradation, a natural growth process, can be boosted by beneficial microbes. Plants' endogenous enzymes can be influenced by microbial signals to disrupt hydrogen bonds between cellulose microfibrils, allowing them to bind and extend cell walls without rupture (Qamar et al. 2020). Enzymes produce root hairs, single-cell extensions that greatly increase root absorption. Complex enzyme interactions that change cell wall composition control root hair initiation and elongation. Seeds with enzyme-producing microbes are distributed to produce crops with deeper root systems. Water and nutrients from deeper soil layers help crops resist climate change-induced drought and nutrient deficiencies (Maleki et al. 2017).

17.3.3 Biotechnological Formulations for Crop Growth

Enzyme-based plant growth promotion requires complex biotechnological formulations to become agricultural inputs. In addition to microbial suspensions, these advanced products ensure enzymatic activity viability, stability, and efficacy. This involves fermenting specific PGPR strains under optimised conditions to produce high yields of target enzymes for the ACC deaminase, phosphatases and encapsulating cells, spores, and extracellular enzymes in talc, lignite, or advanced polymers (Li et al. 2014). These carriers protect bioactive compounds from UV and desiccation. Biodegradable microcapsules or nanofibers containing enzymes or enzyme-producing microbes can be used as carriers to protect bioactive compounds from UV radiation and desiccation while enabling slow and controlled release for plant development. This technology slows enzyme release for plant development. When the plant is stressed, nanoparticles with ACC deaminase and a root-exudate-responsive polymer could release the enzyme in the rhizosphere. Microbes that produce enzymes are used in seed coatings. ACC deaminase-producing bacteria and P-solubilising fungi help the emerging radicle develop roots and access phosphorus (Singhania et al. 2010). Enzymatic synergy between an N-fixer, P-solubiliser, and ACC deaminase producer promotes growth. These next-generation formulations must be rigorously field tested under diverse agro-climatic conditions to give farmers consistent and reliable performance.

17.4 ENZYME-BASED PEST AND DISEASE MANAGEMENT

Sustainable agriculture necessitates a reduction in the use of broad-spectrum chemical pesticides. Enzyme-based strategies offer highly specific, environmentally benign, and effective alternatives for managing agricultural pests and diseases. These approaches can be direct, using enzymes to attack the structural components of pathogens or insects, or indirect, by priming the plant's own innate immune systems (Renge et al. 2012). The deployment of enzymes as biopesticides aligns perfectly with the principles of integrated pest management (IPM), providing powerful tools that are compatible with biological control and cultural practices, thereby minimising resistance development and preserving beneficial organisms.

17.4.1 Pathogen-Degrading Enzymes

Cell walls are essential for the structural integrity and virulence of most plant pathogens, especially fungi and bacteria. Plant wall polymers can be attacked by hydrolytic enzymes, called lytic enzymes (Haroon and Ghazanfar 2016). Notable enzymes in this category include chitinase and β-glucanase. Chitin, a polymer of N-acetylglucosamine, is found in fungal cell walls and insect exoskeletons. Chitinases lyse chitin, weakening insect pests by destroying fungal hyphae. Another structural polysaccharide in fungal cell walls, glucan, is also degraded by β-1,3-glucanase. The synergistic effect of chitinases and glucanases is effective against many phytopathogenic fungi, including *Botrytis cinerea* (grey mould), *Fusarium oxysporum* (wilt), and *Rhizoctonia solani* (damping-off). The glycosidic bonds of peptidoglycan in bacterial cell walls can be broken by enzymes like lysozyme (muramidase), causing cell lysis. As biopesticides, microbial strains that produce these lytic enzymes can colonise the rhizosphere or phyllosphere and constantly produce them to protect against pathogens (Sharma et al. 2017). Recent metagenomic exploration of harsh conditions has found novel, hyper-active, thermostable chitinases and glucanases with a wider pH range, making them more applicable in a variety of field conditions.

17.4.2 Induction of Plant Defence Enzymes

Plants also have an intrinsic immune system capable of being triggered when certain microbial signals are detected, and this effect is referred to as induced systemic resistance (ISR) of the plant. The development of ISR is characterised by a fast and strong activation of the activity of major defence-related enzymes. The initial and rate-limiting step in the phenylpropanoid pathway is phenylalanine ammonia-lyase (PAL), which results in the synthesis of an enormous repertoire of phenolic compounds, such as lignin (strengthening cell walls), phytoalexins (antimicrobial compounds), and salicylic acid (a defence signalling molecule). Other enzymes, such as peroxidase (PO) and polyphenol oxidase (PPO), are involved after the activation of PAL. The final stages of lignin synthesis and production of reactive oxygen species (ROS) capable of attacking pathogens directly are catalysed by POs, phenolics oxidised to quinones that are toxic to many invaders by the action of PPOs. Microbes that are beneficial when used as seed treatment or soil drenches can be used as elicitors of this enzymatic defence arsenal (Mauri et al. 2017). This approach is highly sustainable as it leverages the plant's innate metabolic defences and does not place a direct selection pressure on the population of the pathogen, which helps eliminate the chances of developing resistance.

17.4.3 Enzyme-Based Biopesticides

The development and commercialisation of enzyme-based agricultural technologies require coordinated efforts in research, regulatory approval, industrial-scale production, and market adoption to ensure their effectiveness, safety, and long-term sustainability.. These are defined products whose active ingredient is a specific enzyme or a cocktail of enzymes with pesticidal properties. Unlike microbial biopesticides, where the microbe is the active ingredient and the enzyme is produced *in situ*, these formulations contain the purified or semi-purified enzymes themselves. This offers advantages such as immediate activity, consistent quality, and easier registration and quality control. In insect control, proteases and chitinases are being developed as potent bio-insecticides (Raveendran et al. 2018). Proteases can degrade the proteins in the insect gut lining, leading to mortality, while chitinases, as mentioned, degrade the exoskeleton. Formulations containing a combination of these enzymes have shown efficacy against pests like aphids, whiteflies, and lepidopteran larvae. Significant challenges remain, including the high production costs and poor stability of these enzymes under field conditions. However, advances in recombinant DNA technology and fermentation science are steadily lowering production costs (El-Gendi et al. 2021). Furthermore, formulation science is addressing stability issues through the use of UV protectants, humectants, and adjuvants that improve adhesion to plant surfaces. The integration of these enzyme-based technologies into modern agricultural systems offers a sustainable approach to improving crop productivity, enhancing soil health, and reducing the environmental impact associated with conventional chemical inputs.

17.5 ENZYME-ASSISTED BIOREMEDIATION IN AGRICULTURE

The intensification of agriculture has contaminated agroecosystems with persistent pesticides, heavy metals, and organic pollutants. This contamination endangers soil, water, and human health. Traditional physicochemical remediation is expensive, disruptive to soil structure, and can generate secondary pollutants. Enzyme-assisted bioremediation is a powerful, precise, and eco-friendly alternative. This method uses enzymes extracted or produced in situ by microbes and plants to detoxify, transform, or immobilise environmental contaminants in the soil matrix (Rajasree and Pugalendhi 2021). Since enzymes target pollutants without harming native soil biota, they are a sustainable soil management strategy.

17.5.1 Detoxification of Pesticides and Pollutants

One of the most advanced agricultural applications of environmental biotechnology is enzymatic pesticide and organic pollutant detoxification. Oxidoreductases (laccases, POs, cytochrome) and hydrolases (phosphotriesterases, carboxylesterases, organophosphorus acid anhydrolases) are enzymes that can break down toxic compounds into harmless metabolites. Microbial phosphotriesterases can hydrolyse acetylcholinesterase inhibitor organophosphorus pesticides (OPs) by cleaving their phosphate ester bonds, making them non-toxic (Kumar et al. 2020). A novel bacterial laccase immobilised on magnetic nanoparticles degrades multiple pesticide classes, including atrazine and chlorpyrifos, with over 90% degradation efficiency under optimised conditions and remarkable reusability and stability (Doan et al. 2021). Engineered rhizosphere bacteria expressing cytochrome P450 monooxygenases have been shown to improve phytoremediation in contaminated soils by degrading herbicides like isoproturon and pyrethroids. In addition to pesticides, enzymatic remediation is being used to address pharmaceutical residues in irrigation wastewater and microplastic accumulation in soils. Hydrolases like cutinases and lipases can degrade polyester-based microplastics, while lignin-modifying enzymes can break down endocrine disruptors (Edoamodu and Nwodo 2024). Strategic enzyme use allows in-situ remediation of agricultural lands, preventing toxins from leaching into groundwater and entering the food chain, closing the pesticide pollution loop and protecting ecosystem integrity (Yang et al. 2017).

17.5.2 Soil Health Restoration Through Enzymes

Agricultural productivity and sustainability depend on soil health for its physical structure, chemical fertility, and, most importantly, biological activity. Enzymes are soil health indicators and restoration agents (Sharma et al. 2013). Degraded and contaminated soils have repressed enzymatic profiles, damaging the soil microbiome. Certain enzymes stimulate biogeochemical processes and restore soil fertility. Cellulases and hemicellulases accelerate crop residue decay to increase soil organic matter and stable humus, which improves water retention and soil aggregation (de Sá and Keshavarz-Moore 2021). When handled properly, urease enzymes can control the hydrolysis of urea fertilisers to reduce ammonia volatilisation and improve nitrogen use, and dehydrogenase activity is a direct indicator of microbial metabolic activity and soil quality. One major development is the use of an enzymatic complex to degrade allelopathic substances and autotoxins, which accumulate in monocropping systems and cause soil sickness and lower yields. Phosphatase enzymes mobilise organic phosphorus, a major soil P source not available to plants. Phytase and organic amendments can liberate phytate-phosphorus and prevent phosphorus deficiency in alkaline and acidic soils (Rajasree and Pugalendhi 2021). They directly act as an enzymatic priming agent in degraded soils, improving nutrient cycling, detoxifying inhibitory compounds, and restoring a healthy soil food web. This proactive enzyme management decontaminates and restores the ecological process that creates healthy, resilient, and productive soil.

17.6 INDUSTRIAL AND COMMERCIAL APPLICATIONS

A major green revolution milestone was the commercialisation of enzyme-based agricultural inputs. The global agricultural enzyme in industrial and commercial environment is growing rapidly to provide sustainable agricultural solutions (El-Gendi et al. 2021). This is to produce stable, cost-effective, easy-to-use enzyme products that can withstand the unpredictable and hostile agricultural environment. Commercialisation involves advanced fermentation technology, large-scale enzyme production, downstream processing, and product formulation that is easily transportable, stored, and used in the field. Bioformulations are used in agriculture to boost crop yields and reduce environmental impact (Sharma and Patel 2017).

17.6.1 Enzyme-Based Bioformulations

Enzyme-based bioformulations are specialised products that entail the use of one or more enzymes incorporated frequently together with useful microorganisms or organic carriers for their desired agronomic effect. These formulations preserve the enzymatic activity at the target site, either in soil or on a plant.

17.6.1.1 Commercial Enzyme Products for Soil and Crop Management

Industrial enzymes are designed for various soil and crop management applications. These include soil conditioners, biofertilisers, biostimulants, and biopesticides. Soil conditioners contain cellulase, xylanase, pectinase, and ligninase complexes to accelerate composting and organic matter breakdown to turn agricultural waste into humus quickly (Yang et al. 2017). Some companies sell enzymatic compost accelerators that can speed up composting by a week. Phosphatase- and phytase-fortified biofertilisers dissolve soil phosphorus, and nitrogenase-rich enzyme complexes fix nitrogen by rhizobia. Biostimulants with superoxide dismutase and catalase enzymes are also popular (Qaim 2020). These enzymes scavenge ROS to help plants survive abiotic stresses like drought, salinity, and heat, not nutrient cycling. In a 2024 field experiment on wheat under water shortages, foliar application of a commercial formulation of antioxidant enzymes and amino acids increased biomass, relative water content, and grain yield. Chitinase and glucanase formulations are also being sold as biopesticides because they degrade fungal pathogen cell walls, a more natural disease control method.

17.6.1.2 Encapsulation and Immobilisation Techniques for Field Stability

Enzymatic products' catalytic properties in the presence of environmental stressors like ultraviolet radiation, extreme pH, temperature variations, and microbial proteolytic activities in the critical period between application and action are the biggest challenges to their commercialisation in agriculture. The most popular lab-scale biocatalyst methods are encapsulation and immobilisation (Khan and Khan 2010). This encapsulation uses alginate, chitosan, or synthetic polymers to protect enzymes in micro- or nanoscale capsules. A selective physical barrier, this matrix protects the enzyme and allows substrates and products to diffuse. Immobilisation bonds enzymes to insoluble support chemicals like silica nanoparticles, clay minerals, or biochar. This conjugation stabilises the enzyme's tertiary conformation, improving thermal stability, pH resistance, and catalyst recovery and reuse in controlled environments (Kole 2012).

17.7 EMERGING TECHNOLOGIES

There is a fundamental shift in the frontiers of enzyme biotechnology in agriculture with the convergent applications in genetic manipulation and nanotechnology (Nigam 2013). This development will lead to the creation of an era characterised by a high level of resource efficiency, eco-compatibility, and sustainability in the agricultural management systems. Figure 17.2 illustrates that the integration of key technological domains, including genetic engineering, recombinant enzymes, and Clustered Regularly Interspaced Short Palindromic Repeats (CRISPR), is an innovation in crop protection, soil health, and precision farming.

17.7.1 Genetic Engineering of Enzymes for Agriculture

Enzyme catalytic activity has traditionally been recognised as specific or catalytically efficient in large-scale field applications (Brinch-Pedersen et al. 2007). Genetic engineering shifted towards the naturally occurring enzymes to the active design and optimisation of biocatalysts to specific agricultural factors for the direct enhancement of biological system enzyme activity.

17.7.1.1 Recombinant Enzymes with Enhanced Stability and Activity

The large-scale manufacturing of enzyme products using recombinant DNA technology is a major step in enzyme optimisation. Modern protein engineering tools like directed evolution diversify genes and select the best phenotypes. These feed-grade phytases are more thermostable, so they don't lose activity during pelletisation (Hefferon 2015). This stability ensures good hydrolysis of phytic acid in the animal's digestive system, which increases phosphorus bioavailability and reduces manure phosphate levels, reducing eutrophication. Lignin-degrading enzymes like laccases and POs are also being modified to work better in different pH and temperature conditions. These enzymes accelerate the decomposition of crop residues and complex organic matter in the soil, which provides nutrients (Falk et al. 2002). Higher stability and catalytic efficiency improve agronomic benefits like lower application rates, in-situ usage, and economic viability.

17.7.1.2 CRISPR and Synthetic Biology Approaches for Enzyme Optimisation

Complex genetic engineering technologies have greatly improved biological systems reprogramming, and CRISPR-Cas systems and synthetic biology are leading the charge in this field. CRISPR-Cas technology can knock out genes or edit the promoters and coding sequences that regulate plant enzymes, unlike other recombinant technologies that require gene isolation (Mansour et al. 2012). This accuracy has led to crop varieties that produce digestive or protective enzymes. Improving nitrogen-absorbing enzymes in cereal crops to make them more efficient at using nitrogen fertilisers to reduce nitrogen runoff and its environmental impacts is an example. Synthetic biology allows multi-level and de novo genetic circuit design. Engineering soil bacteria to produce desirable phosphatase enzymes only in certain environmental conditions, such as low soil pH or root exudate signal,

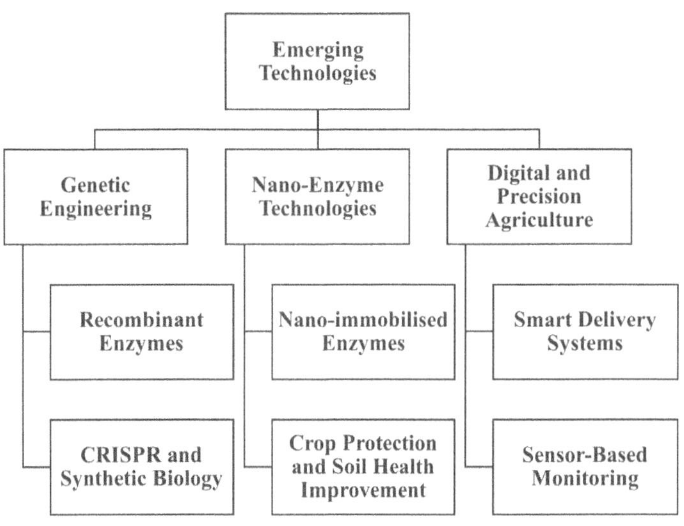

FIGURE 17.2 Advanced biotechnology tools in modern agriculture.

is an interesting innovation. Instead of passive enzyme use, development is essential to active, responsive biological systems. Recent studies have shown that engineered rhizobia colonising legume roots synthesise chitinases, which quickly control soil-borne fungal pathogens (Joshi et al. 2011). Genetic specificity and systems biology can redefine farming by developing an internal source of enzyme production in the agricultural microbiome, which would lead to sustainability and efficiency in the agricultural process.

17.7.2 Nano-Enzyme Technologies

The convergence of enzyme biotechnology and nanotechnology has given rise to nano-enzyme technologies, which aim to address persistent challenges such as limited shelf-life across different environments and insufficient delivery systems (Renge et al. 2012). The strategic combination of biocatalysts with engineered nanomaterials represents a promising agro-biotechnological approach that improves functional stability, enhances catalytic efficiency, and expands the practical applications of enzyme-based technologies in sustainable agriculture..

17.7.2.1 Nano-immobilised Enzymes for Slow-Release Applications

Enzyme encapsulation on solid surfaces improves enzyme stability and reusability, especially in harsh conditions (Mamiro et al. 2011). Challenges include nanoporous silica nanoparticles, chitosan-based carriers, magnetic nanoparticles, and carbon nanotubes. Big surface area to volume ratios, surface chemistry, and unique physicochemical properties make such materials suitable for nano-immobilisation. This immobilisation method protects biocatalysts from protein degradation, pH extremes, and temperature changes, extending their lifespan. Biodegradable polymeric nanocarriers can limit enzymes and release them gradually, potentially triggered by environmental factors like soil moisture, microbial activity, and root exudates. This gradual release process prolongs catalytic effects at important phenological processes like germination (Haroon and Ghazanfar 2016). This method matches crop demand with nitrogen availability.

17.7.2.2 Crop Protection and Soil Health Improvement

Nano-immobilisation of enzyme technology revolutionises nutrient control, especially crop protection and soil health. The technology breaks fungal cell walls with chitinases and glucanases. After nano-encapsulation, these enzymes become biodegradable fungicides that enhance hyaline fungi adherence. Nano-immobilised oxidoreductases like POs and laccases can be used in soil remediation to decontaminate complex pesticides and herbicides. Strategic use of enzyme-loaded nanoparticles improves soil microbial communities (Ma et al. 2019). These metabolic priming nano-bio-conjugates break down complex organic substances like cellulose and lignin into bioavailable forms. The process increases positive microorganisms, biomass, and functional diversity, which boosts soil fertility (Jin et al. 2021; Nadeem et al. 2021). Nano-enzyme technology can be used to develop novel integrated soil health management approaches that combine pathogen management, pollutant degradation, and microbiome engineering.

17.7.3 Integration with Digital and Precision Agriculture

Novel enzyme technologies hold radical potential for integration with precision agriculture. This development enables enzymes to move beyond the biochemical feeds and become core elements of the advanced agronomic system, hence enabling more ecologically suitable measures in the soil ecosystem (Gagaoua et al. 2014).

17.7.3.1 Smart Delivery Systems for Enzymes

Smart delivery in agricultural enzymology focuses on strategic use of an enzyme, dose, and time of day to increase agricultural efficiency (Ranjha et al. 2021). This approach is possible thanks to formulation science, Internet of Things, and predictive data analytics innovations. Advanced designs, like hydrogel-based capsules, release enzymes when in-field sensor networks detect environmental stimuli. A phosphorus-solubilising enzyme capsule may release material when the soil matric potential is optimal for plant growth. Precision delivery technologies are being modified for enzymatic delivery. Variable-rate technology (VRT) drones, originally used for fertilisers and pesticides, can dispense enzyme formulations using multispectral satellite imagery or proximal sensing (Pooja et al. 2023). They are cellulose-degrading enzymes with high organic matter or residue content, making them more efficient at decomposition and resource use.

17.7.3.2 Sensor-Based Monitoring of Enzyme Activity and Soil Health

The viability of agricultural interventions based on the use of enzymes depends on the capacity to observe their biochemical impacts in real time and in situ. Their electrochemical and optical biosensors can measure particular enzymes and their metabolites on-site on soil, thus they can be analysed by the soil sensors, making it possible to have an essential indicator of soil health, including pH, electrical conductivity, macro-nutrients (nitrogen, phosphorus, potassium), and rates of microbial respiration (Medyantseva et al. 2015). They have the high resolution that can be processed into a cloud-based analytics system, enabling people to model the relationship between the applications of enzymes and the health of the soil.

17.8 CHALLENGES AND LIMITATIONS

Enzyme-based biotechnologies have great potential for sustainable agriculture, but complex scientific, economic, environmental, and regulatory challenges prevent their commercialisation. Enzymes must be stable and effective

TABLE 17.2
Challenges and Impacts of Agricultural Enzyme Applications

Categories	Limitations	Impacts	References
Production and formulation	High production costs	Limits accessibility for smallholder farmers.	Mauri et al. (2017)
	Formulation stability and shelf-life	Enzymes can degrade, making products ineffective before use.	Raveendran et al. (2018)
Efficacy and application	Specific environmental conditions	Performance is variable and unpredictable across different fields.	Ma et al. (2019)
	Short duration of action	Requires applications, increasing cost and labour.	Singhania et al. (2010)
	Interaction with soil matrix	Enzymes can bind to soil particles, reducing their activity.	Jain et al. (2021)
Economic and regulatory	Cost-effectiveness	More expensive per unit area than conventional agrochemicals.	Doan et al. (2021)
	Lack of standardised regulations	Creates uncertainty and delays product approval.	El-Gendi et al. (2021)
	Farmer awareness and adoption	Low adoption due to scepticism and lack of results.	Ferreira et al. (2013)
Environmental and ecological	Non-target effects	May disrupt native soil microbial communities and cycles.	Fasim et al. (2021)
	Potential for allergenicity	Enzymes can act as allergens during manufacturing or application.	Jin et al. (2021)

in open fields, where pH fluctuations, extreme temperatures, UV radiation, and soil compounds can denaturate protein and reduce catalytic activity. The difficulty of precise enzyme delivery risks premature degradation before reaching plant or soil sites (Semenova et al. 2006). Large-scale enzyme production is expensive due to fermentation, purification, and formulation processes needed to produce high-purity, stable enzyme products. Due to cost, enzymes are usually used on high-value crops, limiting their impact on staple foods needed by the global population. Exogenous enzymes may disrupt native soil microbial communities' metabolic balance, causing undesirable plant species or pathogen resistance, requiring careful ecological assessment. Global biotechnology regulations are inconsistent, making it difficult to assess the unique risks of genetically engineered enzymes. Small and medium-sized enterprises face lengthy and expensive approval processes. Despite technological advances, public apprehension about genetically modified organisms (GMOs) can hinder market acceptance (Ezhilan et al. 2017; Maleki et al. 2017). Enzyme performance varies from traditional synthetic chemicals, making universal application protocols difficult to develop and the adoption of these technologies. Enzyme-driven agriculture has great potential, but molecular biologists, ecologists, economists, policymakers, and farmers must work together to overcome the many challenges to create practical, scalable, and equitable agricultural solutions (Francisco et al. 2010). Table 17.2 shows the key challenges in agricultural enzyme use, such as stability issues and high costs.

17.9 FUTURE PROSPECTS

The use of enzyme-based biotechnological techniques in sustainable agriculture will substitute input-intensive farming with ecologically balanced, precision-driven farming. Enzymes improve crop yields, soil health, and waste management, ensuring global food security and environmental protection. Biopesticides and fertilisers use hydrolytic enzymes. This economy envisages "smart" seed coatings with micro-encapsulated enzymes to improve germination and reduce allelopathic effects in the rhizosphere, promoting chemical-free seedlings (Singhania et al. 2010). Phosphatases dissolve phosphorus, nitrilases supply nitrogen, and chitinases regulate fungal pathogens and mycorrhizal relationships. Self-sustaining soil ecosystems with crop-specific nutrients should result from the innovations. Enzymatic technologies could affect agricultural drought and salinity. Enzymes detoxify soil ions and salts, and engineered carbonic anhydrases improve water efficiency (Khan et al. 2017).

Precision, environmentally safe enzymatic biopesticides like engineered chitinases, proteases, and lipases that target insect pests and fungal pathogens while minimising resistance and residues that harm pollinators and soil biodiversity, may reduce chemical pesticide use. Food-grade enzymes break down ethylene or strengthen plant cell walls against microbes to reduce harvest spoilage. In farm biorefineries, cellulases and hemicellulases convert agricultural waste into fermentable sugars for biofuels and bioplastics, closing nutrient and carbon loops (Patel et al. 2017). This vision needs synthetic biology and computational protein design. Artificial Intelligence-driven algorithms should develop stable, specific enzymes with unique catalytic functions that selectively cleave herbicides or produce growth-promoting compounds. The improvements in fermentation technologies using genetically altered but safe organisms like *Trichoderma reesei* and *Bacillus subtilis*, enzymatic solution production costs should compete with bulk chemicals (Lee et al. 2006). Based on sensor data on soil moisture, nutrient concentrations, and pests, drones can apply enzyme formulations to reduce waste and maximise agricultural efficacy. The development of effective regulatory frameworks to assess the environmental impacts of newly engineered enzymes and their microbial producers, address public concerns about GMOs, ensure product durability and effectiveness across different climates, and maintain enzyme stability during storage and application remains a key challenge. (Mulualem 2015). Enzyme-based biotechnologies promise gradual improvements and agricultural transformation. Intelligent, selective, and renewable enzymes will reduce petroleum use, improve crop resilience to climate change, restore degraded soils, and create cyclical bioeconomies from linear waste streams (Montagnac et al. 2009). This progress allows a sustainable and productive agricultural system to feed a growing global population without harming the environment (Paine et al. 2005).

17.10 CONCLUSION

In the enzyme-based biotechnological strategies, enzymes play a crucial role in promoting sustainable agriculture. They enhance soil health by improving nutrient cycling and organic matter decomposition, thereby supporting plant growth through hormonal modulation and nutrient solubilisation. This leads to efficient resource utilisation, particularly through biofertilisers that reduce the environmental impacts of synthetic fertilisers, addressing issues like eutrophication and soil acidification. In pest and disease management, enzymes offer targeted and eco-friendly alternatives to chemical pesticides by using pathogen-degrading lytic enzymes, such as chitinases and glucanases, and enhancing plant resistance according to IPM principles. This approach not only lowers the use of chemicals but also mitigates pesticide resistance risks. Enzyme-assisted bioremediation further contributes to ecological restoration by detoxifying pollutants like persistent pesticides and heavy metals, thereby protecting water and food quality. Advancements in commercial enzyme technologies have improved enzyme stability and efficacy through innovative formulation techniques such as encapsulation and nano-immobilisation, leading to reliable bioformulations. The integration of enzyme biotechnology with cutting-edge fields, including genetic engineering, CRISPR, and nanotechnology, paves the way for precision biocatalysis to produce more stable and active recombinant enzymes. Intelligent delivery systems and nano-enzyme conjugates are also being developed for targeted interventions. Moreover, enzymatic solutions with digital agricultural technologies, alongside biosensors and data analytics, enhance resource efficiency by enabling real-time monitoring and precise interventions. Despite challenges such as production costs and regulatory frameworks, enzyme-based biotechnologies are poised for significant evolution, contributing to a circular bioeconomy that transforms agricultural waste into valuable products such as biofuels. Agriculture can shift toward a more productive, intelligent, and ecologically sustainable framework that secures food production and ecological integrity for future generations. The integration of biological creativity with technological advancements is vital for the sustainable agricultural future.

REFERENCES

Alcalde, M., Ferrer, M., Plou, F.J. and Ballesteros, A. 2006. Environmental biocatalysis: From remediation with enzymes to novel green processes. *Trends Biotechnol.* 24: 281–287.

Brinch-Pedersen, H., Borg, S., Tauris, B. and Holm, P.B. 2007. Molecular genetic approaches to increasing mineral availability and vitamin content of cereals. *J. Cereal Sci.* 46: 308–326.

Carvalho, J.C., Oishi, B.O., Woiciechowski, A.L., Pandey, A., Babitha, S. and Soccol, C.R. 2007. Effect of substrates on the production of *Monascus* biopigments by solid-state fermentation and pigment extraction using different solvents. *Indian J. Biotechnol.* 6: 194–199.

de Sá, M.S. and Keshavarz-Moore, E. 2021. Pichia pastoris (*Komagataella phaffii*) as a cost-effective tool for vaccine production for low- and middle-income countries (LMICs). *Bioengineering* 8: 119.

Doan, C.T., Chen, C.L., Nguyen, V.B., Tran, T.N., Nguyen, A.D. and Wang, S.L. 2021. Conversion of pectin-containing by-products to pectinases by *Bacillus amyloliquefaciens* and its applications on hydrolysing banana peels for prebiotics production. *Polymers* 13: 1483.

Dutt, K. and Meghwanshi, G.K. 2024. Advances in fungal enzymes and their applications. In *Applied mycology for agriculture and foods.* Apple Academic Press. pp. 383–419.

Edoamodu, C.E. and Nwodo, U.U. 2024. Optimisation and physicochemical characterisation of a thermo-alkali stable laccase produced by wastewater-associated *Bacillus* sp. NU2. *Environ. Technol.* 45: 4441–4456.

El-Gendi, H., Saleh, A.K., Badierah, R., Redwan, E.M., El-Maradny, Y.A. and El-Fakharany, E.M. 2021. A comprehensive insight into fungal enzymes: Structure, classification, and their role in mankind's challenges. *J. Fungi.* 8: 23.

Ezhilan, M., Gumpu, M.B., Ramachandra, B.L., Nesakumar, N., Babu, K.J., Krishnan, U.M. and Rayappan, J.B.B. 2017. Design and development of an electrochemical biosensor for the simultaneous detection of melamine and urea in adulterated milk samples. *Sens. Actuators B Chem.* 238: 1283–1292.

Falk, M.C., Chassy, B.M., Harlander, S.K., Hoban, T.J., McGloughlin, M.N. and Akhlaghi, A.R. 2002. Food biotechnology: Benefits and concerns. *J. Nutr.* 132: 1384–1390.

Fasim, A., More, V.S. and More, S.S. 2021. Large-scale production of enzymes for biotechnology uses. *Curr. Opin. Biotechnol.* 69: 68–76.

Fernandes, L.P., Ventorim, R.Z., de Oliveira, M.G., Almeida, L.F., Guimarães, V.M., Maitan-Alfenas, G.P. et al. 2024. Xylooligosaccharides from pretreated rice bran produced by immobilised xylanase. *Bioenergy Res.* 17: 1–10.

Ferreira, L.R., Macedo, J.A., Ribeiro, M.L. and Macedo, G.A. 2013. Improving the chemopreventive potential of orange juice by enzymatic biotransformation. *Food Res. Int.* 51: 526–535.

Francisco, É.C., Neves, D.B., Jacob-Lopes, E. and Franco, T.T. 2010. Microalgae as feedstock for biodiesel production: Carbon dioxide sequestration, lipid production and biofuel quality. *J. Chem. Technol. Biotechnol.* 85: 395–403.

Fuentes-Zaragoza, E., Riquelme-Navarrete, M.J., Sánchez-Zapata, E. and Pérez-Álvarez, J.A.2010. Resistant starch as a functional ingredient: A review. *Food Res. Int.* 43: 931–942.

Gagaoua, M., Boucherba, N., Bouanane-Darenfed, A., Sancha, S., Hafid, K. and Ziane, F. 2014. Three-phase partitioning as an efficient method for the purification and recovery of ficin from Mediterranean (*Ficus carica* L.) latex. *Sep. Purif. Technol.* 132: 461–467.

Haroon, F. and Ghazanfar, M. 2016. Applications of food biotechnology. *J. Ecosyst. Ecography.* 6: 1000215.

Hefferon, K.L. 2015. Nutritionally enhanced food crops: Progress and perspectives. *Int. J. Mol. Sci.* 16: 3895–3914.

Jain, A., Singh, N., Kumari, S., Khan, S., Agarwal, S., Kumari, S. et al. 2021. Bioentrepreneurship in agricultural biotechnology. In *Bioentrepreneurship and transferring technology into product development.* IGI Global. pp. 183–200.

Jin, Y.H., Robledo, D., Hickey, J.M. and Houston, R.D. 2021. Surrogate broodstock to enhance biotechnology research and applications in aquaculture. *Biotechnol. Adv.* 49: 107756.

Joshi, V.K., Parmar, M. and Rana, N. 2011. Purification and characterisation of pectinase produced from apple pomace and evaluation of its efficiency in fruit juice extraction and clarification. *Indian J. Nat. Prod. Resour.* 2: 189–197.

Khan, M.I., Rashid, B., Tariq, M. and Khan, A.A. 2017. Crop improvement: New approaches and modern techniques. *Plant Gene Trait* 8.

Khan, S. and Khan, J. 2010. Drought-tolerant wheat cultivars for rainfed areas of KPK, Pakistan. *Pak. J. Agric. Sci.* 47: 355–359.

Kole, C. 2012. *Wild crop relatives: Genomic and breeding resources: Cereals.* Springer Science & Business Media.

Kumar, K., Gambhir, G., Dass, A., Tripathi, A.K., Singh, A., Jha, A.K. et al. 2020. Genetically modified crops: Current status and prospects. *Planta* 251: 91.

Lee, W.C., Yusof, S., Hamid, N.S.A. and Baharin, B.S. 2006. Optimising conditions for enzymatic clarification of banana juice using Response Surface Methodology (RSM). *J. Food Eng.* 73: 55–63.

Li, G., Wang, Y., Liu, B. and Zhang, G. 2014. Transgenic *Bacillus thuringiensis* (Bt) rice is safer to aquatic ecosystems than its non-transgenic counterpart. *PLoS One* 9: e104270.

Ma, L., He, Y., Wang, Y., Wang, Y., Li, R., Huang, Z., Jiang, Y. and Gao, J. 2019. Nanocomposites of Pt nanoparticles anchored on UiO66-NH_2 as carriers to construct acetylcholinesterase biosensors for organophosphorus pesticide detection. *Electrochim. Acta.* 318: 525–533.

Mahdieh, N. and Rabbani, B. 2013. An overview of mutation detection methods in genetic disorders. *Iran. J. Pediatr.* 23: 375–388.

Maleki, N., Kashanian, S., Maleki, E. and Nazari, M. 2017. A novel enzyme-based biosensor for catechol detection in water samples using an artificial neural network. *Biochem. Eng. J.* 128: 1–11.

Mamiro, P., Mbwaga, A., Mamiro, D., Chiwona-Karltun, L. and Mgoba, C. 2011. Nutritional quality and utilisation of local and improved cowpea varieties in some regions in Tanzania. *Afr. J. Food Agric. Nutr. Dev.* 11: 4490–4506.

Mansour, H.B., Houas, I., Montassar, F., Ghedira, K., Barillier, D., Mosrati, R. and Chekir-Ghedira, L. 2012. Alteration of *in vitro* and acute *in vivo* toxicity of textile dyeing wastewater after chemical and biological remediation. *Environ. Sci. Pollut. Res.* 19: 2634–2643.

Mauri, E., Chincaroli, M.F., Rigamonti, R., Magagnin, L., Sacchetti, A. and Rossi, F.2017. Modulation of electrostatic interactions to improve controlled drug delivery from nanogels. *Mater. Sci. Eng. C.* 72: 308–315.

Medyantseva, E.P., Brusnitsyn, D.V., Varlamova, R.M., Beshevets, M.A., Budnikov, H.C. and Fattakhova, A.N. 2015. Capabilities of amperometric monoamine oxidase biosensors based on screen-printed graphite electrodes modified with multiwall carbon nanotubes in the determination of some antidepressants. *J. Anal. Chem.* 70: 535–539.

Montagnac, J.A., Davis, C.R. and Tanumihardjo, S.A. 2009. Nutritional value of cassava for use as a staple food and recent advances for improvement. *Compr. Rev. Food Sci. Food Saf.* 8: 181–194.

Mulualem, T. 2015. Application of bio-fortification through plant breeding to improve the value of staple crops. *Biomed. Biotechnol.* 3: 11–19.

Nadeem, M., Ghaffar, A., Hashim, M.M. and Ullah, I. 2021. Sonication and microwave processing of phalsa drink: A synergistic approach. *Int. J. Fruit Sci.* 21: 993–1007.

Nigam, P.S. 2013. Microbial enzymes with special characteristics for biotechnological applications. *Biomolecules* 3: 597–611.

Paine, J.A., Shipton, C.A., Chaggar, S., Howells, R.M., Kennedy, M.J., Vernon, G.. 2005. Improving the nutritional value of Golden Rice through increased provitamin A content. *Nat. Biotechnol.* 23: 482–487.

Patel, A.K., Singhania, R.R., Pandey, A. and Brahmachari, G. 2017. Production, purification, and application of microbial enzymes. In *Biotechnology of microbial enzymes: Production, biocatalysis, and industrial applications.* Academic Press. pp. 13–41.

Pooja, S.G., Hongal, S., Vasudeva, K.R., Ugalath, J. and Suresh, G.J. 2023. Optimisation of wood apple (*Limonia acidissima* L.) juice extraction for improved yield and quality using commercial pectinase. *Pharma Innov. J.* 12: 1304–1311.

Qaim, M. 2020. Bt cotton, yields and farmers' benefits. *Nat. Plants.* 6: 1318–1319.

Qamar, S., Tantray, A.Y., Bashir, S.S., Ali, T., Roy, A., Elshiekh, M.Z. et al. 2020. Golden rice: Genetic engineering, promises, present status and prospects. In *Rice research for quality improvement: Genomics and genetic engineering.* Springer. pp. 581–604.

Rajasree, V. and Pugalendhi, L. 2021. Breeding vegetables for nutritional security. In *Veganism – A fashion trend or food as a medicine.* IntechOpen.

Ranjha, M.M.A.N., Irfan, S., Lorenzo, J.M., Shafique, B., Kanwal, R., Pateiro, M., Arshad, R.N., Wang, L., Nayik, G.A., Roobab, U. and Aadil, R.M. 2021. Sonication, a potential technique for the extraction of phytoconstituents: A systematic review. *Processes* 9: 1406.

Raveendran, S., Parameswaran, B., Ummalyma, S.B., Abraham, A., Mathew, A.K., Madhavan, A., Rebello, S. and Pandey, A. 2018. Applications of microbial enzymes in the food industry. *Food Technol. Biotechnol.* 56: 16–30.

Renge, V.C., Khedkar, S.V. and Nandurkar, N.R. 2012. Enzyme synthesis by fermentation method: A review. *Sci. Rev. Chem. Commun.* 2: 585–590.

Semenova, M.V., Sinitsyna, O.A., Morozova, V.V., Fedorova, E.A., Gusakov, A.V., Okunev, O.N., Sokolova, L.M., Koshelev, A.V., Bubnova, I.P. and Sinitsyn, A.P. 2006. Use of a preparation from fungal pectin lyase in the food industry. *Prikl. Biokhim. Mikrobiol.* 42: 681–685.

Sharma, H.P. and Patel, H. 2017. Enzymatic added extraction and clarification of fruit juices – A review. *Crit. Rev. Food Sci. Nutr.* 57: 1215–1227.

Sharma, N., Rathore, M. and Sharma, M. 2013. Microbial pectinase: Sources, characterisation and applications. *Rev. Environ. Sci. Biotechnol.* 12: 45–60.

Singhania, R.R., Sukumaran, R.K., Patel, A.K., Larroche, C. and Pandey, A. 2010. Advancement and comparative profiles in the production technologies using solid-state and submerged fermentation for microbial cellulases. *Enzyme Microb. Technol.* 46: 541–549.

Traore, S.B., Carlson, R.E., Pilcher, C.D. and Rice, M.E. 2000. Bt and non-Bt maize growth and development as affected by temperature and drought stress. *Agron. J.* 92: 1027–1035.

Yang, H., Li, J., Du, G., Liu, L. and Brahmachari, G. 2017. Microbial production and molecular engineering of industrial enzymes: Challenges and strategies. In *The biotechnology of microbial enzymes.* Academic Press. pp. 151–165.

18 Plant Enzymes in Bioremediation and Environmental Cleanup

L. Jahnavi Reddy, Kavisa Ghosh, and Riyaz Sayyed

18.1 INTRODUCTION

Environmental pollution from industrialization, urban expansion, and agricultural practices has resulted in the accumulation of toxic substances, including heavy metals, pesticides, and persistent organic pollutants (POPs), in soil and water systems. These contaminants pose a threat to biodiversity, food safety, and human health (Nuruzzaman et al., 2025). While traditional cleanup methods, such as incineration, chemical neutralization, and landfilling, are expensive and environmentally harmful, bioremediation offers a sustainable and cost-effective solution that utilizes biological systems to restore polluted areas (Evode et al., 2021).

Bioremediation is a sustainable approach to addressing environmental contamination by utilizing microorganisms, plants, and their enzymes to break down or transform toxic pollutants. Unlike traditional cleanup methods, bioremediation lowers pollutant levels while supporting a circular bioeconomy by converting waste into harmless byproducts (Mokrani et al., 2024). Plant-based methods, particularly phytoremediation, have garnered attention for their scalability and cost-effectiveness. Phytoremediation utilizes plants to remove, stabilize, or break down contaminants such as heavy metals, pesticides, and hydrocarbons, thereby helping to restore ecosystems and improve damaged sites (Bansal et al., 2024; Kafle et al., 2022). Plant enzymes play a crucial role in bioremediation by breaking down toxic pollutants into less harmful compounds. Enzymes like peroxidases, laccases, and dehalogenases degrade pesticides, dyes, and heavy metals (Narayanan et al., 2023a). Recent studies show the benefits of plant-based bioremediation. Phytoremediation helps restore soil fertility while limiting the transfer of heavy metals into the food chain, making it essential for environmental management (Aryal, 2024).

Compared to microbial and fungal enzymes, plant-derived enzymes have unique advantages for bioremediation and environmental cleanup. Their broader pH tolerance and environmental stability allow them to work effectively in diverse conditions found in contaminated soils and water (Bittencourt et al., 2023). Unlike microbial enzymes, which require controlled settings, plant enzymes are naturally adapted to handle abiotic stresses such as temperature changes, salinity, and oxidative stress (Luciński & Adamiec, 2023). Their in-situ production within plant tissues reduces the need for outside support or fermentation, ensuring continuous enzyme release for self-sustained remediation (Chauhan et al., 2023). Plant enzymes interact with root exudates and microbial communities in the root zone, where enzymes and metabolites work together to break down pollutants. This interaction transforms xenobiotics into simpler forms while improving soil fertility (Chen & Liu, 2024). Plant enzymes also have low immunogenicity and minimal environmental toxicity, making them safer for use than some microbial enzymes (Dkhar et al., 2025). Among the most extensively studied plant enzymes, peroxidases, laccases, and dehalogenases play a crucial role in bioremediation. Peroxidases break down phenolic compounds, dyes, and pesticides, while laccases play a key role in degrading lignin and transforming xenobiotics (Khatua, 2025). Dehalogenases facilitate the detoxification of halogenated organic pollutants, which are typically highly resistant to degradation (Sadiq et al., 2025). Besides degrading pollutants, these enzymes also mitigate oxidative stress within plants, enabling them to survive in contaminated areas. Overall, these traits position plant enzymes as effective, eco-friendly tools for sustainable environmental management (Rathore et al., 2022).

Plant enzymes have emerged as potent biocatalysts in the remediation of environmental pollutants, offering eco-friendly and sustainable alternatives to conventional methods. This chapter explores how plant-derived enzymes degrade xenobiotics, heavy metals, and POPs, highlighting technologies such as synthetic biology and enzyme immobilization that enhance efficacy. Despite their promise, challenges include enzyme stability and scalability. The chapter evaluates these limitations and proposes strategies to overcome them, positioning plant enzymes as key players in green biotechnology.

18.2 MECHANISMS OF PLANT ENZYME-MEDIATED BIOREMEDIATION

Plant enzymes play an important role in bioremediation. They help change complex, toxic pollutants into simpler, less harmful compounds through various biochemical pathways. These enzymes are involved in oxidation, reduction, hydrolysis, and dehalogenation reactions. They target a wide range of environmental contaminants, such as pesticides, dyes, hydrocarbons, and heavy metals (Abu-Tahon et al., 2025). Root exudates and interactions within the rhizosphere enhance the activity of enzymes such as peroxidases, laccases, and dehalogenases. In this zone, plants and microbes collaborate to accelerate the breakdown of pollutants. By combining enzymatic detoxification with natural plant defence and metabolic processes, plant enzyme-mediated bioremediation offers an effective, eco-friendly,

DOI: 10.1201/9781003669920-18

and sustainable approach to restoring contaminated ecosystems (Chakraborty et al., 2025; Kafle et al., 2022).

18.2.1 Key Enzymes and Their Functions

Plant enzymes play a central role in the bioremediation process, as they catalyse reactions that break down or convert toxic pollutants into less harmful products. The enzymes operate under ambient conditions, providing environmentally friendly and sustainable remediation solutions (Kuppan et al., 2024). The major enzymes include peroxidases, laccases, dehalogenases, and nitrilases, which target a broad spectrum of pollutants, thus making them effective tools for ecosystem restoration.

18.2.1.1 Peroxidases

Peroxidases are versatile oxidative enzymes that use hydrogen peroxide (H_2O_2) as an electron acceptor to oxidize phenolic compounds, aromatic amines, and structures similar to lignin. In bioremediation, plant peroxidases play a crucial role in breaking down polycyclic aromatic hydrocarbons (PAHs), synthetic dyes, and pesticide residues that persist in soil and water (Twala et al., 2020). For example, horseradish peroxidase (HRP) has been widely studied for its ability to decolourize textile dyes and to turn phenolic contaminants into stable, less toxic polymers. In contaminated agricultural soils, peroxidases enhance the breakdown of herbicides and insecticides, thereby reducing their environmental impact. Their ability to act on a wide variety of phenolic and aromatic substances makes them crucial for managing industrial waste and areas polluted with pesticides (Bakar et al., 2024).

18.2.1.2 Laccases

Laccases are copper-containing oxidases that promote the one-electron oxidation of aromatic rings while turning molecular oxygen into water. These enzymes are very effective at breaking down phenols, polyphenols, and chlorinated organic pollutants. Plant laccases are particularly important for degrading POPs, such as chlorophenols and phenolic derivatives, found in industrial waste. For instance, laccase activity helps detoxify chlorinated organics and decompose synthetic dyes used in the textile and paper industries (Curran et al., 2021). Their ability to function without needing extra cofactors makes them affordable biocatalysts for large-scale cleanup projects. In soil systems, laccases also help transform compounds derived from lignin, connecting pollution control with natural carbon cycling (Scheibel et al., 2024; Sondhi et al., 2023).

18.2.1.3 Dehalogenases

Dehalogenases are specialized enzymes that break carbon–halogen bonds to detoxify halogenated xenobiotics. These pollutants, including organochlorines and polychlorinated biphenyls (PCBs), are among the most persistent and hazardous contaminants due to their stability and potential for bioaccumulation (Ang et al., 2018). Plant dehalogenases play a crucial role in converting these compounds into less halogenated, more biodegradable forms. For example, in soils contaminated with organochlorine pesticides like dichlorodiphenyl-trichloroethane (DDT), dehalogenases initiate detoxification by removing chlorine atoms, which facilitates microbial degradation. This enzymatic action is particularly valuable for cleaning up industrial sites and agricultural land where long-lasting halogenated compounds present ecological and health risks (Mahalle et al., 2025).

18.2.1.4 Nitrilases

Nitrilases catalyse the hydrolysis of nitrile groups (–C≡N) into carboxylic acids and ammonia, providing a direct method for detoxifying nitrile-based pollutants. These compounds are commonly found in industrial solvents, cyanides, and synthetic pesticide intermediates (Bhatt et al., 2021). Plant nitrilases play a crucial role in converting highly toxic cyanides into less harmful organic acids, thereby reducing environmental toxicity. For example, nitrilase activity has been seen in plants that detoxify industrial waste containing nitrile compounds. Additionally, nitrilases aid in the biodegradation of specific herbicides and fungicides, which helps restore soil quality in heavily farmed areas. Their role in both treating industrial waste and remediating agricultural land highlights their significance as biocatalysts in environmental cleanup (Egelkamp et al., 2019).

18.2.2 Plant Enzymes-Based Bioremediation: Two-Phase Detoxification System

Plant enzymes involved in bioremediation generally operate through a two-phase detoxification system (Chia et al., 2024). This system allows for the efficient breakdown and removal of various pollutants.

i. *Phase I Reactions – Functionalization (Oxidation, Reduction, and Hydrolysis)*: In Phase I, enzymes catalyse reactions such as oxidation, reduction, and hydrolysis. These reactions introduce or expose functional groups (–OH, $–NH_2$, –COOH) on hydrophobic pollutants, thereby rendering them more hydrophilic. These changes increase the solubility and reactivity of the compounds, making them easier to detoxify (Chia et al., 2024). For example, peroxidases and laccases oxidize aromatic hydrocarbons and phenolic pollutants. Esterases and hydrolases cleave ester bonds in pesticides, producing metabolites that are more polar and less likely to accumulate in living organisms. Such reactions turn POPs into intermediates that plants or associated microbes can metabolize in the rhizosphere (Dong et al., 2023; Suresh et al., 2025).
ii. *Phase II Reactions – Conjugation and Detoxification*: After Phase I, the activated or functionalized pollutants undergo Phase II reactions. In this phase, they are conjugated with molecules from the plant, such as sugars (e.g., glucose, xylose), amino acids

(e.g., glycine, glutamine), or peptides (e.g., glutathione) (Chia et al., 2024). Enzymes such as glycosyltransferases and glutathione-S-transferases (GSTs) catalyse this conjugation. The resulting conjugates are usually non-toxic, more water-soluble, and less likely to re-enter the food chain. For instance, cyanide metabolites might be conjugated to amino acids for safe storage. Pesticide derivatives are glycosylated to form stable, less mobile complexes (Mazari et al., 2023).

Together, these two phases ensure that pollutants are transformed and rendered harmless. They can be sequestered in vacuoles or bound to components of the cell wall. This dual mechanism helps plants survive in contaminated environments while also restoring ecological balance (Oubohssaine & Dahmani, 2024).

18.2.3 Pathways of Pollutant Degradation

Plant enzymes employ various biochemical pathways to break down and detoxify environmental pollutants. These processes not only reduce toxicity but also improve the availability of contaminants for further breakdown, ensuring effective bioremediation (Lee et al., 2025) (Figure 18.1).

18.2.3.1 Oxidative Cleavage of Aromatic Rings

One of the most important pathways is the oxidative cleavage of aromatic rings. This process targets complex and stable structures, such as phenols, dyes, and PAHs. Enzymes such as laccases and peroxidases initiate reactions that destabilize and open aromatic rings, thereby facilitating their subsequent reactions. This leads to smaller, more biodegradable compounds that can be further processed and utilized.

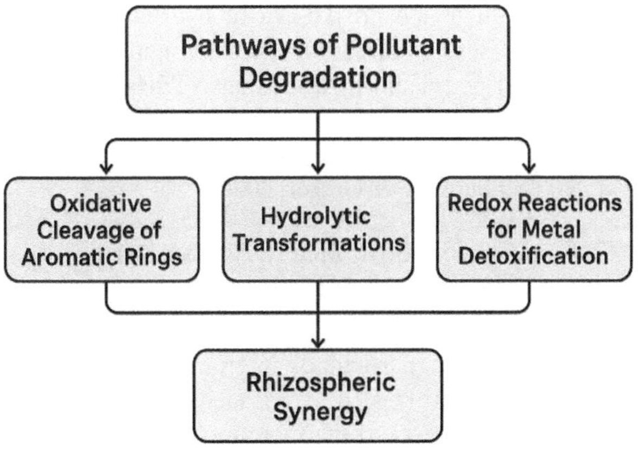

FIGURE 18.1 Pathways of pollutant degradation. This figure shows important biochemical processes that break down pollutants. Pollutants go through oxidative cleavage of aromatic rings, hydrolytic transformations, and redox reactions for metal detoxification, which convert them into less toxic forms. These processes work together in the rhizospheric synergy, where interactions between plants and microbes improve degradation efficiency and soil detoxification.

These oxidative processes are commonly used to detoxify pesticide residues and industrial effluents rich in phenolic compounds (Gundlapalli et al., 2024; Singh et al., 2024a).

18.2.3.2 Hydrolytic Transformations

Another key mechanism involves hydrolytic enzymes, such as nitrilases and esterases. These enzymes act on nitriles and esters, breaking them down into carboxylic acids and alcohols. This reaction reduces toxicity while increasing the solubility and mobility of pollutants, allowing plants to either break them down or store them. Hydrolytic pathways are particularly useful in treating nitrile-based industrial solvents and agrochemical residues typically found in agricultural soils and wastewater (Mousavi et al., 2021).

18.2.3.3 Redox Reactions for Metal Detoxification

Plant enzymes also participate in redox reactions that alter the chemical states of toxic metals. For example, hexavalent chromium [Cr(VI)], which is highly toxic and mobile, can be reduced enzymatically to trivalent chromium [Cr(III)], which is much less harmful and bioavailable. Such changes help lower heavy metal toxicity and prevent their entry into the food chain (Boros-Lajszner et al., 2023; Qureshi et al., 2024).

18.2.3.4 Rhizospheric Synergy

These enzymatic pathways become even more effective in the rhizosphere, where root exudates promote microbial growth and co-metabolism. The combined action of plant enzymes, root-released substances, and microbial communities speeds up pollutant breakdown, making the rhizosphere a key area for effective bioremediation (Karnwal et al., 2025).

18.2.4 Plant-Assisted Pollutant Remediation: Global Case Studies and Mechanistic Insights

Phytoremediation has been widely used in both experimental and real-world applications. Numerous success stories have been documented worldwide. Various plant species, with their own enzymes, can absorb, detoxify, or break down a broad range of environmental pollutants, including heavy metals, organic compounds, and radionuclides (Oubohssaine & Dahmani, 2024; Wang & Delavar, 2024). The case studies showcased in Table 18.1 are important examples that illustrate the variety, effectiveness, and practical applications of phytoremediation methods.

18.2.4.1 Case Study 1: Vetiver Grass for Heavy Metal Remediation

Vetiveria zizanioides has shown great promise in cleaning up heavy metal-contaminated soils. Its enzymatic activity, especially peroxidases, is key in reducing oxidative stress and improving the uptake of metals like zinc (Zn) and copper (Cu). Field studies have demonstrated that this plant can effectively remove Zn and Cu, confirming its efficacy in restoring polluted environments and its potential for

TABLE 18.1
Summary of Plant-Based Pollutant Remediation Case Studies

Case Study	Plant Species	Pollutants	Key Enzymes/Mechanisms	Outcome	Key Citations
Vetiver grass	*Vetiveria zizanioides*	Cu, Fe, Zn	Peroxidase activity linked to metal uptake	Significant removal of Zn and Cu	Gu et al. (2025) and Kumar et al. (2025)
Poplar trees	Populus deltoides	TCE, benzene	Laccase, dehalogenase	70% reduction of TCE in 6 months	Banan et al. (2024) and Jassal et al. (2025)
Indian mustard	Brassica juncea	Cd, Pb	Peroxidase, phytochelatin synthase	Used in soils affected by mining in Jharkhand	Bortoloti and Baron (2022) and Motshumi et al. (2025)
Sunflower	Helianthus annuus	Cs, Sr	Peroxidases, antioxidant enzymes	Reduced radionuclides after Chernobyl	Ejaz et al. (2023) and Sharma et al. (2025)
Willow	*Salix viminalis*	PAHs, diesel	Oxidoreductases, laccases	Substantial reduction of TPH in 12 months	Košnář et al. (2020) and Landberg and Greger (2022)
Rice	Oryza sativa	Arsenic	Arsenate reductase, phytochelatin synthase	Effective at root sequestration, grain risk remains	He et al. (2020) and Ullah et al. (2024)
Water hyacinth	*Eichhornia crassipes*	Pb, Cd, dyes	Peroxidase, catalase, extracellular enzymes	Reduced COD and metal load in effluents	Monroy-Licht et al. (2024) and Narayanan et al. (2023b)
Alfalfa	Medicago sativa	Atrazine	Glutathione-S-transferase (GST)	Lowered atrazine persistence in soils	Yang et al. (2022)

rehabilitating metalliferous lands (Gu et al., 2025; Kumar et al., 2025).

18.2.4.2 Case Study 2: Poplar Trees for Organic Pollutants

Populus deltoides (poplar trees) have been thoroughly researched for their ability to remediate organic pollutants, such as trichloroethylene (TCE) and benzene. Their roots produce laccase and dehalogenase enzymes that help break down chlorinated solvents. Additionally, root exudates stimulate microbial activity in the rhizosphere, thereby enhancing biodegradation. Field applications have reported up to a 70% reduction in TCE levels in contaminated soils and groundwater within six months, demonstrating poplar's strong potential as a sustainable phytoremediation option (Banan et al., 2024; Jassal et al., 2025).

18.2.4.3 Case Study 3: Indian Mustard (*Brassica juncea*) for Cadmium and Lead

Brassica juncea (Indian mustard) is a well-known hyperaccumulator plant that effectively remediates soils contaminated with heavy metals, particularly cadmium (Cd) and lead (Pb). Its high levels of peroxidase and phytochelatin synthase activity enable the effective chelation and sequestration of toxic metals, thereby reducing their bioavailability. Field applications in mining-affected soils in Jharkhand, India, have demonstrated significant reductions in Cd and Pb levels, thereby improving soil quality and highlighting the potential for sustainable phytoremediation in contaminated areas (Bortoloti & Baron, 2022; Motshumi et al., 2025).

18.2.4.4 Case Study 4: Sunflower (*Helianthus annuus*) for Radioactive Metals

Helianthus annuus (sunflower) has proven effective in cleaning up radioactive metals like caesium (Cs) and strontium (Sr). Its ability to take up these metals is supported by peroxidases and antioxidant enzymes that help reduce radiation-induced oxidative stress. This enables the plant to tolerate and accumulate radionuclides. Following the Chernobyl disaster, sunflowers were deployed on floating rafts in contaminated waters, successfully reducing Cs and Sr concentrations, thereby showcasing their value in managing radioactive pollution (Ejaz et al., 2023; Sharma et al., 2025).

18.2.4.5 Case Study 5: Willow (*Salix viminalis*) for Petroleum Hydrocarbons

Salix viminalis (willow) has shown strong potential in remediating petroleum hydrocarbons, including PAHs and diesel residues. The root exudates of the plant enhance microbial activity in the rhizosphere, while its oxidoreductases and laccases facilitate the breakdown of hydrocarbons. Field trials across Europe have demonstrated significant reductions in total petroleum hydrocarbons (TPHs) within just 12 months of planting, making willow a valuable candidate for phytoremediation of oil-contaminated sites (Košnář et al., 2020; Landberg & Greger, 2022).

18.2.4.6 Case Study 6: Rice (*Oryza sativa*) for Arsenic Contamination

Oryza sativa (rice) has a dual role in managing arsenic (As) contamination, particularly in groundwater and paddy fields. The plant utilizes enzymes, such as arsenate reductase, to convert As(V) to the more mobile As(III). It also utilizes phytochelatin synthase to sequester arsenic in root vacuoles, thereby limiting its movement to other parts of the plant. Field studies in West Bengal and Bangladesh have demonstrated rice's ability to sequester arsenic in its roots. However, concerns about bioaccumulation in edible grains highlight both the potential and limitations of this approach in addressing areas contaminated with arsenic (He et al., 2020; Ullah et al., 2024).

18.2.4.7 Case Study 7: Water Hyacinth (*Eichhornia crassipes*) for Wastewater Treatment

Eichhornia crassipes (water hyacinth) is widely utilized in wastewater treatment due to its ability to absorb and detoxify a range of pollutants, including lead (Pb), cadmium (Cd), and textile dyes. Its high levels of peroxidase and catalase activity assist in metal detoxification, while extracellular enzymes break down complex dye molecules. Applications in tannery and textile effluents have shown significant decreases in chemical oxygen demand (COD) and heavy metal concentrations, establishing water hyacinth as a cost-effective and sustainable choice in constructed wetland systems (Monroy-Licht et al., 2024; Narayanan et al., 2023b).

18.2.4.8 Case Study 8: Alfalfa (*Medicago sativa*) for Atrazine (Herbicide)

Medicago sativa (alfalfa) has been used in the phytoremediation of atrazine and other chlorinated herbicides often found in agricultural soils. The plant employs GST enzymes to bind atrazine residues with glutathione, enhancing detoxification and reducing toxicity. Agricultural trials have demonstrated that growing alfalfa effectively reduces atrazine persistence in soil, thereby lowering the risk of groundwater contamination and providing a sustainable approach to managing herbicide pollution (Yang et al., 2022).

18.3 EMERGING TECHNOLOGIES ENHANCING PLANT ENZYME EFFICACY

Plant enzymes have great potential in bioremediation, industrial processing, and sustainable agriculture. However, their practical use often faces challenges like low stability, narrow substrate specificity, and decreased activity in harsh conditions (Dkhar et al., 2025). Recent advancements in new technologies are addressing these issues, leading to improvements in enzyme efficiency, stability, and scalability. Innovations such as genetic engineering, nanotechnology, immobilization methods, and computational enzyme design are enhancing catalytic performance while keeping costs low and maintaining environmental friendliness. Together, these methods are changing how we use plant-derived enzymes and opening new opportunities for large-scale applications in environmental and industrial biotechnology (Gkantzou et al., 2021; Tadesse & Liu, 2025).

18.3.1 Synthetic Biology and Enzyme Engineering

Synthetic biology provides new approaches to redesign biological systems, enhancing enzyme performance in plants. By utilizing modern molecular tools, this field enhances enzyme specificity, catalytic efficiency, and stress tolerance, enabling plants to break down a wide range of pollutants more effectively (Hayakawa et al., 2024). Key strategies include directed evolution, CRISPR/Cas9 genome editing, and the creation of artificial biosensors.

18.3.1.1 Directed Evolution

Directed evolution is a laboratory technique that accelerates natural selection to create more effective enzyme variants. Through repeated rounds of mutagenesis and screening, researchers obtain enzymes with enhanced stability, improved catalytic activity, and increased abilities to degrade pollutants. For example, evolved peroxidases and laccases show greater resistance to temperature and pH changes, which enhances their effectiveness in breaking down aromatic hydrocarbons, dyes, and other hard-to-remove organic pollutants (García-García et al., 2021).

18.3.1.2 CRISPR/Cas9 Genome Editing

Genome editing with CRISPR/Cas9 provides a precise method for altering or introducing genes that encode detoxifying enzymes in plants. Species such as *Arabidopsis thaliana* and *Populus* have been genetically modified to produce more GSTs and peroxidases, thereby enhancing their ability to neutralize pesticides, heavy metals, and industrial toxins. This targeted approach ensures the consistent expression of pollutant-degrading enzymes, thereby improving phytoremediation efficiency while maintaining plant health (Meng et al., 2023; Oubohssaine & Dahmani, 2024).

18.3.1.3 Artificial Biosensors

Synthetic biology also enables the creation of artificial biosensors that integrate pollutant detection with the expression of enzymes. These biosensors can sense contaminants such as herbicides or heavy metals and activate specific detoxifying enzymes in response. This programmable reaction ensures efficient degradation while minimizing unnecessary metabolic strain on plants when pollutants are absent (Aminian-Dehkordi et al., 2023; Chen et al., 2023).

Overall, synthetic biology and enzyme engineering provide robust methods for enhancing plant-based bioremediation. By combining directed evolution, precise genome editing, and biosensor development, researchers can create plants with customized enzyme systems that are well-equipped to tackle various and complex environmental challenges (García-García et al., 2021; Meng et al., 2023; Oubohssaine & Dahmani, 2024).

18.3.2 AI/ML-Assisted Enzyme Modelling

The integration of artificial intelligence (AI) and machine learning (ML) into enzyme research is transforming the design, optimization, and utilization of plant enzymes in bioremediation (Ndochinwa et al., 2024). Synzymes are poised to be the future of modern enzyme engineering (Palabiyik, 2025). These computational methods enable the accurate prediction of enzyme performance, reduce experimental costs, and facilitate the faster application of laboratory results in s settings.

18.3.2.1 Predictive Modelling

AI-driven predictive models simulate enzyme–substrate interactions. This helps identify mutations that boost activity, stability, and pollutant specificity. For example, an ML-based algorithm successfully predicted mutations in laccase enzymes. These mutations enhanced their ability to break down phenolic pollutants and synthetic dyes. Such methods avoid lengthy experimental screening and provide clear pathways for creating high-performance enzymes (Cui et al., 2025; Salas-Nuñez et al., 2024).

18.3.2.2 Digital Twins

Digital twins create virtual replicas of contaminated sites. They simulate environmental factors like soil composition, pH, temperature, and pollutant concentrations. This enables researchers to evaluate enzyme performance before conducting field trials (Silva et al., 2023). A case study featured a digital twin model of a textile effluent site. It tested peroxidase-based remediation strategies under varying pollutant loads. The simulations accurately predicted enzyme degradation rates, guiding effective deployment strategies and reducing ecological risks (Kim et al., 2024).

18.3.2.3 Real-Time Monitoring

AI-powered biosensors and Internet of Things (IoT) systems enable real-time tracking of enzyme activity and pollutant levels on-site (Nashruddin et al., 2024). For instance, researchers have developed AI-integrated biosensors for detecting atrazine. These biosensors monitored herbicide residues in soil and triggered the expression of detoxifying enzymes in engineered *Arabidopsis*. This adaptive response system enhanced resource utilization and increased degradation efficiency (Singh et al., 2024b).

By combining predictive modelling, digital twins, and real-time biosensing, AI/ML tools provide a useful framework for enzyme-based remediation. These methods speed up enzyme design and enable responsive, data-driven actions in field applications. This marks a significant step towards sustainable and precision-focused environmental biotechnology (Akintola, 2024; Shi et al., 2025).

18.3.3 Immobilization and Nanotechnology

The combination of immobilization methods and nanotechnology has significantly enhanced the application of plant-based enzymes in environmental and industrial processes. Plant enzymes, such as HRP, laccases, polyphenol oxidases, and ascorbate oxidase, are effective biocatalysts. However, they often exhibit limited stability, can be inhibited, and have short operational lifetimes (Abdel-Mageed, 2025). Immobilization enhances their stability, reusability, and resistance to degradation. Meanwhile, nanotechnology boosts catalytic efficiency, pollutant specificity, and controlled delivery (Figure 18.2). Together, these methods connect laboratory studies with large-scale environmental cleanup and industrial use (Tadesse & Liu, 2025). Some of the methods that are being explored are further discussed below.

18.3.3.1 Nanozymes

Nanozymes are engineered nanoparticles that mimic the activities of natural enzymes. Although they do not come from plants, they complement natural plant enzymes when there are issues with supply or stability. For example, iron oxide nanozymes can break down phenolic compounds in wastewater, and cerium oxide nanozymes show peroxidase-like activity in removing textile dyes (Alizadeh et al., 2020; Nagendran et al., 2024; Zhang et al., 2022). When used in conjunction with plant peroxidases or laccases, nanozymes enhance the overall degradation efficiency. This creates a hybrid system that combines the natural targeting of plant enzymes with the durability of nanomaterials.

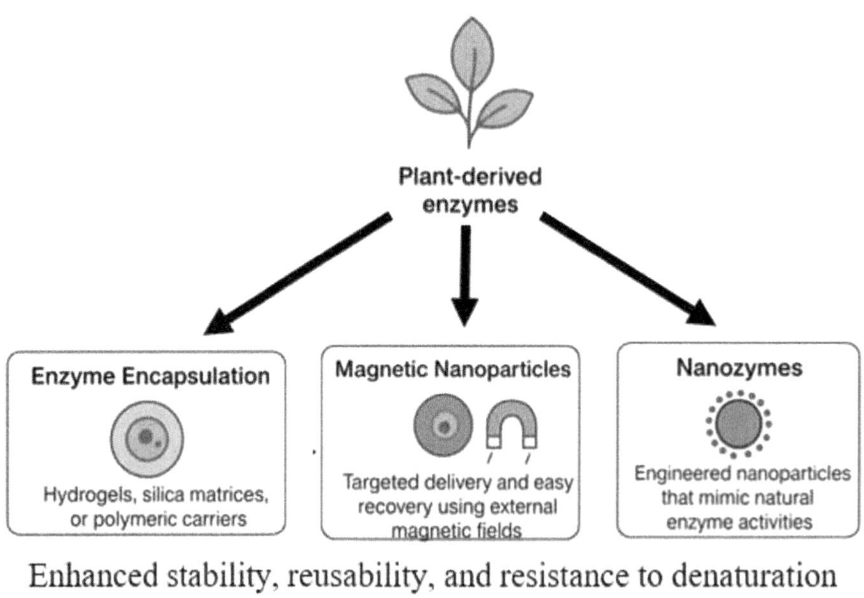

FIGURE 18.2 Plant-derived enzymes combined with nanotechnology. Nanozymes, encapsulation carriers, and magnetic nanoparticles improve the stability, reusability, and catalytic efficiency of plant enzymes for environmental and industrial uses.

18.3.3.2 Enzyme Encapsulation

Encapsulation methods that utilize hydrogels, silica matrices, or polymer carriers have proven highly effective for plant enzymes. For example, HRP enclosed in silica matrices resists inactivation by heavy metals during soil cleanup. Similarly, plant-derived laccases entrapped in alginate hydrogels have maintained their activity over multiple reuse cycles while degrading textile dyes (Bakar et al., 2024; Colin et al., 2024; Hussain et al., 2024). These protective carriers preserve the natural catalytic activity of plant enzymes, prolong their lifespan, and make them more suitable for ongoing treatment systems.

18.3.3.3 Magnetic Nanoparticles

Magnetic nanoparticles provide the benefits of targeted enzyme delivery and easy recovery using external magnetic fields. Plant enzymes attached to magnetic nanoparticles combine eco-friendliness with operational efficiency (Cavalcante et al., 2024). For instance, HRP attached to magnetic nanoparticles has been utilized to break down phenolic pollutants, while laccase–magnetic nanocomposites have effectively removed bisphenol A from wastewater, demonstrating high reusability (Du et al., 2023). These systems reduce enzyme loss, lower operational costs, and enhance the utilization of plant enzyme-based nanobiocatalysts for environmental cleanup.

Immobilization and nanotechnology offer complementary methods for addressing the natural limitations of plant enzymes. By enhancing stability, reusability, and catalytic effectiveness, these techniques reveal the potential for affordable, scalable, and sustainable cleanup solutions (Fahim et al., 2025).

18.3.4 Integration with Phytoremediation Systems

Integrating plant enzymes into phytoremediation systems provides a combined approach to improve pollutant removal from contaminated areas. Plants naturally absorb, store, and break down toxins. Adding engineered enzymes significantly increases degradation rates, pollutant targeting, and resistance to harsh conditions. This strategy not only overcomes the limits of traditional phytoremediation but also enhances the ability to manage large-scale environmental issues (Aryal, 2024; Islam et al., 2024; Oubohssaine & Dahmani, 2024).

18.3.4.1 Transgenic Plants

Genetic engineering has led to the creation of transgenic plants that produce detoxifying enzymes directly in their tissues. For example, poplar trees engineered to express fungal laccases have shown better degradation of chlorinated solvents in groundwater (Jassal et al., 2025). Similarly, *A. thaliana* modified to overexpress peroxidase genes has shown improved breakdown of phenolic pollutants in soil. These transgenic systems function as self-sustaining biocatalysts, enabling long-term remediation with minimal external assistance (Riyazuddin et al., 2019).

18.3.4.2 Rhizosphere Engineering

The rhizosphere, the soil area influenced by plant roots, is a key spot for microbe–plant interactions. Engineering root exudates to boost microbial co-metabolism can enhance pollutant degradation (Afzal et al., 2025). For instance, willow plants have been shown to encourage rhizospheric bacteria that can break down PAHs (Landberg & Greger, 2022). This collaborative action between plant enzymes and microbial communities accelerates detoxification while promoting soil health.

18.3.4.3 Smart Sensors

Integrating smart sensors and IoT devices into phytoremediation systems enables real-time monitoring and control. AI-driven biosensors can detect pollutant levels and trigger enzyme production in transgenic plants when contaminant levels exceed a certain threshold (Popescu et al., 2024). A case study demonstrated the use of IoT-enabled monitoring to detect atrazine in agricultural soils, where real-time detection activated GST production, resulting in rapid detoxification. These systems lessen metabolic stress on plants while ensuring effective remediation (Li et al., 2025).

By merging transgenic expression, rhizosphere engineering, and sensor-driven control, phytoremediation becomes a highly adaptable system integrated with enzymes. These innovations offer scalable and sustainable solutions for cleaning up complex mixtures of pollutants in various ecosystems (Ayub et al., 2025).

18.4 LIMITATIONS AND CHALLENGES OF PLANT ENZYME-BASED BIOREMEDIATION

The use of plant enzymes in bioremediation and environmental cleanup comes with several challenges as well as limitations. One of the main issues is the low operational stability and catalytic efficiency of free enzymes, which necessitate improvements in these areas for effective bioremediation. Over time, the low stability of enzymes decreases, posing a significant challenge to maintaining the effectiveness of bioremediation processes. To overcome this problem, techniques such as enzyme immobilization on an inert material are employed, which aid in enhancing operational stability and reducing the overall remediation costs (Abdelhamid et al., 2024; Yamaguchi & Miyazaki, 2024). Though several enzyme immobilization techniques have been developed to enhance stability and reduce the costs, identifying the most suitable immobilization technologies and materials remains a major challenge due to the high cost as well as limited availability of effective enzyme carriers for field use (Wang et al., 2023; Yamaguchi & Miyazaki, 2024). Further, the in-situ application of these methods remains limited due to environmental factors like temperature, pH, as well as the presence of microbial species that can affect the enzyme activity and stability. Additionally, the complex environmental matrices may hinder the complete degradation or mineralization of pollutants (Yang & Liu, 2025). To overcome these challenges, microbe–plant interactions and the integration of nanoparticles have been explored to enhance bioremediation efforts. However, these approaches are associated with challenges, such as ensuring compatibility and optimizing synergistic effects to achieve significant bioremediation of pollutants (Unimke et al., 2024).

Another challenge that hinders the application of plant-based enzymes for bioremediation is associated with the production and application costs of these enzymes. High production costs impede the widespread implementation of enzymes in fields. Furthermore, the complexity of enzyme recovery and reusability poses a barrier, as it can be intricate and resource-intensive (Makam et al., 2022). Additionally, in terms of application, there are limitations in the biodegradation efficiency of certain contaminants. For instance, the heterogeneity of contaminated environments and the presence of complex mixtures of pollutants can complicate enzyme activity and effectiveness. Moreover, specific enzyme activities could be affected by variations in environmental conditions (pH, temperature, etc.) and the presence of inhibitors, which can impede optimal performance (Wang et al., 2023). Future research should focus on optimizing enzyme performance in natural conditions and integrating biotechnological advances to enhance the economic feasibility of enzyme-based solutions for environmental cleanup.

Another challenge is scaling bioremediation processes from laboratory to field applications. The efficiency seen in controlled environments often diminishes in real-world conditions due to unpredictable environmental variables. Moreover, economic considerations like the cost of producing enzymes at a large scale may hinder the wide-scale adoption of enzyme-based bioremediation technology. Techniques such as the integration of enzyme systems with advanced technological solutions, like bio-electrochemical systems, aim to enhance the effectiveness and scalability of enzymatic bioremediation (Yang & Liu, 2025). However, scaling these solutions while maintaining efficacy and cost-effectiveness remains a considerable challenge. Thus, ongoing research and development efforts should focus on improving enzyme stability, catalytic efficiency, and adaptability to various environmental conditions.

Further, the adoption of bioremediation strategies, including those using plant enzymes, is subject to regulatory scrutiny due to concerns about biosafety and genetic pollution. The deployment of genetically modified organisms or the release of enzymes into natural environments raises ecological and ethical questions that regulators must address. Biosafety assessments are crucial for ensuring that bioremediation processes do not lead to unintended ecological consequences (Maglione et al., 2024; Mani & Kumar, 2013). Public perception can also significantly influence the deployment of bioremediation technologies. Misgivings about the safety and ecological consequences of such technologies could impede their acceptance and regulatory approval (Afshar et al., 2020). Furthermore, the complexity of compliance associated with various regulatory standards across different jurisdictions worldwide can also delay as well as prevent the implementation of new bioremediation technologies. Thus, addressing these regulatory hurdles, while ensuring public and environmental safety, is essential for the broader acceptance and implementation of enzyme-based bioremediation methods.

While plant enzymes offer a promising avenue for bioremediation and environmental cleanup, their field application is limited by several technical, regulatory, and ethical challenges. Overcoming these limitations is crucial for the effective application of these technologies in environmental cleanup. Moreover, further research and development are warranted to enhance enzyme stability, reduce production costs, and optimize application strategies, thereby improving the feasibility and efficiency of plant enzyme-based bioremediation.

18.5 FUTURE DIRECTIONS

The future directions for utilizing plant enzymes in bioremediation and environmental cleanup are promising, with ongoing research and technological advancements aimed at enhancing their efficiency and applicability. The most innovative approaches include modular bioreactor systems, integration with circular bioeconomy models, and fostering global collaboration to achieve scalable solutions.

18.5.1 Modular Bioreactor Systems

Modular bioreactor systems enhance the scale and flexibility of enzyme-based bioremediation processes by allowing for the fine-tuning of environmental conditions to optimize enzymatic activity and pollutant degradation. This flexibility enables the efficient conversion of pollutants into non-toxic forms, facilitating effective environmental cleanup. Although plant enzyme modular systems are not widely available, this concept is utilized in microbial and algal bioreactor applications and can be adapted for plant enzyme-based models. These systems optimize enzymatic activity and pollutant degradation by controlling environmental conditions, resulting in the efficient conversion of pollutants into non-toxic forms for environmental cleanup (Maglione et al., 2024).

Further, the operational stability and catalytic efficiency of plant enzymes can be enhanced through immobilization techniques. This method involves anchoring enzymes onto supports, improving their stability and reusability while reducing costs. Immobilized enzyme systems in bioreactors are crucial for the degradation of pollutants and various industrial applications (Abdelhamid et al., 2024; Yamaguchi & Miyazaki, 2024). The integration of sensors ensures precise bioprocess control for optimal enzyme activity. Innovations in bioreactor technology and real-time monitoring capabilities are expanding opportunities for environmental cleanup (Palladino et al., 2024). The coupling of enzymes with nanomaterials to form nanobiocatalysts improves enzyme durability, efficiency, and recyclability. These approaches use green synthesis methods to enhance biocatalytic processes in bioreactors (Khafaga et al., 2024). Research indicates that combining endophytic microorganisms with nanoparticles enhances water remediation by augmenting pollutant degradation and supporting the health of aquatic plants through the production of bioactive compounds. These strategies can address water pollution in modular bioreactor systems (Manganyi et al., 2024). Future research directions include enzyme engineering and machine learning to optimize the use of immobilized enzymes for environmental remediation purposes. These technologies can aid in designing improved enzymes for effective bioremediation in modular bioreactors (Abdelhamid et al., 2024). Thus, the development and optimization of modular bioreactor systems for enzyme deployment are key to advancing plant enzyme bioremediation. Integrating immobilization, nanotechnology, and bioreactor designs provides efficient solutions to environmental pollution challenges.

18.5.2 Integration with Circular Bioeconomy Models

Circular bioeconomy frameworks prioritize the sustainable utilization of resources through the implementation of the "Reduce, Reuse and Recycle" (3R) strategies. This integration is crucial in addressing global challenges, including climate change, pollution, and resource depletion. Within this model, bioremediation involves employing biological agents, including plants and enzymes, to convert waste into valuable resources, thereby contributing to a low-carbon and zero-waste economy. The adoption of bioeconomy measures facilitates the sustainable management of waste, supporting economic growth while maintaining ecological balance (Dahiya et al., 2022; Kuppan et al., 2023; Leong et al., 2021). This approach aligns with the principles of waste reduction, resource recycling, and material reuse, offering a comprehensive solution to environmental challenges. Consequently, integrating plant enzyme bioremediation with circular bioeconomy models presents a viable and sustainable avenue for environmental remediation. By focusing on waste reduction, recycling, and resource recovery, these strategies not only address current environmental challenges but also promote a more sustainable future.

18.5.3 Funding and Global Collaboration for Scalable Solutions

Developing scalable bioremediation solutions requires significant investment and international collaboration, aligning research, funding, and policy frameworks to promote biotechnology innovations globally. Effective collaboration facilitates the sharing of resources and the transfer of technology to address ecological challenges. Research indicates an increasing global funding for bioeconomy and bioremediation initiatives, which are crucial for innovation and deployment at scale (Abad-Segura et al., 2021; Dhiman et al., 2024; Holden et al., 2022; Kostas et al., 2021).

Funding is crucial for advancing the stabilization and catalytic efficiency of enzyme-based systems, necessitating significant investment in research and development. The high cost of enzyme immobilization technologies necessitates support for developing inexpensive carriers and methods for eco-remediation (Wang et al., 2023). Countries like the U.S. and Europe lead global collaborations in environmental sustainability and scalable bioremediation solutions. Collaboration enables sharing knowledge and technology to overcome current methodological limitations in environmental cleanup (Zhan et al., 2025). Enzyme engineering, including directed evolution, enhances enzyme properties and ensures scalability in bioremediation processes, thereby facilitating the detoxification of micro-pollutants (Radley et al., 2023). The integration of computational methodologies and biotechnological strategies can enhance the capabilities of enzymes in environmental cleanup, thereby optimizing their application for sustainability (Abdelhamid et al., 2024). Innovations such as 3D bioprinting for precision bioremediation demonstrate the synergy between biological systems and engineered solutions, enabling efficient strategies that align with sustainability goals (Finny, 2024). International collaboration fosters scalable solutions through shared technologies, thereby enhancing the implementation of enzymatic bioremediation across various environmental contexts (Yamaguchi & Miyazaki, 2024).

Overall, future directions in plant enzymes for bioremediation emphasize cross-disciplinary partnerships and technological advancements for sustainable environmental cleanup. These directions highlight the potential of plant enzymes in integrated environmental management systems, which combine novel technologies and sustainable models to address global challenges effectively.

18.6 CONCLUSION

Plant enzymes have emerged as robust biocatalysts for environmental remediation, offering sustainable and eco-friendly alternatives to conventional cleanup methods. This chapter has explored the diverse mechanisms by which plant-derived enzymes degrade xenobiotics, heavy metals, and POPs, highlighting their potential in addressing complex environmental challenges. Key enzymes, such as peroxidases, laccases, dehalogenases, and nitrilases, play crucial roles in transforming toxic pollutants into less harmful compounds through oxidation, reduction, hydrolysis, and conjugation reactions. The two-phase detoxification system in plants, involving functionalization and conjugation, further enhances the efficacy of these enzymatic processes. Case studies from around the world have demonstrated the practical applications of plant-based remediation strategies across various pollutants and ecosystems. From heavy metal accumulation by vetiver grass to the breakdown of organic pollutants by poplar trees, these examples prove the versatility and effectiveness of plant enzymes in environmental cleanup. Emerging technologies are revolutionizing the field of plant enzyme-based bioremediation. Synthetic biology and enzyme engineering techniques, such as directed evolution and CRISPR/Cas9 genome editing, are enhancing the performance and specificity of enzymes. AI/ML-assisted enzyme modelling is accelerating the design and optimization process, while immobilization and nanotechnology are enhancing enzyme stability and reusability. The integration of plant enzymes with phytoremediation systems, including transgenic plants and rhizosphere engineering, offers promising avenues for large-scale environmental restoration. While plant enzymes offer promising solutions for bioremediation and environmental cleanup, challenges such as enzyme stability, scalability, and regulatory hurdles need to be addressed for widespread implementation. Future directions indicate the development of modular bioreactor systems, integration with circular bioeconomy models, and enhanced global collaboration to achieve scalable solutions. As research progresses and technologies advance, plant enzymes are evident to play an increasingly vital role in addressing environmental pollution and promoting sustainable ecosystem management. In conclusion, plant enzymes represent a powerful tool in the arsenal of environmental biotechnology. Their continued development and application hold great promise for tackling some of the most pressing ecological challenges of our time, paving the way for cleaner, healthier environments worldwide.

REFERENCES

Abad-Segura, E, Belmonte-Ureña, L J, Batlles-Delafuente, A, & González-Zamar, M-D (2021). Implications for sustainability of the joint application of bioeconomy and circular economy: A worldwide trend study. *Sustainability*, *13*(13), 7182. https://doi.org/10.3390/su13137182

Abdelhamid, M A A, Khalifa, H O, Yoon, H J, Ki, M-R, & Pack, S P (2024). Microbial immobilized enzyme biocatalysts for multipollutant mitigation: Harnessing nature's toolkit for environmental sustainability. *International Journal of Molecular Sciences*, *25*(16), 8616. https://doi.org/10.3390/ijms25168616

Abdel-Mageed, H M. (2025). Frontiers in nanoparticles redefining enzyme immobilization: A review addressing challenges, innovations, and unlocking sustainable future potentials. *Micro and Nano Systems Letters*, *13*(1). https://doi.org/10.1186/s40486-025-00228-2

Abu-Tahon, M A, Housseiny, M M, Aboelmagd, H I, Daifalla, N, Khalili, M, Isichei, A C, Ramadan, A S, El-Saad, A M A, Seddek, N H, Ebrahim, D, Ali, Y H, Saeed, I K, Rikabi, H A, & Eltaib, L (2025). A holistic perspective on the efficiency of microbial enzymes in bioremediation process: Mechanism and challenges: A review. *International Journal of Biological Macromolecules*, *142278*. https://doi.org/10.1016/j.ijbiomac.2025.142278

Afshar, L, Hashemi, S-M, Rezania Moallem, M-R, Aghayan, H-R, Shariatinasab, S, Azin, S-M, Hajizadeh-Saffar, E, Basiri, M, Baharvand, H, Sadighi, J, Ashtiani, M K, Hamidieh, A-A, Soleymani-Goloujeh, M, Azin, S-A, Gooshki, E S, Samani, R O, & Arjmand, B. (2020). Ethics of research on stem cells and regenerative medicine: Ethical guidelines in the Islamic Republic of Iran. *Stem Cell Research & Therapy*, *11*(1). https://doi.org/10.1186/s13287-020-01916-z

Afzal, M R, Naz, M, Yu, Y, Yan, L, Wang, P, Mohotti, J, Hao, G, Zhou, J, Chen, Z, Zhang, L, & Wang, Q (2025). Root exudates: The rhizospheric frontier for advancing sustainable plant protection. *Resources Environment and Sustainability*, *21*, 100249. https://doi.org/10.1016/j.resenv.2025.100249

Akintola, A A (2024). AI-driven monitoring systems for bioremediation: Real-time data analysis and predictive modelling. *World Journal of Advanced Research and Reviews*, *24*(1), 788–803. https://doi.org/10.30574/wjarr.2024.24.1.3099

Alizadeh, N, Salimi, A, Sham, T, Bazylewski, P, & Fanchini, G (2020). Intrinsic enzyme-like activities of cerium oxide nanocomposite and its application for extracellular H_2O_2 detection using an electrochemical microfluidic device. *ACS Omega*, *5*(21), 11883–11894. https://doi.org/10.1021/acsomega.9b03252

Aminian-Dehkordi, J, Rahimi, S, Golzar-Ahmadi, M, Singh, A, Lopez, J, Ledesma-Amaro, R, & Mijakovic, I (2023). Synthetic biology tools for environmental protection. *Biotechnology Advances*, *68*, 108239. https://doi.org/10.1016/j.biotechadv.2023.108239

Ang, T, Maiangwa, J, Salleh, A B, Normi, Y M, & Leow, T C (2018). Dehalogenases: From improved performance to potential microbial dehalogenation applications. *Molecules*, *23*(5), 1100. https://doi.org/10.3390/molecules23051100

Aryal, M (2024). Phytoremediation strategies for mitigating environmental toxicants. *Heliyon*, *10*(19), e38683. https://doi.org/10.1016/j.heliyon.2024.e38683

Ayub, A, Wani, A K, Malik, S M, Ayub, M, Chopra, C, Singh, R, & Malik, T (2025). Harnessing microbes and plants for bioremediation of heavy metal contaminants: Current paradigms and future perspectives. *Environmental Challenges*, *101220*. https://doi.org/10.1016/j.envc.2025.101220

Bakar, B, Akbulut, M, Ulusal, F, Ulu, A, Özdemir, N, & Ateş, B (2024). Horseradish peroxidase immobilized onto mesoporous magnetic hybrid nanoflowers for enzymatic decolorization of textile dyes: A highly robust bioreactor and boosted enzyme stability. *ACS Omega*, *9*(23), 24558–24573. https://doi.org/10.1021/acsomega.4c00703

Banan, D, Sher, A, Doty, S L, & Kim, S (2024). Endophyte mediated *Populus trichocarpa* water use efficiency is dependent on time of day and plant water status. *Phytobiomes Journal*, *8*(3), 309–317. https://doi.org/10.1094/pbiomes-11-22-0077-r

Bansal, M, Santhiya, D, & Sharma, J G (2024). Mechanistic understanding on the uptake of micro-nano plastics by plants and its phytoremediation. *Environmental Science and Pollution Research*, *31*(6), 8354–8368. https://doi.org/10.1007/s11356-023-31680-5

Bhatt, A, Prajapati, D, & Gupte, A (2021). Current status and future of nitrile catalysis using key nitrilases enzymes and their biotechnological impact. *The Open Biotechnology Journal*, *15*(1), 71–81. https://doi.org/10.2174/1874070702115010071

Bittencourt, G A, De Souza Vandenberghe, L P, Martínez-Burgos, W J, Valladares-Diestra, K K, De Mello, A F M, Maske, B L, Brar, S K, Varjani, S, De Melo Pereira, G V, & Soccol, C R (2023). Emerging contaminants bioremediation by enzyme and nanozyme-based processes – A review. *iScience*, *26*(6), 106785. https://doi.org/10.1016/j.isci.2023.106785

Boros-Lajszner, E, Wyszkowska, J, & Kucharski, J (2023). Evaluation and assessment of trivalent and hexavalent chromium on *Avena sativa* and soil enzymes. *Molecules*, *28*(12), 4693. https://doi.org/10.3390/molecules28124693

Bortoloti, G A, & Baron, D (2022). Phytoremediation of toxic heavy metals by Brassica plants: A biochemical and physiological approach. *Environmental Advances*, *8*, 100204. https://doi.org/10.1016/j.envadv.2022.100204

Cavalcante, A L G, Dari, D N, Da Silva Aires, F I, De Castro, E C, Santos, K M D, & Santos, J C S D (2024). Advancements in enzyme immobilization on magnetic nanomaterials: Toward sustainable industrial applications. *RSC Advances*, *14*(25), 17946–17988. https://doi.org/10.1039/d4ra02939a

Chakraborty, S, Talukdar, A, Dey, S, & Bhattacharya, S. (2025). Role of fungi, bacteria and microalgae in bioremediation of emerging pollutants with special reference to pesticides, heavy metals and pharmaceuticals. *Discover Environment*, *3*(1). https://doi.org/10.1007/s44274-025-00217-7

Chauhan, J, Prathibha, M, Singh, P, Choyal, P, Mishra, U N, Saha, D, Kumar, R, Anuragi, H, Pandey, S, Bose, B, Mehta, B, Dey, P, Dwivedi, K, Gupta, N, & Singhal, R K (2023). Plant photosynthesis under abiotic stresses: Damages, adaptive, and signaling mechanisms. *Plant Stress*, *10*, 100296. https://doi.org/10.1016/j.stress.2023.100296

Chen, L, & Liu, Y (2024). The function of root exudates in the root colonization by beneficial soil rhizobacteria. *Biology*, *13*(2), 95. https://doi.org/10.3390/biology13020095

Chen, S, Chen, X, Su, H, Guo, M, & Liu, H (2023). Advances in synthetic-biology-based whole-cell biosensors: Principles, genetic modules, and applications in food safety. *International Journal of Molecular Sciences*, *24*(9), 7989. https://doi.org/10.3390/ijms24097989

Chia, X K, Hadibarata, T, Kristanti, R A, Jusoh, M N H, Tan, I S, & Foo, H C Y (2024). The function of microbial enzymes in breaking down soil contaminated with pesticides: A review. *Bioprocess and Biosystems Engineering*, *47*(5), 597–620. https://doi.org/10.1007/s00449-024-02978-6

Colin, C, Akpo, E, Perrin, A, Cornu, D, & Cambedouzou, J (2024). Encapsulation in alginates hydrogels and controlled release: An overview. *Molecules*, *29*(11), 2515. https://doi.org/10.3390/molecules29112515

Cui, X, Zheng, Y, Liu, Y, Yuchi, Z, & Yuan, Y (2025). AI-driven de novo enzyme design: Strategies, applications, and future prospects. *Biotechnology Advances*, *108603*. https://doi.org/10.1016/j.biotechadv.2025.108603

Curran, L M L K, Pham, L T M, Sale, K L, & Simmons, B A (2021). Review of advances in the development of laccases for the valorization of lignin to enable the production of lignocellulosic biofuels and bioproducts. *Biotechnology Advances*, *54*, 107809. https://doi.org/10.1016/j.biotechadv.2021.107809

Dahiya, D, Sharma, H, Rai, A K, & Nigam, P S (2022). Application of biological systems and processes employing microbes and algae to Reduce, Recycle, Reuse (3Rs) for the sustainability of circular bioeconomy. *AIMS Microbiology*, *8*(1), 83–102. https://doi.org/10.3934/microbiol.2022008

Dhiman, S, Kaur, P, Narang, J, Mukherjee, G, Thakur, B, Kaur, S, & Tripathi, M (2024). Fungal bioprocessing for circular bioeconomy: Exploring lignocellulosic waste valorization. *Mycology*, *15*(4), 1–26. https://doi.org/10.1080/21501203.2024.2316824

Dkhar, D S, Swain, R P, Dubey, R, Patel, G K, & Chandra, P (2025). Plant-derived enzymes as sustainable biocatalysts for biosensing and industrial applications. *Industrial Crops and Products*, *233*, 121336. https://doi.org/10.1016/j.indcrop.2025.121336

Dong, C, Tiwari, A, Anisha, G S, Chen, C, Singh, A, Haldar, D, Patel, A K, & Singhania, R R (2023). Laccase: A potential biocatalyst for pollutant degradation. *Environmental Pollution*, *319*, 120999. https://doi.org/10.1016/j.envpol.2023.120999

Du, M, Liu, J, Wang, Q, Wang, F, Bi, L, Ma, C, Song, M, & Jiang, G (2023). Immobilization of laccase on magnetic PEGDA–CS inverse opal hydrogel for enhancement of bisphenol A degradation in aqueous solution. *Journal of Environmental Sciences*, *147*, 74–82. https://doi.org/10.1016/j.jes.2023.10.017

Egelkamp, R, Zimmermann, T, Schneider, D, Hertel, R, & Daniel, R (2019). Impact of nitriles on bacterial communities. *Frontiers in Environmental Science*, *7*. https://doi.org/10.3389/fenvs.2019.00103

Ejaz, U, Khan, S M, Khalid, N, Ahmad, Z, Jehangir, S, Rizvi, Z F, Lho, L H, Han, H, & Raposo, A (2023). Detoxifying the heavy metals: A multipronged study of tolerance strategies against heavy metals toxicity in plants. *Frontiers in Plant Science*, *14*. https://doi.org/10.3389/fpls.2023.1154571

Evode, N, Qamar, S A, Bilal, M, Barceló, D, & Iqbal, H M (2021). Plastic waste and its management strategies for environmental sustainability. *Case Studies in Chemical and Environmental Engineering*, *4*, 100142. https://doi.org/10.1016/j.cscee.2021.100142

Fahim, Y A, Ragab, W M, Hasani, I W, & El-Khawaga, A M. (2025). Biomedical and environmental applications via nanobiocatalysts and enzyme immobilization. *European Journal of Medical Research*, *30*(1). https://doi.org/10.1186/s40001-025-02782-2

Finny, A S (2024). 3D bioprinting in bioremediation: A comprehensive review of principles, applications, and future directions. *PeerJ*, *12*, e16897. https://doi.org/10.7717/peerj.16897

García-García, J D, Van Gelder, K, Joshi, J, Bathe, U, Leong, B J, Bruner, S D, Liu, C C, & Hanson, A D (2021). Using continuous directed evolution to improve enzymes for plant applications. *Plant Physiology*, *188*(2), 971–983. https://doi.org/10.1093/plphys/kiab500

Gkantzou, E, Chatzikonstantinou, A V, Fotiadou, R, Giannakopoulou, A, Patila, M, & Stamatis, H (2021). Trends in the development of innovative nanobiocatalysts and their application in biocatalytic transformations. *Biotechnology Advances*, *51*, 107738. https://doi.org/10.1016/j.biotechadv.2021.107738

Gu, Y, Wang, H, Yang, Y, Chen, H, Chen, C, & Cheng, W. (2025). Metabonomics reveals the mechanism of stress resistance in *Vetiveria zizanioides* inoculated with AMF under copper stress. *Scientific Reports*, *15*(1). https://doi.org/10.1038/s41598-025-90595-y

Gundlapalli, M, Sivagami, K, Gopalakrishnan, M, Harshini, P, Janjaroen, D, & Ganesan, S (2024). Biodegradation of low molecular weight polycyclic aromatic hydrocarbons in soil: Insights into bacterial activities and bioremediation techniques. *Sustainable Chemistry for the Environment*, *7*, 100146. https://doi.org/10.1016/j.scenv.2024.100146

Hayakawa, T, Suzuki, H, Yamamoto, H, & Mitsuda, N (2024). Review article for *Plant Biotechnology*. *Plant Biotechnology*, *41*(3), 173–193. https://doi.org/10.5511/plantbiotechnology.24.0630b

He, S, Wang, X, Wu, X, & Yin, Y, & Q, L., MA. (2020). Using rice as a remediating plant to deplete bioavailable arsenic from paddy soils. *Environment International*, *141*, 105799. https://doi.org/10.1016/j.envint.2020.105799

Holden, N M, Neill, A M, Stout, J C, O'Brien, D, & Morris, M A (2022). Biocircularity: A framework to define sustainable, circular bioeconomy. *Circular Economy and Sustainability*, *3*(1), 77–91. https://doi.org/10.1007/s43615-022-00180-y

Hussain, N, Shahbaz, A, Malik, H A, Ehsan, F, Santos, J C S D, & Balčiūnaitė, A (2024). Porous nanomaterials for enzyme immobilization and bioremediation applications. *Microbes Based Approaches for the Management of Hazardous Contaminants*, 146–161. https://doi.org/10.1002/9781119851158.ch9

Islam, M M, Saxena, N, & Sharma, D (2024). Phytoremediation as a green and sustainable prospective method for heavy metal contamination: A review. *RSC Sustainability*, *2*(5), 1269–1288. https://doi.org/10.1039/d3su00440f

Jassal, P S, Kudave, P S, Wani, A K, & Yadav, T. (2025). Prospects of phytoremediation in degradation of environmental contaminants: Recent advances, challenges and way forward. *International Journal of Phytoremediation*, 1–18. https://doi.org/10.1080/15226514.2025.2500643

Kafle, A, Timilsina, A, Gautam, A, Adhikari, K, Bhattarai, A, & Aryal, N (2022). Phytoremediation: Mechanisms, plant selection and enhancement by natural and synthetic agents. *Environmental Advances*, *8*, 100203. https://doi.org/10.1016/j.envadv.2022.100203

Karnwal, A, Kumar, G, Mahmoud, A E D, Dutta, J, Singh, R, Al-Tawaha, A R M S, & Malik, T (2025). Eco-engineered remediation: Microbial and rhizosphere-based strategies for heavy metal detoxification. *Current Research in Biotechnology*, *100297*. https://doi.org/10.1016/j.crbiot.2025.100297

Khafaga, D S R, Muteeb, G, Elgarawany, A, Radwan, M G, Allam, S, Farhan, M, Aatif, M, & Almatar, B A (2024). Green nanobiocatalysts: Enhancing enzyme immobilization for industrial and biomedical applications. *PeerJ*, *12*, e17589. https://doi.org/10.7717/peerj.17589

Khatua, S. (2025). Myco-enzymes as a potent tool for the degradation of xenobiotics. In *Industrial crops and products* (pp. 549–584). https://doi.org/10.1007/978-981-96-8260-7_21

Kim, M, Shim, J Y, Lim, S, Lee, H, Kwon, S C, Hong, S, & Ryu, S. (2024). Reduction of greenhouse gas emissions by optimizing the textile dyeing process using digital twin technology. *Fashion and Textiles*, *11*(1). https://doi.org/10.1186/s40691-024-00384-w

Košnář, Z, Mercl, F, & Tlustoš, P (2020). Long-term willows phytoremediation treatment of soil contaminated by fly ash polycyclic aromatic hydrocarbons from straw combustion. *Environmental Pollution*, *264*, 114787. https://doi.org/10.1016/j.envpol.2020.114787

Kostas, E T, Adams, J M M, Ruiz, H A, Durán-Jiménez, G, & Lye, G J (2021). Macroalgal biorefinery concepts for the circular bioeconomy: A review on biotechnological developments and future perspectives. *Renewable and Sustainable Energy Reviews*, *151*, 111553. https://doi.org/10.1016/j.rser.2021.111553

Kumar, S, Banerjee, S, Dutta, S, Roy, P K, & Bhattacharyya, P (2025). Evaluating the impact of arbuscular mycorrhizal fungi on the phytoremediation potential of *Vetiveria zizanioides* and *Saccharum spontaneum* on coal mine waste, India. *Journal of Environmental Chemical Engineering*, *116593*. https://doi.org/10.1016/j.jece.2025.116593

Kuppan, N, Padman, M, Mahadeva, M, Srinivasan, S, & Devarajan, R (2024). A comprehensive review of sustainable bioremediation techniques: Eco friendly solutions for waste and pollution management. *Waste Management Bulletin*, *2*(3), 154–171. https://doi.org/10.1016/j.wmb.2024.07.005

Kuppan, P, Sudharsanam, A, Venkateswarlu, K, & Megharaj, M. (2023). Solar technology–closed loop synergy facilitates low-carbon circular bioeconomy in microalgal wastewater treatment. *npj Clean Water*, *6*(1). https://doi.org/10.1038/s41545-023-00256-8

Landberg, T, & Greger, M (2022). Phytoremediation using willow in industrial contaminated soil. *Sustainability*, *14*(14), 8449. https://doi.org/10.3390/su14148449

Lee, Y, Cho, K, & Yun, J (2025). Phytoremediaton strategies for co-contaminated soils: Overcoming challenges, enhancing efficiency, and exploring future advancements and innovations. *Processes*, *13*(1), 132. https://doi.org/10.3390/pr13010132

Leong, H Y, Show, P L, Chew, K W, Chang, C-K, Lim, J W, Khoo, K S, Chia, S R, & Chang, J-S. (2021). Waste biorefinery towards a sustainable circular bioeconomy: A solution to global issues. *Biotechnology for Biofuels*, *14*(1). https://doi.org/10.1186/s13068-021-01939-5

Li, X, Tian, J, Ren, X, Wang, J, & Liu, X (2025). Enhancement in the catalytic properties of CotA laccase from *Bacillus pumilus* via high-throughput screening using malachite green as a pressure. *Microorganisms*, *13*(2), 377. https://doi.org/10.3390/microorganisms13020377

Luciński, R, & Adamiec, M (2023). The role of plant proteases in the response of plants to abiotic stress factors. *Frontiers in Plant Physiology*, *1*. https://doi.org/10.3389/fphgy.2023.1330216

Maglione, G, Zinno, P, Tropea, A, Mussagy, C U, Dufossé, L, Giuffrida, D, & Mondello, A (2024). Microbes' role in environmental pollution and remediation: A bioeconomy focus approach. *AIMS Microbiology*, *10*(3), 723–755. https://doi.org/10.3934/microbiol.2024033

Mahalle, S, Bhende, R S, Bokade, P, Bajaj, A, & Dafale, N A (2025). Emerging microbial remediation methods for rejuvenation of pesticide-contaminated sites. *Total Environment Microbiology*, *1*(3), 100026. https://doi.org/10.1016/j.temicr.2025.100026

Makam, P, Yamijala, S S R K C, Bhadram, V S, Shimon, L J W, Wong, B M, & Gazit, E. (2022). Single amino acid bionanozyme for environmental remediation. *Nature Communications*, *13*(1). https://doi.org/10.1038/s41467-022-28942-0

Manganyi, M C, Dikobe, T B, & Maseme, M R (2024). Exploring the potential of endophytic microorganisms and nanoparticles for enhanced water remediation. *Molecules (Basel, Switzerland)*, *29*(12), 2858. https://doi.org/10.3390/molecules29122858

Mani, D, & Kumar, C (2013). Biotechnological advances in bioremediation of heavy metals contaminated ecosystems: An overview with special reference to phytoremediation. *International Journal of Environmental Science and Technology*, *11*(3), 843–872. https://doi.org/10.1007/s13762-013-0299-8

Mazari, A M A, Zhang, L, Ye, Z, Zhang, J, Tew, K D, & Townsend, D M (2023). The multifaceted role of glutathione S-transferases in health and disease. *Biomolecules*, *13*(4), 688. https://doi.org/10.3390/biom13040688

Meng, H, Zhao, J, Yang, Y, Diao, K, Zheng, G, Li, T, Dai, X, & Li, J (2023). PeGSTU58, a glutathione S-transferase from *Populus euphratica*, enhances salt and drought stress tolerance in transgenic *Arabidopsis*. *International Journal of Molecular Sciences*, *24*(11), 9354. https://doi.org/10.3390/ijms24119354

Mokrani, S, Houali, K, Yadav, K K, Arabi, A I A, Eltayeb, L B, AwjanAlreshidi, M, Benguerba, Y, Cabral-Pinto, M M, & Nabil, E (2024). Bioremediation techniques for soil organic pollution: Mechanisms, microorganisms, and technologies – A comprehensive review. *Ecological Engineering*, *207*, 107338. https://doi.org/10.1016/j.ecoleng.2024.107338

Monroy-Licht, A, Carranza-Lopez, L, De La Parra-Guerra, A C, & Acevedo-Barrios, R (2024). Unlocking the potential of *Eichhornia crassipes* for wastewater treatment: Phytoremediation of aquatic pollutants, a strategy for advancing sustainable development goal-06 clean water. *Environmental Science and Pollution Research*, *31*(31), 43561–43582. https://doi.org/10.1007/s11356-024-33698-9

Motshumi, K K, Mbangi, A, Van Der Van Der Watt, E, & Khetsha, Z P (2025). Phytoremediation potential of silicon-treated *Brassica juncea* L. In mining-affected water and soil composites In South Africa: A review. *Agriculture*, *15*(15), 1582. https://doi.org/10.3390/agriculture15151582

Mousavi, S M, Hashemi, S A, Moezzi, S M I, Ravan, N, Gholami, A, Lai, C W, Chiang, W, Omidifar, N, Yousefi, K, & Behbudi, G (2021). Recent advances in enzymes for the bioremediation of pollutants. *Biochemistry Research International, 2021*, 1–12. https://doi.org/10.1155/2021/5599204

Nagendran, V, Goveas, L C, Vinayagam, R, Varadavenkatesan, T, & Selvaraj, R (2024). Nanozymes in environmental remediation: A bibliometric and comprehensive review of their oxidoreductase-mimicking capabilities. *Microchemical Journal, 111748*. https://doi.org/10.1016/j.microc.2024.111748

Narayanan, M, Ali, S S, & El-Sheikh, M (2023a). A comprehensive review on the potential of microbial enzymes in multipollutant bioremediation: Mechanisms, challenges, and future prospects. *Journal of Environmental Management, 334*, 117532. https://doi.org/10.1016/j.jenvman.2023.117532

Narayanan, M, Ma, Y, Obaid, S A, Alfarraj, S, Duc, P A, & Karuppusamy, I (2023b). *Eichhornia crassipes* biochar aided pollutants sorption competence of multi-metal tolerant fungi species on South Pennar river. *Environmental Research, 231*, 116152. https://doi.org/10.1016/j.envres.2023.116152

Nashruddin, S N A B M, Salleh, F H M, Yunus, R M, & Zaman, H B (2024). Artificial intelligence–powered electrochemical sensor: Recent advances, challenges, and prospects. *Heliyon, 10*(18), e37964. https://doi.org/10.1016/j.heliyon.2024.e37964

Ndochinwa, O. G., Wang, Q., Amadi, O. C., Nwagu, T. N., Nnamchi, C. I., Okeke, E. S., & Moneke, A. N. (2024). Current status and emerging frontiers in enzyme engineering: An industrial perspective. *Heliyon, 10*(11), e32673. https://doi.org/10.1016/j.heliyon.2024.e32673

Nuruzzaman, M, Bahar, M M, & Naidu, R (2025). Diffuse soil pollution from agriculture: Impacts and remediation. *The Science of the Total Environment, 962*, 178398. https://doi.org/10.1016/j.scitotenv.2025.178398

Oubohssaine, M, & Dahmani, I (2024). Phytoremediation: Harnessing plant power and innovative technologies for effective soil remediation. *Plant Stress, 14*, 100578. https://doi.org/10.1016/j.stress.2024.100578

Palabiyik, A A (2025). Synzymes: The future of modern enzyme engineering. *Applied Biochemistry and Biotechnology*. https://doi.org/10.1007/s12010-025-05305-1

Palladino, F, Scholl, A E, José, Á H M, Rosa, C A, Marceline, P R F, Santos, I J B, Fibrin, D L, & Rodrigues, R D C L B (2024). Bioreactors: Applications and innovations for a sustainable and healthy future—A critical review. *Applied Sciences, 14*(20), 9346. https://doi.org/10.3390/app14209346

Popescu, S M, Mansour, S, Wani, O A, Kumar, S S, Sharma, V, Sharma, A, Arya, V M, Kirkham, M B, Hou, D, Bolan, N, & Chung, Y S (2024). Artificial intelligence and IoT driven technologies for environmental pollution monitoring and management. *Frontiers in Environmental Science, 12*. https://doi.org/10.3389/fenvs.2024.1336088

Qureshi, F F, Ashraf, M A, Rasheed, R, Hussain, I, Rizwan, M, Iqbal, M, & Yong, J W H (2024). Microbial-assisted alleviation of chromium toxicity in plants: A critical review. *Plant Stress, 11*, 100394. https://doi.org/10.1016/j.stress.2024.100394

Radley, E, Davidson, J, Foster, J, Obexer, R, Bell, E L, & Green, A P (2023). Engineering enzymes for environmental sustainability. *Angewandte Chemie (International Ed. in English), 62*(52). https://doi.org/10.1002/anie.202309305

Rathore, S, Varshney, A, Mohan, S, & Dahiya, P (2022). An innovative approach of bioremediation in enzymatic degradation of xenobiotics. *Biotechnology and Genetic Engineering Reviews, 38*(1), 1–32. https://doi.org/10.1080/02648725.2022.2027628

Riyazuddin, R, Bela, K, Horváth, E, Rigó, G, Gallé, Á, Szabados, L, Fehér, A, & Csiszár, J (2019). Overexpression of the Arabidopsis glutathione peroxidase-like 5 gene (AtGPXL5) resulted in altered plant development and redox status. *Environmental and Experimental Botany, 167*, 103849. https://doi.org/10.1016/j.envexpbot.2019.103849

Sadiq, F K, Sadiq, A A, Matsika, T A, & Momoh, B A (2025). Sustainable remediation of persistent organic pollutants: A review on recent innovative technologies. *Current Research in Biotechnology, 100293*. https://doi.org/10.1016/j.crbiot.2025.100293

Salas-Nuñez, L F, Barrera-Ocampo, A, Caicedo, P A, Cortes, N, Osorio, E H, Villegas-Torres, M F, & Barrios, A F G (2024). Machine learning to predict enzyme–substrate interactions in elucidation of synthesis pathways: A review. *Metabolites, 14*(3), 154. https://doi.org/10.3390/metabo14030154

Scheibel, D M, Gitsov, I P I, & Gitsov, I (2024). Enzymes in "Green" synthetic chemistry: Laccase and lipase. *Molecules, 29*(5), 989. https://doi.org/10.3390/molecules29050989

Sharma, S, Dadhwal, R, & Banerjee, R (2025). Nanoparticle assisted phytoremediation: An eco-friendly approach for removal of heavy metals from the environment. *Journal of Environmental Sciences*. https://doi.org/10.1016/j.jes.2025.05.034

Shi, Z, Xu, S, Xue, S, Chen, K, Lu, Y, Wang, F, Long, S, Tian, Y, Zhang, P, Wang, J, Gu, Y, Zhou, J, Zhou, H, Meng, S, & Cui, H (2025). From machine learning to multimodal models: The AI revolution in enzyme engineering. *BioDesign Research, 100044*. https://doi.org/10.1016/j.bidere.2025.100044

Silva, L, Rodríguez-Sedano, F, Baptista, P, & Coelho, J P (2023). The digital twin paradigm applied to soil quality assessment: A systematic literature review. *Sensors, 23*(2), 1007. https://doi.org/10.3390/s23021007

Singh, A K, Fernandez-Lafuente, R, Schmidt, J E, Boczkaj, G, & Bilal, M (2024a). Biocatalytic functionalities of lignin peroxidase-based systems in lignin depolymerization and pollutants removal from environmental matrices. *Current Pollution Reports, 10*(3), 345–361. https://doi.org/10.1007/s40726-024-00310-0

Singh, S, Kaur, N P, Varshney, H, Khan, R, Kumar, N A, Sharma, R, Singh, A K, & Ramamurthy, J, P C. (2024b). Enzyme-based sensor for the real-time detection of atrazine: Evidence from electrochemical and docking studies. *Scientific Reports, 14*(1). https://doi.org/10.1038/s41598-024-65801-y

Sondhi, S, Chopra, N K, Kumar, A, & Gupta, N (2023). Laccase: A green solution for environmental problems. *Advances in Environmental and Engineering Research, 04*(02), 1–32. https://doi.org/10.21926/aeer.2302030

Suresh, V, Shams, R, Dash, K K, Shaikh, A M, & Béla, K (2025). Comprehensive review on enzymatic polymer degradation: A sustainable solution for plastics. *Journal of Agriculture and Food Research, 101788*. https://doi.org/10.1016/j.jafr.2025.101788

Tadesse, M, & Liu, Y (2025). Recent advances in enzyme immobilization: The role of artificial intelligence, novel nanomaterials, and dynamic carrier systems. *Catalysts, 15*(6), 571. https://doi.org/10.3390/catal15060571

Twala, P P, Mitema, A, Baburam, C, & Feto, N A (2020). Breakthroughs in the discovery and use of different peroxidase isoforms of microbial origin. *AIMS Microbiology, 6*(3), 330–349. https://doi.org/10.3934/microbiol.2020020

Ullah, I, Anwar, Y, Siddiqui, M F, Alsulami, N, & Ullah, R (2024). Phytoremediation of arsenic (AS) in rice plants mediated by *Bacillus subtilis* strain IU31 through alleviation of antioxidant responses and gibberellins and indole-3-acetic acid synthesis. *Environmental Pollution*, *355*, 124207. https://doi.org/10.1016/j.envpol.2024.124207

Unimke, A, Iya, N I D, Fatunla, O K, Ofon, U A, Ayanda, O S, Abdullahi, A, Olim, D M, Yahaya, S M, Abdulahi, S, Muhammad, A B, Galadima, A I, Mmuoegbulam, A O, Mohammed, S E, Okezie, O, Ibrahim, M M, & Badamasi, H (2024). Microbe-plant-nanoparticle interactions: Role in bioremediation of petroleum hydrocarbons. *Water Science and Technology: A Journal of the International Association on Water Pollution Research*, *90*(10), 2870–2893. https://doi.org/10.2166/wst.2024.362

Wang, J, & Delavar, M A (2024). Modelling phytoremediation: Concepts, methods, challenges and perspectives. *Soil & Environmental Health*, *2*(1), 100062. https://doi.org/10.1016/j.seh.2024.100062

Wang, L, Du, X, Li, Y, Bai, Y, Tang, T, Wu, J, Liang, H, & Gao, D (2023). Enzyme immobilization as a sustainable approach toward ecological remediation of organic-contaminated soils: Advances, issues, and future perspectives. *Critical Reviews in Environmental Science and Technology*, *53*(18), 1684–1708. https://doi.org/10.1080/10643389.2023.2180285

Yamaguchi, H, & Miyazaki, M (2024). Bioremediation of hazardous pollutants using enzyme-immobilized reactors. *Molecules*, *29*(9), 2021. https://doi.org/10.3390/molecules29092021

Yang, H, & Liu, J. (2025). Recent advances in microbial and bioelectrochemical strategies for degradation of per- and polyfluoroalkyl substances: Mechanisms, limitations, and research opportunities. *Biotechnology Letters*, *47*(3). https://doi.org/10.1007/s10529-025-03593-5

Yang, X, He, Y, Liu, B, Guo, H, Xue, L, Duan, Y, Hu, H, Gao, F, Zhou, L, & Zhang, J J (2022). Alfalfa's response to atrazine stress and its secreted atrazine metabolites. *Ecotoxicology and Environmental Safety*, *241*, 113780. https://doi.org/10.1016/j.ecoenv.2022.113780

Zhan, M, Xu, M, Lin, W, He, H, & He, C (2025). Graphene oxide research: Current developments and future directions. *Nanomaterials (Basel, Switzerland)*, *15*(7), 507. https://doi.org/10.3390/nano15070507

Zhang, R, Zhao, H, & Fan, K (2022). Structure-activity mechanism of iron oxide nanozymes. In *ACS symposium series* (pp. 1–35). https://doi.org/10.1021/bk-2022-1422.ch001

19 Interaction Between Plant Enzymes and Hormonal Regulation in Growth

Muhammad Imran, Dan-Dan Zhao, Shifa Shaffique, Ibrahim Khan, Odongkara Peter, Md. Injamum-Ul-Hoque, Sang-Mo Kang, Abdul Latif Khan, and In-Jung Lee

19.1 INTRODUCTION

Plant development and growth involves a web of biochemical, molecular, and complex interactions between enzymes and hormones rather than simple cell division and elongation. Starting from seed germination until the plant develops flowers and fruit, the hormonal and enzymatic processes and signals constantly communicate with each other. This forms a significant coordinating system that allows plants to react accordingly to both their internal development and changes in their environment (Santner and Estelle, 2009, Mazzoni-Putman *et al.*, 2021). Among these phytohormones, cytokinins, ethylene, auxins (IAA), gibberellins (GAs), brassinosteroids (BRs), abscisic acid (ABA), and jasmonates (JAs) play a significant role in coordinating developmental processes across stages of growth and tissues. These hormonal signals rarely act on their influence and are frequently channeled through specific enzymes responsible for metabolic fluxes, cell wall remodeling, and downstream signaling cascades. At the same time, enzymes are responsible for making hormones, breaking them down, modifying them, and executing the metabolic changes that actually make growth happen. This two-way interplay creates a dynamic feedback system in which enzymes keep hormone levels in check, while hormones tell enzymes when and how much to work (Ismail *et al.*, 2021). The biosynthesis of plant hormones highlights this complex interplay between enzymes and hormones, which work together rather than operating independently. Recent advances in molecular genetics, proteomics, and metabolomics have illuminated many of these interactions. Take auxin, for example; its biosynthesis involves flavin monooxygenases and tryptophan aminotransferases, while GA production requires the sequential action of cytochrome P450 monooxygenases, 2-oxoglutarate-dependent dioxygenases, and terpene synthases (Zhao, 2010, Ismail *et al.*, 2021).

Furthermore, auxin-regulated cell wall loosening enzymes are shown to influence tissue elongation by switching on genes for these wall-modifying enzymes through proteins called auxin response factors. In contrast, cytokinin oxidase/dehydrogenase enzymes fine-tune cytokinin levels during shoot development (Perrot-Rechenmann, 2010). Similarly, GAs promote stem growth; they do so by destroying DELLA proteins, which normally act as brakes on genes that encode growth-stimulating enzymes, and the GA 20-oxidases and GA 2-oxidases tightly control the GA metabolism, which adjust hormone availability in response to environmental stimuli (Hedden and Thomas, 2012, Bhatla and Lal, 2023). The connection between enzyme and hormone goes well beyond just making and breaking down hormones. There's a whole suite of enzymes that modify cell walls, like xyloglucan endotransglucosylases/hydrolases (XTHs), expansins, and pectin methylesterases, which are absolutely crucial for putting hormonal growth signals into action (Robichaux, 2024). This interaction between hormones and enzymes becomes important, especially in stress response. Hormones such as JA and ABA adjust the enzyme activity which protect against oxidative damage, regulate water balance, and produce defensive compounds. Interestingly, a transcription factor phytochrome-interacting factor 4 (PIF4) integrates signals from multiple hormones and acts like a master switch to help plants adjust their shape in response to temperature changes and shade avoidance responses (Koini *et al.*, 2009, Casal and Fankhauser, 2023). Similarly, the receptor proteins such as BR insensitive 1 (BRI1) that detect BRs and directly activate other proteins through phosphorylation, setting off cascades of enzymatic reactions that ultimately change which genes are expressed and how the cell's metabolism should operate (Wang *et al.*, 2012). Collectively, these findings underscore that plant growth regulation is not the result of linear pathways but rather an integrated network of feedback loops linking enzymatic activity to hormonal balance. Understanding how enzymes and hormones interact can have profound implications for agriculture and biotechnology. Manipulating these regulatory networks offers strategies for improving stress tolerance, crop yield, and developmental plasticity. Recent technological advances in metabolomics, genomics, and proteomics have shown unprecedented complexity in these interactions, revealing novel cross-talk mechanisms between different hormonal pathways and regulatory nodes (Singhal *et al.*, 2021). In this chapter, we will explore the multifaceted interactions of enzymes and hormonal regulation that work together to control growth by examining specific examples from different stages of plant development processes and emerging concepts in the field. By integrating current knowledge about how hormone signals work, how enzymes function, and their reciprocal regulation, we aim to provide a comprehensive framework for understanding how plants manage

DOI: 10.1201/9781003669920-19

to develop with such impressive precision while remaining adaptable to their environment.

19.2 ENZYME AND HORMONE INTERACTIONS IN PLANT GROWTH REGULATION

19.2.1 Gibberellins (GAs) and Growth Enzymes

Gibberellins (GAs) constitute a large family of diterpenoid phytohormones that regulate stem elongation, fruit development, seed germination, and flowering. GA biosynthesis proceeds via the terpenoid pathway in endoplasmic reticulum and plastids, where GA3-oxidases (GA3ox) and GA20-oxidases (GA20ox) catalyze the successive conversion of bioactive GAs such as GA_1 and GA_4 (Hedden and Thomas, 2012, He *et al.*, 2019). Inactivation is mediated by GA2-oxidases (GA2ox), which convert active GAs through hydroxylation to less active or inactive derivatives. This biosynthesis–catabolism balance allows deactivation and biosynthesis, which allows plants to adjust GA levels in response to developmental or environmental signals. One of the most extensively studied showed that during seed germination, GA responses signaling in cereal cells stimulate transcription of α-amylase and associated hydrolytic enzyme production which enables the storage of proteins and mobilization of starch from the endosperm to fuel seedling growth (He *et al.*, 2019). The same GA pathway upregulates proteases and β-glucanases to loosen cell wall polymers and degrade storage reserves to facilitate reserve mobilization and radicle protrusion (Zondo and Mafa, 2025) and regulates expansins, xyloglucan endotransglycosylases, and endoglucanases that modify the reducing wall rigidity and cellulose–hemicellulose network and enable turgor-driven expansion (Lee and Kende, 2002, Zondo and Mafa, 2025). This enzymatic remodeling contributes to the GA-induced elongation phenotype. The GA action is tightly controlled by DELLA proteins, a key repressor in the GA signaling pathway, which links environmental inputs such as temperature and light to growth-promoting enzyme activity (Zondo and Mafa, 2025). A recent study shows that auxin can indirectly influence GA homeostasis through the upregulation of GA2ox gene expression in roots, therefore, crosstalking through enzyme control hubs such as GA2ox6 and GA2ox8 (Kubalová *et al.*, 2025). Moreover, glycosyltransferases (which conjugate GAs with glucose), such as GA β-D-glucosyltransferase, add another layer of enzymatic control over GA compartmentation and bioavailability (Avezum *et al.*, 2023). Thus, the GA enzyme network is deeply enmeshed with other hormonal pathways and ensures responsive coordination of growth (Figure 19.1).

19.2.2 Abscisic Acid (ABA) and Stress Enzymes

Abscisic acid (ABA) is a vital hormone integrating environmental stress responses and developmental regulation. Its biosynthesis originates from carotenoids in plastids, with *NCED* (9-cis-epoxycarotenoid dioxygenase) catalyzing the rate-limiting cleavage that produces xanthoxin, a direct

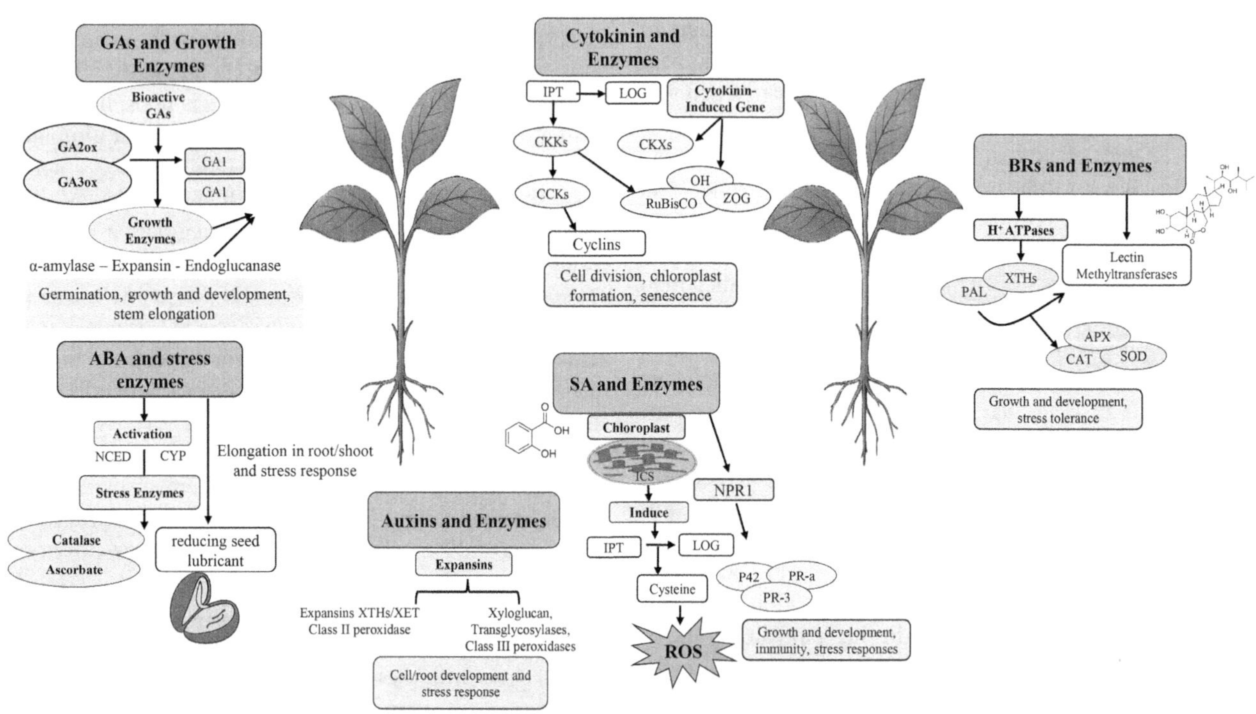

FIGURE 19.1 Schematic representation of plant hormone-regulating enzymatic pathways controlling plant development, growth, and stress responses. Gibberellins, ABA, auxins, cytokinins, SA, and BRs regulate specific enzymes such as peroxidases, oxidases, and transferases that drive cell elongation, chloroplast formation, seed germination, and antioxidant defense. Coordinating crosstalk between hormonal and enzymatic activities is essential for growth, immunity, and stress tolerance.

ABA precursor. Under salinity, drought, and cold stress, the *NCED* transcription is rapidly upregulated, allowing plants to elevate ABA levels and adjust their physiology to environmental conditions (Cutler *et al.*, 2010, Yang *et al.*, 2021a). ABA plays a central role in inducing antioxidant defense systems to protect cellular components from oxidative damage. It enhances the transcription or activity of catalase (CAT), ascorbate peroxidase (APX), and superoxide dismutase (SOD), thereby scavenging reactive oxygen species (ROS) during stress (Mittler and Blumwald, 2015, Rao *et al.*, 2025). Through these enzymes, ABA signaling minimizes oxidative injury to membranes, proteins, and nucleic acids, enabling plants to maintain cellular homeostasis under environmental challenges (Figure 19.1). In the seeds, ABA functions as an antagonist to GA-driven hydrolytic enzymes by repressing hydrolytic enzyme synthesis, such as α-amylases, in aleurone cells (Zhu, 2016). This suppression helps avoid premature mobilization of starch reserves and enforces seed dormancy until favorable germination conditions. The ABA and GA antagonism thus acts as a checkpoint, integrating environmental cues of germination timing. At the signaling level, ABA-regulated protein phosphatases such as *SnRK2* kinases and *PP2Cs* modulate downstream transcription factors and enzymes, linking stress perception to metabolic reprogramming (Finkelstein, 2013). Emerging studies show that in tomato plants under cold stress, ABA and BR signaling can positively regulate ABA biosynthesis genes, enhancing chilling tolerance (An *et al.*, 2023). Similarly, BR signaling via BZR1 can downregulate ABA levels to promote growth in *Arabidopsis* plants (Moon *et al.*, 2021). Moreover, the ABA and BR crosstalk modulates antioxidant enzyme activities, balancing defense and growth under water deficit (Zhu, 2016). These interactions imply that ABA control over stress enzymes is not isolated but integrated into a broader hormonal network that dynamically balances growth and resilience.

19.2.3 Auxins and Enzymatic Modulation

Auxins are crucial plant hormones that sculpt plant architecture, growth responses, and organ patterning. Auxin biosynthesis predominantly follows the IPA (indole-3-pyruvic acid) pathway, where TAA1/TAR (tryptophan aminotransferases) convert tryptophan to IPA, and flavin monooxygenases of the YUCCA family oxidize IPA to IAA (indole-3-acetic acid), a bioactive auxin (Luo and Di, 2023). Spatial auxin gradients are dynamically regulated by auxin-conjugating enzymes (GH3 family) and hydrolases that release active IAA from stored conjugates, thereby fine-tuning hormonal homeostasis under shifting developmental or environmental conditions. Auxin signaling orchestrates a battery of cell wall-modifying enzymes essential for cell expansion (Luo *et al.*, 2023). Expansins, XTHs/XETs (xyloglucan endotransglycosylases/hydrolases), and certain class III peroxidases restructure the cell wall matrix, reducing mechanical resistance to turgor-driven growth. Auxin also upregulates plasma membrane H^+-ATPases and apoplastic hydrolases, resulting in cell wall acidification, a key feature for the acid growth hypothesis (Takahashi *et al.*, 2012, Luo *et al.*, 2023). Acidification activates the expansins, which loosen the hemicellulose and cellulose, enabling elongation of roots and shoots. Moreover, auxin-mediated control regulation of pectin methylesterases modulates pectin porosity and crosslinking, providing another layer of enzymatic control over the cell wall (Bajguz and Piotrowska-Niczyporuk, 2023). Moreover, recent evidence emphasizes feedback coupling between the mechanical trigger of these enzymes and auxin distribution. For instance, cell wall stiffness or mechanical strain can influence the trafficking or polarity of PIN auxin efflux carriers, such as mechanosensitive signaling pathways that can modulate auxin transporters (PIN proteins), thereby altering local auxin flux and biosynthetic enzymes (Luo *et al.*, 2023). Thereby, the mechanical status of the cell wall can influence auxin transport or biosynthesis, enabling growth to self-adjust to mechanical constraints. This emergent feedback ensures that auxin transporter localization, enzyme activity, and gradients remain harmonized with tissue mechanics as a dynamic interplay important for stable morphogenesis and tropic responses. This tight coupling of hormone gradients and enzyme activity allows plants to adapt growth direction and rate in response to gravity, light, and mechanical stress.

19.2.4 Jasmonates (JAs) and Defense/Development Enzymes

Jasmonic acid (JA) and its derivatives, such as JA-Ile (jasmonoyl-isoleucine) and MeJA (methyl jasmonate), are a vital class of phytohormones that coordinate plant responses to biotic and abiotic stresses while regulating developmental processes (Chini *et al.*, 2023). The jasmonate biosynthesis begins in chloroplasts through the octadecanoid pathway, where AOC (allene oxide cyclase), AOS (allene oxide synthase), and LOX (lipoxygenases) sequentially convert α-linolenic acid into OPDA (12-oxo-phytodienoic acid), which is subsequently reduced and undergoes β-oxidation in peroxisomes to yield JA (Yang *et al.*, 2013, Chini *et al.*, 2023). When the COI1–JAZ co-receptor complex senses the JA-Ile, it triggers ubiquitin-mediated degradation of JAZ repressor proteins; therefore, transcription factors such as *MYC4*, *MYC3*, and *MYC2* are activated to stimulate jasmonate-responsive gene expression (Hodnik *et al.*, 2016, Gan *et al.*, 2025). Jasmonates significantly influence numerous enzyme activities and expressions that are involved in defense responses, secondary metabolism, and developmental processes. In response to pathogen attack and herbivory, JA signaling triggers the expression of defensive enzymes such as proteinase inhibitors (PIs), polyphenol oxidases (PPO), and various biosynthetic enzymes (Wasternack and Feussner, 2018, Chini *et al.*, 2023). For instance, as a result of JA treatment, the activity of phenylalanine ammonia-lyase (PAL) (the enzyme that limits the rate of phenylpropanoid biosynthesis) is upregulated, resulting in the accumulation of defensive compounds such as phytoalexins, flavonoids, and lignins (Kumar *et al.*, 2023). Furthermore, jasmonates regulate the expression of hydroperoxide lyases and lipoxygenase enzymes, which generate organic volatile compounds acting as airborne signals for plant–plant communication and indirect defense by attracting natural enemies of herbivores (Corpas *et al.*, 2018). Beyond defense, jasmonates play crucial roles through enzymatic modulation in developmental processes. JA regulates the activity of cellulases and peroxidases during root development, influencing the root hair formation and cell wall remodeling (Zhao *et al.*, 2015, Kumar *et al.*, 2023). Jasmonates influence the pollen viability and stamen maturation by regulation of enzymes involved in pollen coat formation and lipid metabolism, with JA-deficient mutants exhibiting male sterility (Huang *et al.*, 2022). The jasmonates crosstalk with other hormones, such as GAs and auxins, involves complex enzymatic networks that calibrate growth–defense trade-offs, wherein JAZ proteins can suppress GAs and auxin signaling during JA-mediated activation by interfering with their respective transcriptional regulators (Ye and Cui, 2019).

19.2.5 Salicylic Acid (SA) and Pathogenesis-Related Enzymes

Salicylic acid (SA) is a key hormone in plant immunity, coordinating both systemic and local defense responses against hemibiotrophic and biotrophic pathogens. SA biosynthesis in plants involves two primary pathways: one is the PAL pathway in the cytosol and the other is ICS (isochorismate synthase) in chloroplasts, which predominates during pathogen infection (Figure 19.1) (Dempsey *et al.*, 2011). The ICS uses ICS1 (also known as SID2 in *Arabidopsis*) pathway to convert chorismate to isochorismate, followed by conversion to SA through enzymatic steps that remain partially characterized (Rekhter *et al.*, 2019). SA perception involves the nonexpressor of pathogenesis-related genes 1 (NPR1) protein, which functions as a transcriptional co-activator and undergoes nuclear translocation and redox-dependent oligomerization upon SA accumulation (Bothma *et al.*, 2018). Moreover, during the SA-mediated immune response, several enzymatic reprogramming activities, such as the induction of pathogenesis-related (PR) proteins, possess direct antimicrobial enzymatic activities. SA increases the accumulation of PR-1, a systemic marker of acquired resistance (SAR), PR-3 (chitinases), and PR-2 (β-1,3-glucanases), which enzymatically degrade fungal cell wall components (Xu *et al.*, 2018). Additionally, SA induces PR-9 (peroxidases) and thaumatin-like proteins (PR-5) with antifungal properties that generate ROS and catalyze lignin biosynthesis for pathogen containment (Sharma *et al.*, 2021). The enzymatic cascade enhanced the activity of CAT, SOD, and APX, which maintain ROS homeostasis during the hypersensitive response (Herrera-Vásquez *et al.*, 2017). SA significantly influences primary metabolism through enzymatic regulation by inhibiting the activity of CAT, which allows accumulation of hydrogen peroxide that serves as a signaling molecule for defense gene activation (Ku *et al.*, 2018). Furthermore, during immune responses, SA triggers the activity of alternative oxidase (AOX) in mitochondria, affecting cellular redox status and energy metabolism (Vishwakarma *et al.*, 2020). Moreover, the antagonistic crosstalk between SA and jasmonate signaling involves enzymatic nodes, where SA interacts with transcription factors to induce the NPR1, which can suppress JA-responsive gene expression, thereby prioritizing defense against biotrophic over necrotrophic pathogens (Roychowdhury *et al.*, 2025). Recent studies have revealed that SA also modulates enzymes involved in DNA repair and nucleotide metabolism, contributing to enhanced genome stability and stress tolerance during pathogen attack (Roychowdhury *et al.*, 2024).

19.2.6 Cytokinins and Enzymes

Cytokinins are adenine-derived hormones that promote shoot initiation, delay senescence, cell division, and integrate nutrient signals with growth regulation (Wu *et al.*, 2021). Their biosynthesis is primarily initiated by IPTs (isopentenyl transferases), which transfer the isopentenyl moiety from dimethylallyl diphosphate to adenosine nucleotides. Subsequent hydroxylation by *CYP735A*, a cytochrome P450 monooxygenase, and activation via LOG (LONELY GUY) enzymes refine cytokinin activity and localization (Argueso and Kieber, 2024). Cytokinin homeostasis is further maintained by cytokinin oxidase/dehydrogenase (CKX) enzymes, which degrade active molecules, providing a mechanism to regulate local cytokinin concentration in specific tissues. Cytokinins influence control over cell cycle–related and metabolic enzymes. Within chloroplasts, they enhance photosynthetic capacity by upregulating RuBisCO activase (ribulose-1,5-bisphosphate carboxylase/oxygenase) and stabilizing RuBisCO subunits, thereby boosting carbon fixation (Cortleven and Schmülling, 2015, Wu *et al.*, 2021). They modulate carbohydrate metabolism, which is involved in starch breakdown and sucrose synthesis, helping to coordinate source–sink relationships. Critically, cytokinins induce cyclins and CDKs (cyclin-dependent kinases) to regulate the cell cycle by facilitating the transitions of G1 to S and G2 to M in meristematic tissues (Yang *et al.*, 2021b). Cross-regulation between cytokinin signaling components, including the CDK–cyclin machinery and type-A/B response regulators, enables the dynamic adjustment of division rates in response to developmental or environmental cues (Sizani, 2019). This enzymatic modulation underlies processes such as leaf initiation, shoot branching, and vascular differentiation. Importantly, cytokinins are involved in crosstalk between hormones by modulating ethylene biosynthesis (Figure 19.1). In *Arabidopsis* roots, cytokinins spatially regulate ACC synthase (ACS) and ACC oxidase (ACO) genes, controlling ethylene production in defined tissue zones, which influences apical meristem size and root elongation (Yamoune *et al.*, 2024). Moreover, cytokinins can affect ACS protein stability post-translationally via modulation of

ACS turnover, in cooperation with BRs, linking cytokinin signaling to ethylene regulation (Werner and Schmülling, 2009, Klein *et al.*, 2018). This integration of cytokinin and ethylene pathways adds further enzymatic layers to developmental control and helps the plant coordinate growth, differentiation, and stress responses.

19.2.7 Ethylene and Enzyme Modulation

Ethylene is a gaseous hormone in organ abscission, senescence, fruit ripening, and responses to mechanical stress. Its biosynthesis involves the Yang cycle, where 1-aminocyclopropane-1-carboxylic acid (ACC) converts S-adenosyl-L-methionine to ACC, and ACC oxidase (ACO) oxidizes ACC to ethylene (Li *et al.*, 2025). Both ACS and ACO are tightly regulated at the post-translational, hormonal, and transcriptional levels, enabling rapid adjustment in ethylene output in response to environmental stimuli and development. Ethylene activates cell wall-degrading enzymes that drive tissue softening. During fruit floral organ abscission and ripening, it induces pectin methylesterases, polygalacturonases, cellulases, and pectate lyases, which disassemble pectic substances and cellulose microfibrils in the primary cell wall and middle lamella (Chaves *et al.*, 2006, Li *et al.*, 2025). Such enzymatic remodeling, facilitating organ separation, reduces cell adhesion and creates the textural changes characteristic of ripened fruit. Additionally, ethylene signaling also interfaces with oxidative enzymes by modulating oxidases and class III peroxidases involved in lignin polymerization; ethylene contributes to secondary cell wall strengthening or loosening depending on context (Vanholme *et al.*, 2019). For example, in stress conditions or wounding, ethylene can enhance lignin deposition to reinforce structural barriers, and sometimes, by promoting elongation in tissues, it may limit lignification to preserve flexibility. Moreover, crosstalk with auxin and jasmonate and ethylene fine-tunes the induction of these enzymes, integrating its actions into broader developmental programs and stress responses (Gan *et al.*, 2025). The auxin and jasmonate crosstalk further refines the enzyme induction, such as auxin may prime tissues for ethylene sensitivity by increasing ACS transcript levels, while jasmonate signals may amplify ethylene-induced cell wall enzyme expression under stress (González Ortega-Villaizán *et al.*, 2024). EIN3/EIL transcription factors, an ethylene signaling component, regulate downstream transcription of both oxidative enzymes and hydrolases to embed ethylene's enzymatic effects into broader hormonal networks.

19.2.8 Brassinosteroids and Enzyme Regulation

BRs are a class of polyhydroxy steroids essential for plant stress adaptation and growth processes (Clouse, 2015). Acting through receptors such as BRI1 (BR insensitive 1) initiates phosphorylation cascades, which regulate transcription factors that include BES1 and BZR1 (Li *et al.*, 2018). Activation of plasma membrane H^+-ATPases, which acidify the apoplast and activate the cell wall modification enzymes such as xyloglucan endotransglycosylases/hydrolases (XTHs), expansins, and pectin methylesterases, is one of the primary growth-promoting actions of BRs (Delesalle *et al.*, 2024). This enzymatic activity enables turgor-driven cell elongation and cell wall rigidity in stems and roots. BRs also reinforce stress tolerance pathways through upregulating antioxidant enzymes such as CAT, SOD, and APX, scavenging ROS (Figure 19.1), detoxifying enzymes such as glutathione S-transferases, and minimizing oxidative damage under salinity, drought, and extreme temperatures (Planas-Riverola *et al.*, 2019). Additionally, BR signaling modulates carbohydrate metabolism enzymes and detoxifying enzymes like glutathione S-transferases to maintain energy balance under stress. By coordinating defense-related growth-promoting enzymes, BRs ensure that plants can both protect tissues and expand under fluctuating environmental conditions. Importantly, under cold stress, the ABA and BR crosstalk can upregulate ABA biosynthetic genes (An *et al.*, 2023), while, through BZR1-mediated repression, BR may suppress ABA levels in non-stress conditions, working antagonistically. This is also demonstrated through the regulation of remorin proteins that mediate ABA and BR balance (Clouse, 2015). The BR and GA cross-talk converge, where BZR1 interacts with DELLA proteins to regulate expression of GA-responsive genes, therefore modulating shared downstream enzymes (Allen and Ptashnyk, 2017). Thus, BRs activate cell wall enzymes to bolster stress defenses, drive growth, and coordinate with ABA and GA pathways to fine-tune enzymatic regulation across developmental and environmental contexts (Figure 19.1).

19.3 ENZYME AND HORMONE CROSSTALK IN PLANT DEVELOPMENT

Plant development involves a constant exchange of feedback between enzyme activities and hormone signaling, resulting in a highly adaptable regulatory framework. Rather than acting alone, these two components operate in a highly flexible system that governs stress adaptation, growth, and differentiation (Avezum *et al.*, 2023). Hormones such as GA, ethylene, ABA, jasmonates, cytokinins, and SA function as signaling hubs that modulate enzymatic pathways across tissues and developmental stages. Through this coordination, plants optimize developmental transitions in response to environmental conditions, ensuring physiological flexibility and survival (Kurepa and Smalle, 2025). GAs and ABA exemplify antagonistic yet tightly regulated action during germination. In cereal grains, GA perception in the peripheral endosperm initiates the transcription of α-amylase and other hydrolase enzymes responsible for the mobilization of starch, supplying sugars and energy to emerging seedlings (Bethke *et al.*, 1997, Kurepa and Smalle, 2025). This process is regulated through the GID1–DELLA signaling, where GA binds to its receptor, leading to DELLA degradation, activating

GA-responsive transcription factors that activate hydrolytic enzyme genes. Conversely, ABA represses the same amylase expression and maintains seed dormancy under unfavorable conditions (Zhu, 2016). This crosstalk between ABA and GA modulates the activities of GA2-oxidases and GA20-oxidases (key enzymes in GA biosynthesis and deactivation), *NCED*, and related enzymes, modulating the local hormone balance, allowing seeds to respond flexibly to environmental conditions such as drought, temperature, and moisture (Zhou *et al.*, 2023). Thus, the environmental fluctuations or imbibition alter the transcription of these enzymes, leading to local shifts in GA/ABA ratios that determine whether a seed remains dormant or germinates. Current studies also reveal that ROS act as intermediaries in this hormonal crosstalk. During early imbibition the ROS generation can promote GA biosynthesis via oxidizing ABA or inhibiting *NCED* expression, effectively coupling redox status to hormonal balance (Tardast *et al.*, 2023). This integration of enzyme and hormone activities ensures that seed germination occurs only when metabolic readiness aligns with environmental opportunity (Figure 19.2).

Fruit ripening represents another significant process that involves hormone signaling with enzymatic modifications to produce characteristic changes in flavor, aroma, and texture. In climacteric fruits, ethylene is the key regulatory hormone, where ripening is marked by an increase in ethylene biosynthesis, catalyzed by ACC oxidase and 1-aminocyclopropane-1-carboxylic acid (ACC) synthase, whose expression is transcriptionally activated in response to rising ethylene levels, forming a positive feedback loop (Li *et al.*, 2025). Thetranscriptional activation of cell wall-degrading enzymes such as pectate lyases, polygalacturonases, and pectin methylesterases, which demethylated pectins and depolymerize in the middle lamella, resulting in cell separation and softening of fruit tissue, and upregulate cellulases and expansins, which further loosen the cellulose hemicellulose network to complete textural changes (Cosgrove, 2015, Li *et al.*, 2025). Ethylene signaling through transcription factors such as EIN3/EIL1 also upregulates genes involved in pigment and volatile biosynthesis, enhancing fruit flavor and coloration. In non-climacteric and climacteric fruit species, ABA acts synergistically, particularly during the early stages of ripening (Finkelstein, 2013, Dolgikh *et al.*, 2019). Moreover, ABA accumulation catalyzes sucrose cleavage and hexose phosphorylation, increasing sugar availability and sweetness through preceding ethylene induction, and stimulates enzymes such as invertases and hexokinases. ABA also regulates the expression of *CYP707A* and *NCED* genes to control its own degradation and synthesis while influencing ACC synthase to reinforce ethylene production. Recent studies suggest that Ca^{2+}-ATPases and calmodulin-like proteins (a calcium-dependent signaling) act downstream of ethylene and ABA to regulate enzyme gene expression during ripening (Iqbal *et al.*, 2020). This reveals that fruit ripening is not solely hormonal but integrates transcriptional, ionic, and enzymatic networks, ensuring precise coordination and timing of metabolic transitions (Figure 19.2).

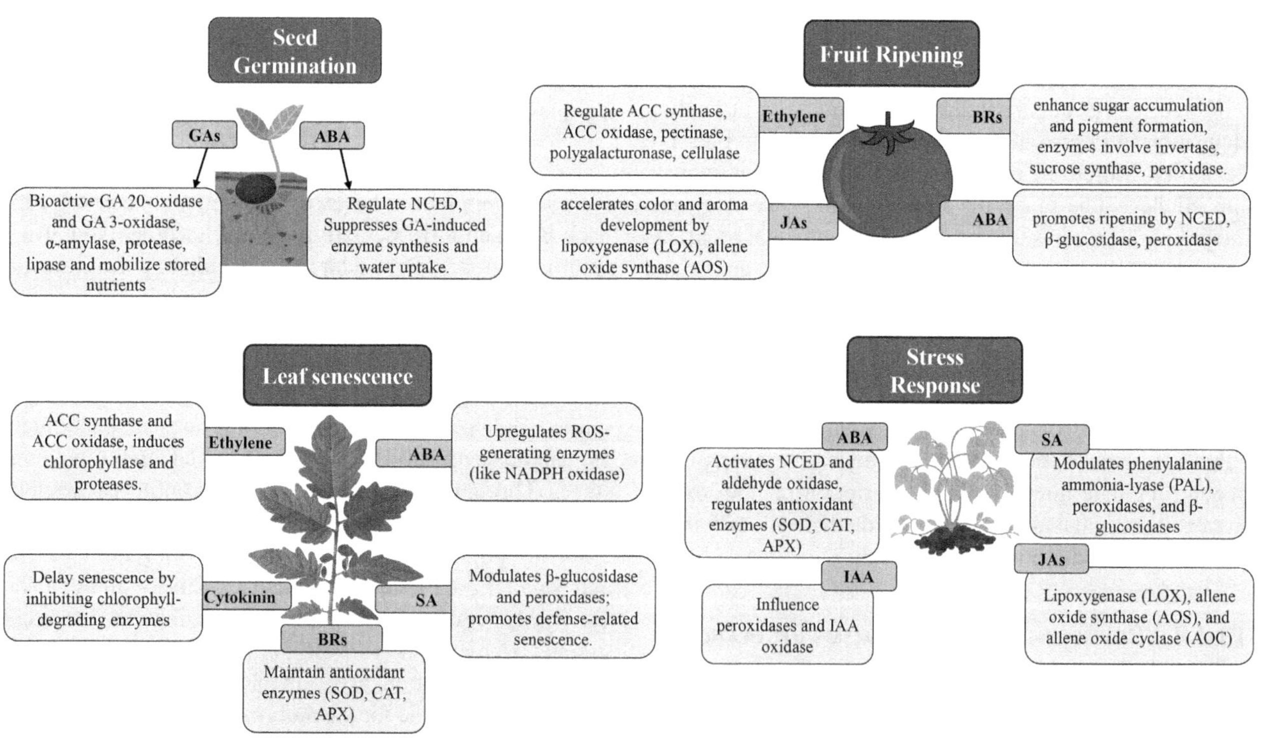

FIGURE 19.2 Hormonal regulation of enzymes during plant developmental and stress responses. The illustrations represent the interactions between major plant hormones and their associated enzymes across different physiological processes such as seed germination, leaf senescence, fruit ripening, and stress response.

Leaf and organ senescence exemplify a genetically programmed process driven by hormonal and enzymatic reprogramming that reallocates resources from aging tissues to developing or storage organs. ABA and ethylene act as key promoters of senescence by stimulating serine proteases and cysteine proteases that dismantle chloroplast and cytoplasmic proteins into transportable amino acids (Wang and Schippers, 2019). During this phase, chlorophyll degrades and enzymes such as pheophytinase, chlorophyllase, and red chlorophyll catabolite reductase are upregulated, resulting in visible leaf yellowing and the breakdown of photosynthetic apparatus components. SA and jasmonates contribute to senescence progression, particularly under stress conditions, through the activation of lipoxygenases, phospholipases, and PAL, which break down membrane lipids, releasing fatty acids for energy or signaling (He *et al.*, 2023). In contrast, cytokinins antagonize these pathways and act as anti-senescence signals by delaying the activation of catabolic enzymes and repressing senescence-associated genes (SAGs) (Werner and Schmülling, 2009, Altaf *et al.*, 2024). Furthermore, the *WRKY* transcription factors work as convergence nodes for multiple hormonal signals, such as *WRKY57* and *WRKY75*, which integrate ethylene, JA, and ABA signals, modulating the expression of catabolic enzymes involved in senescence (Huang *et al.*, 2022). Through the activation of allene oxide synthase, ethylene can further enhance JA biosynthesis, forming a reinforcing loop that accelerates senescence once initiated. Thus, enzyme activation during senescence is a hormonally coordinated process ensuring the orderly remobilization of lipid, nitrogen, and carbon reserves from aging to developing organs rather than uncontrolled tissue decay (Figure 19.2).

Under unfavorable conditions such as salinity, high temperature, drought, or pathogen attack, plants undergo a dramatic shift, where hormone and enzyme interplay determine redefined cellular priorities between defense and growth. In this regard, ABA serves as a central mediator, by upregulating antioxidant enzymes such as SOD, CAT, and APX, to help plants detoxify ROS (González Ortega-Villaizán *et al.*, 2024). Through this ABA minimizes cellular injury while activating osmoprotectant metabolism. Furthermore, enzymes such as proline dehydrogenase, pyrroline-5-carboxylate synthetase (P5CS), and sorbitol dehydrogenase are upregulated to accumulate osmolytes such as sugar alcohols and proline, which stabilize proteins and membranes under osmotic stress (Sellamuthu *et al.*, 2024). Moreover, the SA and jasmonates further modify stress responses through the activation of PAL and lipoxygenases, which direct the carbon flux toward secondary metabolites to strengthen defense barriers (He *et al.*, 2023). Ethylene acts synergistically with ABA or antagonistically to modulate growth inhibition and enhance ROS detoxification. Recently, multi-omics analyses reveal that enzyme and hormone interactions under stress are also post-transcriptionally controlled through redox modifications and protein phosphorylation cascades, ensuring rapid adaptation (Mei & Wang, 2024). These interconnected networks allow plants to preserve metabolic homeostasis and survive harsh environments by balancing enzymatic defense activation with hormone-regulated energy management (Figure 19.2).

19.4 MOLECULAR MECHANISMS OF ENZYME–HORMONE INTEGRATION

The intricate integration of enzymatic activities with hormonal signaling is a key aspect of plant biology, allowing coordinated responses in development and environmental challenges. This incorporation operates through a variety of molecular mechanisms, including transcriptional regulation to post-translational modifications, which produce dynamic regulatory networks that regulate cellular metabolism and physiology (Kurepa and Smalle, 2025). These mechanisms are essential to understand how plants achieve the remarkable plasticity required for sessile existence in fluctuating environments (Bajguz and Piotrowska-Niczyporuk, 2023). Plant hormones influence the control of enzyme gene expression at the transcriptional level through HREs (hormone-responsive cis-regulatory elements) in promoter regions. Hormone signaling pathways modulated the activity of these transcription factors at the binding sites (Figure 19.3). For instance, AuxREs (auxin-responsive elements) containing the conserved TGTCTC motif, which are perceived by ARFs (auxin response factors), become transcriptionally active upon degradation of Aux/IAA repressor proteins triggered by auxin perception through the TIR1/AFB receptor complex (Freire-Rios *et al.*, 2020). This mechanism directly links auxin perception to the expression of enzymes involved in expansins, cell wall modification, and xyloglucan endotransglucosylases, including pectinases, which collectively facilitate cell organogenesis and expansion (Luo and Di, 2023). Similarly, through DELLA protein degradation, GA signaling liberates transcription factors such as PIFs (phytochrome-interacting factors) and other bHLH proteins, which activate the expression of enzymes involved in GA biosynthesis, downstream developmental processes, and catabolism (Zondo and Mafa, 2025) (Figure 19.3). This transcriptional integration strategy is demonstrated by the presence of GA-responsive elements (GAREs) and DELLA-binding sites in the promoters of genes encoding hydrolytic enzymes, such as α-amylases, during seed germination (Zentella *et al.*, 2007).

The *bZIP* transcription factors, such as *ABFs/AREBs*, bind the *ABREs* (ABA-responsive elements) with the core sequence ACGTGG/TC, whose activity is regulated through ABA-triggered phosphorylation cascades involving SnRK2 protein kinases. This transcriptional mechanism controls the expression of enzymes involved in ROS detoxification, metabolic adjustments, and osmolyte biosynthesis during drought stress, creating coordinated metabolic responses that enhance stress tolerance (Ayub *et al.*, 2025). Similarly, ERFs (ethylene-responsive element binding factors) recognize GCC-box elements and activate transcription of genes encoding cell wall-degrading enzymes during senescence and fruit ripening, while also regulating enzymes involved in ethylene biosynthesis (Sriskantharajah *et al.*, 2021). The

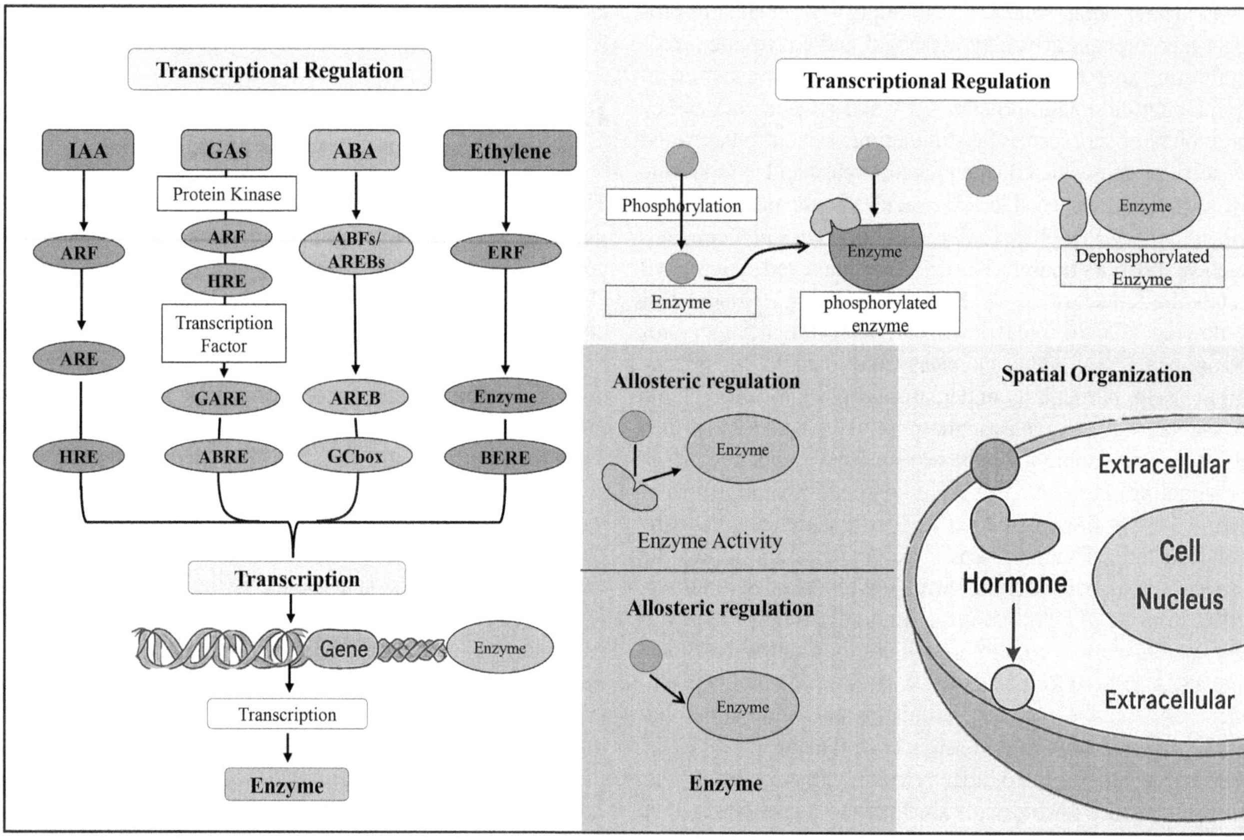

FIGURE 19.3 Schematic representation of hormone-mediated transcriptional and enzymatic regulation in plants. Illustrations represent a complex regulation of enzyme activity by hormones through transcriptional, post-translational, and spatial mechanisms. The left side (transcriptional regulation) represents major plant hormones such as IAA (auxin), GAs (gibberellins), ABA (abscisic acid), and ethylene, which regulate gene expression by activating specific transcription factors that bind to hormone-responsive cis-elements in promoters. The right side represents the post-translational modifications such as phosphorylation and dephosphorization, converting inactive enzymes to active phosphorylated forms along with allosteric regulation and spatial organization.

BRREs (BR-responsive elements) and their cognate transcription factors, *BES1/BZR1*, which undergo dephosphorylation upon BR perception through the BRI1 receptor kinase, have revealed how BR signaling coordinates expression of enzymes involved in cell wall modification, steroid biosynthesis, and growth processes (Cao *et al.*, 2024). Furthermore, the post-translational modifications represent quicker and reversible mechanisms for modulating enzyme activity in response to hormonal signals (Figure 19.3). The hormone-activated protein kinases catalyze the phosphorylation, which can dramatically alter enzyme activity, subcellular localization, and stability. For example, to coordinate stomatal closure with metabolic adjustments, the ABA activates SnRK2 kinases that phosphorylate multiple target enzymes, including metabolic enzymes such as sucrose-phosphate synthase and NADPH oxidases (RBOHs) that produce ROS signals (Peixoto and Baena-González, 2022). Similarly, phosphorylation-dependent inactivation of transcription factors and metabolic enzymes is a part of BR signaling via the GSK3-like kinase BIN2. BIN2 inactivation is triggered by BR perception, which in turn causes target proteins to become dephosphorylated (Delesalle *et al.*, 2024). The MAPK (mitogen-activated protein kinase) cascades, on the other hand, serve as vital integration nodes where different hormones activate specific MAPKs that phosphorylate overlapping sets of target enzymes (Xu Juan and Zhang ShuQun, 2015).

Through allosteric mechanisms, the enzyme activity can also be directly modulated, involving hormones and their derivatives as ligands, although such direct interactions remain less well-characterized than transcriptional mechanisms (Bajguz and Piotrowska-Niczyporuk, 2023). Certain enzymes possess regulatory domains that detect changes at cellular metabolite levels altered by hormone action, generating feedback loops. For example, transcriptional regulation and post-translational control via phosphorylation and calcium-dependent protein binding affect ACC synthase, the rate-limiting enzyme in ethylene production, establishing several integration sites for hormone crosstalk (Bajguz and Piotrowska-Niczyporuk, 2023). An additional degree of regulatory complexity is introduced by the spatial organization of enzyme–hormone integration. The biosynthesis, perception, and signal transduction of hormones often occur in specific subcellular compartments, generating localized domains of enzymatic activity. The compartmentalization of enzyme activities and hormone metabolism in different cell types within tissues regulates cell-specific responses to hormonal signals, contributing to the complex patterns of plant development and stress responses (Takatsuka and Umeda, 2014).

19.5 FUTURE PERSPECTIVES

The combination of transcriptomics, proteomics, and metabolomics offers unrivaled opportunities to coordinate complex networks connecting phytohormones and enzymes. Through integration of these datasets with network analysis and computational modeling, researchers can predict regulatory hubs and feedback loops that underlie development, growth, and stress adaptation. These approaches will help to identify unknown enzyme and hormone interactions and provide integrated knowledge into plant physiology. Furthermore, the manipulation of enzyme-responsive genes within hormone signaling and biosynthetic pathways can be revealed through advances in molecular breeding. Additionally, translating knowledge of hormone and enzyme regulation into practical applications can enhance stress resistance, fruit shelf life, and seed vigor in major crops. For instance, in climacteric fruits, the post-harvest softening can be delayed through moderating ethylene-responsive cell wall-degrading enzymes and protecting against oxidative stress in drought-prone regions by boosting antioxidant enzyme activity through ABA pathways. Understanding enzyme and hormone interaction can be vital for engineering climate-resilient plants with stable productivity in a global agriculture that faces unpredictable climate shifts. The genome editing, traditional breeding strategies, and integration of systems biology insights will accelerate the creation of cultivars capable of sustaining yield and quality under diverse and challenging environmental conditions.

19.6 CONCLUSION

The integration of plant enzymes and hormones constitutes a balanced regulatory network that supports every aspect of plant life, from seed germination to senescence. Through the regulation of enzymes involved in cell wall dynamics, redox balance, and metabolism, hormones orchestrate developmental and stress-responsive processes. Conversely, enzymes influence the signaling of hormones, biosynthesis, and degradation, creating feedback loops that ensure precise regulation under different internal and external conditions. Advances in systems biology, biochemistry, and molecular genetics have unveiled the complexity of these interactions and have been able to identify key regulatory networks. As food security and climate variability challenges escalate, applying this knowledge through genome editing, modern breeding, and biotechnological innovations offers a promising path to develop crops with enhanced resilience, productivity, and post-harvest quality. Ultimately, unraveling the hormone and enzyme crosstalk also deepens our understanding of plant biology and equips agriculture with tools to meet the demands of rapid climate change.

REFERENCES

Allen, H.R. & Ptashnyk, M., 2017. Mathematical modelling and analysis of the brassinosteroid and gibberellin signalling pathways and their interactions. *Journal of Theoretical Biology*, *432*, 109–131.

Altaf, F., Parveen, S., Farooq, S., Lone, M.L., Haq, A.U. & Tahir, I., 2024. Enigmas of senescence: A reappraisal on the hormonal crosstalk and the molecular mechanisms. *Theoretical and Experimental Plant Physiology*, *36*, 51–81.

An, S., Liu, Y., Sang, K., Wang, T., Yu, J., Zhou, Y. & Xia, X., 2023. Brassinosteroid signaling positively regulates abscisic acid biosynthesis in response to chilling stress in tomato. *Journal of Integrative Plant Biology*, *65*, 10–24.

Argueso, C.T. & Kieber, J.J., 2024. Cytokinin: From autoclaved DNA to two-component signaling. *The Plant Cell*, *36*, 1429–1450.

Avezum, L., Rondet, E., Mestres, C., Achir, N., Madode, Y., Gibert, O., Lefevre, C., Hemery, Y., Verdeil, J.-L. & Rajjou, L., 2023. Improving the nutritional quality of pulses via germination. *Food Reviews International*, *39*, 6011–6044.

Ayub, A., Javed, T., Nayab, A., Nan, Y., Xie, Y., Hussain, S., Shafiq, Y., Tian, H., Hui, J. & Gao, Y., 2025. AREB/ABF/ABI5 transcription factors in plant defense: Regulatory cascades and functional diversity. *Critical Reviews in Biotechnology*, *45*(7), 1414–1434.

Bajguz, A. & Piotrowska-Niczyporuk, A., 2023. Biosynthetic pathways of hormones in plants. *Metabolites*, *13*, 884.

Bethke, P.C., Schuurink, R. & Jones, R.L.J.J.O.E.B., 1997. Hormonal signalling in cereal aleurone. *Journal of Experimental Botany*, *48*, 1337–1356.

Bhatla, S.C. & Lal, M.A., 2023. *Plant physiology, development and metabolism*. Springer Nature.

Bothma, J.P., Norstad, M.R., Alamos, S. & Garcia, H.G., 2018. LlamaTags: A versatile tool to image transcription factor dynamics in live embryos. *Cell*, 173, 1810–1822.e16.

Cao, X., Wei, Y., Shen, B., Liu, L. & Mao, J., 2024. Interaction of the transcription factors BES1/BZR1 in plant growth and stress response. *International Journal of Molecular Sciences*, *25*, 6836.

Casal, J.J. & Fankhauser, C., 2023. Shade avoidance in the context of climate change. *Plant Physiology*, *191*, 1475–1491.

Chaves, A.L.S., Mello-Farias, P.C.D.J.G. & Biology, M., 2006. Ethylene and fruit ripening: From illumination gas to the control of gene expression, more than a century of discoveries. *Genetics and Molecular Biology*, *29*, 508–515.

Chini, A., Monte, I., Zamarreño, A.M., García-Mina, J.M. & Solano, R., 2023. Evolution of the jasmonate ligands and their biosynthetic pathways. *New Phytologist*, *238*, 2236–2246.

Clouse, S.D., 2015. A history of brassinosteroid research from 1970 through 2005: Thirty-five years of phytochemistry, physiology, genes, and mutants. *Journal of Plant Growth Regulation*, *34*, 828–844.

Corpas, F. J., Freschi, L., Rodríguez-Ruiz, M., Mioto, P.T., González-Gordo, S. & Palma, J.M., 2018. Nitro-oxidative metabolism during fruit ripening. *Journal of Experimental Botany*, *69*, 3449–3463.

Cortleven, A. & Schmülling, T.J.J.O.E.B., 2015. Regulation of chloroplast development and function by cytokinin. *Journal of Experimental Botany*, *66*, 4999–5013.

Cosgrove, D.J.J.C.O.I.P.B., 2015. Plant expansins: Diversity and interactions with plant cell walls. *Current Opinion in Plant Biology*, *25*, 162–172.

Cutler, S.R., Rodriguez, P.L., Finkelstein, R.R. & Abrams, S.R.J.a.R.O.P.B., 2010. Abscisic acid: Emergence of a core signaling network. *Annual Review of Plant Biology*, *61*, 651–679.

Delesalle, C., Vert, G. & Fujita, S., 2024. The cell surface is the place to be for brassinosteroid perception and responses. *Nature Plants*, *10*, 206–218.

Dempsey, D.M.A., Vlot, A.C., Wildermuth, M.C. & Klessig, D.F., 2011. Salicylic acid biosynthesis and metabolism. *The Arabidopsis book/American Society of Plant Biologists*, *9*, e0156.

Dolgikh, V.A., Pukhovaya, E.M. & Zemlyanskaya, E.V., 2019. Shaping ethylene response: The role of EIN3/EIL1 transcription factors. *Frontiers in Plant Science*, *10*, 1030.

Finkelstein, R., 2013. Abscisic acid synthesis and response. *The Arabidopsis Book/American Society of Plant Biologists*, *11*, e0166.

Freire-Rios, A., Tanaka, K., Crespo, I., Van Der Wijk, E., Sizentsova, Y., Levitsky, V., Lindhoud, S., Fontana, M., Hohlbein, J. & Boer, D.R., 2020. Architecture of DNA elements mediating ARF transcription factor binding and auxin-responsive gene expression in Arabidopsis. *Proceedings of the National Academy of Sciences*, *117*, 24557–24566.

Gan, H., Wang, S., Yang, Z. & Ma, P., 2025. Molecular decoding of phytohormone crosstalk: JA-mediated key regulatory nodes and signal integration. *Plants*, *14*, 2647.

González Ortega-Villaizán, A., King, E., Patel, M.K. & Pollmann, S., 2024. Plant hormone crosstalk under abiotic stress conditions. In: *Progress in botany vol. 85* (pp. 167–194). Springer.

He, H., Liang, G., Lu, S., Wang, P., Liu, T., Ma, Z., Zuo, C., Sun, X., Chen, B. & Mao, J., 2019. Genome-wide identification and expression analysis of GA2ox, GA3ox, and GA20ox are related to gibberellin oxidase genes in grape (*Vitis vinifera* L.). *Genes*, *10*, 680.

He, S., Zhi, F., Min, Y., Ma, R., Ge, A., Wang, S., Wang, J., Liu, Z., Guo, Y. & Chen, M., 2023. The MYB59 transcription factor negatively regulates salicylic acid-and jasmonic acid-mediated leaf senescence. *Plant Physiology*, *192*, 488–503.

Hedden, P. & Thomas, S.G., 2012. Gibberellin biosynthesis and its regulation. *Biochemical Journal*, *444*, 11–25.

Herrera-Vásquez, A., Salinas, P. & Holuigue, L., 2017. Corrigendum: Salicylic acid and reactive oxygen species interplay in the transcriptional control of defense genes expression. *Frontiers in Plant Science*, *8*, 964.

Hodnik, N., Baldizzone, C., Polymeros, G., Geiger, S., Grote, J.-P., Cherevko, S., Mingers, A., Zeradjanin, A. & Mayrhofer, K.J., 2016. Platinum recycling going green via induced surface potential alteration enabling fast and efficient dissolution. *Nature Communications*, *7*, 13164.

Huang, P., Li, Z. & Guo, H., 2022. New advances in the regulation of leaf senescence by classical and peptide hormones. *Frontiers in Plant Science*, *13*, 923136.

Iqbal, Z., Shariq Iqbal, M., Singh, S.P. & Buaboocha, T., 2020. Ca2+/calmodulin complex triggers CAMTA transcriptional machinery under stress in plants: Signaling Cascade and molecular regulation. *Frontiers in Plant Science*, *11*, 598327.

Ismail, M.A., Amin, M.A., Eid, A.M., Hassan, S.E.-D., Mahgoub, H.A., Lashin, I., Abdelwahab, A.T., Azab, E., Gobouri, A.A. & Elkelish, A., 2021. Comparative study between exogenously applied plant growth hormones versus metabolites of microbial endophytes as plant growth-promoting for *Phaseolus vulgaris* L. *Cells*, *10*, 1059.

Klein, T., Eckhard, U., Dufour, A., Solis, N. & Overall, C.M., 2018. Proteolytic cleavage mechanisms, function, and "Omic". *Chemical Reviews*, *118*, 1137–1168.

Koini, M.A., Alvey, L., Allen, T., Tilley, C.A., Harberd, N.P., Whitelam, G.C. & Franklin, K.A., 2009. High temperature-mediated adaptations in plant architecture require the bHLH transcription factor PIF4. *Current Biology*, *19*, 408–413.

Ku, Y.-S., Sintaha, M., Cheung, M.-Y. & Lam, H.-M., 2018. Plant hormone signaling crosstalks between biotic and abiotic stress responses. *International Journal of Molecular Sciences*, *19*, 3206.

Kubalová, M., Griffiths, J., Müller, K., Levenets, L., Tylová, E., Tarkowská, D., Jones, A.M. & Fendrych, M., 2025. Gibberellin-deactivating GA2OX enzymes act as a hub for auxin–gibberellin cross talk in Arabidopsis thaliana root growth regulation. *Proceedings of the National Academy of Sciences*, *122*, e2425574122.

Kumar, A.P., Bhasker, K., Nikhil, B. & Srinivas, P., 2023. Role of phenylpropanoids and flavonoids in plant defense mechanism. *International Journal of Environment and Climate Change*, *13*, 2951–2960.

Kurepa, J. & Smalle, J., 2025. The evolution of plant hormones: From metabolic byproducts to regulatory hubs. *International Journal of Molecular Sciences*, *26*, 7190.

Lee, Y. & Kende, H.J.P.P., 2002. Expression of α-expansin and expansin-like genes in deepwater rice. *Plant Physiology*, *130*, 1396–1405.

Li, J., Cheng, K., Lu, Y., Wen, H., Ma, L., Zhang, C., Suprun, A.R. & Zhu, H., 2025. Regulation of 1-aminocyclopropane-1-carboxylic acid synthase (ACS) expression and its functions in plant life. *Plant Hormones*, 1.

Li, Q.-F., Lu, J., Yu, J.-W., Zhang, C.-Q., He, J.-X. & Liu, Q.-Q., 2018. The brassinosteroid-regulated transcription factors BZR1/BES1 function as a coordinator in multisignal-regulated plant growth. *Biochimica et Biophysica Acta (BBA)-Gene Regulatory Mechanisms*, *1861*, 561–571.

Luo, P. & Di, D.-W., 2023. Precise regulation of the TAA1/TAR-YUCCA auxin biosynthesis pathway in plants. *International Journal of Molecular Sciences*, *24*, 8514.

Luo, P., Li, T.-T., Shi, W.-M., Ma, Q. & Di, D.-W., 2023. The roles of GRETCHEN HAGEN3 (GH3)-dependent auxin conjugation in the regulation of plant development and stress adaptation. *Plants*, *12*, 4111.

Mazzoni-Putman, S.M., Brumos, J., Zhao, C., Alonso, J.M. & Stepanova, A.N., 2021. Auxin interactions with other hormones in plant development. *Cold Spring Harbor Perspectives in Biology*, *13*, a039990.

Mei, Y. & Wang, N.N., 2024. New insights into the regulation of ethylene biosynthesis during leaf senescence in Arabidopsis. *New Phytologist*, 244.

Mittler, R. & Blumwald, E., 2015. The roles of ROS and ABA in systemic acquired acclimation. *The Plant Cell*, *27*, 64–70.

Moon, J., Park, C.-H., Son, S.-H., Youn, J.-H. & Kim, S.-K., 2021. Endogenous level of abscisic acid down-regulated by brassinosteroids signaling via BZR1 to control the growth of Arabidopsis thaliana. *Plant Signaling & Behavior*, *16*, 1926130.

Peixoto, B. & Baena-González, E., 2022. Management of plant central metabolism by SnRK1 protein kinases. *Journal of Experimental Botany*, *73*, 7068–7082.

Perrot-Rechenmann, C., 2010. Cellular responses to auxin: Division versus expansion. *Cold Spring Harbor Perspectives in Biology*, *2*, a001446.

Planas-Riverola, A., Gupta, A., Betegón-Putze, I., Bosch, N., Ibañes, M. & Caño-Delgado, A.I.J.D., 2019. Brassinosteroid signaling in plant development and adaptation to stress. *Development*, *146*, dev151894.

Rao, M.J., Duan, M., Zhou, C., Jiao, J., Cheng, P., Yang, L., Wei, W., Shen, Q., Ji, P. & Yang, Y., 2025. Antioxidant defense system in plants: Reactive oxygen species production, signaling, and scavenging during abiotic stress-induced oxidative damage. *Horticulturae*, *11*, 477.

Rekhter, D., Lüdke, D., Ding, Y., Feussner, K., Zienkiewicz, K., Lipka, V., Wiermer, M., Zhang, Y. & Feussner, I., 2019. Isochorismate-derived biosynthesis of the plant stress hormone salicylic acid. *Science*, *365*, 498–502.

Robichaux, K.J., 2024. *Elucidation and characterization of glycosyltransferases implicated in pectic polysaccharide biosynthesis in Arabidopsis thaliana*. University of Nevada.

Roychowdhury, R., Hada, A., Biswas, S., Mishra, S., Prusty, M.R., Das, S.P., Ray, S., Kumar, A. & Sarker, U., 2025. Jasmonic acid (JA) in plant immune response: Unravelling complex molecular mechanisms and networking of defence signalling against pathogens. *Journal of Plant Growth Regulation*, *44*, 89–114.

Roychowdhury, R., Mishra, S., Anand, G., Dalal, D., Gupta, R., Kumar, A. & Gupta, R., 2024. Decoding the molecular mechanism underlying salicylic acid (SA)-mediated plant immunity: An integrated overview from its biosynthesis to the mode of action. *Physiologia Plantarum*, *176*, e14399.

Santner, A. & Estelle, M., 2009. Recent advances and emerging trends in plant hormone signalling. *Nature*, *459*, 1071–1078.

Sellamuthu, G., Tarafdar, A., Jasrotia, R.S., Chaudhary, M., Vishwakarma, H. & Padaria, J.C., 2024. Introgression of Δ 1-pyrroline-5-carboxylate synthetase (PgP5CS) confers enhanced resistance to abiotic stresses in transgenic tobacco. *Transgenic Research*, *33*, 131–147.

Sharma, A., Sharma, A., Kumar, R., Sharma, I. & Vats, A.K., 2021. PR proteins: Key genes for engineering disease resistance in plants. *Crop improvement*. CRC Press, 81–98.

Singhal, R.K., Saha, D., Skalicky, M., Mishra, U.N., Chauhan, J., Behera, L.P., Lenka, D., Chand, S., Kumar, V. & Dey, P., 2021. Crucial cell signaling compounds crosstalk and integrative multi-omics techniques for salinity stress tolerance in plants. *Frontiers in Plant Science*, *12*, 670369.

Sizani, B.L., 2019. *Functional analysis of cell cycle inhibitors and crosstalk between auxin and drought stress during leaf development of Arabidopsis thaliana*. University of Antwerp.

Sriskantharajah, K., El Kayal, W., Torkamaneh, D., Ayyanath, M.M., Saxena, P.K., Sullivan, A.J., Paliyath, G. & Subramanian, J., 2021. Transcriptomics of improved fruit retention by hexanal in 'Honeycrisp' reveals hormonal crosstalk and reduced cell wall degradation in the fruit abscission zone. *International Journal of Molecular Sciences*, *22*, 8830.

Takahashi, K., Hayashi, K.-I. & Kinoshita, T.J.P.P., 2012. Auxin activates the plasma membrane H+-ATPase by phosphorylation during hypocotyl elongation in Arabidopsis. *Plant Physiology*, *159*, 632–641.

Takatsuka, H. & Umeda, M., 2014. Hormonal control of cell division and elongation along differentiation trajectories in roots. *Journal of Experimental Botany*, *65*, 2633–2643.

Tardast, Z., Iranbakhsh, A., Ebadi, M. & Oraghi Ardebili, Z., 2023. Carboxylic acid-functionalized multiwalled carbon nanotubes (COOH-MWCNTs) improved production of atropine in callus of *Datura inoxia* by influencing metabolism, gene regulation, and DNA cytosine methylation; an in vitro biological assessment. *Plant Physiology and Biochemistry*, *202*, 107975.

Vanholme, R., De Meester, B., Ralph, J. & Boerjan, W.J.C.O.I.B., 2019. Lignin biosynthesis and its integration into metabolism. *Current Opinion in Biotechnology*, *56*, 230–239.

Vishwakarma, K., Kumar, N., Shandilya, C., Mohapatra, S., Bhayana, S. & Varma, A., 2020. Revisiting plant–microbe interactions and microbial consortia application for enhancing sustainable agriculture: A review. *Frontiers in Microbiology*, *11*, 560406.

Wang, H. & Schippers, J.H., 2019. The role and regulation of autophagy and the proteasome during aging and senescence in plants. *Genes*, *10*, 267.

Wang, Z.-Y., Bai, M.-Y., Oh, E. & Zhu, J.-Y., 2012. Brassinosteroid signaling network and regulation of photomorphogenesis. *Annual Review of Genetics*, *46*, 701–724.

Wasternack, C. & Feussner, I., 2018. The oxylipin pathways: Biochemistry and function. *Annual Review of Plant Biology*, *69*, 363–386.

Werner, T. & Schmülling, T., 2009. Cytokinin action in plant development. *Current Opinion in Plant Biology*, *12*, 527–538.

Wu, W., Du, K., Kang, X. & Wei, H., 2021. The diverse roles of cytokinins in regulating leaf development. *Horticulture Research*, 8.

Xu, P., Ali, A., Han, B. & Wu, X., 2018. Current advances in molecular basis and mechanisms regulating leaf morphology in rice. *Frontiers in Plant Science*, *9*, 1528.

Xu Juan, X.J. & Zhang Shuqun, Z.S., 2015. Mitogen-activated protein kinase cascades in signaling plant growth and development.

Yamoune, A., Zdarska, M., Depaepe, T., Rudolfova, A., Skalak, J., Berendzen, K.W., Mira-Rodado, V., Fitz, M., Pekarova, B. & Mala, K.L.N., 2024. Cytokinins regulate spatially specific ethylene production to control root growth in Arabidopsis. *Plant Communications*, 5.

Yang, G., Chen, Y., Yu, H., Zhang, H., Han, D., Guo, X., Yan, E., Quan, H. & Li, T., 2021a. Raspberry (*Rubus idaeus* L.) NCED1 gene enhances high salinity and cold tolerance in Arabidopsis. *In Vitro Cellular & Developmental Biology-Plant*, *57*, 811–819.

Yang, W., Cortijo, S., Korsbo, N., Roszak, P., Schiessl, K., Gurzadyan, A., Wightman, R., Jönsson, H. & Meyerowitz, E., 2021b. Molecular mechanism of cytokinin-activated cell division in Arabidopsis. *Science*, *371*, 1350–1355.

Yang, X.Y., Wang, Y., Jiang, W.J., Liu, X.L., Zhang, X.M., Yu, H.J., Huang, S.W. & Liu, G.Q., 2013. Characterization and expression profiling of cucumber kinesin genes during early fruit development: Revealing the roles of kinesins in exponential cell production and enlargement in cucumber fruit. *Journal of Experimental Botany*, *64*, 4541–4557.

Ye, J. & Cui, X., 2019. Next-generation crop breeding methods. *Molecular Plant*, *12*, 470–471.

Zentella, R., Zhang, Z.-L., Park, M., Thomas, S.G., Endo, A., Murase, K., Fleet, C.M., Jikumaru, Y., Nambara, E. & Kamiya, Y., 2007. Global analysis of DELLA direct targets in early gibberellin signaling in Arabidopsis. *The Plant Cell*, *19*, 3037–3057.

Zhao, W., Li, Z., Fan, J., Hu, C., Yang, R., Qi, X., Chen, H., Zhao, F. & Wang, S., 2015. Identification of jasmonic acid-associated microRNAs and characterization of the regulatory roles of the miR319/TCP4 module under root-knot nematode stress in tomato. *Journal of Experimental Botany*, *66*, 4653–4667.

Zhao, Y., 2010. Auxin biosynthesis and its role in plant development. *Annual Review of Plant Biology*, *61*, 49–64.

Zhou, Y., Zhao, C., Du, T., Li, A., Qin, Z., Zhang, L., Dong, S., Wang, Q. & Hou, F., 2023. Overexpression of 9-cis-epoxycarotenoid dioxygenase gene, IbNCED1, negatively regulates plant height in transgenic sweet potato. *International Journal of Molecular Sciences*, *24*, 10421.

Zhu, J.-K., 2016. Abiotic stress signaling and responses in plants. *Cell*, *167*, 313–324.

Zondo, S.N. & Mafa, M.S., 2025. Peroxidase and β-1, 3-glucanase synergistic functions strengthen plant cell wall and protect wheat against *Diuraphis noxia* infestation. *Planta*, *262*, 41.

20 Enzymatic Mechanisms in Plant Senescence and Aging

Md. Parvez Kabir, Falguni Akter Shraboni, Faridul Alam, Ahmad Numery Ashfaqul Haque, Md. Selim Reza, and Akbar Hossain

20.1 INTRODUCTION

Senescence is the final stage of plant life, which results in programmed degradation and recycling of cellular components, leading to death of either individual parts or whole plants (Baseer et al., 2024). This is more than a modest or passive loss but pre-programmed in intricate biochemical and molecular pathways (Guo et al., 2021). Senescence is involved in several aspects of aging and performs many functions, like nutrient remobilization, seed development and response to environmental stress (Bhat et al., 2019). It's also essential to ecosystem equilibrium and for successful breeding.

The enzymatic control of plant age-dependent retardation or senescence represents a key hint toward the elucidation of complex pathways related to cellular degradation, nutrient re-use and defense responses of plants against environmental stresses (Doan et al., 2024). Recent literatures reported that protease, nuclease and hydrolases are active in hormonal regulation during the degradation of macromolecules, such as proteins, lipids, nucleic acid, in development of senescence (Ahmad & Tahir, 2016; Haq et al., 2024; Jhanji et al., 2025). This selective decline of cell-material reserves essential nutrients for the re-use and prepares the plant at same time to its next phase.

Crosstalk in senescence with plant aging or development is complicated (Woo et al., 2018). The precise control of senescence is orchestrated during normal plant growth, which is regulated by exogenous factors such as light, temperature, water supply and internal signals (e.g., hormones: ethylene, jasmonic acid (JA) and abscisic acid (ABA)) (Bayanati et al., 2021; Zhao et al., 2022). These hormones and their crosstalk with intricate transcriptional networks determine the duration and severity of senescence to enable plants to adjust life cycle during development or in response to environmental stimuli.

The biological significance of senescence should not be limited to the end of plant life. It is the ecosystem level that is necessary for the recycling of resources e.g., nitrogen, phosphorous, potassium etc., to mobilize it in such condition that other plants or organisms can be able to use them (Estiarte et al., 2023; Nascimento et al., 2024). Senescence is an important adaptive response not only to biotic (drought, pathogen attacks) and abiotic stresses (high or low temperature) but also for normal developmental processes (Häffner et al., 2015; Hu et al., 2023; Raza et al., 2020). Importantly, on stress-induced senescence, the use of such treatments lessens the effects of them and could even support plants live under adverse ecological conditions, resulting in a success to species (Sade et al., 2018).

The following chapter covers the enzymatic aspect of plant senescence and aging, with focus on molecular mechanisms underpinning these processes or particular enzymes or biochemical pathways. These phenomena are not only relevant for further knowledge in plant biology but also to improve breeding strategies with which agricultural productivity and, particularly, tolerance of crops toward global climate change conditions and resource shortness can be increased.

20.2 MOLECULAR MECHANISMS OF PLANT SENESCENCE AND AGING

Senescence is a programmed developmental stage of vegetative plant organs characterized by the degradation of cellular constituents with the eventual remobilization of nutrients to developing reproductive structures. It is induced by oxidative stress that results in macromolecular cell damage mediated by the reactive oxygen species (ROS), protein misfolding, lipid peroxidation and DNA damage leads to an acceleration of cellular aging, which are illustrated in Figure 20.1. The incidence of oxidative damage and shortened telomeres increase with age, increasing the plants' susceptibility to genotoxicity and hampering their DNA repair capacity.

The main characteristic of senescence is programmed cell death (PCD) that removes the unhealthy cells disposing a complete nutrient recycling within an organism, and it regulates the transferring of nutrients to growing tissues. The aging is orchestrated by senescence-related genes and three different types of transcription factors, NAC, WRKY and MYB, which coordinate the relocation of nutrients into sink organs. Finally, autophagy is yet another important constituent of senescence which involves eliminating damaged organelles and proteins and thereby providing for dynamic cellular homeostasis on the one hand as well as nutrients redistribution on the other hand.

Therefore, the molecular mechanisms of senescence (i.e., oxidative stress, DNA damage signaling, PCD and autophagy) are finely tuned processes that rather than causing a waste of resources may act to redistribute nutrients in order to optimize the plant transition into the final phase of its own life cycle. These mechanisms can be exploited

DOI: 10.1201/9781003669920-20

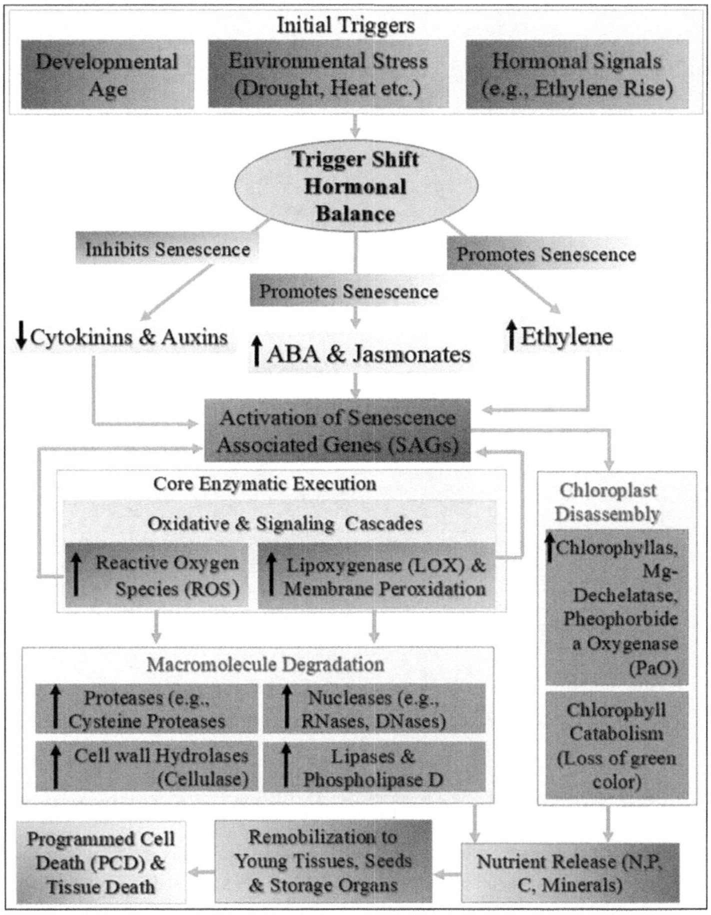

FIGURE 20.1 Hormonal regulation and enzymatic pathways in plant senescence.

to enhance high yields, long life duration and tolerance to stress in agriculture.

20.3 ENZYMES INVOLVED IN PLANT SENESCENCE

Senescence in plants is defined as the regulated breakdown of cellular components, a crucial process for nutrient remobilization and stress acclimation (Baseer et al., 2024). The breakdown of proteins, lipids, carbohydrates and cell wall materials allows for efficient resource recycling and also primes the plant toward death or dormancy (Barnes & Anderson, 2018). The key classes of enzymes that are responsible for plant senescence, including proteases, hydrolases, antioxidant enzymes, cell wall–degrading enzymes and chlorophyll-degrading enzymes, are discussed below.

20.3.1 Proteases

Proteases, such as the 26S proteasome and autophagy pathway, are central in protein turnover during senescence (Wang & Schippers, 2019). This enzyme which is involved in nitrogen recycling by degrading the damaged or unnecessary proteins and maintaining cellular homeostasis (Luciński & Adamiec, 2023). Autophagy also plays similar role in the degradation of organelles and damaged cellular components for nutrient recycling (Masclaux-Daubresse et al., 2017).

20.3.2 Hydrolases

Enzymes, like β-glucosidases, amylases and lipases, are important in the degradation of macromolecules i.e., starch, cellulose and lipid (Akimova et al., 2024). β-Glucosidases hydrolyze β-glucosidic linkages in polysaccharides to release simple sugar, such as glucose (Muradova et al., 2023). Amylases hydrolyze starch into glucose and maltose to release energy, and lipases hydrolyze lipids into fatty acids and glycerol for cell membrane turnover and energy mobilization (Choi et al., 2022; Saini et al., 2017).

20.3.3 Antioxidant Enzymes

ROS are regulated by enzymes, including superoxide dismutases (SODs), catalases (CATs), peroxidases, glutathione reductase (Ighodaro & Akinloye, 2018). They detoxify ROS and safeguard cellular components from the oxidative damage to maintain cellular integrity during senescence (Davalli et al., 2016). These enzymes play up a role in the retardation of senescence and mitigate oxidative

TABLE 20.1
Enzymatic Activity and Antioxidant Reactions in Plants during Senescence under Environmental Stress Conditions

Antioxidant	Reaction	Plant Species	Enzyme Activity Status	Key Reference
Superoxide dismutase (SOD)	$2O_2^{\cdot-} + 2H^+ \Rightarrow O_2 + H_2O_2$	Cluster bean	↓	Deo *et al.* (2006)
		Tobacco	↕	Dertinger *et al.* (2003)
		Bean	↕	Procházková and Wilhelmová (2007)
		Maize	↕	Procházková *et al.* (2001)
		Wheat	↓	Srivalli and Khanna-Chopra (2001)
Ascorbate peroxidase (APX)	$H_2O_2 + 2Asc. \Rightarrow 2H_2O + 2MDHA$	Pea	↓	Jimenéz *et al.* (1998)
		Cucumber	↑	Kanazawa *et al.* (2000)
		Bean	↕	Procházková and Wilhelmová (2004)
		Maize	↕	Procházková *et al.* (2001)
Catalase (CAT)	$2H_2O_2 \Rightarrow 2H_2O + O_2$	Cucumber	↓	Kanazawa *et al.* (2000)
		Soybean	↕	Fu *et al.* (2000)
		Maize	↕	Procházková *et al.* (2001)
		Bean	↓	Procházková and Wilhelmová (2004)
		Arabidopsis	↓	Zimmermann *et al.* (2006)
Glutathione reductase (GR)	$GSSG + NAD(P)H + H^+ \Rightarrow 2GSH + NAD(P)^+$	Cucumber	↕	Kanazawa *et al.* (2000)
		Bean	↕	Procházková and Wilhelmová (2004)
		Tobacco	↓	Dertinger *et al.* (2003)
Dehydroascorbate reductase (DHAR)	$DHA + 2GSH \Rightarrow Asc. + GSSG$	Cucumber	↓	Kanazawa *et al.* (2000)
Glutathione peroxidase (GPX)	$H_2O_2 + 2GSH \Rightarrow 2H_2O + GSSG$	Cucumber	↑	Kanazawa *et al.* (2000)

Asc.: ascorbate; DHA: dehydroascorbate; GSH: reduced glutathione; GSSG: oxidized glutathione; MDHA: monodehydroascorbate radical; NADP: nicotinamide adenine dinucleotide phosphate; ↑: continuous increase in activity of enzymes during senescence; ↓: continuous decrease in activity of enzymes during senescence; ↕: initial increase followed by decrease activity of enzymes during senescence.

stress (Varesi et al., 2022). The activity of these antioxidant enzymes during the senescence process of different plant species under environmental stress conditions is summarized in Table 20.1.

20.3.4 Cell Wall-Degrading Enzymes

Structural components like cellulose, pectin and hemicellulose are degraded by enzymes such as cellulases, pectinases and xylanases (Toushik et al., 2017). These enzymes break down and induce degradation of plant tissues that leads to drop off, particularly during leaf and fruit abscission, thereby allowing relocation of nutrients to developing tissues (Barnes & Anderson, 2018).

20.3.5 Chlorophyll-Degrading Enzymes

Chlorophyllase and pheophytinase initiate the breakdown of chlorophyll during leaf senescence (Syvash & Zolotareva, 2017). The degradation of chlorophyll is required for the recycling of nitrogen and carbon, which are essential requirements for survival of the plant (Tanaka & Ito, 2025). Jointly, the action of these enzymes enables efficient cellular material degradation, directly feeding into nutrient remobilization during aging (War et al., 2022).

These enzymes are involved in the coordinated regulation of biochemical events occurring during plant senescence to ensure efficient nutrient remobilization, adaptation to environmental challenges and thus, successful preparation for plant death or dormancy.

20.4 HORMONAL REGULATION OF ENZYMATIC ACTIVITY IN SENESCENCE

Hormonal regulation plays an important role in controlling the activity of enzymes in plant senescence because hormones such as ABA, ethylene, JA and the cytokinins are involved in many processes (e.g., protein degradation, resistance to oxidative stress and initiation of senescence) (Panozzo et al., 2025).

The senescence-regulated hormones like ethylene induced the degradation of protein content by increasing proteolytic and antioxidant enzyme (for counteracting against oxidative stress) expression level (Iqbal et al., 2017). This in turn contributes to delay irreversible cellular damage resulting in a programmed aging process (Alonso-Salinas et al., 2024). In contrast, cytokinins-retard senescence by suppressing the activities of senescence-associated proteases and increasing antioxidant enzyme activity to maintain cellular integrity and delay aging (Hönig et al., 2018).

TABLE 20.2
Phytohormonal Regulation and Enzyme Activity during Senescence

Phytohormones	Plant/Species	Regulation	Effect	Key Reference
Cytokinins (CKs)	Soybean, corn, rice, *Arabidopsis*	Negative (−)	Endogenous CK levels decrease during senescence, with increased expression of CK oxidase, an enzyme that degrades CK.	Lim et al. (2007) and Buchanan-Wollaston et al. (2005)
Auxins (IAA)	*Arabidopsis*	Negative (−)	Exogenous IAA application reduces the expression of genes from the SAG family during senescence.	Sarwat et al. (2013)
	Arabidopsis	Positive (+)	The gene SAUR 36, a positive regulator of senescence, shows increased levels.	Hou et al. (2013)
	Arabidopsis	Positive (+)	During senescence, the expression of genes involved in auxin biosynthesis (TSA1, AO1, NIT13) is increased.	Hou et al. (2013)
Abscisic acid (ABA)	Soybean, corn, rice, *Arabidopsis*	Positive (+)	Exogenous ABA application promotes senescence and regulates the expression of SAG family genes.	Lim et al. (2007) and Zhang et al. (2012)
Ethylene	*Arabidopsis thaliana*	Positive (+)	The expression of the gene encoding ACC is decreased.	Khan et al. (2013)
	Arabidopsis	Positive (+)	The EIN2 protein influences ore1 gene transcription, leading to increased expression of multiple SAG family genes.	Kim et al. (2009)
Jasmonic acid (JA)	*A. thaliana*	Positive (+)	The expression of several key enzymes involved in chlorophyll degradation is induced.	Reinbothe et al. (2009)
	Arabidopsis	Positive (+)	JA represses and degrades Rubisco activase.	Shan et al. (2011)
Polyamines (PAs)	Barley	Positive (+)	During senescence, transcript levels and activation of PA catabolic enzymes increase.	Sobieszczuk Nowicka et al. (2016)
	Arabidopsis	Positive (+)	Arabidopsis polyamine back-conversion oxidase mutants, which are deficient in converting spermine to spermidine, show delayed entry into dark-induced leaf senescence.	Sequera-Mutiozabal et al. (2016)
	Barley	Positive (+)	Spermidine and spermine are produced and transported into the apoplast, where they generate H_2O_2 and diamino propane, both of which contribute to senescence-dependent degradation processes.	Sobieszczuk-Nowicka et al. (2016)
Sugars	*Arabidopsis*	Positive (+)	High sugar levels have been reported in petals exhibiting visible symptoms of senescence.	van Doorn (2004) and van Doorn (2008)
	Tobacco (*Nicotiana tabacum*)	Negative (−)	Sugar accumulation can inhibit the effects of CKs.	Wingler et al. (1998)

SAG: Senescence-associated gene; IAA: Indole-3-acetic acid; Positive (+): Promotion of senescence; Negative (−): Inhibition of senescence.

ABA, stress hormone, enhances senescence under unfavorable conditions by inducing oxidative enzymes through accumulation of free radicals, and breakdown of chlorophyll, and other macromolecules in the cell (Sağlam et al., 2024). This triggers aging, especially in case of stress (drought or salinity) (Ma et al., 2020). Similarly, JA also participates in controlling the defense response to biotic stress such as pathogen attacks (Roychowdhury et al., 2025). JA induces the upregulation of proteases and antioxidants that are able to counteract oxidative stress, besides its involvement in nutrient remobilization (Ali & Baek, 2020). Table 20.2 summarizes the regulatory roles of various phytohormones and other factors in modulating senescence-related enzyme activity, highlighting their influence on gene expression and senescence progression.

The connection between ABA and JA functions seems to be complicated and could be context-specific. Yet their effect can act in a positive or a negative direction under certain circumstances (Khan, 2025). This feedback loop provides a means by which the plant might be able to tune their aging responses rapidly in relation to environmental stimuli.

Information on hormonal interactions in the regulation of enzyme activity during senescence is important to improve crop tolerance and prolong productivity. Manipulation of these pathways are used to increase crop life span, nutrient recycling and resilience to stress which helps in the sustainability in agriculture.

20.5 GENETIC REGULATION OF ENZYMATIC ACTIVITIES IN PLANT SENESCENCE

The activities of enzymes related to senescence, especially in plants, are almost genetically controlled and are therefore very important for the coordination of decomposition of cellular components and nutrient mobilization (Zhang et al., 2020). These varied genes encode proteins with activities in protein catabolism, oxidation stress tolerance, macromolecule degradation and nutrition cycling (Krishnamurthy et al., 2024). The main enzymes tend to be proteases (e.g., 26S proteasome and cysteine protease), hydrolases (β-glucosidases, lipases), antioxidant enzymes (e.g., SODs) and chlorophyll-degrading enzymes (Girondé et al., 2015; Kaabel et al., 2020; Tanaka & Ito, 2025; Wang et al., 2016).

Several transcription factors, including NAC, WRKY, MYB and AP2/ERF, regulate the expression of these enzymes to guarantee their proper activation during senescence (Cao et al., 2023). Moreover, epigenetic changes – such as histone acetylation – are indeed changing the structure of chromatin and promoting gene expression necessary for senescence (Crouch et al., 2022). These changes are responsible for the regulation of senescence at the genetic level.

The molecular–genetic networks integrated by hormonal signaling pathways (ethylene, ABA and cytokinins) also receive developmental signals that allow to incorporate the environmental inputs via a coordinated acting of senescence (Yetgin et al., 2025). Mutations in key genes of these systems may disrupt the activity of enzymes and culminate in altered senescence timing, and fertilization imbalanced between uptake and recycling nutrients (Zhang et al., 2020). For example, mutants with defects in ethylene signaling or NAC transcription factors may exhibit either delayed or accelerated senescence (Forlani, 2021; Podzimska-Sroka et al., 2015). Likewise, knocked-out mutant of enzymes such as chlorophyllase and antioxidant enzymes can result in premature or impaired senescence, highlighting the importance of these enzyme activities for keeping efficient senescence (Tamary et al., 2019).

The genetic regulation of these enzyme activities is insight into the molecular basis of senescence. It also offers potential options for increasing agricultural production and stress tolerance of plants, by imposing selection on these genetic networks to better control the timing and severity of senescence. The genes and transcription factors involved in plant senescence, as outlined in Table 20.3, provide a comprehensive overview of the genetic regulation of enzymatic activities during this process, highlighting their roles in macromolecule degradation, nutrient recycling and the overall progression of senescence.

20.6 ENZYMATIC MECHANISMS IN ENVIRONMENTAL STRESS AND PLANT SENESCENCE

Biotic (e.g., pathogens) and abiotic (e.g., drought and heat) environmental stresses have a significant impact on plant senescence by activating specific enzymatic responses to reduce potential damage and control aging process (Barna, 2022; Rasool et al., 2019). Under drought conditions, several antioxidant enzymes, including SOD and CAT, are activated in order to scavenge ROS level, as a result senescence delay is prolonged and efficient nutrient recycling is enhanced (Bhattacharjee, 2019; Jajic et al., 2015). In response to heat, the induction of HSPs (heat shock proteins) is associated with the activation of enzymes for scavenging ROS and maintaining cellular integrity and minimizing damage induced by stress (Driedonks et al., 2015). Pathogen pressure, however, triggers an increase in ROS production, which in turn activates both antioxidant and cell wall-degrading enzymes (Sahu et al., 2022). Such responses not only facilitate the control of pathogen spread but also induce premature senescence as a defense mechanism.

To cope with these stresses, plants increase the activities of antioxidant enzymes, proteases and autophagy to manage ROS and protein turnover for survival by regulating the aging process (Salehi et al., 2020; Su et al., 2020). Hydrolases as well as amylases and lipases are also induced to favor mobilization of stored macromolecules as well as offering key metabolites during stressful periods (Arora et al., 2017).

The senescence-stress response crosstalk is complex and stress can either promote or retard the senescence depending on the capability of plants to balance enzymes activities (Kalinin et al., 2023). Long-term stress can lead to premature senescence (Schippers, 2015), while efficient enzymatic regulation may delay the aging process and make plants better adapted to stresses (Cannea & Padiglia, 2025). Another level of control is hormonal crosstalk, especially with stress hormones, ABA and JA and with senescence hormone (ethylene), which also contribute toward these enzymatic adaptations (Verma et al., 2016). These contacts modulate enzymatic activities regarding the quality and degree of stress encountered, suggesting that perception of environmental cues can dynamically control plant responses.

20.7 AGRICULTURE AND CROP IMPROVEMENT PROSPECTS

Enzymatic senescence holds the potential to enhance crop yield, stress tolerance and post-harvest quality (Zhao et al., 2022). The elucidations of the aging-related mechanism about these enzymes are useful ways to regulate the finding of models for delaying senescence, and prolonging crop shelf life (Zhou & Yang, 2003). For instance, senescence in fruits and vegetables can be retarded by downregulating the expression of proteolytic enzymes or enzymes involved in quenching oxidative stress that may extend productivity as well as post-harvest quality (Jhanji et al., 2025; Lal et al., 2023).

Modification of enzymes such as chlorophyllase may play a positive role in preserving color and nutritive value (Pareek et al., 2017), and suppression of the level of cell wall-degrading enzymes can inhibit the softening during fruit ripening, which not only can make from hard to more marketable fruit but also can decrease waste (Li et al., 2022). Furthermore, enzyme modulation may benefit for resource and protein turnover in achieving higher crop yield (Wang et al., 2024), stress tolerance as a consequence of regulative antioxidative enzymes leads to tolerate the plant under abiotic stresses like drought, heat (Fujita & Hasanuzzaman, 2022).

In horticulture, it could result in regulation of senescence-related enzymes either to extend post-harvest life or to lengthen plant life for an open petal and long floral

TABLE 20.3
Genes and Transcription Factors Involved in Plant Senescence across Various Species

Genes or Transcription Factors Involved in Senescence	Plant Species	Function	Key References
Red chlorophyll catabolite reductase (RCCR)	*Arabidopsis*	Chlorophyll degradation	Pruzinska et al. (2007)
Cysteine proteases	Peruvian lily	Protein degradation in petals	Wagstaff et al. (2002)
Serine proteases	Daffodil	Protein degradation in petals	Hunter et al. (2002)
Protein modification gene (GNT1)	Rice	Negative response in senescence	Fanata et al. (2013)
Reductase (NYC1)	Rice	Positive response in senescence	Sato et al. (2009)
NAC family (NYC4)	Rice	Chlorophyll degradation	Yamatani et al. (2013)
Transcription factor gene, NAC family	Daffodil	Chlorophyll degradation	Hunter et al. (2002)
Stay green (Sgr1)	*Arabidopsis*	Chlorophyll degradation	Rodoni et al. (1997)
Sucrose non-fermenting related kinase 1 (SnRK1)	*Physcomitrella patens*	Sugar-mediated regulation of aging	Thelander et al. (2004)
Allene oxide synthetase	Petunia	Phospholipids and fatty acids degradation in petals	Xu et al. (2006)
Lipases	*Dianthus*	Phospholipids and fatty acids degradation in petals	Hoeberichts et al. (2007)
Hexokinase 1 (HXK1)	–	Sugar-mediated regulation of aging	Smeekens et al. (2010)
Telomerase reverse transcriptase (TERT)	*Arabidopsis*	Telomere attrition	Watson and Riha (2010)
BFN1	*Arabidopsis*	Nucleic acids degradation in leaves	Matallana-Ramirez et al. (2013)
NUC1	Petunia	Nucleic acids degradation in petals	Langston et al. (2005)
Senescence-associated genes 12 (SAG12)	*Arabidopsis*	Senescence specific, proteolysis	Lohman et al. (1994)
Pheophorbide A oxygenase (PAO)	*Arabidopsis*	Positive regulator in leaf senescence	Sakuraba et al. (2012)
Lipoxygenase (OsLOX2)	Rice	Polyunsaturated fatty acid degradation	Huang et al. (2014)
Transcription factor genes (Ghd2)	Rice	Positive response in senescence	Liu et al. (2016)
Chitinase (CHI)	*Arabidopsis*	Senescence-enhanced gene	Masclaux-Daubresse et al. (2007)
Transporter gene (OsPSL50)	Rice	Negative response in leaf senescence	He et al. (2021)
Cytokinin receptor1/Arabidopsis histidine kinase4 (CRE1/AHK4)	*Arabidopsis*	Programmed cell death	Vescovi et al. (2012)
Autophagy-related genes (ATG18, ATG7, ATG8a, ATG 8e, ATG9)	*Arabidopsis*	Breakdown of cell and cellular organelles	Xiong et al. (2005); van der Graaff et al. (2006)
Light signaling gene (Lhca4)	Rice	Positive regulator of plant senescence	Yamatani et al. (2018)
Transcription factor gene (OsHox33)	Rice	Positive regulator of plant senescence	Luan et al. (2013)
Soybean (*Glycine max*) senescence-associated receptor-like kinase (GmSARK)	Soybean	Leaf senescence	Xu et al. (2011)
Senescence-associated E3 ubiquitin ligase 1 (SAUL1)	*Arabidopsis*	Negative regulator of plant senescence	Raab et al. (2009)
Transcription factors NAC (NAM, ATAF and CUC)	Petunia	Control organ development and response to pathogens	Souer et al. (1996)
NAC family (OsNAC054)	Rice	Positive response during senescence	Sakuraba et al. (2020)
NAC family (OsNAC106)	Rice	Negative response during senescence	Sakuraba et al. (2015)
NAC-like activated by AP3 (AtNAP)	*Arabidopsis*	Delay leaf senescence	Guo and Gan (2006)
NAC family (OsNAC2)	Rice	Positive regulator	Mao et al. (2017)
Jungbrunnen1 (JUB1)	*Arabidopsis*	Negative regulator of leaf senescence	Wu et al. (2012)
WRKY transcription factor genes (WRKY53, WRKY6)	*Arabidopsis*	Leaf senescence	Miao et al. (2004)
WRKY transcription factor genes (OsWRKY5)	Rice	Leaf senescence	Kim et al. (2019)
WRKY family	Switchgrass	Leaf senescence	Rinerson et al. (2015)
WRKY family	*Gardenia*	Petal senescence	Tsanakas et al. (2014)
Zn finger family *Solanum lycopersicum* Zinc Finger2 (SlZF2)	Tomato	Negative role in leaf senescence	Hichri et al. (2014)
MYB family Myeloblastosis-related protein R1/Myb-related protein 44 (MYBR1/MYB44)	*Arabidopsis*	Negative role in leaf senescence	Jaradat et al. (2013)
MYB family (OsMYB102)	Rice	Negative role in senescence process	Piao et al. (2019)
Ethylene-insensitive3 (EIN3)	*Arabidopsis*	Accelerates age-dependent leaf senescence	Li et al. (2013)

TABLE 20.4
Implication of Plant-Delayed Senescence Strategies in Combination with Modern Biotechnological Processes for Agricultural Crop Improvement

Plant Species	Strategies Implied for Senescence	Yield Increase (%)	Key References
Rice	pSARK::IPT	144	Peleg et al. (2011)
	Introgression of japonica OsSGR allele	10.6–12.7	Shin et al. (2020)
	OsGATA12 overexpression	30	Lu et al. (2017)
	Cytokinin-mediated increasing of the root sink strength	4–8	Zhu et al. (2022)
	OsNAP RNAi	6.3–10.3	Liang et al. (2014)
	ONAC096 T-DNA insertion	16	Kang et al. (2019)
	OsCKX11 Crispr/Cas9	7.5	Zhang et al. (2021)
Wheat and barley	Cytokinin-mediated increasing of the root sink strength	2–15	Werner et al. (2003), Yang et al. (2016)
Wheat	TaNAM RNAi	Not affected	Uauy et al. (2006)
	pAtMYB32xs::IPT	40–67	Joshi et al. (2019)
	pSAG12::IPT	Not affected	Sýkorová et al. (2008)
	Abscisic acid inhibition of stomatal closure to induce senescence	3	Zhang et al. (2012), Yang et al. (2003)
	EMS mutagenesis of GPC-A1 and GPC-D1	Not affected	Avni et al. (2014)
	TaNAC-S overexpression	Not affected	Zhao et al. (2015)
	Ethylene activation of nutrient remobilization	Delayed chlorophyll degradation and grain filling	Beltrano et al. (1999)
	NAM RNAi	Not affected	Borrill et al. (2015)
Maize	HaHB11 overexpression	28	Raineri et al. (2019)
	Cytokinin-mediated increasing of the root sink strength	3–30	Ren et al. (2016), Gao et al. (2017)
	pSARK::IPT	47–66	Oneto et al. (2016)
	nac7 RNAi	2.2–3.5	Zhang et al. (2019)
Tomato	SlORE1S02 RNAi	43–95	Lira et al. (2017)
	SlNAP2 RNAi	19	Ma et al. (2018)
Canola	pAtMYB32xs::IPT	16–23	Kant et al. (2015)
Peanut	pSARK::IPT	51	Qin et al. (2011)
Cassava	pSAG12::IPT	–23.33	Zhang et al. (2010)

maturation stage which would enhance visual appeal and add value in ornamental plants (Jhanji et al., 2025; Serek & Reid, 2024).

Modulation of these enzymatic mechanisms would provide the potential strategies to improve agricultural sustainability, decrease post-harvest losses and guarantee food security under fluctuating climatic conditions (Nishad et al., 2024). Such innovations seem to be bringing more robust crops and generating economic value throughout the marketplace. The strategies for delaying plant senescence, as outlined in Table 20.4, offers a valuable implication for agricultural improvement.

20.8 FUTURE DIRECTIONS AND RESEARCH

Studies on the enzymatic mechanism of plant senescence are rapidly being done because of the cutting-edge technologies and innovative methodologies (Xiao et al., 2024). These days also the modern methodologies, such as proteomics and metabolomics, and that of CRISPR-based gene editing offer unique possibilities to unravel the complex biochemical scenario of what is aging (Ali et al., 2025). While proteome provides insight to senescence-induced enzymes (Xu et al., 2018), the depiction of footprints left in metabolic networks during this progression is starkly revealing by comparison (Shen et al., 2023). CRISPR has made it possible to fine-tune the genetics of enzymes that foster senescence; and this will expand understanding of how these pathways regulate aging and stress responses (Zhang et al., 2023).

Although these recent progresses, still some challenges remain, like the identification of all senescence enzymes and the complete role played by environment on their enzymatic activities (Lozano-Torres et al., 2019). Moreover, the control of enzymatic activity by plant turnover materials (PTMs) (such as phosphorylation and acetylation) remains elusive (Friso & van Wijk, 2015). Further work along these lines may have potential for the development of crops with delayed senescence aspect, improved tolerance and enhanced post-harvest quality which would ensure increased agricultural productivity with decreased post-harvest loss (Ninama et al., 2024).

Crops may have increased yield and better environmental tolerance in relation to climate change through the manipulation of enzymes involved in protein degradation and nutrient recycling (Soares et al., 2019). They can also enhance nutrient-use efficiency, which is beneficial in low-nutrient soils (Hazrati et al., 2025). Applying these findings to crop breeding and biotechnology could revolutionize modern agriculture as a system that is more flexible, resilient and better equipped to feed the world's expanding population.

20.9 CONCLUSION

The mechanisms of enzymatic system involved in plant senescence are critical for nutrient recycling, stressed environment, controlled aging of plants. Enzymes (proteases, hydrolases, antioxidants) are key players in the breakdown cellular material, and their activity is regulated by hormones (ethylene, ABA, cytokinins and JA). Knowledge of genetic control and environmental sensitivity of gene expression is critical for aging research in plants. The knowledge of the study also provides applied opportunities in agriculture, such as retardation of senescence, increasing yield, stress tolerance and prolonging post-harvest quality. Now, armed with new tools like proteomics, gene editing and many others, these enigmas will also be on a crash course to revolutionize crop breeding, food production and the overall sustainability of agriculture.

REFERENCES

Ahmad, S. S., & Tahir, I. (2016). How and why of flower senescence: Understanding from models to ornamentals. *Indian Journal of Plant Physiology*, *21*(4), 446–456.

Akimova, D., Kakimov, A., Suychinov, A., Urazbayev, Z., Zharykbasov, Y., Ibragimov, N., …, & Utegenova, A. (2024). Enzymatic hydrolysis in food processing: Biotechnological advancements, applications, and future perspectives. *Slovak Journal of Food Sciences/Potravinarstvo*, *18*.

Ali, M. S., & Baek, K. H. (2020). Jasmonic acid signaling pathway in response to abiotic stresses in plants. *International Journal of Molecular Sciences*, *21*(2), 621.

Ali, Z., Ullah, Z., Iqbal, J., Abbasi, B. A., Ijaz, S., Ibrahim, M. A., …, & Mahmood, T. (2025). Abiotic stress resilience of CRISPR-edited crops and underlying mechanisms dissected by high-throughput omics technologies. In: *Stress-resilient crops: Coordinated omics-CRISPR-nanotechnology strategies* (pp. 189–233). CABI.

Alonso-Salinas, R., López-Miranda, S., Pérez-López, A. J., & Acosta-Motos, J. R. (2024). Strategies to delay ethylene-mediated ripening in climacteric fruits: Implications for shelf life extension and postharvest quality. *Horticulturae*, *10*(8), 840.

Arora, K., Panda, K. K., Mittal, S., Mallikarjuna, M. G., Rao, A. R., Dash, P. K., & Thirunavukkarasu, N. (2017). RNAseq revealed the important gene pathways controlling adaptive mechanisms under waterlogged stress in maize. *Scientific Reports*, *7*(1), 10950.

Avni, R., Zhao, R., Pearce, S., Jun, Y., Uauy, C., Tabbita, F., …, & Distelfeld, A. (2014). Functional characterization of GPC-1 genes in hexaploid wheat. *Planta*, *239*(2), 313–324.

Barna, B. (2022). Manipulation of senescence of plants to improve biotic stress resistance. *Life*, *12*(10), 1496.

Barnes, W. J., & Anderson, C. T. (2018). Release, recycle, rebuild: Cell-wall remodeling, autodegradation, and sugar salvage for new wall biosynthesis during plant development. *Molecular Plant*, *11*(1), 31–46.

Baseer, A. Q., Niazi, P., Monib, A. W., Hassand, M. H., Hejran, A. B., Sarwari, A., …, & Fahmawi, S. M. S. (2024). Lifecycle transitions in plant development: Ripening, senescence, & cell death. *Journal of Pharma Insights and Research*, *2*(2), 169–179.

Bayanati, M., Lajayer, B. A., & Pessarakli, M. (2021). Senescence: The final phase of plant life. In: *Handbook of plant and crop physiology* (pp. 98–109). CRC Press.

Beltrano, J., Ronco, M. G., & Montaldi, E. R. (1999). Drought stress syndrome in wheat is provoked by ethylene evolution imbalance and reversed by rewatering, aminoethoxyvinylglycine, or sodium benzoate. *Journal of Plant Growth Regulation*, *18*(2), 59–64.

Bhat, M. A., Lone, H. A., & Mehraj, S. S. (2019). Nutrient remobilization during senescence. In: *Senescence signalling and control in plants* (pp. 227–237). Academic Press.

Bhattacharjee, S. (2019). ROS in aging and senescence. In: S. Bhattacharjee (Ed.), *Reactive oxygen species in plant biology* (pp. 65–79). Springer New Delhi. https://doi.org/10.1007/978-81-322-3941-3_3

Borrill, P., Fahy, B., Smith, A. M., & Uauy, C. (2015). Wheat grain filling is limited by grain filling capacity rather than the duration of flag leaf photosynthesis: A case study using NAM RNAi plants. *PloS One*, *10*(8), e0134947.

Buchanan-Wollaston, V., Page, T., Harrison, E., Breeze, E., Lim, P. O., Nam, H. G., …, & Leaver, C. J. (2005). Comparative transcriptome analysis reveals significant differences in gene expression and signalling pathways between developmental and dark/starvation-induced senescence in Arabidopsis. *The Plant Journal*, *42*(4), 567–585.

Cannea, F. B., & Padiglia, A. (2025). Antioxidant defense systems in plants: Mechanisms, regulation, and biotechnological strategies for enhanced oxidative stress tolerance. *Life*, *15*(8), 1293.

Cao, J., Liu, H., Tan, S., & Li, Z. (2023). Transcription factors-regulated leaf senescence: Current knowledge, challenges and approaches. *International Journal of Molecular Sciences*, *24*(11), 9245.

Choi, Y. J., Zaikova, K., Yeom, S. J., Kim, Y. S., & Lee, D. W. (2022). Biogenesis and lipase-mediated mobilization of lipid droplets in plants. *Plants*, *11*(9), 1243.

Crouch, J., Shvedova, M., Thanapaul, R. J. R. S., Botchkarev, V., & Roh, D. (2022). Epigenetic regulation of cellular senescence. *Cells*, *11*(4), 672.

Davalli, P., Mitic, T., Caporali, A., Lauriola, A., & D'Arca, D. (2016). ROS, cell senescence, and novel molecular mechanisms in aging and age-related diseases. *Oxidative Medicine and Cellular Longevity*, *2016*(1), 3565127.

Deo, P. M., Biswal, U. C., & Biswal, B. (2006). Water stress-sensitized photoinhibition in senescing cotyledons of clusterbean: Changes in thylakoid structures and inactivation of photosystem 2. *Photosynthetica*, *44*(2), 187–192.

Dertinger, U., Schaz, U., & Schulze, E. D. (2003). Age-dependence of the antioxidative system in tobacco with enhanced glutathione reductase activity or senescence-induced production of cytokinins. *Physiologia Plantarum*, *119*(1), 19–29.

Doan, P. P. T., Vuong, H. H., & Kim, J. (2024). Genetic foundation of leaf senescence: Insights from natural and cultivated plant diversity. *Plants*, *13*(23), 3405.

Driedonks, N., Xu, J., Peters, J. L., Park, S., & Rieu, I. (2015). Multi-level interactions between heat shock factors, heat shock proteins, and the redox system regulate acclimation to heat. *Frontiers in Plant Science*, *6*, 999.

Estiarte, M., Campioli, M., Mayol, M., & Penuelas, J. (2023). Variability and limits of nitrogen and phosphorus resorption during foliar senescence. *Plant Communications*, *4*(2).

Fanata, W. I. D., Lee, K. H., Son, B. H., Yoo, J. Y., Harmoko, R., Ko, K. S., …, & Lee, K. O. (2013). N-glycan maturation is crucial for cytokinin-mediated development and cellulose synthesis in *Oryza sativa*. *The Plant Journal*, *73*(6), 966–979.

Forlani, S. (2021). Investigating plant senescence: The role of NAC transcription factors in *Solanum lycopersicum* and *Arabidopsis thaliana*. PhD course in Environmental Sciences XXXIII cycle, Dipartimento di Bioscienze, Università degli Studi di Milano, Italy, pp. 150. Available weblink: https://tesidottorato.depositolegale.it/handle/20.500.14242/126564

Friso, G., & van Wijk, K. J. (2015). Posttranslational protein modifications in plant metabolism. *Plant Physiology*, *169*(3), 1469–1487.

Fu, J., Huang, B., & Zhang, G. (2000). Physiological and biochemical changes during seed filling in relation to leaf senescence in soybean. *Biologia Plantarum*, *43*(4), 545–548.

Fujita, M., & Hasanuzzaman, M. (2022). Approaches to enhancing antioxidant defense in plants. *Antioxidants*, *11*(5), 925.

Gao, Z., Liang, X. G., Zhang, L., Lin, S., Zhao, X., Zhou, L. L., …, & Zhou, S. L. (2017). Spraying exogenous 6-benzyladenine and brassinolide at tasseling increases maize yield by enhancing source and sink capacity. *Field Crops Research*, *211*, 1–9.

Girondé, A., Poret, M., Etienne, P., Trouverie, J., Bouchereau, A., Le Cahérec, F., …, & Avice, J. C. (2015). A comparative study of proteolytic mechanisms during leaf senescence of four genotypes of winter oilseed rape highlighted relevant physiological and molecular traits for NRE improvement. *Plants*, *5*(1), 1.

Guo, Y., & Gan, S. (2006). AtNAP, a NAC family transcription factor, has an important role in leaf senescence. *The Plant Journal*, *46*(4), 601–612.

Guo, Y., Ren, G., Zhang, K., Li, Z., Miao, Y., & Guo, H. (2021). Leaf senescence: Progression, regulation, and application. *Molecular Horticulture*, *1*(1), 5.

Häffner, E., Konietzki, S., & Diederichsen, E. (2015). Keeping control: The role of senescence and development in plant pathogenesis and defense. *Plants*, *4*(3), 449–488.

Haq, A. U., Farooq, S., Lone, M. L., Parveen, S., Altaf, F., & Tahir, I. (2024). Flower senescence coordinated by ethylene: An update and future scope on postharvest biology in the "buttercup" family. *Journal of Plant Growth Regulation*, *43*(2), 402–422.

Hazrati, R., Asghari-Zakaria, R., & Zare, N. (2025). Crop genome editing for nutrient use efficiency. In: *Omics and genome editing: Revolution in crop improvement for sustainable agriculture* (pp. 87–106). Springer Nature Switzerland.

He, Y., Zhang, X., Shi, Y., Xu, X., Li, L., & Wu, J. L. (2021). Premature senescence leaf 50 promotes heat stress tolerance in rice (*Oryza sativa* L.). *Rice*, *14*(1), 53.

Hichri, I., Muhovski, Y., Žižková, E., Dobrev, P. I., Franco-Zorrilla, J. M., Solano, R., …, & Lutts, S. (2014). The *Solanum lycopersicum* zinc Finger2 cysteine-2/histidine-2 repressor-like transcription factor regulates development and tolerance to salinity in tomato and Arabidopsis. *Plant Physiology*, *164*(4), 1967–1990.

Hoeberichts, F. A., Van Doorn, W. G., Vorst, O., Hall, R. D., & van Wordragen, M. F. (2007). Sucrose prevents up-regulation of senescence-associated genes in carnation petals. *Journal of Experimental Botany*, *58*(11), 2873–2885.

Hönig, M., Plíhalová, L., Husičková, A., Nisler, J., & Doležal, K. (2018). Role of cytokinins in senescence, antioxidant defence and photosynthesis. *International Journal of Molecular Sciences*, *19*(12), 4045.

Hou, K., Wu, W., & Gan, S. S. (2013). SAUR36, a small auxin up RNA gene, is involved in the promotion of leaf senescence in Arabidopsis. *Plant Physiology*, *161*(2), 1002–1009.

Hu, J., Zhao, X., Gu, L., Liu, P., Zhao, B., Zhang, J., & Ren, B. (2023). The effects of high temperature, drought, and their combined stresses on the photosynthesis and senescence of summer maize. *Agricultural Water Management*, *289*, 108525.

Huang, J., Cai, M., Long, Q., Liu, L., Lin, Q., Jiang, L., …, & Wan, J. (2014). OsLOX2, a rice type I lipoxygenase, confers opposite effects on seed germination and longevity. *Transgenic Research*, *23*(4), 643–655.

Hunter, D. A., Steele, B. C., & Reid, M. S. (2002). Identification of genes associated with perianth senescence in daffodil (*Narcissus pseudonarcissus* L. 'Dutch Master'). *Plant Science*, *163*(1), 13–21.

Ighodaro, O. M., & Akinloye, O. A. (2018). First line defence antioxidants-superoxide dismutase (SOD), catalase (CAT) and glutathione peroxidase (GPX): Their fundamental role in the entire antioxidant defence grid. *Alexandria Journal of Medicine*, *54*(4), 287–293.

Iqbal, N., Khan, N. A., Ferrante, A., Trivellini, A., Francini, A., & Khan, M. I. R. (2017). Ethylene role in plant growth, development and senescence: Interaction with other phytohormones. *Frontiers in Plant Science*, *8*, 475.

Jajic, I., Sarna, T., & Strzalka, K. (2015). Senescence, stress, and reactive oxygen species. *Plants*, *4*(3), 393–411.

Jaradat, M. R., Feurtado, J. A., Huang, D., Lu, Y., & Cutler, A. J. (2013). Multiple roles of the transcription factor AtMYBR1/AtMYB44 in ABA signaling, stress responses, and leaf senescence. *BMC Plant Biology*, *13*(1), 192.

Jhanji, S., Kaur, G., & Chumber, M. (2025). Deciphering flower senescence physiology: Advancements in post-harvest storage and preservation techniques for enhancing longevity. *The Journal of Horticultural Science and Biotechnology*, *100*(2), 164–195.

Jiménez, A., Hernández, J. A., Pastori, G., del Rıo, L. A., & Sevilla, F. (1998). Role of the ascorbate-glutathione cycle of mitochondria and peroxisomes in the senescence of pea leaves. *Plant Physiology*, *118*(4), 1327–1335.

Joshi, S., Choukimath, A., Isenegger, D., Panozzo, J., Spangenberg, G., & Kant, S. (2019). Improved wheat growth and yield by delayed leaf senescence using developmentally regulated expression of a cytokinin biosynthesis gene. *Frontiers in Plant Science*, *10*, 1285.

Kaabel, S., Friščić, T., & Auclair, K. (2020). Mechanoenzymatic transformations in the absence of bulk water: A more natural way of using enzymes. *ChemBioChem*, *21*(6), 742–758.

Kalinin, A., Zubkova, E., & Menshikov, M. (2023). Integrated stress response (ISR) pathway: Unraveling its role in cellular senescence. *International Journal of Molecular Sciences*, *24*(24), 17423.

Kanazawa, S., Sano, S., Koshiba, T., & Ushimaru, T. (2000). Changes in antioxidative enzymes in cucumber cotyledons during natural senescence: Comparison with those during dark-induced senescence. *Physiologia Plantarum*, *109*(2), 211–216.

Kang, K., Shim, Y., Gi, E., An, G., & Paek, N. C. (2019). Mutation of ONAC096 enhances grain yield by increasing panicle number and delaying leaf senescence during grain filling in rice. *International Journal of Molecular Sciences*, *20*(20), 5241.

Kant, S., Burch, D., Badenhorst, P., Palanisamy, R., Mason, J., & Spangenberg, G. (2015). Regulated expression of a cytokinin biosynthesis gene IPT delays leaf senescence and improves yield under rainfed and irrigated conditions in canola (*Brassica napus* L.). *PLoS One*, *10*(1), e0116349.

Khan, M., Rozhon, W., & Poppenberger, B. (2013). The role of hormones in the aging of plants—A mini-review. *Gerontology*, *60*(1), 49–55.

Khan, N. (2025). Molecular insights into ABA-mediated regulation of stress tolerance and development in plants. *International Journal of Molecular Sciences*, *26*(16), 7872.

Kim, J. H., Woo, H. R., Kim, J., Lim, P. O., Lee, I. C., Choi, S. H., …, & Nam, H. G. (2009). Trifurcate feed-forward regulation of age-dependent cell death involving miR164 in Arabidopsis. *Science*, *323*(5917), 1053–1057.

Kim, T., Kang, K., Kim, S. H., An, G., & Paek, N. C. (2019). OsWRKY5 promotes rice leaf senescence via senescence-associated NAC and abscisic acid biosynthesis pathway. *International Journal of Molecular Sciences*, *20*(18), 4437.

Krishnamurthy, H. K., Rajavelu, I., Pereira, M., Jayaraman, V., Krishna, K., Wang, T., …, & Rajasekaran, J. J. (2024). Inside the genome: Understanding genetic influences on oxidative stress. *Frontiers in Genetics*, *15*, 1397352.

Lal, M. K., Tiwari, R. K., Lal, P., Kumar, A., Kumar, R. (2023). Regulatory Role of Melatonin in Post-harvest Management of Vegetables and Fruits. In: R. Kumar, M. A. Altaf, M. K. Lal, & R. K. Tiwari (Eds.), *Melatonin in plants: A regulator for plant growth and development* (pp. 219–244). Springer, Singapore. https://doi.org/10.1007/978-981-99-6745-2_10

Langston, B. J., Bai, S., & Jones, M. L. (2005). Increases in DNA fragmentation and induction of a senescence-specific nuclease are delayed during corolla senescence in ethylene-insensitive (etr1-1) transgenic petunias. *Journal of Experimental Botany*, *56*(409), 15–23.

Li, Y., He, H., Hou, Y., Kelimu, A., Wu, F., Zhao, Y., …, & Zhu, X. (2022). Salicylic acid treatment delays apricot (*Prunus armeniaca* L.) fruit softening by inhibiting ethylene biosynthesis and cell wall degradation. *Scientia Horticulturae*, *300*, 111061.

Li, Z., Peng, J., Wen, X., & Guo, H. (2013). ETHYLENE-INSENSITIVE3 is a senescence-associated gene that accelerates age-dependent leaf senescence by directly repressing miR164 transcription in *Arabidopsis*. *The Plant Cell*, 25(9), 3311–3328. https://doi.org/10.1105/tpc.113.113340

Liang, C., Wang, Y., Zhu, Y., Tang, J., Hu, B., Liu, L., …, & Chu, C. (2014). OsNAP connects abscisic acid and leaf senescence by fine-tuning abscisic acid biosynthesis and directly targeting senescence-associated genes in rice. *Proceedings of the National Academy of Sciences*, *111*(27), 10013–10018.

Lim, P. O., Kim, H. J., & Gil Nam, H. (2007). Leaf senescence. *Annual Review of Plant Biology*, *58*, 115–136.

Lira, B. S., Gramegna, G., Trench, B. A., Alves, F. R., Silva, E. M., Silva, G. F., …, & Rossi, M. (2017). Manipulation of a senescence-associated gene improves fleshy fruit yield. *Plant Physiology*, *175*(1), 77–91.

Liu, J., Shen, J., Xu, Y., Li, X., Xiao, J., & Xiong, L. (2016). Ghd2, a CONSTANS-like gene, confers drought sensitivity through regulation of senescence in rice. *Journal of Experimental Botany*, *67*(19), 5785–5798.

Lohman, K. N., Gan, S., John, M. C., & Amasino, R. M. (1994). Molecular analysis of natural leaf senescence in *Arabidopsis thaliana*. *Physiologia Plantarum*, *92*(2), 322–328.

Lozano-Torres, B., Estepa-Fernández, A., Rovira, M., Orzaez, M., Serrano, M., Martinez-Manez, R., & Sancenon, F. (2019). The chemistry of senescence. *Nature Reviews Chemistry*, *3*(7), 426–441.

Lu, G., Casaretto, J. A., Ying, S., Mahmood, K., Liu, F., Bi, Y. M., & Rothstein, S. J. (2017). Overexpression of OsGATA12 regulates chlorophyll content, delays plant senescence and improves rice yield under high density planting. *Plant Molecular Biology*, *94*(1), 215–227.

Luan, W., Shen, A., Jin, Z., Song, S., Li, Z., & Sha, A. (2013). Knockdown of OsHox33, a member of the class III homeodomain-leucine zipper gene family, accelerates leaf senescence in rice. *Science China Life Sciences*, *56*(12), 1113–1123.

Luciński, R., & Adamiec, M. (2023). The role of plant proteases in the response of plants to abiotic stress factors. *Frontiers in Plant Physiology*, *1*, 1330216.

Ma, X., Zhang, Y., Turečková, V., Xue, G. P., Fernie, A. R., Mueller-Roeber, B., & Balazadeh, S. (2018). The NAC transcription factor SlNAP2 regulates leaf senescence and fruit yield in tomato. *Plant Physiology*, *177*(3), 1286–1302.

Ma, Y., Dias, M. C., & Freitas, H. (2020). Drought and salinity stress responses and microbe-induced tolerance in plants. *Frontiers in Plant Science*, *11*, 591911.

Mao, C., Lu, S., Lv, B., Zhang, B., Shen, J., He, J., …, & Ming, F. (2017). A rice NAC transcription factor promotes leaf senescence via ABA biosynthesis. *Plant Physiology*, *174*(3), 1747–1763.

Masclaux-Daubresse, C., Chen, Q., & Havé, M. (2017). Regulation of nutrient recycling via autophagy. *Current Opinion in Plant Biology*, *39*, 8–17.

Masclaux-Daubresse, C., Purdy, S., Lemaitre, T., Pourtau, N., Taconnat, L., Renou, J. P., & Wingler, A. (2007). Genetic variation suggests interaction between cold acclimation and metabolic regulation of leaf senescence. *Plant Physiology*, *143*(1), 434–446.

Matallana-Ramirez, L. P., Rauf, M., Farage-Barhom, S., Dortay, H., Xue, G. P., Dröge-Laser, W., …, & Mueller-Roeber, B. (2013). NAC transcription factor ORE1 and senescence-induced BIFUNCTIONAL NUCLEASE1 (BFN1) constitute a regulatory cascade in Arabidopsis. *Molecular Plant*, *6*(5), 1438–1452.

Miao, Y., Laun, T., Zimmermann, P., & Zentgraf, U. (2004). Targets of the WRKY53 transcription factor and its role during leaf senescence in Arabidopsis. *Plant Molecular Biology*, *55*(6), 853–867.

Muradova, M., Proskura, A., Canon, F., Aleksandrova, I., Schwartz, M., Heydel, J. M., …, & Neiers, F. (2023). Unlocking flavor potential using microbial β-glucosidases in food processing. *Foods*, *12*(24), 4484.

Nascimento, D. B. D., Lopes, M. L. S., Izidro, J. L. P. S., Bezerra, R. C. A., Gois, G. C., Amaral, T. N. E. D., …, & Coêlho, J. J. (2024). Nitrogen, phosphorus, and potassium cycling in pasture ecosystems. *Ciência Animal Brasileira*, *25*, e-76743.

Ninama, N., Gangal, L., Khayum, A., & Singh, A. (2024). Post-harvest biotechnology or genetic engineering solutions: Extending shelf life and reducing food waste. *Journal of Advances in Biology & Biotechnology*, *27*(4), 1–26.

Nishad, D. C., Mishra, H., Tiwari, A. K., & Mishra, D. (2024). Post-harvest management: Enhancing food security and sustainability. *Advances in Agriculture Sciences Volume II*, *24*(4), 136.

Oneto, C. D., Otegui, M. E., Baroli, I., Beznec, A., Faccio, P., Bossio, E., …, & Lewi, D. (2016). Water deficit stress tolerance in maize conferred by expression of an isopentenyltransferase (IPT) gene driven by a stress-and maturation-induced promoter. *Journal of Biotechnology*, *220*, 66–77.

Panozzo, A., Bolla, P. K., Barion, G., Botton, A., & Vamerali, T. (2025). Phytohormonal regulation of abiotic stress tolerance, leaf senescence and yield response in field crops: A comprehensive review. *BioTech*, *14*(1), 14.

Pareek, S., Sagar, N. A., Sharma, S., Kumar, V., Agarwal, T., González, Aguilar, G. A., & Yahia, E. M. (2017). Chlorophylls: Chemistry and biological functions. In: E. M. Yahia (Ed.), *Fruit and vegetable phytochemicals: Chemistry and human health* (2nd edn., pp. 269–284). John Wiley & Sons, Ltd. https://doi.org/10.1002/9781119158042.ch14

Peleg, Z., Reguera, M., Tumimbang, E., Walia, H., & Blumwald, E. (2011). Cytokinin-mediated source/sink modifications improve drought tolerance and increase grain yield in rice under water-stress. *Plant Biotechnology Journal*, *9*(7), 747–758.

Piao, W., Kim, S. H., Lee, B. D., An, G., Sakuraba, Y., & Paek, N. C. (2019). Rice transcription factor OsMYB102 delays leaf senescence by down-regulating abscisic acid accumulation and signaling. *Journal of Experimental Botany*, *70*(10), 2699–2715.

Podzimska-Sroka, D., O'Shea, C., Gregersen, P. L., & Skriver, K. (2015). NAC transcription factors in senescence: From molecular structure to function in crops. *Plants*, *4*(3), 412–448.

Prochazkova, D., Sairam, R. K., Srivastava, G. C., & Singh, D. V. (2001). Oxidative stress and antioxidant activity as the basis of senescence in maize leaves. *Plant Science*, *161*(4), 765–771.

Procházková, D., & Wilhelmová, N. (2004). Changes in antioxidative protection in bean cotyledons during natural and continuous irradiation-accelerated senescence. *Biologia Plantarum*, *48*(1), 33–39.

Procházková, D., & Wilhelmová, N. (2007). The capacity of antioxidant protection during modulated ageing of bean (*Phaseolus vulgaris* L.) Cotyledons. 1. The antioxidant enzyme activities. *Cell Biochemistry and Function*, *25*(1), 87–95.

Pruzinska, A., Anders, I., Aubry, S., Schenk, N., Tapernoux-Lüthi, E., Müller, T., …, & Hörtensteiner, S. (2007). In vivo participation of red chlorophyll catabolite reductase in chlorophyll breakdown. *The Plant Cell*, *19*(1), 369–387.

Qin, H., Gu, Q., Zhang, J., Sun, L., Kuppu, S., Zhang, Y., …, & Zhang, H. (2011). Regulated expression of an isopentenyltransferase gene (IPT) in peanut significantly improves drought tolerance and increases yield under field conditions. *Plant and Cell Physiology*, *52*(11), 1904–1914.

Raab, S., Drechsel, G., Zarepour, M., Hartung, W., Koshiba, T., Bittner, F., & Hoth, S. (2009). Identification of a novel E3 ubiquitin ligase that is required for suppression of premature senescence in Arabidopsis. *The Plant Journal*, *59*(1), 39–51.

Raineri, J., Campi, M., Chan, R. L., & Otegui, M. E. (2019). Maize expressing the sunflower transcription factor HaHB11 has improved productivity in controlled and field conditions. *Plant Science*, *287*, 110185.

Rasool, S., Mir, B. A., Rehman, M. U., Amin, I., Mir, M. U. R., & Ahmad, S. B. (2019). Abiotic stress and plant senescence. In: *Senescence signalling and control in plants* (pp. 15–27). Academic Press.

Raza, A., Ashraf, F., Zou, X., Zhang, X., & Tosif, H. (2020). Plant adaptation and tolerance to environmental stresses: Mechanisms and perspectives. In: *Plant ecophysiology and adaptation under climate change: Mechanisms and perspectives I: General consequences and plant responses* (pp. 117–145). Springer Singapore.

Reinbothe, C., Springer, A., Samol, I., & Reinbothe, S. (2009). Plant oxylipins: Role of jasmonic acid during programmed cell death, defence and leaf senescence. *The FEBS Journal*, *276*(17), 4666–4681.

Ren, B., Zhu, Y., Zhang, J., Dong, S., Liu, P., & Zhao, B. (2016). Effects of spraying exogenous hormone 6-benzyladenine (6-BA) after waterlogging on grain yield and growth of summer maize. *Field Crops Research*, *188*, 96–104.

Rinerson, C. I., Rabara, R. C., Tripathi, P., Shen, Q. J., & Rushton, P. J. (2015). The evolution of WRKY transcription factors. *BMC Plant Biology*, *15*(1), 66.

Rodoni, S., Muhlecker, W., Anderl, M., Krautler, B., Moser, D., Thomas, H., …, & Hortensteiner, S. (1997). Chlorophyll breakdown in senescent chloroplasts (cleavage of pheophorbide a in two enzymic steps). *Plant Physiology*, *115*(2), 669–676.

Roychowdhury, R., Hada, A., Biswas, S., Mishra, S., Prusty, M. R., Das, S. P., …, & Sarker, U. (2025). Jasmonic acid (JA) in plant immune response: Unravelling complex molecular mechanisms and networking of defence signalling against pathogens. *Journal of Plant Growth Regulation*, *44*(1), 89–114.

Sade, N., del Mar Rubio-Wilhelmi, M., Umnajkitikorn, K., & Blumwald, E. (2018). Stress-induced senescence and plant tolerance to abiotic stress. *Journal of Experimental Botany*, *69*(4), 845–853.

Sağlam, N. G., Yaycılı, O., & Arıcı, C. D. (2024). Plant growth regulators and senescence. In: *Plant growth regulators to manage biotic and abiotic stress in agroecosystems* (pp. 37–53). CRC Press.

Sahu, P. K., Jayalakshmi, K., Tilgam, J., Gupta, A., Nagaraju, Y., Kumar, A., …, & Rajawat, M. V. S. (2022). ROS generated from biotic stress: Effects on plants and alleviation by endophytic microbes. *Frontiers in Plant Science*, *13*, 1042936.

Saini, R., Saini, H. S., & Dahiya, A. (2017). Amylases: Characteristics and industrial applications. *Journal of Pharmacognosy and Phytochemistry*, *6*(4), 1865–1871.

Sakuraba, Y., Kim, D., Han, S. H., Kim, S. H., Piao, W., Yanagisawa, S., …, & Paek, N. C. (2020). Multilayered regulation of membrane-bound ONAC054 is essential for abscisic acid-induced leaf senescence in rice. *The Plant Cell*, *32*(3), 630–649.

Sakuraba, Y., Kim, Y. S., Han, S. H., Lee, B. D., & Paek, N. C. (2015). The Arabidopsis transcription factor NAC016 promotes drought stress responses by repressing AREB1 transcription through a trifurcate feed-forward regulatory loop involving NAP. *The Plant Cell*, *27*(6), 1771–1787.

Sakuraba, Y., Schelbert, S., Park, S. Y., Han, S. H., Lee, B. D., Andres, C. B., …, & Paek, N. C. (2012). STAY-GREEN and chlorophyll catabolic enzymes interact at light-harvesting complex II for chlorophyll detoxification during leaf senescence in Arabidopsis. *The Plant Cell*, *24*(2), 507–518.

Salehi, B., Azzini, E., Zucca, P., Maria Varoni, E., Anil Kumar, N., Dini, L., …, & Sharifi-Rad, J. (2020). Plant-derived bioactives and oxidative stress-related disorders: A key trend towards healthy aging and longevity promotion. *Applied Sciences*, *10*(3), 947.

Sarwat, M., Naqvi, A. R., Ahmad, P., Ashraf, M., & Akram, N. A. (2013). Phytohormones and microRNAs as sensors and regulators of leaf senescence: Assigning macro roles to small molecules. *Biotechnology Advances*, *31*(8), 1153–1171.

Sato, Y., Morita, R., Katsuma, S., Nishimura, M., Tanaka, A., & Kusaba, M. (2009). Two short-chain dehydrogenase/reductases, non-yellow COLORING 1 and NYC1-LIKE, are required for chlorophyll b and light-harvesting complex II degradation during senescence in rice. *The Plant Journal*, *57*(1), 120–131.

Schippers, J. H. (2015). Transcriptional networks in leaf senescence. *Current Opinion in Plant Biology*, *27*, 77–83.

Sequera-Mutiozabal, M. I., Erban, A., Kopka, J., Atanasov, K. E., Bastida, J., Fotopoulos, V., …, & Tiburcio, A. F. (2016). Global metabolic profiling of *Arabidopsis* polyamine oxidase 4 (AtPAO4) loss-of-function mutants exhibiting delayed dark-induced senescence. *Frontiers in Plant Science*, *7*, 173.

Serek, M., & Reid, M. S. (2024). Role of growth regulators in the postharvest life of ornamentals. In: *Plant growth regulators in agriculture and horticulture* (pp. 147–174). CRC Press.

Shan, X., Wang, J., Chua, L., Jiang, D., Peng, W., & Xie, D. (2011). The role of Arabidopsis Rubisco activase in jasmonate-induced leaf senescence. *Plant Physiology, 155*(2), 751–764.

Shen, S., Zhan, C., Yang, C., Fernie, A. R., & Luo, J. (2023). Metabolomics-centered mining of plant metabolic diversity and function: Past decade and future perspectives. *Molecular Plant, 16*(1), 43–63.

Shin, D., Lee, S., Kim, T. H., Lee, J. H., Park, J., Lee, J., ..., & Nam, H. G. (2020). Natural variations at the stay-green gene promoter control lifespan and yield in rice cultivars. *Nature Communications, 11*(1), 2819.

Smeekens, S., Ma, J., Hanson, J., & Rolland, F. (2010). Sugar signals and molecular networks controlling plant growth. *Current Opinion in Plant Biology, 13*(3), 273–278.

Soares, J. C., Santos, C. S., Carvalho, S. M., Pintado, M. M., & Vasconcelos, M. W. (2019). Preserving the nutritional quality of crop plants under a changing climate: Importance and strategies. *Plant and Soil, 443*(1), 1–26.

Sobieszczuk-Nowicka, E., Kubala, S., Zmienko, A., Małecka, A., & Legocka, J. (2016). From accumulation to degradation: Reprogramming polyamine metabolism facilitates dark-induced senescence in barley leaf cells. *Frontiers in Plant Science, 6*, 1198.

Souer, E., van Houwelingen, A., Kloos, D., Mol, J., & Koes, R. (1996). The no apical meristem gene of Petunia is required for pattern formation in embryos and flowers and is expressed at meristem and primordia boundaries. *Cell, 85*(2), 159–170.

Srivalli, B., & Khanna-Chopra, R. (2001). Induction of new isoforms of superoxide dismutase and catalase enzymes in the flag leaf of wheat during monocarpic senescence. *Biochemical and Biophysical Research Communications, 288*(4), 1037–1042.

Su, T., Li, X., Yang, M., Shao, Q., Zhao, Y., Ma, C., & Wang, P. (2020). Autophagy: An intracellular degradation pathway regulating plant survival And stress response. *Frontiers in Plant Science, 11*, 164.

Sýkorová, B., Kurešová, G., Daskalova, S., Trčková, M., Hoyerová, K., Raimanová, I., ..., & Kamínek, M. (2008). Senescence-induced ectopic expression of the *A. tumefaciens* ipt gene in wheat delays leaf senescence, increases cytokinin content, nitrate influx, and nitrate reductase activity, but does not affect grain yield. *Journal of Experimental Botany, 59*(2), 377–387.

Syvash, O. O., & Zolotareva, O. K. (2017). Regulation of chlorophyll degradation in plant tissues. *Biotechnologia Acta, 10*(3), 20–30.

Tamary, E., Nevo, R., Naveh, L., Levin-Zaidman, S., Kiss, V., Savidor, A., ..., & Adam, Z. (2019). Chlorophyll catabolism precedes changes in chloroplast structure and proteome during leaf senescence. *Plant Direct, 3*(3), e00127.

Tanaka, A., & Ito, H. (2025). Chlorophyll degradation and its physiological function. *Plant and Cell Physiology, 66*(2), 139–152.

Thelander, M., Olsson, T., & Ronne, H. (2004). Snf1-related protein kinase 1 is needed for growth in a normal day-night light cycle. *European Molecular Biology Organization Journal, 23*, 1900–1910.

Toushik, S. H., Lee, K. T., Lee, J. S., & Kim, K. S. (2017). Functional applications of lignocellulolytic enzymes in the fruit and vegetable processing industries. *Journal of Food Science, 82*(3), 585–593.

Tsanakas, G. F., Manioudaki, M. E., Economou, A. S., & Kalaitzis, P. (2014). De novo transcriptome analysis of petal senescence in *Gardenia jasminoides* Ellis. *BMC Genomics, 15*(1), 554.

Uauy, C., Distelfeld, A., Fahima, T., Blechl, A., & Dubcovsky, J. (2006). A NAC gene regulating senescence improves grain protein, zinc, and iron content in wheat. *Science, 314*(5803), 1298–1301.

van der Graaff, E., Schwacke, R., Schneider, A., Desimone, M., Flugge, U. I., & Kunze, R. (2006). Transcription analysis of *Arabidopsis* membrane transporters and hormone pathways during developmental and induced leaf senescence. *Plant Physiology, 141*(2), 776–792.

van Doorn, W. G. (2004). Is petal senescence due to sugar starvation? *Plant Physiology, 134*(1), 35–42.

van Doorn, W. G. (2008). Is the onset of senescence in leaf cells of intact plants due to low or high sugar levels? *Journal of Experimental Botany, 59*(8), 1963–1972.

Varesi, A., Chirumbolo, S., Campagnoli, L. I. M., Pierella, E., Piccini, G. B., Carrara, A., ..., & Pascale, A. (2022). The role of antioxidants in the interplay between oxidative stress and senescence. *Antioxidants, 11*(7), 1224.

Verma, V., Ravindran, P., & Kumar, P. P. (2016). Plant hormone-mediated regulation of stress responses. *BMC Plant Biology, 16*(1), 86.

Vescovi, M., Riefler, M., Gessuti, M., Novák, O., Schmülling, T., & Lo Schiavo, F. (2012). Programmed cell death induced by high levels of cytokinin in Arabidopsis cultured cells is mediated by the cytokinin receptor CRE1/AHK4. *Journal of Experimental Botany, 63*(7), 2825–2832.

Wagstaff, C., Leverentz, M. K., Griffiths, G., Thomas, B., Chanasut, U., Stead, A. D., & Rogers, H. J. (2002). Cysteine protease gene expression and proteolytic activity during senescence of Alstroemeria petals. *Journal of Experimental Botany, 53*(367), 233–240.

Wang, F., Liu, J., Zhou, L., Pan, G., Li, Z., Zaidi, S. H. R., & Cheng, F. (2016). Senescence-specific change in ROS scavenging enzyme activities and regulation of various SOD isozymes to ROS levels in psf mutant rice leaves. *Plant Physiology and Biochemistry, 109*, 248–261.

Wang, H., & Schippers, J. H. (2019). The role and regulation of autophagy and the proteasome during aging and senescence in plants. *Genes, 10*(4), 267.

Wang, H., Yang, Y., Yao, C., Feng, Y., Wang, H., Kong, Y., ..., & Deng, G. (2024). The correct combination and balance of macronutrients nitrogen, phosphorus and potassium promote plant yield and quality through enzymatic and antioxidant activities in potato. *Journal of Plant Growth Regulation, 43*(12), 4716–4734.

War, A. F., Sheergojri, I. A., Nanda, S. A., Khan, M. A., Rehmaan, I. U., Reshi, Z. A., & Rashid, I. (2022). Recycling and remobilization of nitrogen during senescence. In: *Advances in plant nitrogen metabolism* (pp. 96–117). CRC Press.

Watson, J. M., & Riha, K. (2010). Comparative biology of telomeres: Where plants stand. *FEBS Letters, 584*(17), 3752–3759.

Werner, T., Motyka, V., Laucou, V., Smets, R., Van Onckelen, H., & Schmülling, T. (2003). Cytokinin-deficient transgenic Arabidopsis plants show multiple developmental alterations indicating opposite functions of cytokinins in the regulation of shoot and root meristem activity. *The Plant Cell, 15*(11), 2532–2550.

Wingler, A., von Schaewen, A., Leegood, R. C., Lea, P. J., & Paul Quick, W. (1998). Regulation of leaf senescence by cytokinin, sugars, and light: Effects on NADH-dependent hydroxypyruvate reductase. *Plant Physiology, 116*(1), 329–335.

Woo, H. R., Masclaux-Daubresse, C., & Lim, P. O. (2018). Plant senescence: How plants know when and how to die. *Journal of Experimental Botany*, *69*(4), 715–718.

Wu, A., Allu, A. D., Garapati, P., Siddiqui, H., Dortay, H., Zanor, M. I., …, & Balazadeh, S. (2012). JUNGBRUNNEN1, a reactive oxygen species–responsive NAC transcription factor, regulates longevity in Arabidopsis. *The Plant Cell*, *24*(2), 482–506.

Xiao, R., Hu, S., Du, X., Wang, Y., Fang, K., Zhu, Y., …, & Yang, J. (2024). Revolutionizing senescence detection: Advancements from traditional methods to cutting-edge techniques. *Aging and Disease*, *16*(3), 1285.

Xiong, Y., Contento, A. L., & Bassham, D. C. (2005). AtATG18a is required for the formation of autophagosomes during nutrient stress and senescence in *Arabidopsis thaliana*. *The Plant Journal*, *42*(4), 535–546.

Xu, F., Meng, T., Li, P., Yu, Y., Cui, Y., Wang, Y., …, & Wang, N. N. (2011). A soybean dual-specificity kinase, GmSARK, and its Arabidopsis homolog, AtSARK, regulate leaf senescence through synergistic actions of auxin and ethylene. *Plant Physiology*, *157*(4), 2131–2153.

Xu, X., Guo, K., Liang, W., Chen, Q., Shi, J., & Shen, B. (2018). Quantitative proteomics analysis of proteins involved in leaf senescence of rice (*Oryza sativa* l.). *Plant Growth Regulation*, *84*(2), 341–349.

Xu, Y., Ishida, H., Reisen, D., & Hanson, M. R. (2006). Upregulation of a tonoplast-localized cytochrome P450 during petal senescence in *Petunia inflata*. *BMC Plant Biology*, *6*(1), 8.

Yamatani, H., Kohzuma, K., Nakano, M., Takami, T., Kato, Y., Hayashi, Y., …, & Kusaba, M. (2018). Impairment of Lhca4, a subunit of LHCI, causes high accumulation of chlorophyll and the stay-green phenotype in rice. *Journal of Experimental Botany*, *69*(5), 1027–1035.

Yamatani, H., Sato, Y., Masuda, Y., Kato, Y., Morita, R., Fukunaga, K., …, & Kusaba, M. (2013). NYC4, the rice ortholog of *Arabidopsis* THF1, is involved in the degradation of chlorophyll–protein complexes during leaf senescence. *The Plant Journal*, *74*(4), 652–662.

Yang, D., Li, Y., Shi, Y., Cui, Z., Luo, Y., Zheng, M., …, & Wang, Z. (2016). Exogenous cytokinins increase grain yield of winter wheat cultivars by improving stay-green characteristics under heat stress. *PLoS One*, *11*(5), e0155437.

Yang, J. C., Zhang, J. H., Wang, Z. Q., Zhu, Q. S., & Liu, L. J. (2003). Involvement of abscisic acid and cytokinins in the senescence and remobilization of carbon reserves in wheat subjected to water stress during grain filling. *Plant, Cell & Environment*, *26*(10), 1621–1631.

Yetgin, A., Srivastava, R.K., Mandal, N. (2025). Insights into Plant Hormone Signaling Networks for Environmental Responses. In: R. K. Srivastava, & A. Chakraborty (Eds.), *Mitigation and adaptation strategies against climate change in natural systems. Environmental earth sciences* (pp. 505–523). Springer, Cham. https://doi.org/10.1007/978-3-031-75968-0_26

Zhang, J., Fengler, K. A., Van Hemert, J. L., Gupta, R., Mongar, N., Sun, J., …, & Shen, B. (2019). Identification and characterization of a novel stay-green QTL that increases yield in maize. *Plant Biotechnology Journal*, *17*(12), 2272–2285.

Zhang, K., & Gan, S. S. (2012). An abscisic acid-AtNAP transcription factor-SAG113 protein phosphatase 2C regulatory chain for controlling dehydration in senescing Arabidopsis leaves. *Plant Physiology*, *158*(2), 961–969.

Zhang, K., Xia, X., Zhang, Y., & Gan, S. S. (2012). An ABA-regulated and Golgi-localized protein phosphatase controls water loss during leaf senescence in Arabidopsis. *The Plant Journal*, *69*(4), 667–678.

Zhang, P., Wang, W. Q., Zhang, G. L., Kaminek, M., Dobrev, P., Xu, J., & Gruissem, W. (2010). Senescence-inducible expression of isopentenyl transferase extends leaf life, increases drought stress resistance and alters cytokinin metabolism in cassava. *Journal of Integrative Plant Biology*, *52*(7), 653–669.

Zhang, W., Peng, K., Cui, F., Wang, D., Zhao, J., Zhang, Y., …, & Zhang, K. (2021). Cytokinin oxidase/dehydrogenase OsCKX11 coordinates source and sink relationship in rice by simultaneous regulation of leaf senescence and grain number. *Plant Biotechnology Journal*, *19*(2), 335–350.

Zhang, Y., Huang, D., & Miao, Y. (2023). Epigenetic control of plant senescence and cell death and its application in crop improvement. *Frontiers in Plant Science*, *14*, 1258487.

Zhang, Y., Wang, H. L., Li, Z., & Guo, H. (2020). Genetic network between leaf senescence and plant immunity: Crucial regulatory nodes and new insights. *Plants*, *9*(4), 495.

Zhao, D., Derkx, A. P., Liu, D. C., Buchner, P., & Hawkesford, M. J. (2015). Overexpression of a NAC transcription factor delays leaf senescence and increases grain nitrogen concentration in wheat. *Plant Biology*, *17*(4), 904–913.

Zhao, W., Zhao, H., Wang, H., & He, Y. (2022). Research progress on the relationship between leaf senescence and quality, yield and stress resistance in horticultural plants. *Frontiers in Plant Science*, *13*, 1044500.

Zhou, M., & Yang, J. (2023). Delaying or promoting? Manipulation of leaf senescence to improve crop yield and quality. *Planta*, *258*(3), 48.

Zhu, K., Ren, W., Yan, J., Zhang, Y., Zhang, W., Xu, Y., …, & Yang, J. (2022). Grain yield and nitrogen use efficiency are increased by exogenous cytokinin application through the improvement in root physiological traits of rice. *Plant Growth Regulation*, *97*(1), 157–169.

Zimmermann, P., Heinlein, C., Orendi, G., & Zentgraf, U. (2006). Senescence-specific regulation of catalases in *Arabidopsis thaliana* (L.) Heynh. *Plant, Cell & Environment*, *29*(6), 1049–1060.

21 Plant–Microbe Interaction Modulating Enzymatic Response Under Stress Conditions

Hina Jabeen, Akhtar Rasool, Nadia Sharif, Rabiha Seboussi, and Zainul Abideen

21.1 INTRODUCTION

Modern agriculture is under immense pressure to produce more food on less land, with fewer resources, and under increasingly unpredictable environmental conditions. This urgent need has sparked growing interest in nature-friendly or bio-rational solutions which can be innovative and sustainable. In this regard, plant-associated microorganisms have been proved to be a promising path toward resilient and sustainable crop production even in the periods of environmental and pollutant stress (Murugesan et al., 2024). The challenge is significant; rapid urbanization and industrialization have resulted in the enhanced contamination of agricultural soils with chemical pollutants and heavy metals (HMs) (e.g., Ni, Cd, Pb, Cr, As). This contamination has threatened human health, plant growth, and food security on a global scale (Khilji et al., 2025). Agricultural lands are exposed to increased accumulation of pesticides, fertilizers, polyaromatic hydrocarbons, and polychlorinated biphenyls (Xia et al., 2024). Exposure to various stresses and toxic levels of pollutants severely hampers plant growth and triggers a cascade of physiological and molecular disturbances across different plant varieties and species leading to the detrimental effects on plant growth and physiology. To cope with these and other stresses, including abiotic factors like heat, drought, and salinity, and biotic threats like pathogenic invasions, plants must mount complex defense mechanisms (Mujeeb et al., 2023).

Scientists have been increasingly drawn to the subtle but powerful partnerships between plants and the microorganisms that live alongside them. These microscopic allies ranging from beneficial bacteria and fungi to hidden entophytic microbes are emerging as essential players in helping plants adapt to and cope with the stress conditions of climate/environment (Abideen et al., 2022; Hanif et al., 2023). Whether it is extreme heat, drought, nutrient-poor soils, or pathogen attack, plant-associated microbes act like invisible support teams, fine-tuning plant enzyme activity, optimizing nutrient uptake, and strengthening natural defense systems (Adeleke et al., 2024). This relationship is especially exciting because the microbes actively train plants for resilience, often by modulating hormone levels and boosting antioxidant capacity even before a stress event occurs. This suggests a powerful, sustainable alternative to chemical fertilizers and pesticides.

Plants are constantly challenged by climatic changes or ecological stress factors as the environment is more like a dynamic platform rather than a static entity which necessitates strong competition for the available resources. These stresses include (1) abiotic stresses like heat, cold, drought, nutrients deficiency, salinity, HMs toxicity, pollutants toxicity and (2) biotic stresses such as pathogenic or parasitic invasion and presence of pests (insects, worms, birds, animals, etc.). All the stress factors elicit complex physiological responses in plants which disturb the regular metabolic processes. One of such physiological responses is the activation of enzymatic network which regulates the signaling pathways and evokes the defense mechanisms. Moreover, human interventions include the use of pesticides, fertilizers, or other chemical means which are applied with the aim of restoring soil fertility and nutrient balance for better agriculture productivity. Unfortunately, these chemicals pose a severe threat to plants and their consumers (Muhammad et al., 2024; Upadhyay, 2025).

These beneficial partnerships are far from accidental; over time, plants and microbes have evolved to cooperate, especially under tough conditions like drought, poor soil, or disease pressure. Certain microbes help plants grow better by improving nutrient availability, producing plant hormones, or helping the plant handle stress more efficiently (Mikiciuk et al., 2024). One example of this comes from studies on the native shrub *Ceanothus velutinus*, where researchers have found plant growth-promoting rhizobacteria (PGPR) with some impressive abilities: they produce the growth hormone indole-3-acetic acid (IAA), release iron-binding molecules called siderophores, break down complex proteins, detoxify harmful compounds with catalase (CAT), and even help fix nitrogen and unlock phosphate from the soil (Zhang et al., 2025). When soil from this plant natural habitat was added to a propagation mix, it helped improve rooting and growth, and some bacterial strains even boosted growth in *Arabidopsis*, a model plant used in research. Plant growth-promoting microorganisms (PGPMs) include a wide variety of bacteria and fungi that boost plant health in multiple ways, such as improving nutrient availability, producing growth hormones, and helping plants survive harsh conditions (Rafique et al., 2025). In sorghum, certain bacteria have proven especially effective at enhancing drought resistance and growth. For example,

DOI: 10.1201/9781003669920-21

Azospirillum brasilense fixes nitrogen and *Bacillus subtilis* makes phosphorus easier to absorb by releasing organic acids like gluconic, citric, and oxalic acid, which dissolve otherwise unavailable phosphate compounds. Other beneficial microbes such as *Pseudomonas*, *Burkholderia*, *Penicillium*, and mycorrhizal fungi use similar strategies: balancing hormones, producing antioxidants, and improving root development (Ansabayeva et al., 2025; Laishram et al., 2025).

Phosphate-solubilizing microorganisms (PSMs) are particularly important for phosphorus nutrition. They release acids that lower soil pH, dissolve locked up phosphates, and use enzymes like phosphatases and phytases to unlock organic phosphorus sources. On top of that, they produce growth hormones, fix nitrogen, and reduce plant stress by breaking down the ethylene (ET) precursor 1-aminocyclopropane-1-carboxylic acid (ACC). They can even protect plants from diseases by producing compounds such as hydrogen cyanide and lytic enzymes (Ahash et al., 2025; Pang et al., 2024). Furthermore, more studies highlight the significance of PGPMs, mycorrhizal fungi, nitrogen-fixing bacteria, and endophytes in boosting crop resilience, facilitating viable substitutes to chemical inputs and abiotic stressors. For example, sorghum was reported to show enhanced effect after the use of PGPMs by Srivastava et al. (2025). They further quoted a vast literature to support the significant role of PGPMs in the plant growth maintenance under stress. This chapter is going to focus on (1) different enzymatic retorts to stress conditions and (2) the role of plant–microbe interactions on enzymatic responses.

21.2 ROLE OF MICROBES IN MODULATING STRESS RESPONSES

All major abiotic stresses (drought, salinity, heat, cold, pollutant exposure) induce oxidative stress via overproduction of reactive oxygen species (ROS). Plants rapidly upregulate the antioxidant enzyme network to maintain redox homeostasis. Important antioxidant enzymes include CAT, superoxide dismutase (SOD), ascorbate peroxidase (APX), glutathione peroxidase (GPX), glutathione reductase (GR), peroxidases (POD), monodehydroascorbate reductase (MDHAR), dehydroascorbate reductase (DHAR). SOD converts superoxide to H_2O_2, which is then removed by CAT, APX, and POD; GR regenerates reduced glutathione, sustaining the ascorbate–glutathione cycle. The coordinated activity of these enzymes determines whether a plant tolerates or succumbs to stress (Rao et al., 2025). Table 21.1

TABLE 21.1
Impact of Plant–Microbe Interaction on Plant Enzyme Mediated Stress Tolerance

Enzyme Group/ Function	Stress Type	Without Plant–Microbe Interaction	With Plant–Microbe Interaction	Reports	References
Antioxidant enzymes (SOD, catalase, APX, GR)	Drought, salinity, heat, cold, pollution	ROS accumulation and oxidative damage; reduced photosynthesis	ROS scavenged efficiently, membrane stability maintained, higher photosynthetic rate	*Pseudomonas fluorescens*	Azeem et al. (2025) and Khan et al. (2025)
Pathogenesis-related enzymes (chitinase, PAL, PPO)	Pathogen attack	Delayed or weak defense activation; higher disease severity	Enhanced defense mechanisms by increased induction of related enzymes	*Trichoderma asperellum*	Hasna et al. (2025)
Osmoprotectant enzymes (P5CS, trehalose synthase)	Drought, salinity	Low osmolyte production and cell dehydration and wilting	High osmolyte production/ trehalose levels and turgor pressure maintenance	*Acinetobacter calcoaceticus* and *Bacillus amyloliquefaciens*	Eswaran et al. (2024)
Heat shock proteins/ chaperones	Heat stress	Protein misfolding and enzyme inactivation	Improved protein folding and repair, higher thermal tolerance	Endophytes and rhizobacteria (*Burkholderia* sp. and *Saitozyma podzolica*)	Dlamini et al. (2025) and Liu et al. (2025)
Cold-response enzymes (adenosine triphosphates, AFPs, cold metabolism)	Cold/freezing	Ice crystal damage to cells	Increased AFPs and sugars prevent ice damage, improved cold survival	Psychrotolerant endophytes	Licciardello et al. (2025)
Detoxification enzymes (PCS, GSTs)	HM, pollutants	Slow detox metal toxicity, chlorosis	Faster detox, reduced heavy metal uptake, restored growth	Metal-tolerant PGPR	Liu et al. (2024) and Yao et al. (2024)
Nitrogen metabolism enzymes (NR, GS)	Nutrient imbalance	Poor N assimilation under stress	Better N uptake and assimilation, improved growth and yield	AMF, N-fixing bacteria	Li et al. (2025) and Zhao et al. (2025)
Hormone-modulating enzyme (ACC deaminase)	Cold, heat, salt, or flooding	Ethylene buildup and growth inhibition	Reduced ethylene, better root growth, and recovery	PGPRs and *Bacillus safensis*	Kaleh et al. (2025) and Karimian et al. (2025)

further summarizes the impact of plant–microbe interactions on these key antioxidants and other stress-tolerance enzymes, contrasting plant performance with and without their microbial partners.

Plants exposed to HMs and xenobiotics rely on glutathione (GSH)-dependent pathways, phytochelatin synthase (PCS), and metal-chelating peptides. An established cascade demonstrates that PCS polymerizes the glutathione to produce phytochelatins which act against HMs by vacuolar sequestration. Glutathione-S-transferases (GSTs) conjugate the electrophilic toxins to glutathione facilitating their transport and sequestration. These detox systems are inducible by metal exposure and oxidative cues; manipulating PCS/GSH pathways improves tolerance in model and crop plants (Seregin and Kozhevnikova, 2023). Metabolism maintenance under stress conditions frequently perturb both carbon fixation and nitrogen assimilation, necessitating metabolic reprogramming to maintain cellular homeostasis. The primary nitrogen-assimilatory enzymes nitrate reductase (NR), glutamine synthetase (GS), and glutamate synthase (GOGAT) are central to sustaining amino acid biosynthesis and osmolyte precursors, particularly proline and γ-amino-butyric acid (GABA), under drought, salinity, heat, or nutrient-limiting environments. Stress typically suppresses NR/GS/GOGAT activities, although adaptive signaling and symbiotic interactions may counteract this downregulation (Rivero-Marcos et al., 2026). Concomitantly, sucrose catabolism undergoes a stress-induced shift: sucrose synthase (SUS) is often favored over invertases due to its lower adenosine triphosphate (ATP) demand, thereby channeling carbon skeletons toward maintenance of respiration and osmolyte synthesis rather than growth. These adjustments are tightly regulated by carbon–nitrogen (C/N) signaling networks and cellular energy sensors, notably the SnRK1–TOR axis, which integrates sugar status, redox balance, and phytohormonal signals to optimize metabolic allocation under energy-limited conditions (Artins et al., 2024).

When attacked by pathogens, plants rapidly induce PR enzymes. One of such enzymes are chitinases and β-1,3-glucanases that can degrade fungal cell walls. Other enzymes include phenylalanine ammonia-lyase (PAL) and polyphenol oxidase (PPO) that channel carbon into phenolics and lignin for cell-wall reinforcement and antimicrobial compound synthesis. Moreover, a list of pathogenesis-protective proteins is reported which help in plant defense against pathogenic attacks. As shown in Table 21.1, the presence of beneficial microbes like *Trichoderma asperellum* significantly enhances this defense mechanism through the increased induction of these related enzymes. These enzymes are under the control of immune signaling networks such as ET, salicylic acid (SA), and jasmonic acid (JA) pathways. Further, these are actively regulated by transcription factors such as WRKYs (Peng et al., 2024).

Ying et al. (2025) evaluated the contributions and limitations of *Arabidopsis* and *Oryza* in cold stress research with the fishing out of the non-coding RNAs in ROS modulation. Reviews and experimental studies across crops repeatedly identify SOD, CAT, and APX as reliable markers of stress tolerance and targets of improvement (Ying et al., 2025). Beneficial rhizobacteria and endophytes frequently stimulate antioxidant enzyme activity in hosts by increasing the enzyme induction under drought or salinity which leads to the lowered levels of ROS and enhance the protection of photosynthetic machinery and cellular membranes. El-Saadony et al. (2024) demonstrated the effectiveness of PGPRs in augmenting the plant tolerance to drought stress. Moreover, decrease in the ROS concentration was reported with subsequence increase in malondialdehyde (MDA, antioxidant enzyme) levels and growth in *Mentha pulegium* (peppermint) and *Mentha spicata* (spearmint) under drought conditions (Azab et al., 2025; Souda Koraichi et al., 2025).

Recent studies highlight the significance of biological control agents and beneficial fungi (e.g., *Trichoderma* spp.) in regulating the PR enzyme responses leading to robust protection against pathogenic challenges and a substantial reduction in the effect of diseases in the treated crops. One fascinating discovery from recent research is that plant-associated microbes such as PGPR, endophytes, and mycorrhizal fungi can "train" a plant's antioxidant system to be stronger and faster. Microbial inoculation often boosts the levels and activity of SOD, CAT, and APX, reduces membrane damage from lipid peroxidation, and helps keep photosynthesis humming under tough conditions like drought, salinity, and HM stress. They do this by triggering host antioxidant genes, improving water and nutrient uptake, producing their own antioxidant compounds, and tweaking hormonal signaling to stabilize redox balance. This plant–microbe teamwork has become a hot topic in sustainable agriculture because it offers a natural way to prepare plants for future stress (González et al., 2024).

21.3 PLANT GROWTH-PROMOTING RHIZOBACTERIA (PGPR) AND ENZYMATIC MODULATION

Under water stress, drought-tolerant PGPR are thought to be a good substitute for conventional farming methods. PGPR is seen as a long-term, synergistic biological strategy to address the negative consequences of drought. Microbial communities are gaining more attention these days for boosting crop output and offering resilience to abiotic stress because of their capacity for colonization and symbiotic relationships with plants, either on the surface of the rhizosphere or inside plant cells. Under controlled conditions, a number of advantageous drought-resistant rhizobacteria are being isolated and tested to help plants become more tolerant of water stress. Numerous studies have demonstrated that PGPR, including *Burkholderia*, *Arthrobacter*, *Bacillus*, *Achromobacter* sp., *A. brasilense*, *Pseudomonas* sp., and *Paenibacillus polymyxa*, enhance drought tolerance in *Arabidopsis thaliana* (El-Saadony et al., 2024), tomato (Jofré et al., 2024), wheat (Noureen et al., 2024), and maize

TABLE 21.2
Role of Different Microbial Groups in Modulating Plant Enzyme Responses to Abiotic and Biotic Stress

Microbial Group	Example Species/ Strain	Target Enzyme(s)	Type of Stress	Mechanism of Modulation	Plant Response/ Benefit	References
PGPR	*Bacillus subtilis*, *Pseudomonas fluorescens*	SOD, catalase, POD	Drought, salinity	Induce antioxidant enzyme activity via phytohormone signaling (IAA, ABA)	Reduced oxidative stress, improved root growth	Kumawat et al. (2023)
Endophytic bacteria	*Enterobacter cloacae*, *Azospirillum brasilense*	GR, APX	HM toxicity	Enhance glutathione–ascorbate cycle enzymes for detoxification	Increased metal tolerance and biomass	Kaushal and Pati (2025)
Arbuscular mycorrhizal fungi (AMF)	*Rhizophagus irregularis*, *Funneliformis mosseae*	Acid phosphatase, NR, POD	Phosphorus deficiency, drought	Increase nutrient uptake and induce defense-related enzymes	Improved nutrient status and water-use efficiency	Talaat and Shawky (2017)
Endophytic fungi	*Trichoderma harzianum*, *Piriformospora indica*	PAL, chitinase, β-1,3-glucanase	Pathogen attack	Activate defense enzyme pathways (SA/JA signaling)	Enhanced systemic resistance	Poveda et al. (2020)
Cyanobacteria	*Anabaena variabilis*, *Nostoc* spp.	NR, SOD	Nutrient limitation, oxidative stress	Improve nitrogen assimilation and ROS scavenging	Higher photosynthetic efficiency	Singh and Prasad (2024)

(Agunbiade et al., 2024). PGPR have amazing potential to control the physiological response to water deprivation, therefore maintaining plant life (Table 21.2).

Metal-tolerant rhizobacteria and certain endophytes alter metal speciation in the rhizosphere, stimulate host GST and PCS activity, and can lower translocation of toxic metals into shoots and edible tissues an important route for phytoremediation and safer crop production on marginal soils (Gupta et al., 2024). GSH, a powerful natural antioxidant and excellent source of thion in cells, can help plants better withstand the toxic effects of HMs and soil/chemical pollutants by regulating their internal defense systems (Roy et al., 2023). Research has shown that giving plants a boost of GSH from outside sources ramps up the activity of key antioxidant enzymes, which in turn helps them cope with metal stress. For example, HM-affected Mexican cotton showed significantly lowered level of cadmium (Cd) in leaves following the supplementation with glutathione from outside source. Moreover, increased activity of protective enzymes like SOD, APX, POD, and GR was observed which augment metal tolerance in plants by fighting ROS. Similarly, many researchers demonstrated the role of glutathione in HMs and pollutant stress tolerance in plants by lowering harmful oxidative stress markers such as MDA and H_2O_2, while also enhancing the activity of SOD, CAT, POD, APX, and GR. These results highlight GSH as a promising ally for helping crops survive and thrive in HM-contaminated soils.

Beneficial microbes (PGPR and some endophytes) further increase host osmolyte levels by stimulating P5CS/TPS expression or by supplying compatible solutes, producing a measurable improvement in plant water status and recovery after stress (Eswaran et al., 2024). The levels of stress hormones continue to increase under the extensive stress tenures which need to be regulated anyways. However, many PGPR produce an enzyme called ACC deaminase (*acdS*). This enzyme breaks down the ET precursor ACC into ammonia and α-ketobutyrate, helping to keep ET levels in check and enabling roots to keep growing and gathering resources even during adversity (Buragohain et al., 2024). Figure 21.1 visually captures this intricate process, demonstrating how microbial signals like ACC deaminase trigger a coordinated response in the plant, upregulating stress-tolerance enzymes (SOD and CAT) while downregulating stress-induced hormones (like ET via ACC synthase/ ACC oxidase (ACS/ACO)), ultimately leading to enhanced survival and growth.

This figure illustrates how PGPR/arbuscular mycorrhizal fungi (AMF) in the rhizosphere mediate the plant response to abiotic stress (red waves). The microbes release key signaling molecules (e.g., IAA, ACC deaminase) which interact with the roots. This interaction results in the upregulation of antioxidant genes (SOD, catalase) to enhance tolerance, and the downregulation of stress-related genes (e.g., ACS/ACO for ethylene) to fine-tune the production of key enzymes that improve survival and growth under stress.

21.4 MYCORRHIZAL FUNGI AND ENZYMATIC MODULATION

There is a tight relationship between the functioning of AMF and soil enzymatic activities (Matinizadeh et al., 2024). Soil enzymes produced by saprobic fungi, roots, and bacteria are the main factor in controlling biochemical processes such as decomposition of organic matter and soil nutrient cycles (Wang et al., 2024a). By producing enzymes, fungi promote the breakdown of organic matter and the recovery of nutrients for their own benefit and that of the ecosystem. Enzymes such as arylsulfatase and acid phosphatase are the main catalytic enzymes breaking down

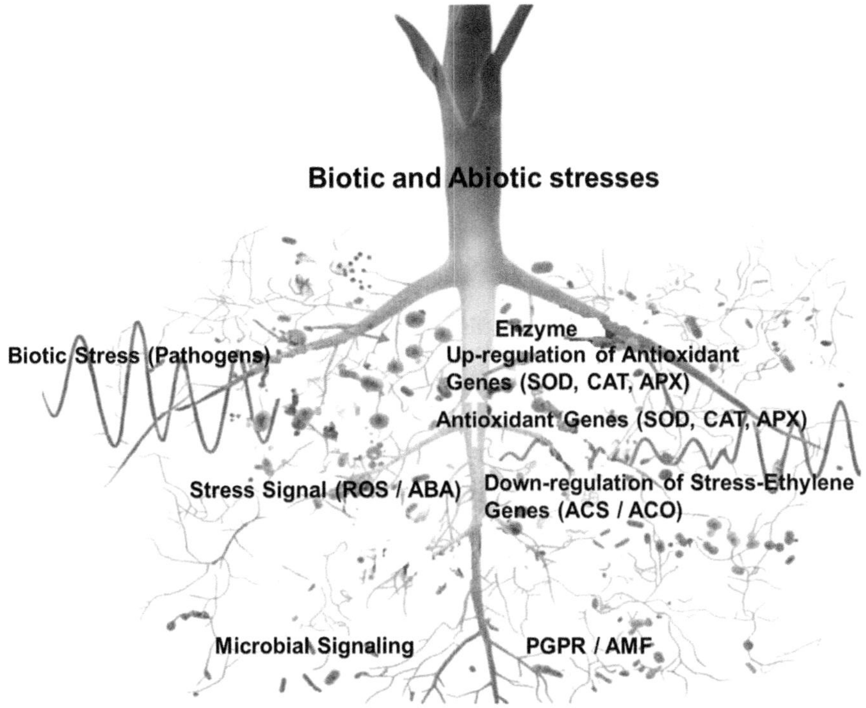

FIGURE 21.1 Plant–microbe interaction modulating enzymatic response under stress conditions.

lignin and cellulose as the main constituents of organic matter (Daunoras et al., 2024). Increased enzymatic activity in soils often indicates an increase in fungal abundance and diversity (Bogati et al., 2025).

Root-associated AMF help to expand the soil volume for inorganic P acquisition and activate root phosphatase to hydrolyze organic P. On the other hand, enzymatic activities are sensitive to changes in plant cover, soil chemistry, and grazing intensity and are linked to soil nutrient cycles. Herbivores alter nutrient distribution and fluxes, thereby affecting soil enzyme activities in semi-arid ecosystems. For example, a localized nutrient input from animal excreta has been shown to promote urease, alkaline phosphatase, and dehydrogenase activities under high grazing intensity in arid and semi-arid grasslands (Mencel et al., 2025). Recently, natural and biological agricultural practices have attracted the attention of researchers to ensure the sustainability of agricultural production. AMF have proven effective as a safe and practical approach for plant production under normal and stress conditions; AMF and plant roots create a symbiotic relationship that gives extension to the root system, increasing the absorption surface, and enhances nutrient and water uptake. However, this symbiotic interaction is the most widespread symbiotic relationship in nature, as about 80% of plants are involved in this relationship; therefore, AMF is considered an environmentally friendly biological fertilizer that positively affects the growth and production of host plants (Samanta et al., 2025). As detailed in Table 21.2, AMF like *Rhizophagus irregularis* primarily impact acid phosphatase and NR, enhancing nutrient uptake and water-use efficiency under deficiency or drought.

Mycorrhizal inoculation is a sustainable way to reduce the use of mineral fertilizers and enhance plant tolerance to biotic and abiotic stresses; however, most vegetable crops form a symbiotic relationship with AMF except Brassicaceae crops (Shalaby and Ramadan, 2024). Previous studies showed that mycorrhizal colonization exhibited a significant increase in metabolic activity and tolerance to abiotic stresses, as AMF inoculation alleviated oxidative damage caused by salt stress in *Valeriana officinalis* plants by activating the antioxidant defense system, reducing MDA and Na accumulation, and maintaining membrane integrity, osmolytes content, and water status, leading to increased nutrient content, photosynthetic pigments, and growth parameters (Begum et al., 2023), improved morphological traits (root length and shoot weight), biochemical traits (chlorophyll, protein, and proline), nutrient content (N, P, and Mg), and antioxidant enzyme activities (APX, SOD, and CAT) in salt-stressed safflower (*Carthamus tinctorius* L.) plants (Gürsoy, 2023), increased antioxidant enzyme activity, eliminated ROS, enhanced photosynthetic activity, and decreased H_2O_2, MDA, and electrolyte leakage, which resulted in increased size and weight of watermelon (*Citrullus lanatus*) fruits under salt stress (Ma et al., 2025). Under drought stress, mycorrhizal fungi improved the ability of tomato plants to resist drought stress by increasing antioxidant enzyme activity (CAT and APX), photosynthetic pigments (chlorophyll and carotenoids), nutrient content (K, Ca, and Mg), and reducing the accumulation of H_2O_2 and MDA, which increases root length, fresh and dry weight of shoot and root, and tomato yield (Dolatmand-Shahri et al., 2025), increased shoot and root dry weight, stomatal conductance, photosynthesis rate, water use efficiency, glycyrrhizin content, decreased abscisic acid (ABA) in liquorice (*Glycyrrhiza glabra*) plants (Jia et al., 2024), increased non-enzymatic antioxidants (phenol, flavonoid, and glycine betaine), ATP production, and nutrient content (N, P, and K), and improved hormonal status (increased gibberellin, cytokinins, and IAA, and decreased ABA), thus improving soybean (*Glycine max* L.) growth and yield parameters (plant height, fresh weight, seed weight, and oil content) (Hernandez-Leon and Valenzuela-Soto, 2023).

Under heat stress, AMF improved physiological and biochemical traits, including leaf water potential, stomatal conductance, photosynthetic activity, photochemical efficiency, antioxidant enzyme activity (POD, SOD, CAT), phytohormones (ABA, SA, JA, IAA), and reduced H_2O_2, MDA accumulation in tomato plants. Under Cd stress, mycorrhizal inoculation significantly improved photosynthetic parameters (net photosynthetic rate, transpiration rate, intercellular CO_2 concentration, and stomatal conductance), chlorophyll content, and carbohydrate content, while reducing the accumulation of Cd, H_2O_2, and MDA, which led to an increase in the dry matter of roots, stems, and leaves, and weight of wheat (*Triticum aestivum* L.) grains (Valenzuela-Aragon et al., 2025). Mutualistic interactions provide an additional regulatory layer. Arbuscular mycorrhizal (AM) fungi and nitrogen-fixing or associative bacteria enhance nutrient availability and modulate host signaling, frequently upregulating NR and GS expression, improving amino acid pools, and promoting recovery and biomass accumulation under stress or nutrient-deficient conditions (Bhardwaj et al., 2023). Such symbioses thus mitigate stress-induced repression of N-assimilation and improve plant resilience (Wang et al., 2024b).

21.5 ENDOPHYTES AND ENZYMATIC MODULATION

Endophytes play a pivotal role in modulating enzymatic responses in plants under various abiotic stresses, contributing significantly to plant tolerance and adaptation mechanisms. These microbial symbionts, both fungal and bacterial, enhance host plants' antioxidative defense systems by regulating key enzymes involved in ROS scavenging, thereby alleviating oxidative damage caused by environmental stresses (Ameen et al., 2024). This regulatory function is broadly summarized in Table 21.2, which highlights their role in enhancing enzymes like GR and defense-related enzymes such as chitinase and PAL to manage HM and pathogen stress, respectively. Under HM stress, endophytic consortia containing fungi like *Paecilomyces formosus* and bacteria such as *Sphingomonas* sp. LK11 have been shown to reduce metal uptake and oxidative stress markers in *G. max*. Co-inoculation led to lower hydrogen peroxide, lipid peroxidation, and an optimized activity of antioxidant enzymes, including CAT and SOD, while simultaneously improving nutrient uptake and modulating soil enzymatic activities critical for stress resilience (Das et al., 2025). Similarly, inoculation with *P. formosus* also alleviated nickel-induced oxidative stress by modulating enzymatic antioxidants such as peroxidase, PPO, CAT, and SOD, coupled with stress-responsive phytohormonal regulation (Mahmoud and Fouad, 2024).

In drought-stressed plants, fungal endophytes like *Diaporthe atlantica* and consortia of *Microdochium majus*, *Meyerozyma guilliermondii*, and *Aspergillus aculeatus* enhance enzymatic antioxidant responses by upregulating CAT, APX, and total antioxidant capacity. This enzymatic modulation confers protection against ROS accumulation, sustaining photosynthetic capacity and osmotic balance in species such as tomato and *Moringa oleifera* under water deficits. ACC deaminase-producing endophytic fungi, including *Trichoderma gamsii* and *Fusarium proliferatum*, further mitigate drought stress by reducing ET and hydrogen peroxide production, simultaneously inducing antioxidant enzyme activities like peroxidase, CAT, and APX along with enhanced gene expression linked to oxidative stress tolerance (Rehman et al., 2022).

Salinity stress responses are similarly modulated by endophytes such as *Sphingomonas* sp. LK11 and bacterial strains, including *Pseudomonas koreensis*. These microbes improve enzymatic antioxidant defense by regulating CAT, peroxidase, PPO, and SOD activities. The balanced enzymatic activity mitigates ROS damage and ionic toxicity while enhancing chlorophyll content and nutrient uptake. Notably, nitro-fatty acids mediate nitroalkylation post-translational modifications that fine-tune CAT activity during salt stress, providing a novel molecular mechanism for enzymatic modulation by microbial interaction (Chaki et al., 2025). Endophyte-mediated enzymatic modulation also involves transcriptional regulation where beneficial microbes influence the expression of stress-responsive genes encoding antioxidants and phytohormone biosynthesis enzymes, thereby orchestrating a comprehensive defense network (Taheri, 2025). For instance, melatonin co-applied with *Bacillus* sp. strain IPR-4 enhances peroxidase, CAT, APX, SOD, and GR activities in soybean under drought, simultaneously modulating hormonal crosstalk to optimize stress responses (Peter et al., 2024). Furthermore, endophytic fungi like *Aspergillus nomiae* and *Aspergillus fumigatus* improve waterlogging tolerance by modulating ET metabolism and antioxidant enzyme activities, including CAT and peroxidase in maize, indicating a broad spectrum of enzymatic adaptation across diverse stress conditions. Endophytic *Serendipita vermifera* exerts indirect enzymatic modulation by maintaining redox homeostasis and repressing pathogen virulence factors, thus contributing to plant defense priming (Amin and Zaman, 2025).

21.6 MICROBE-INDUCED HORMONAL REGULATION

Under stressful conditions, plants often produce high levels of phytohormones such as ET and ABA which are produced under different stress circumstances and support plant to maintain their growth during stress periods. It has been established that ET production is associated with metal, flood, or drought stress while ABA is produced as a response to salt stress (Qin et al., 2024). Excessive production of ET ends up subduing the root growth and production. The enzymes ACSs and ACOs are considered to be the key biosynthetic enzymes that are responsible for ET biosynthesis (Khan et al., 2024).

The level of ABA rises during the period of drought, salinity, high heat, or chilling cold, triggering a cascade of signaling events. This hormone activates specific transcription factors, which in turn switch on stress-responsive genes, including those involved in antioxidant defense, osmolyte accumulation, and protective protein production. By fine-tuning these responses, ABA enables plants to maintain cellular homeostasis, safeguard photosynthesis, and survive adverse conditions (Qin et al., 2025).The beneficial microbes also pinch the plant broader hormonal networks governed by the auxin, ABA, JA, and SA. They achieve this through secreted metabolites and effector molecules that modify the expression of stress-responsive genes. Hormonal modulation by the responsive enzymes in plants make them stronger for tougher challenges, setting the stage for heightened defenses by promoting antioxidant activity, osmolyte buildup, and other protective pathways. Recent studies on controlled environment discover the significant potential of ACC deaminase-producing strains to enhance tolerance to drought, salinity, and even flooding. For example, in tomato and maize under drought stress, inoculation with ACC deaminase-producing strains improved oxidative stress tolerance and sustained growth (Buragohain et al., 2024). In grapevine, *Burkholderia* sp. reduced the ET high level and corresponding inhibitory effect on root development, thus supporting better resilience under water deficit (Yin et al., 2025). Maize treated with native ACC deaminase PGPR in saline soils demonstrated improved growth performance highlighting these strains as promising bio-inoculants for saline environments (Mohamed et al., 2025).

21.7 CONCLUSION

This chapter confirms that the symbiotic relationship between plants and their associated microbes is not merely supportive but functions as a highly sophisticated regulatory circuit that primes and optimizes the plant enzymatic defense capacity under stress. Future research must move toward functional precision, requiring large-scale genomic and metabolomic studies to identify the specific microbial effector molecules and PGPM genes responsible for targeted enzymatic enhancement. The crucial next step involves transitioning from generalized microbial applications to rational, site-specific bio-inoculation strategies. This evolution demands the establishment of standardized delivery platforms and robust field-testing to validate the reliable persistence and enzymatic modulatory efficacy of these microbes across diverse environmental interactions, such as varying soil temperature, salinity, or drought by strategically leveraging the PGPM-induced optimization of plant enzyme systems, especially the antioxidant machinery and stress-signaling pathways.

REFERENCES

Abideen, Z., Cardinale, M., Zulfiqar, F., Koyro, H.-W., Rasool, S. G., Hessini, K., Darbali, W., Zhao, F., Siddique, K. H., 2022. Seed endophyte bacteria enhance drought stress tolerance in *Hordeum vulgare* by regulating, physiological characteristics, antioxidants and minerals uptake. Frontiers in Plant Science. 13, 980046.

Adeleke, B. S., Chaudhary, P., Ayilara, M. S., Ojo, F. M., Erinoso, S. M., Upadhayay, V. K., Adeyemo, A. I., Akinola, S. A., 2024. Rhizosphere microbiomes mediating abiotic stress mitigation for improved plant nutrition. Ecologies. 5, 375–401.

Agunbiade, V. F., Fadiji, A. E., Agbodjato, N. A., Babalola, O. O., 2024. Isolation and characterization of plant-growth-promoting, drought-tolerant rhizobacteria for improved maize productivity. Plants. 13, 1298.

Ahash, S., Manikandan, K., Sivasankari Devi, T., Elamathi, S., Maragatham, S., Subrahmaniyan, K., 2025. Phosphate-solubilizing microorganisms for sustainable phosphorus management in rice. Rhizosphere. 34, 101096.

Ameen, M., Mahmood, A., Sahkoor, A., Zia, M. A., Ullah, M. S., 2024. The role of endophytes to combat abiotic stress in plants. Plant Stress. 12, 100435.

Amin, A., Zaman, W., 2025. Integrative perspectives on multi-level mechanisms in plant-pathogen interactions: From molecular defense to ecological resilience. Phyton-International Journal of Experimental Botany. 94, 1973–1996.

Ansabayeva, A., Makhambetov, M., Rebouh, N. Y., Abdelkader, M., Saudy, H. S., Hassan, K. M., Nasser, M. A., Ali, M. A. A., Ebrahim, M., 2025. Plant growth-promoting microbes for resilient farming systems: Mitigating environmental stressors and boosting crops productivity—A review. Horticulturae. 11, 260.

Artins, A., Martins, M. C. M., Meyer, C., Fernie, A. R., Caldana, C., 2024. Sensing and regulation of C and N metabolism – Novel features and mechanisms of the TOR and SnRK1 signaling pathways. The Plant Journal. 118, 1268–1280.

Azab, M. S., Aboul-Nasr, A. M., Behiry, S. I., Sabra, M. A., 2025. Enhancing growth, biochemical traits, and stress resilience of *Mentha pulegium* L. in calcareous soil using PGPR. Egyptian Academic Journal of Biological Sciences, H. Botany. 16, 17–27.

Azeem, M., Sultana, R., Ahmed, N., Abbasi, M. W., Hasan, K. A., Dong, R., Alamri, S., Alfagham, A. T., 2025. Salinity stress resilience in *Sorghum bicolor* through Pseudomonas-mediated modulation of growth, antioxidant system, and eco-physiological adaptations ACS Omega. 10, 940–954.

Begum, N., Xiao, Y., Wang, L., Li, D., Irshad, A., Zhao, T., 2023. Arbuscular mycorrhizal fungus *Rhizophagus irregularis* alleviates drought stress in soybean with overexpressing the GmSPL9d gene by promoting photosynthetic apparatus and regulating the antioxidant system. Microbiological Research. 273, 127398.

Bhardwaj, A. K., Chandra, K. K., Kumar, R., 2023. Mycorrhizal inoculation under water stress conditions and its influence on the benefit of host microbe symbiosis of *Terminalia arjuna* species. Bulletin of the National Research Centre. 47, 89.

Bogati, K. A., Sewerniak, P., Walczak, M., 2025. Unraveling the effect of soil moisture on microbial diversity and enzymatic activity in agricultural soils. Microorganisms. 13.

Buragohain, K., Tamuly, D., Sonowal, S., Nath, R., 2024. Impact of drought stress on plant growth and its management using plant growth promoting rhizobacteria. Indian Journal of Microbiology. 64, 287–303.

Chaki, M., Aranda-Caño, L., Begara-Morales, J. C., Sánchez-Calvo, B., López-Jaramillo, F. J., Padilla, M. N., Valderrama, R., Pedrajas, J. R., Barroso, J. B., 2025. Nitro-fatty acids-mediated nitroalkylation modulates fine-tuning catalase antioxidant function during salinity stress in plants. Protein Science. 34, 70076.

Das, D., Sharma, P. L., Paul, P., Baruah, N. R., Choudhury, J., Begum, T., Karmakar, R., Khan, T., Kalita, J., 2025. Harnessing endophytes: Innovative strategies for sustainable agricultural practices. Discover Bacteria. 2, 1.

Daunoras, J., Kačergius, A., Gudiukaitė, R., 2024. Role of soil microbiota enzymes in soil health and activity changes depending on climate change and the type of soil ecosystem. Biology. 13, 85.

Dlamini, W. N., Lai, W.-A., Chen, W.-C., Shen, F.-T., 2025. Unveiling the thermotolerance and growth-promoting attributes of endophytic bacteria derived from *Oryza sativa*: Implications for sustainable agriculture. Microorganisms. 13, 766.

Dolatmand-Shahri, N., Modarres-Sanavy, S. A. M., Mirjalili, M. H., Mokhtassi-Bidgoli, A., 2025. Phosphorus fertilizer and arbuscular mycorrhizal fungi application improves bitter melon fruit yield and some phytochemical compounds under irrigation deficit stress Current Plant Biology. 42, 100446.

El-Saadony, M. T., Saad, A. M., Mohammed, D. M., Fahmy, M. A., Elesawi, I. E., Ahmed, A. E., Algopishi, U. B., Elrys, A. S., Desoky, E.-S. M., Mosa, W. F. A., Abd El-Mageed, T. A., Alhashmi, F. I., Mathew, B. T., AbuQamar, S. F., El-Tarabily, K. A., 2024. Drought-tolerant plant growth-promoting rhizobacteria alleviate drought stress and enhance soil health for sustainable agriculture: A comprehensive review. Plant Stress. 14, 100632.

Eswaran, S. U. D., Sundaram, L., Perveen, K., Bukhari, N. A., Sayyed, R., 2024. Osmolyte-producing microbial biostimulants regulate the growth of *Arachis hypogaea* L. under drought stress. BMC Microbiology. 24, 165.

González, M. C., Roitsch, T., Pandey, C., 2024. Antioxidant responses and redox regulation within plant-beneficial microbe interaction. Antioxidants. 13.

Gupta, R., Khan, F., Alqahtani, F. M., Hashem, M., Ahmad, F., 2024. Plant growth–promoting Rhizobacteria (PGPR) assisted bioremediation of heavy metal toxicity. Applied Biochemistry and Biotechnology. 196, 2928–2956.

Gürsoy, M., 2023. Morphological and biochemical changes with hormone and hydro-priming applications in safflower (*Carthamus tinctorius* L.) seedlings under salinity stress conditions. Notulae Botanicae Horti Agrobotanici Cluj-Napoca. 51, 13282–13282.

Hanif, A., Abideen, Z., Hasan, K. A., Ali, B., El-Keblawy, A., Umar, M., Rao, Y., Siddiqui, Z. S., 2023. Application of microbial antagonists as alternatives to chemical pesticides: A novel approach for enhancing plant growth and resilience of agricultural land. Land Degradation & Development. 34, 5486–5498.

Hasna, M. K., Paul, N. R., Haque, M. M., Bir, M. S. H., Ali, M. A., Chong, K. P., 2025. Biocontrol efficacy of *Trichoderma asperellum* against fusarium wilt in tomato plants by induction of the host defense genes. Discover Plants 2, 136.

Hernandez-Leon, S. G., Valenzuela-Soto, E. M., 2023. Glycine betaine is a phytohormone-like plant growth and development regulator under stress conditions. Journal of Plant Growth Regulation. 42, 5029–5040.

Jia, T., Gu, J., Ma, M., 2024. La (NO(3))(3) substantially fortified *Glycyrrhiza uralensis*'s resilience against salt stress by interconnected pathways. BMC Plant Biology 24, 024–05644.

Jofré, M. F., Mammana, S. B., Pérez-Rodriguez, M. M., Silva, M. F., Gomez, F. J. V., Cohen, A. C., 2024. Native rhizobacteria improve drought tolerance in tomato plants by increasing endogenous melatonin levels and photosynthetic efficiency. Scientia Horticulturae. 329, 112984.

Kaleh, A. M., Singh, P., Ooi Chua, K., Harikrishna, J. A., 2025. Modulation of plant transcription factors and priming of stress tolerance by plant growth-promoting bacteria: A systematic review. Annals of Botany. 135, 387–402.

Karimian, A., Bahramnejad, B., Siosemardeh, A., Salehi, H., Ahmad, N. S., 2025. Boosting drought resilience: The role of endophytic *Bacillus safensis* in enhancing melatonin production in chickpea cultivars. BMC Plant Biology 25, 789.

Kaushal, P., Pati, A. M., 2025. Unleashing rhizobacteria for sustainable soil remediation: PGPR roles in heavy metal tolerance, detoxification, and plant productivity. Frontiers in Microbiology. 16, 1662000.

Khan, S., Alvi, A. F., Saify, S., Iqbal, N., Khan, N. A., 2024. The ethylene biosynthetic enzymes, 1-aminocyclopropane-1-carboxylate (ACC) synthase (ACS) and ACC oxidase (ACO): The less explored players in abiotic stress tolerance. Biomolecules. 14, 90.

Khan, V., PrinciMubashshir, M., Umar, S., Iqbal, N., 2025. Methyl jasmonate and *Pseudomonas fluorescens* synergistically boost antioxidative defense, secondary metabolites, and osmolyte production to enhance drought resilience in mustard. Journal of Plant Growth Regulation. 44, 198–216.

Khilji, S. A., Tariq, R., Aziz, I., Munir, N., Javed, A., Uppal, T., Abideen, Z., El-Sheikh, M. A., Keblawy, A. E., 2025. Phytoremediation potential and ecophysiological responses of *Pistia stratiotes* L. for removal of cadmium and lead from polluted water: A viable option for agricultural resilience. Environmental Sustainability. 8, 235–248.

Kumawat, K. C., Sharma, B., Nagpal, S., Kumar, A., Tiwari, S., Nair, R. M., 2023. Plant growth-promoting rhizobacteria: Salt stress alleviators to improve crop productivity for sustainable agriculture development. Frontiers in Plant Science. 13, 1101862.

Laishram, B., Devi, O. R., Dutta, R., Senthilkumar, T., Goyal, G., Paliwal, D. K., Panotra, N., Rasool, A., 2025. Plant-microbe interactions: PGPM as microbial inoculants/biofertilizers for sustaining crop productivity and soil fertility. Current Research in Microbial Sciences. 8, 100333.

Li, J., Wang, Y., Fang, Y., Lyu, X., Zhu, Z., Wu, C., Xu, Z., Li, W., Liu, N., Du, C., 2025. Phycospheric bacteria alleviate the stress of erythromycin on *Auxenochlorella pyrenoidosa* by regulating nitrogen metabolism. Plants. 14, 121.

Licciardello, G., Antonielli, L., Sicher, C., Larini, I., Perazzolli, M., 2025. Two Antarctic endophytic bacteria of *Colobanthus quitensis* show functional and genomic characteristics potentially responsible for plant growth promotion and cold tolerance. Polar Biology. 48, 42.

Liu, H., Su, Y., Ye, C., Zuo, D., Wang, L., Mei, X., Deng, W., Liu, Y., Huang, H., Hao, J., Zhao, J., Wang, D., Zhang, X., Zhu, Y., Liu, J., Yang, M., Zhu, S., 2025. Nucleotides enriched under heat stress recruit beneficial rhizomicrobes to protect plants from heat and root-rot stresses. Microbiome. 13, 160.

Liu, L., Xiao, C., Gao, Y., Jiang, T., Xu, K., Chen, J., Lin, Z., Chen, J., Tian, S., Lu, L., 2024. Inoculation of multi-metal-resistant

Bacillus sp. to a hyperaccumulator plant Sedum alfredii for facilitating phytoextraction of heavy metals from contaminated soil. Chemosphere. 366, 143464.

Ma, X., Wang, Y., Li, X., Liu, Y., Luo, H., Shang, W., Di, Y., Gao, N., Kang, L., Piao, F., Guo, Z., Dong, H., Zhao, W., Zhang, T., 2025. Integrative analysis of different low-light-tolerant watermelon lines in response to low-light stress. BMC Plant Biology 25, 25–7180.

Mahmoud, R. M., Fouad, M. S., 2024. Approaches to antioxidant defence system: An overview of coping mechanism against lithium/nickel exposure in plants. Lithium and Nickel Contamination in Plants and the Environment (pp. 95–138). World Scientific.

Matinizadeh, M., Nouri, E., Bayranvand, M., Kolarikova, Z., Janoušková, M., 2024. Arbuscular mycorrhiza and rhizosphere soil enzymatic activities as modulated by grazing intensity and plant species identity in a semi-arid grassland. Rhizosphere. 30, 100893.

Mencel, J., Wojciechowska, A., Mocek-Płóciniak, A., 2025. Effect of grassland vegetation units on soil biochemical properties and the abundance of selected microorganisms in the Obra River Valley. Agronomy. 15, 1573.

Mikiciuk, G., Miller, T., Kisiel, A., Cembrowska-Lech, D., Mikiciuk, M., Łobodzińska, A., Bokszczanin, K., 2024. Harnessing beneficial microbes for drought tolerance: A reviewofecologicalandagriculturalinnovations. Agriculture. 14, 2228.

Mohamed, Z. A., Purayil, G. P., Aljar, F. K., Alhosani, S. M., Alblooshi, A. K., Alnuaimi, S. T., Alhemeiri, N. S., El-Tarabily, K. A., AbuQamar, S. F., 2025. Rhizosphere-competent actinobacteria with ACC deaminase activity mitigate heavy metal stress in corn (*Zea mays*). Biology and Fertility of Soils. 61(1), 1077–1095.

Muhammad, M., Waheed, A., Wahab, A., Majeed, M., Nazim, M., Liu, Y.-H., Li, L., Li, W.-J., 2024. Soil salinity and drought tolerance: An evaluation of plant growth, productivity, microbial diversity, and amelioration strategies. Plant Stress. 11, 100319.

Mujeeb, A., Abideen, Z., Aziz, I., Sharif, N., Hussain, M. I., Qureshi, A. S., Yang, H.-H., 2023. Phytoremediation of potentially toxic elements from contaminated saline soils using *Salvadora persica* L.: Seasonal evaluation. Plants. 12, 598.

Murugesan, P., Krishanan, A., Ramasamy, P., Chandrasekaran, K., Rajendran, J., Paramasivam, S., Chinnaiyan, U., Panneerselvam, C., Aziz, A. T., Alasmari, A., 2024. Biorational methods for effective pest control management in stored products for agricultural sustainability. Entomological Research. 54, e12697.

Noureen, S., Iqbal, A., Muqeet, H. A., 2024. Potential of drought tolerant rhizobacteria amended with biochar on growth promotion in wheat. Plants. 13, 1183.

Pang, F., Li, Q., Solanki, M. K., Wang, Z., Xing, Y. X., Dong, D. F., 2024. Soil phosphorus transformation and plant uptake driven by phosphate-solubilizing microorganisms. Frontiers in Microbiology. 15.

Peng, X., Hu, J., Duan, X., Chai, M., Wen, J., Wang, Z., Xie, H., 2024. Genome-wide identification and analysis of the WRKY transcription factor family associated with leaf senescence in Alfalfa. Plants (Basel). 13.

Peter, O., Imran, M., Shaffique, S., Kang, S.-M., Rolly, N. K., Felistus, C., Bilal, S., Dan-Dan, Z., Injamum-Ul-Hoque, M., Kwon, E.-H., Mong, M. N., Gam, H.-J., Kim, W.-C., Lee, I.-J., 2024. Combined application of melatonin and Bacillus sp. strain IPR-4 ameliorates drought stress tolerance via hormonal, antioxidant, and physiomolecular signaling in soybean. Frontiers in Plant Science. 15.

Poveda, J., Barquero, M., González-Andrés, F., 2020. Insight into the microbiological control strategies against *Botrytis cinerea* using systemic plant resistance activation. Agronomy. 10, 1822.

Qin, H., Xiao, M., Li, Y., Huang, R., 2024. Ethylene modulates rice root plasticity under abiotic stresses. Plants (Basel). 13.

Qin, Z., Lyu, J., Teng, Z., Meng, S., Peng, Y., Yuan, D., Duan, M., Zhang, J., Ye, N., 2025. ABA biosynthesis rather than ABA catabolism is induced by low temperature and inhibits seed germination by activating OsTPP3. The Crop Journal. 13(3), 752–763.

Rafique, N., Khalil, S., Cardinale, M., Rasheed, A., Zhao, F., Abideen, Z., 2025. A comprehensive evaluation of the potential of plant growth-promoting rhizobacteria for applications in agriculture in stressed environments. Pedosphere. 35, 229–248.

Rao, M. J., Duan, M., Zhou, C., Jiao, J., Cheng, P., Yang, L., Wei, W., Shen, Q., Ji, P., Yang, Y., Conteh, O., Yan, D., Yuan, H., Rauf, A., Ai, J., Zheng, B., 2025. Antioxidant defense system in plants: Reactive oxygen species production, signaling, and scavenging during abiotic stress-induced oxidative damage. Horticulturae. 11, 477.

Rehman, B., Javed, J., Rauf, M., Khan, S. A., Arif, M., Hamayun, M., Gul, H., Khilji, S. A., Sajid, Z. A., Kim, W.-C., Lee, I.-J., 2022. ACC deaminase-producing endophytic fungal consortia promotes drought stress tolerance in *M. oleifera* by mitigating ethylene and H_2O_2. Frontiers in Plant Science. 13.

Rivero-Marcos, M., Calleja-Satrustegui, A., Ariz, I., 2026. The counteracting role of nitrate during ammonium toxicity in plants. Horticultural Plant Journal. 12(1), 49–61.

Roy, S., Naimuzzaman, M., Rahman, F., 2023. Glutathione: A key frontier of heavy-metal detoxification and tolerance in plants. Plant Trends. 1, 33–42.

Samanta, I., Ghosh, K., Saikia, R., SavitaMaity, P. J., Chowdhary, G., 2025. Arbuscular mycorrhizal fungi – A natural tool to impart abiotic stress tolerance in plants. Plant Signaling & Behavior. 20, 2525843.

Seregin, I. V., Kozhevnikova, A. D., 2023. Phytochelatins: Sulfur-containing metal(loid)-chelating ligands in plants. International Journal of Molecular Sciences. 24, 2430.

Shalaby, O. A., Ramadan, M. E.-S., 2024. Mycorrhizal colonization and calcium spraying modulate physiological and antioxidant responses to improve pepper growth and yield under salinity stress. Rhizosphere. 29, 100852.

Singh, G., Prasad, S. M., 2024. Crosstalk mechanism of hydrogen sulfide and nitric oxide in regulating the nickel induced toxicity of cyanobacteria *Nostoc muscorum* and *Anabaena* sp. Acta Physiologiae Plantarum. 46, 82.

Souda Koraichi, S. I., Cherradi, O., Ziane, S. O., El Abed, S., El Alaoui Talibi, Z., El Ghachtouli, N., El Modafar, C., 2025. Improving growth and stimulating natural defenses in spearmint with four compatible PGPR strains. Journal of Pure & Applied Microbiology. 19.

Srivastava, A. K., Riaz, A., Jiang, J., Li, X., Uzair, M., Mishra, P., Zeb, A., Zhang, J., Singh, R. P., Luo, L., Chen, S., Yang, S., Zhao, Y., Xie, X., 2025. Advancing climate-resilient sorghum: The synergistic role of plant biotechnology and microbial interactions. Rice. 18, 41.

Taheri, P., 2025. Endophytic fungi as regulators of phytohormones production: Cytomolecular effects on plant growth, stress protection and importance in sustainable agriculture. Plant Stress. 17, 100978.

Talaat, N. B., Shawky, B. T., 2017. Microbe-mediated induced abiotic stress tolerance responses in plants. Microbe-mediated induced abiotic stress tolerance responses in plants. Interactions in Agro-Ecological Perspectives: Volume 2: Microbial Interactions and Agro-Ecological Impacts. Springer, pp. 101–133.

Upadhyay, S. K., 2025. Relevance of cross talk between root exudates, hormones, and root-associated microbes in developing sustainable phytoremediation strategies: A comprehensive review. Physiology and Molecular Biology of Plants. 31, 1629–1649.

Valenzuela-Aragon, B., Cardinale, M., Rolli, E., Rustioni, L., Francioli, D., 2025. The role of arbuscular mycorrhizal fungi in abiotic stress management in viticulture under climatic shifts. Plant Stress. 16, 100863.

Wang, X., Chi, Y., Song, S., 2024a. Important soil microbiota's effects on plants and soils: A comprehensive 30-year systematic literature review. Frontiers in Microbiology. 15.

Wang, Z., Lian, J., Liang, J., Wci, H., Chen, H., Hu, W., Tang, M., 2024b. Arbuscular mycorrhizal symbiosis modulates nitrogen uptake and assimilation to enhance drought tolerance of *Populus cathayana*. Plant Physiology and Biochemistry. 210, 108648.

Xia, F., Zhao, Z., Niu, X., Wang, Z., 2024. Integrated pollution analysis, pollution area identification and source apportionment of heavy metal contamination in agricultural soil. Journal of Hazardous Materials. 465, 133215.

Yao, X., Ren, J., Fang, L., Sun, K., He, W., 2024. The role and mechanism of *Bacillus megaterium* strain A14 in inhibiting the cadmium uptake by peanut plants in acidic red soil. Journal of Applied Microbiology. 135, lxae120.

Yin, X., Wang, C., Yuan, L., Wei, Y., Li, T., Liu, Q., Han, X., Wu, X., Wang, C., Jiang, X., 2025. Complete genome analysis and antimicrobial mechanism of *Burkholderia gladioli* ZBSF BH07 reveal its dual role in the biocontrol of grapevine diseases and growth promotion in grapevines. Microorganisms. 13, 1756.

Ying, Z., Fu, S., Yang, Y., 2025. Signaling and scavenging: Unraveling the complex network of antioxidant enzyme regulation in plant cold adaptation. Plant Stress. 16, 100833.

Zhang, H., Lang, Z., Zhu, J.-K., Wang, P., 2025. Tackling abiotic stress in plants: Recent insights and trends. Stress Biology. 5, 8.

Zhao, Z., Fernie, A. R., Zhang, Y., 2025. Engineering nitrogen and carbon fixation for next-generation plants. Current Opinion in Plant Biology. 85, 102699.

22 Future Perspectives on Plant Enzyme Systems in Biotechnology and Environmental Science

Amit B. Verma, Fatima Zahar Kamal, Said Kewedar, Nimisha Patel, Ahmad Ali, and Hina Alim

22.1 INTRODUCTION

Enzymes are biological catalysts that speed up chemical reactions in living organisms without undergoing in it. They have advantages like low immunogenicity, environmental compatibility, substrate specificity and the ability to work under mild conditions. They are vital in various industries such as the production of sweeteners, modification of antibiotics, manufacturing of detergents and food processing. Enzymes are also essential in analytical tools and tests used in clinical, forensic and environmental fields. While microbial and animal-derived enzymes have been widely used, there are growing concerns about sustainability, biosafety, ethical sourcing and biocompatibility, which have increased interest in plant-based enzymes (Hood 2002; Morton 1955; Sharma and Upadhyay 2020).

Historically, human activities have placed immense pressure on our planet, contributing to an environmental crisis. Our heavy dependence on fossil fuels to generate the energy, chemicals and materials vital for contemporary life has resulted in an unsustainable increase in greenhouse gas emissions (Emily et al. 2023). The growing demand for food is also fueling carbon-intensive agricultural practices, which are causing alarming rates of deforestation and significant biodiversity loss. Hazardous chemicals harmful to health are commonplace in many industrial operations. One technology that is already extensively utilized by major chemical companies and is essential for developing a sustainable chemical industry is biocatalysis. This process employs natural enzymes (Bell et al. 2021; Snajdrova et al. 2023). Plant enzyme systems play a crucial role in the development of biosensors, enabling the real-time detection of substances such as urea, glucose, pesticides and medicines through enhanced electrochemical performance. Many plant enzymes serve as substitutes for harmful chemicals in industries like food, textiles and biofuels, thereby promoting more sustainable production methods. Recent innovations in enzyme engineering and molecular farming have led to increased production and specificity of enzymes, broadening their applications across various fields. Techniques for extracting and analyzing the properties of diverse bioactive enzymes such as amylase, pectinase, cellulase, laccase, asparaginase, galactosidase, actinidin and urease have been developed using various plant tissue compounds.

As demand for environment-friendly biotechnological approaches increases, plants have emerged as promising substitutes for conventional sources of enzyme production like animals and microorganisms. Various plants provide an ample, sustainable and morally sound supply of enzymes boasting superior performance characteristics such as high efficacy, durability and precise substrate affinity. Enzymes derived from plant sources have shown significance as economical, robust and eco-friendly biosensing tools for healthcare analysis, agricultural applications, security checks and pollution surveillance (Seth and Meena 2024). Enzymes' extraction and utilization adhere to green chemistry tenets, minimizing ecological footprint while enhancing biological compatibility. An instance showcasing enzyme extraction techniques derived directly from plants sourced in diverse forms such as seed extracts, leaf components, fruit compounds and root elements through effective and large-scale procedures. These enzymes like amylase, cellobiohydrolase and peroxidases serve practical purposes while fostering innovative sustainability in various sectors.

22.2 PLANT ENZYME IN PHYTOREMEDIATION

The use of plant-based methods in remediation offers potential through their inherent capabilities; it demonstrates efficacy across various pollutants while requiring minimal energy input and being environment-friendly. The classification of this therapy is determined by how it works; for instance, the processes of phytoporation and plant degradation involve the accumulation, storage and metabolism of contaminants. Rhizofiltration involves the use of microbial breakdown in the rhizosphere, and phytostabilization is when plant-produced chemicals immobilize contaminants at the root interface. Phytovolatilization is when xenobiotics are taken up and released via transpiration (Pilon-Smith 2005). Phytoremediation has been considered an economically, sustainably and ecologically sound remedy for diverse environmental contamination problems. Investigating how different types of vegetation can clean contaminated water

DOI: 10.1201/9781003669920-22

bodies reveals success stories involving numerous species, such as those found in swamps (Wang et al. 20020), trees (Luqman et al. 2013); research focused on this topic as well as studies involving aquatic macroscopic plants (Pflugmacher et al. 2015).

22.2.1 Mechanism of Plant Enzymes

The green liver model of plants operates through a three-phase process, utilizing various enzymes to transform hydrophobic substances into hydrophilic metabolites that can be stored within the plant's cells (Figure 22.1).

22.2.1.1 Phase 1 Transformation

During this phase, xenobiotic structures undergo alteration through the incorporation or exposure of functional units, thereby enhancing their polarity significantly. The process takes place via either hydrolysis or oxidation reactions primarily facilitated by CYP450 monooxygenases (Sandermann 1992).

22.2.1.2 Phase 2 Conjugation

During this stage, foreign substances undergo modification through incorporation of exogenous water-soluble compounds into either an intermediary product produced during the initial step or directly onto the xenobiotic itself. This process involves sulfate formation, acetate conversion, methyl group addition, glutathione binding and glucuronic acid attachment mediated through enzyme activities such as glutathione-S-transferases and UDP-glucuronosyltransferases (Sandermann 1992).

22.2.1.3 Phase 3 Compartmentalization/Sequestration

Subsequently, the organism accumulates synthesized compounds within its vacuoles or extracellular spaces through interactions with structural elements such as cellulose and xylan found on the cell walls. Metabolites traverse the cell's plasma membrane or tonoplast via active transport mechanisms, ultimately reaching either the vacuole or extracellular space (Coleman et al. 1997) (Figure 22.1).

Key enzymes in phytoremediation

a. *Peroxidases*

   An enzymatic function involves breaking down contaminants such as those found in nature, livestock diets, industrial chemicals, agricultural products and pulp fibers industries. It is also used in the breakdown of textile dyes, the degradation of lignin in the paper and pulp industry, the removal of color from dyes, sewage treatment and as a biosensor (Bansal and Kanwar 2013; Falade et al. 2017).

b. *Laccases*

   Laccases are multicopper oxidase capable of catalyzing the oxidation of wide range of phenolic and non-phenolic substrates using molecular oxygen with water as the only byproduct (Mayolo-Deloisa et al. 2020). The laccase enzyme effectively removes color and impurities from industrial wastewater by breaking down pollutants like xenobiotics through biological processes known as bioremediation (Nigam 2013; Viswanath et al. 2014).

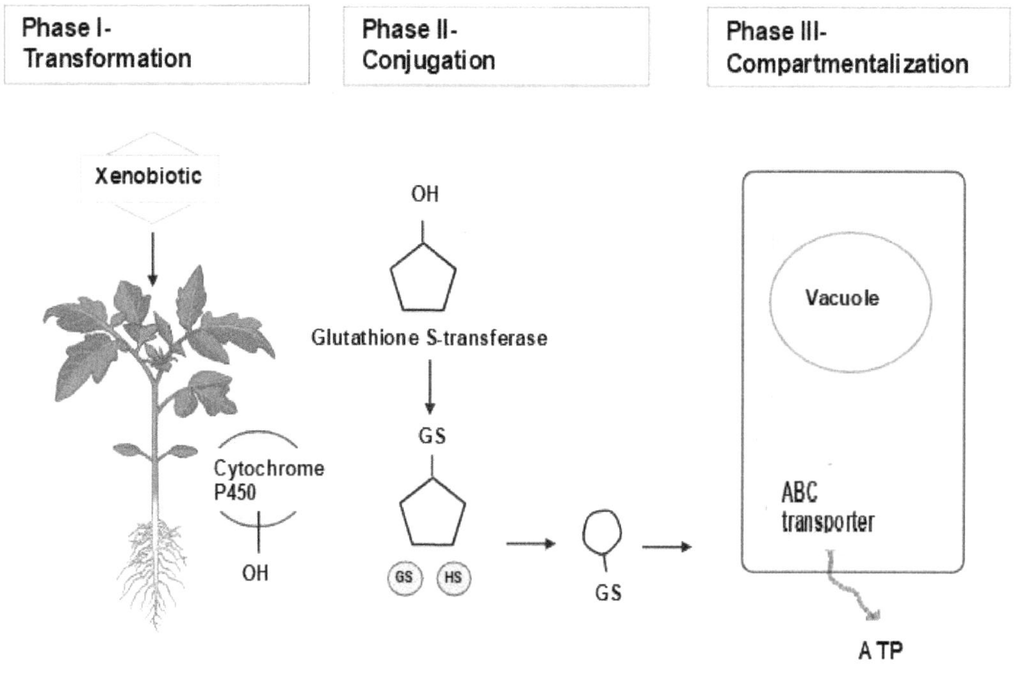

FIGURE 22.1 Mechanism of plant enzyme-mediated detoxification.

c. *Nitro reductase*

These enzymes are involved in the reduction of nitrogen-containing compound, including those compounds that contain nitro as a functional group (De Oliveira et al. 2007; Hecht et al. 1995). Nitroreductase has the ability to metabolize TNT (2,4,6-trinitrotoluene) (Bryant and Deluca 1991).

d. *Glutathione S-transferases (GSTs)*

GSTs have been recognized for their capacity to neutralize herbicides through conjugation with this peptide compound called glutathione. In the metabolism of foreign substances and in protecting against oxidative stress, GST enzymes play a crucial role. These enzymes have been well studied in bacteria, plants and mammals (Pflugmacher et al. 2000). Other genes known as GSTs contribute to the hormone's reaction by aiding in degrading substances like anthocyanins and cinnamic acids found within plants (Marrs 1996) (Table 22.1).

22.2.2 Plant Enzyme as Biocatalysts in Green Chemistry

Green chemistry's primary objective extends far beyond merely managing environmental contamination. Its goal is to develop cleaner, faster and greener substitutes for conventional chemicals. Innovative strategies aim at preventing environmental issues early on, substantially lowering carbon dioxide output and minimizing industrial pollutants such as toxic residues (*United states EPV, 2025*). Biological catalysts called enzymes exhibit remarkable selectivity and efficiency; they significantly contribute to contemporary industrial biochemistry. The capacity of these substances to function in mild environments while minimizing ecological impact positions them as superior alternatives to traditional synthetic catalysts (Farhan et al. 2025).

Incorporating principles and measurements derived from green chemistry toward achieving ecological balance in development via biological catalysts may contribute to

TABLE 22.1
List of Enzymes Extracted from Plant Sources for Industry-Related Application in Green Chemistry

Plant Enzyme	Source	Green Chemistry/Industrial Application	Methodology	Key Reference
Laccase	*Populus trichocarpa*	Demonstrated direct polymerization of sinapyl alcohol (syringyl lignin). Inhibition reduced lignin content and S/G ratio in populus.	Expressed in tobacco, purified and used to catalyze lignin monomer polymerization in vitro. Antisense lines develop in *Populus* to analyze lignin synthesis.	Liu et al. (2021)
Lipase	*Ricinus communis* (castor bean)	Bioremediation of used lubricating oil–contaminated soils.	Seeds were collected and endosperms were separated and defatted with chilled acetone. After filtration and drying lipase extract powder was obtained.	Aguilera Flores et al. (2022)
Urease	1. *Citrullus lanatus* (watermelon seeds) 2. *Vigna umbellata* (rice bean seeds)	Detection and removal of heavy metals from aqueous solutions. Investigated for ureolytic activity and antifungal potential.	Urease catalyzes the hydrolysis of urea to produce ammonia and carbonate ions, resulting in a rise in pH. Isolate from germinated rice bean seeds, homogenize it and centrifuge. Observe or find the enzyme purity on UV-visible spectrophotometry.	Kumar et al. (2024) Siregar et al. (2024)
Actinidin, bromelain	Kiwi, pineapple	Tenderization and microbial safety.	Injecting in to pork foreshank and sous-vide treated.	Chang and Han (2020)
β-amylase	*Arachis hypogaea* (Peanut)	Antioxidant enhancement of food products.	Mix β-amylase with wheat flour. Treat the flour with DPPH. This process improved antioxidant activity due to the presence of phenolic compounds that release when enzymatic hydrolysis occurs.	Ranjana Das and Kayastha (2018)
Papain	Latex of *Carica papaya* (unripe fruit)	Used for tenderizing Chilean abalone.	Impregnation: Injection, immersion and high hydrostatic pressure (HHP).	Pizarro-oteiza et al. (2020)
Ginger protease	*Zingiber officinale* (zingibain)	Tenderization of chicken breast. Resulted in 37.7% reduction in shear force, significant increase in myofibrillar fragmentation.	Ginger extract injected into chicken breast stored at 4°C for 24 h. Cut and analyzed for shear force, myofibrillar fragmentation and water holding capacity.	Cruz et al. (2020)

enhancing an outcome within the realm of molecular biology and biotechnological fields. By employing advanced methods like protein engineering along with aspects in biocatalytic engineering, including substrate selection, media optimization and reactor design, it is possible to enhance productivity while reducing costs (Sheldon and Woodley 2017). In green chemistry, there exist dual strategies for evaluating enzyme performance by assessing their specificity toward substrates. The primary method used is enzyme-based discrimination in its ground-state form to assess the extent of substrate binding and thereby achieve specificity. To illustrate, enzymes like DNA methyltransferases and proteins involved in translation exhibit high precision during their function at rest states. During this transitional phase, efficient coupling occurs between discrimination of substrates and chemical reactions because there's less stringent affinity for specific substrates involved. Precision emerges as an incidental outcome when specific regions optimize their ability to transform substrates into desired outcomes (Tawfik 2014). Biocatalysis can now be considered as a key-enabling technology for the implementation of sustainable chemicals manufacturing, not only in fine chemicals and pharmaceuticals manufacture but also in the production of commodity chemicals and bio-based plastics. Biocatalysis is green and sustainable.

22.2.2.1 Environmental Benefits

Green chemistry yields substantial advantages for our planet's health. Through innovative modifications in manufacturing methods, green chemistry diminishes environmental contamination initially, minimizes byproducts and eliminates hazardous substances that pose risks to both people's well-being and ecosystems (United states EPV, 2025). Benefits in pollution prevention: Reduces harmful emissions and waste at the source instead of treating them after they are produced, Resource conservation: Uses raw materials and energy more efficiently through better processes (United states EPV, 2025), Ecosystem Protection: Eliminates persistent organic pollutants and substances that accumulate in the environment that can be harmful to wildlife and human health and weather protection: Renewable utilization of feedstock and efficient energy processing result in reduction of greenhouse gas emissions (United states EPV, 2025).

22.2.2.2 Economic Benefits

Conventional chemical processes frequently incorporate poisonous metals or harsh acids and bases; biocatalysis employs enzymes that operate under mild condition. Enzymes derived from plants can be reused, so they are eco-friendly in nature. Most common biocatalysts are transaminases, hydrolases and oxidases. Improved understanding and application of enzymes can make chemical production cleaner and more accurate (Singh and Chiphang 2025). Enzymes are environment-friendly, non-toxic in nature and biodegradable; this character plays an essential role in industrial economic benefits, conserving fuel and minimizing emissions by enzymes to reduce/consume less energy in the process as compared to conventional chemical catalyst, and biocatalyst can be sourced from nature and that's why they are renewable as compared to chemical one (Schmid et al. 2001).

22.2.2.3 Types of Enzymes in Green Chemistry Application

Enzymes have strong application in green chemistry due to their accurate selectivity for reaction and activity under extreme environmental conditions. In green chemical processes well-employed enzyme types are transaminases, hydrolases and oxidases/dehydrogenases. These enzymes assist in minimizing waste, conserving energy and evading toxic chemicals (Singh and Chiphang 2025). Transaminases are enzymes that transfer amino group from one molecule to another. Reaction is converted into sitagliptin from ketone, intermediated by an amine donor, with enzyme conferring stereoselectivity (Bornscheuer et al. 2012). Hydrolases are used to catalyze the breakdown of chemical bond with the help of water molecule. In green applications biodiesel and ester undergo hydrolysis process. Another enzyme lipase hydrolyzes fats to glycerol and converts fatty acid esters in biodiesel (Savile et al. 2010). Oxidases and dehydrogenases are enzymes that assist in oxidation–reduction reactions.

22.2.3 Environmental Remediation through Enzyme Engineering

Current remediation methods such as physical, chemical and biological techniques are not enough to clean up environmental pollution. Enzyme-based bioremediation is a simple, quick, eco-friendly and socially acceptable method for removing xenobiotic compounds from the environment. Several plant enzymes with bioremediation abilities have been isolated and studied from various sources, but the low production of these enzymes limits their use. Genetic engineering of enzymes has the potential to produce large quantities of recombinant enzymes (Sharma et al. 2018). Engineered microbial enzymes has proven effective in degrading recalcitrant compounds such as polychlorinated biphenyls (PCBs) and polycyclic aromatic hydrocarbons (PAHs) through microbial bioremediation processes (Ayesha 2024). The breakdown of contaminants using microorganisms is a slow process that can slow down bioremediation (Ghosh et al. 2017). Using enzymes separated from their cell walls for bioremediation is better than using whole plants to overcome some of these limitations. Enzymes are complex biological molecules that act as catalysts in biochemical reactions (Sharma et al. 2018). CRISPR/Cas9 is a modern engineering synthesis endonuclease enzyme that revolutionized genome editing. CRISPR/Cas9 is a powerful tool that has been engineered for applications beyond gene editing (Barakate and Stephens 2016). CRISPR/Cas9's role in enhancing agricultural innovations is to ensure global food security. Researchers can precisely alter the genes of crop plants, particularly important ones like *Triticum aestivum*, *Oryza sativa*, *Zea mays* and *Solanum tuberosum*, by utilizing modern sequencing technology. This allows them to impart traits like pressure tolerance and disorder resistance. The technology is essential for creating next-generation crops (NGCs) that can adjust to shifting climates and provide higher yields and more satisfying products. While ethical and legal considerations are important, responsible genome-enhancing governance holds the potential for significant agricultural improvements. This analysis emphasizes the revolutionary potential of CRISPR/Cas9 in crop development and its critical role in determining the future of food production (Muha-Ud-Din et al. 2024).

Horseradish peroxidase (HRP) is isolated from the plant *Armoracia rusticana* for the bioremediation. HRP cross-links by the cross-linking agent called ethylene glycol-bis (succinic acid N-hydroxysuccinimide) (EGS), which is less toxic molecule compared to glutaraldehyde. The resulting HRP-CLEAs were tested to evaluate their catalytic effectiveness for bioremediation using a newly developed packed bed reactor system (PBRS). These newly developed HRP-CLEAs have a great potential for industrial use in dealing with various industrial dye-based pollutants (Mohammad et al. 2017). Another example is lignin and its derivatives that are being widely used in wastewater treatment because of their advantages such as renewable sources, good performance, biodegradability and no secondary pollution (Wang et al. 2021).

22.2.3.1 Advancements in Enzyme Engineering

An experimental study involving remediation of contaminated aquatic environments utilized lab conditions where five types of native wetland plants were tested: Smooth knotgrass (*Polygonum amphibium*), water milfoil (*Lemna minor*), water lettuce (*Eichhornia crassipes*), swamp milkweed (*Oenanthe javanica*) and cattail (*Typha latifolia*). The focus was on treating nitrogen (N), phosphorus (P) and three pollutants: Cadmium (Cd), mercury (Hg) and lead (Pb). Thickets of sharp docks accumulate nutrients like nitrogen (N) and phosphorus on their leaves; these plants act as super absorbers for cadmium (Cd), such as water hyacinths and duckweeds; they also serve as potent accumulators for mercury (Hg), exemplified by water dropworts; finally, marshes teem with *Calamus* that excel at absorbing lead (Pb) (Wang et al. 2002) (Figure 22.2).

22.2.4 Exploiting Extremophile-Inspired Plant Enzymes

Enzymes from organisms that live in extreme condition (extremozymes) have been studied and used in various industries like food, pharmaceutical, agriculture and molecular biology. These enzymes are adopted to catalyze reactions even under harsh environmental conditions. Enzyme engineering helps in insights from known enzymes to develop or improve catalysts (Chatterjee et al. 2023). Enzymes are becoming more and more important in industries as biocatalysts because they often outperform traditional inorganic catalysts. The use of enzymes for the synthesis of various products is more efficient and greener compared to traditional chemical methods (Sheldon and Woodley 2017). One of the key advantages of enzymes is their high stereoselectivity that allows them a wide range of chemical transformation with high specificity (Kate et al. 2022). Although enzymes bring many benefits, they can be unstable under extreme temperature and pH conditions found in large-scale industrial processes (Littlechild 2015). Recent trends in biocatalysis are exploring the use of extremozymes, enzymes from plants, that thrive in extreme conditions (Herbert 1992); these enzymes are naturally resistant to harsh environment such as high temperatures and pressure, high salinity, water scarcity

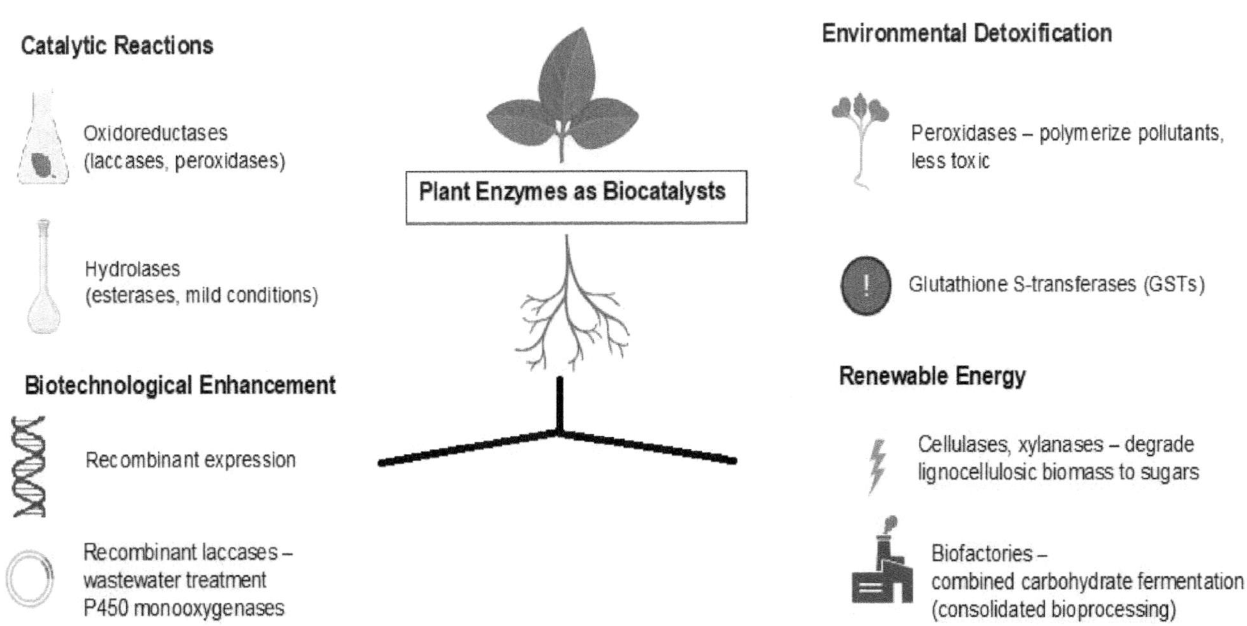

FIGURE 22.2 Engineered plant enzymes in microbial hosts for green biocatalysis.

and extreme pH levels. Because extremozymes can function under such extreme conditions, they are highly sought-after in industrial processes (Liszka et al. 2012). For example, nuclease H, a halophilic enzyme from *Micrococcus varians* subsp. *halophilus*, breaks down RNA at 60°C and 12% salt concentration and is used in the food industry to produce a flavoring agent called 5-guanylic acid (Onishi et al. 1988). Another example is the L-aminoacylase enzyme derived from the thermophilic archaeon *Thermococcus litoralis*, which works best at 85°C in tris-HCl, pH 8 and has a wide range of substrate specificity (Toogood et al. 2002). This enzyme is used by several pharmaceutical companies for the industrial production of various L-amino acids and their analogs.

The extremophiles associated with plants are classified as bacteria, archaea and eukaryotes and further divided into different groups, e.g., *Bacteroidetes*, *Ascomycota*, *Basidiomycota*, *Euryarchaeota*, *Firmicutes*, *Actinobacteria*, *Crenarchaeota, and Proteobacteria*. As the beneficial association, the plant-associated extremophiles display multiple ecological and plant growth promotion (PGP) attributes, positively impacting their plant counterparts (Adil et al. 2023; Bose et al. 2022; Roy et al. 2022). The plant-associated rhizospheric microbes are present in the root zone and key examples include *Paenibacillus*, *Burkholderia*, *Azospirillum*, *Bacillus*, *Methylobacterium* and *Pseudomonas* (Dalmini et al. 2022; Meeta Lavania et al. 2006), while epiphytic microbes are found in various parts of the plants surface and can survive under UV radiation and high temperature (40–50°C); some notable examples include *Agrobacterium*, *Methylobacterium*, *Pantoea* and *Pseudomonas*, which have been found in different crops under both normal and extreme conditions (Nutaratat et al. 2014). Endophytic microbes on the other hand live inside plant tissues and form a mutualistic relationship with the plant (Tiwari et al. 2022). Some examples of these endophytic microbes include *Achromobacter*, *Azozrcus*, *Enterobacter*, *Herbaspirillum*, *Burkholderia*, *Klebsiella, Nocardioides, Pantoea* and others, which have been isolated from various plant species (Ryan et al. 2008; Tiwari et al. 2023).

22.2.5 Enzyme-Mediated Biocontrol and Plant-Mediated Interaction

There are certain enzymes that can break down or destroy the cell walls of harmful plant pathogens. Most frequently found process in rhizosphere is to eliminate harmful pathogens by releasing enzymes that are responsible for the breakdown of the cell wall of microorganisms, which are present around the plant roots where plant growth–promoting microbes are present. Eliminating of pathogens indirectly help in the enhancement of plant growth and development. Research on microbial hydrolases and lytic enzymes prove that they involve to control infectious plant pathogens. By this research we understand the ability of bacteria to control plant pathogens or how to promote plant growth. Experimental process by researcher measures the total production of antimicrobial agents like ammonia and antibiotics, as well as hydrolytic enzymes such as amylase, lipase, proteases and chitinases to check their capacity and dissolve in phosphate (Mota et al. 2017).

Genetically modified microorganisms (GMOs) and genetically engineered microbial systems (GEMS) play an essential role in developing sustainable microbial strains, which live in harsh conditions; this enhancement feature of the microbial strains help to solve agriculture and food security related problems. Advanced technology like DNA synthesis, sequencing and powerful computational tools results in moving of molecular biology toward synthetic biology. One recent example is CRISPR/Cas that is used in various application in agricultural field (Sudheer et al. 2020). Rhizosphere is responsible for providing ideal environment for wide range of microorganisms that influence the plant growth and development. These essential microorganisms boost plant growth by making nutrients available and they have effects on plant hormone regulation and deterring plant pathogens or competitors, competing for harmful microorganisms. Rhizosphere microbiomes might have the opportunity to perform an important job on future green revolution and it might be useful on sustainable and green agriculture. For instance, the knowledge gaps in roots–environment relation are still wide open especially about roots and microorganisms (Solomon et al. 2024). The conversion of sugars into useful products can happen in two main approaches that are used in cellulosic biofuel production (CBP): Improving the yield and ability of microbes to use cellulose, using a different system to break down and use cellulose in microbes that already have high productivity and can withstand various conditions (Lynd et al. 2002).

22.2.6 Machine Learning-Guided Enzyme Design

Biocatalysis has become a promising option for making special chemicals and pharmaceuticals sustainable and helping in the fast reaction and precise control that enzyme catalysts provide under mild and environment-friendly conditions (Bornscheuer et al. 2012; Buller et al. 2023; Zetzsche et al. 2022). In recent decades directed evolution (Bloom and Arnold 2009) has made it possible to develop customized enzymes with high catalytic activity and selectivity, which often outperform small molecule catalysts (Reetz 2011). Despite the success of directed evolution in enzyme engineering (Bloom and Arnold 2009; Packer and Liu 2015), its efficiency in exploring the vast range of possible functions is still limited. Recently, machine learning (ML) methods have emerged as powerful tools for speeding up enzyme engineering (Siedhoff et al. 2020; Yang et al. 2019; Yang et al. 2024). Supervised ML models have been trained to understand the relationship between protein sequences and their properties (Gelman et al. 2021; Hie et al. 2020; Luo and Et 2021; Wu et al. 2019). These models act as alternatives to laboratory screening, quickly predicting and prioritizing useful enzyme variants through in silico fitness prediction, which reduce the overall experimental work required (Hie and Yang 2022; Luo and Et 2021; Romero et al. 2013; Yu et al. 2023).

Recent efforts have focused on advancing ML algorithms to develop enzyme functions that are not found in nature. Inspired by synthetic chemistry, biocatalysis researchers have developed a variety of enzyme activities not seen in natural systems (Brandenberg et al. 2017; Chen and Arnold 2020; Klaus and Hammer 2022; Yang and Arnold 2021), including atom transfer radical cyclase (Zhou et al. 2021) and radical pyridoxal enzyme (Cheng et al. 2023) activities developed. However, general ML strategies to discover new-to-nature enzyme functions remain underdeveloped. Due to the lack of fitness data for these new functions, it's hard to train supervised ML models to guide such directed evolution efforts. Therefore, the key to successfully engineering new-to-nature enzyme lies in designing effective starting libraries without relying on fitness data from experiments, allowing the identification of initial hits and the optimization of highly functional enzyme variants (Ding et al. 2024). Despite the fact that massive-scale random sampling of combinatorial mutants has been used to generate variety, the inclusion of dangerous mutations in these random mixtures ends in a large number of nonfunctional mutants decreasing the usefulness of such libraries. Moreover because of the energy of supervised ML fashions in interpolation in place of extrapolation, a diverse starting library enriched with wonderful practical versions will allow ML models to more correctly map out the health landscape, in the end enhancing the efficiency of downstream ML-guided directed evolution (MLDE), despite the fact that designing an effective starting library with co-optimized health and variety remains a hard assignment (Ding et al. 2024).

MODIFY (ML-optimized library design with improved fitness and diversity) is an ML algorithm introduced for effective beginning library layout in enzyme engineering. Given a hard and fast state-of-the-art residue for enzyme engineering, this design creates libraries that sample variants from the combinatorial series space more likely to be functional even as keeping a high degree trendy range. To address the cold-start assignment in which no experimentally characterized fitness information is available, an ensemble of pretrained unsupervised models can be used to perform zero-shot health predictions. Designed libraries obtain a pareto optimal stability between predicted fitness and sequence diversity in which no metric can be progressed without reducing the other. With a deep mutational scanning at a benchmark of 87%, dataset ALTER offers strong and correct zero-shot fitness predictions, outperforming numerous unsupervised ML strategies. Being further tested on the experimentally characterized fitness landscape ultramodern GB 1 protein, MODIFY produced various combinational libraries enriched with excessive fitness versions. In silico MLDE experiments show that ALTER libraries discover the collection area and identify higher fitness areas more efficaciously, imparting a more informative training set for effective MLDE. Additionally, MODIFY was used to design effective libraries for the fast era trendy, green and stereoselective biocatalysts for the stereoselective formation; modern-day C–B and C-ALTER–designed libraries are more wonderful than those developed experimentally presenting precious insights into enzyme structure hobby relationships (Ding et al. 2024). Moreover, generalist biocatalysts that facilitate each C–B and C–Si bond formation were identified from a library designed using MODIFY, further showcasing the utility ultramodern this library design algorithm in new-to-nature enzyme engineering (Ding et al. 2024).

22.3 CONCLUSION

Enzyme systems originating from plants present a sustainable and biologically harmonious substitute for enzymes derived from animals and microbes, finding utility in diverse areas such as biotechnology, clinical assessments, ecological surveillance and manufacturing processes. Phyto enzymes under mild conditions provide benefits like low immunogenicity, ecological sustainability, accuracy or specificity toward substrates and continuous performance. Advancement and growth in plant biology, genetic plant engineering, genomic alteration and omics-based technologies currently give shape to the innovative strategies for safeguarding food availability and discovering novel methods for biomaterials in response to fluctuation in environmental factors. Plant enzymes are the source of numerous chemical compound that are used in industries to enhance dyes and agrochemicals. In agricultural sector molecular methodologies, detection techniques and financial assessment play a primary role in transformation of plant biotechnology. Plant-derived enzymes have a capacity to challenge global demographics, provision of nourishment and fluctuation in environment. In future plant enzymes will play most essential role in economic stability, industrial processing, pharmaceutical and food safety.

REFERENCES

Adil, M., Tiwari, P., Chen, J. T., & Kanwal, S. (2023). Major bioactive metabolites and antimicrobial potential of Orchidaceae fungal endophytes. In *Advances in orchid biology, biotechnology, and omics*. 978–9819910786. Springer Publishers: Berlin/Heidelberg, Germany

Aguilera Flores, M. M., Sánchez Castro, M. A., Ávila Vázquez, V., Correa Aguado, H. C., & García Torres, J. (2022 Jul 18). Evaluation of the lipase from castor bean (*Ricinus communis* L.) as a potential agent for the remediation of used lubricating oil contaminated soils. *Journal of Environmental Health Science and Engineering*, *20*(2), 657–673. https://doi.org/10.1007/s40201-022-00806-1. PMID: 36406614; PMCID: PMC9672203.

Bansal, N., & Kanwar, S. S. (2013). Peroxidase (s) in environment protection. *The Scientific World Journal*, *2013*(1), 714639.

Barakate, A., & Stephens, J. (2016). An overview of CRISPR-based tools and their improvements: New opportunities in understanding plant–pathogen interactions for better crop protection. *Frontiers in Plant Science*, *7*, 765.

Bell, E. L., Finnigan, W., France, S. P., Green, A. P., Hayes, M. A., Hepworth, L. J., …, & Flitsch, S. L. (2021). Biocatalysis. *Nature Reviews Methods Primers*, *1*(1), 46.

Bloom, J. D., & Arnold, F. H. (2009). In the light of directed evolution: Pathways of adaptive protein evolution. *Proceedings of the National Academy of Sciences*, *106*(supplement_1), 9995–10000.

Bornscheuer, U. T., Huisman, G. W., Kazlauskas, R. J., Lutz, S., Moore, J. C., & Robins, K. (2012). Engineering the third wave of biocatalysis. *Nature*, *485*(7397), 185–194.

Bose, S. K., Bajpai, M., Das, J., Gautam, A., & Tiwari, P. (2022). Actinomycetes endophytes: Overview and significance in the production of bioactive compounds. In Tiwari, P. (Ed.), *Endophytes: Types, potential uses, and mechanisms of action*. Nova Publishers.

Brandenberg, O. F., Fasan, R., & Arnold, F. H. (2017). Exploiting and engineering hemoproteins for abiological carbene and nitrene transfer reactions. *Current Opinion in Biotechnology*, *47*, 102–111.

Bryant, C., & DeLuca, M. (1991). Purification and characterization of an oxygen-insensitive NAD (P) H nitroreductase from *Enterobacter cloacae*. *Journal of Biological Chemistry*, *266*(7), 4119–4125.

Chang, J. H., & Han, J. A. (2020). Synergistic effect of sous-vide and fruit-extracted enzymes on pork tenderization. *Food Science and Biotechnology*, *29*(9), 1213–1222.

Chatterjee, A., Puri, S., Sharma, P. K., Deepa, P. R., & Chowdhury, S. (2023). Nature-inspired Enzyme engineering and sustainable catalysis: Biochemical clues from the world of plants and extremophiles. *Frontiers in Bioengineering and Biotechnology*, *11*, 1229300.

Chen, K., & Arnold, F. H. (2020). Engineering new catalytic activities in enzymes. *Nature Catalysis*, *3*(3), 203–213.

Cheng, L., Li, D., Mai, B. K., Bo, Z., Cheng, L., Liu, P., & Yang, Y. (2023). Stereoselective amino acid synthesis by synergistic photoredox-pyridoxal radical biocatalysis. *Science*, *381*(6656), 444–451.

Coleman, J., Blake-Kalff, M., & Davies, E. (1997). Detoxification of xenobiotics by plants: Chemical modification and vacuolar compartmentation. *Trends in Plant Science*, *2*(4), 144–151.

Cruz, P. L., Panno, P. H. C., Giannotti, J. D. G., de Carvalho, R. V., & Roberto, C. D. (2020). Effect of proteases from ginger rhizome on the fragmentation of myofibrils and tenderness of chicken breast. *LWT*, *120*, 108921.

De Oliveira, I. M., Henriques, J. A. P., & Bonatto, D. (2007). In silico identification of a new group of specific bacterial and fungal nitroreductases-like proteins. *Biochemical and Biophysical Research Communications*, *355*(4), 919–925.

Ding, K., Chin, M., Zhao, Y., Huang, W., Mai, B. K., Wang, H., …, & Luo, Y. (2024). Machine learning-guided co-optimization of fitness and diversity facilitates combinatorial library design in enzyme engineering. *Nature Communications*, *15*(1), 6392.

Dlamini, S. P., Akanmu, A. O., & Babalola, O. O. (2022). Rhizospheric microorganisms: The gateway to a sustainable plant health. *Frontiers in Sustainable Food Systems*, *6*, 925802.

Emily, R., John, D., Jake, F., Richard, O., Elizabeth, L. B., & Anthony, P. G. (2023). Engineering enzymes for environmental sustainability. *Angewandte chemie international edition*, *62*(52). https://doi.org/10.1002/anie.202309305

Falade, A. O., Nwodo, U. U., Iweriebor, B. C., Green, E., Mabinya, L. V., & Okoh, A. I. (2017). Lignin peroxidase functionalities and prospective applications. *MicrobiologyOpen*, *6*(1), e00394.

Farhan, M., Hasani, I. W., Khafaga, D. S., Ragab, W. M., Ahmed Kazi, R. N., Aatif, M., …, & Fahim, Y. A. (2025). Enzymes as catalysts in industrial biocatalysis: Advances in engineering, applications, and sustainable integration. *Catalysts*, *15*(9), 891.

Gelman, S., Fahlberg, S. A., Heinzelman, P., Romero, P. A., & Gitter, A. (2021). Neural networks to learn protein sequence–function relationships from deep mutational scanning data. *Proceedings of the National Academy of Sciences*, *118*(48), e2104878118.

Ghosh, A., Dastidar, M. G., & Sreekrishnan, T. R. (2017). Bioremediation of chromium complex dyes and treatment of sludge generated during the process. *International Biodeterioration & Biodegradation*, *119*, 448–460.

Hecht, H. J., Erdmann, H., Park, H. J., Sprinzl, M., & Schmid, R. D. (1995). Crystal structure of NADH oxidase from *Thermus thermophilus*. *Nature Structural Biology*, *2*(12), 1109–1114.

Herbert, R. A. (1992). A perspective on the biotechnological potential of extremophiles. *Trends in Biotechnology*, *10*, 395–402.

Hie, B., Bryson, B. D., & Berger, B. (2020). Leveraging uncertainty in machine learning accelerates biological discovery and design. *Cell Systems*, *11*(5), 461–477.

Hie, B. L., & Yang, K. K. (2022). Adaptive machine learning for protein engineering. *Current Opinion in Structural Biology*, *72*, 145–152.

Hood, E. E. (2002). From green plants to industrial enzymes. *Enzyme and Microbial Technology*, *30*(3), 279–283.

Kate, A., Sahu, L. K., Pandey, J., Mishra, M., & Sharma, P. K. (2022). Green catalysis for chemical transformation: The need for the sustainable development. *Current Research in Green and Sustainable Chemistry*, *5*, 100248.

Klaus, C., & Hammer, S. C. (2022). New catalytic reactions by enzyme engineering. *Trends in Chemistry*, *4*(5), 363–366.

Kumar, P., Divya, Kayastha, A. M. (2024). Exploring the cata purification, biochemical characterization, and heavy metal precipitation. *International Journal of Biological Macromolecules*, *282*, 136798.

Liszka, M. J., Clark, M. E., Schneider, E., & Clark, D. S. (2012). Nature versus nurture: Developing enzymes that function under extreme conditions. *Annual Review of Chemical and Biomolecular Engineering*, *3*(1), 77–102.

Littlechild, J. A. (2015). Enzymes from extreme environments and their industrial applications. *Frontiers in Bioengineering and Biotechnology*, *3*, 161.

Liu, Y., Cao, S., Liu, X., Li, Y., Wang, B., Sun, Y., …, & Lu, H. (2021). PtrLAC16 plays a key role in catalyzing lignin polymerization in the xylem cell wall of Populus. *International Journal of Biological Macromolecules*, *188*, 983–992.

Luo, Y., Jiang, G., Yu, T., Liu, Y., Vo, L., Ding, H., Su, Y., Wei Qian, W., Zhao, H., & Peng, J. (2021). ECNet is an evolutionary context-integrated deep learning framework for protein engineering. *Nature Communications*, *12*(1), 5743.

Luqman, M., Butt, T. M., Tanvir, A., Atiq, M., Hussan, M. Z. Y., & Yaseen, M. (2013). Phytoremediation of polluted water by tress: A review. *African Journal of Agricultural Research*, *8*(17), 1591–1595.

Lynd, L. R., Weimer, P. J., Van Zyl, W. H., & Pretorius, I. S. (2002). Microbial cellulose utilization: Fundamentals and biotechnology. *Microbiology and Molecular Biology Reviews*, *66*(3), 506–577.

Marrs, K. A. (1996). The functions and regulation of glutathione S-transferases in plants. *Annual Review of Plant Biology*, *47*(1), 127–158.

Mayolo-Deloisa, K., González-González, M., & Rito-Palomares, M. (2020). Laccases in food industry: Bioprocessing, potential industrial and biotechnological applications. *Frontiers in Bioengineering and Biotechnology*, *8*, 222.

Meeta Lavania, M. L., Chauhan, P. S., Chauhan, S. V. S., Singh, H. B., & Nautiyal, C. S. (2006). Induction of plant defense enzymes and phenolics by treatment with plant growth-promoting rhizobacteria Serratia marcescens NBRI1213.

Morton, R. K. (1955). [6] Methods of extraction of enzymes from animal tissues. 25–51.

Mota, M. S., Gomes, C. B., Souza, I. T., & Moura, A. B. (2017). Bacterial selection for biological control of plant disease: Criterion determination and validation. *Brazilian Journal of Microbiology*, *48*(1), 62–70.

Muhammad, B., Hafiz, I., Hongbo, H., & Wei, W. (2017). Development of horseradish peroxidase based crosslinked enzyme aggregates and their environmental exploitation for bioremediation purposes. *Journal of Environmental Management*, *188*, 137–143.

Muha-Ud-Din, G., Ali, F., Hameed, A., Naqvi, S. A. H., Nizamani, M. M., Jabran, M., Sarfraz, S., & Yong, W. (2024). CRISPR/Cas9-based genome editing: A revolutionary approach for crop improvement and global food security. *Physiological and Molecular Plant Pathology*, *129*, 102191.

Nigam, P. S. (2013). Microbial enzymes with special characteristics for biotechnological applications. *Biomolecules*, *3*(3), 597–611.

Nutaratat, P., Srisuk, N., Arunrattiyakorn, P., & Limtong, S. (2014). Plant growth-promoting traits of epiphytic and endophytic yeasts isolated from rice and sugar cane leaves in Thailand. *Fungal Biology*, *118*(8), 683–694.

Onishi, H., Kamekura, M., Yokoi, H., & Kobayashi, T. (1988). Production of 5′ nucleotide by using halophilic nuclease H preferentially adsorbed on flocculated cells of the halophile *Micrococcus varians* subsp. *halophilus*. *Applied and Environmental Microbiology*, *54*(11), 2632–2635.

Packer, M. S., & Liu, D. R. (2015). Methods for the directed evolution of proteins. *Nature Reviews Genetics*, *16*(7), 379–394.

Pflugmacher, S., Kühn, S., Lee, S. H., Choi, J. W., Baik, S., Kwon, K. S., & Contardo-Jara, V. (2015). Green Liver Systems® for water purification: Using the phytoremediation potential of aquatic macrophytes for the removal of different cyanobacterial toxins from water. *American Journal of Plant Sciences*, *6*(9), 1607–1618.

Pflugmacher, S., Schröder, P., & Sandermann, Jr, H. (2000). Taxonomic distribution of plant glutathione S-transferases acting on xenobiotics. *Phytochemistry*, *54*(3), 267–273.

Pilon-Smith, E. (2005). Phytoremediation. *Annual Review Plant Biology*, *56*, 15–39.

Ranjana Das, R. D., & Kayastha, A. M. (2018). An antioxidant rich novel β-amylase from peanuts (*Arachis hypogaea*): Its purification, biochemical characterization and potential applications.

Reetz, M. T. (2011). Laboratory evolution of stereoselective enzymes: A prolific source of catalysts for asymmetric reactions. *Angewandte Chemie International Edition*, *50*(1), 138–174.

Romero, P. A., Krause, A., & Arnold, F. H. (2013). Navigating the protein fitness landscape with Gaussian processes. *Proceedings of the National Academy of Sciences*, *110*(3), E193–E201.

Roy, B., Maitra, D., Ghosh, J., & Mitra, A. K. (2022). Unique extremophilic Bacillus: Their application in plant growth promotion and sustainable agriculture. *Microbes and Microbial Biotechnology for Green Remediation*, 287–304.

Ryan, R. P., Germaine, K., Franks, A., Ryan, D. J., & Dowling, D. N. (2008). Bacterial endophytes: Recent developments and applications. *FEMS Microbiology Letters*, *278*(1), 1–9.

Sandermann, H. (1992). Plant metabolism of xenobiotics. *Trends in Biochemical Sciences*, *17*, 82–84.

Savile, C. K., Janey, J. M., Mundorff, E. C., Moore, J. C., Tam, S., Jarvis, W. R., …, & Hughes, G. J. (2010). Biocatalytic asymmetric synthesis of chiral amines from ketones applied to sitagliptin manufacture. *Science*, *329*(5989), 305–309.

Schmid, A., Dordick, J. S., Hauer, B., Kiener, A., Wubbolts, M., & Witholt, B. (2001). Industrial biocatalysis today and tomorrow. *Nature*, *409*(6817), 258–268.

Seth, R., & Meena, A. (2024). Enzymes-based nanomaterial synthesis: An eco-friendly and green synthesis approach. *Clean Technologies and Environmental Policy*, *27*, 1–24.

Sharma, B., Dangi, A. K., & Shukla, P. (2018). Contemporary enzyme based technologies for bioremediation: A review. *Journal of Environmental Management*, *210*, 10–22.

Sharma, H., & Upadhyay, S. K. (2020). Enzymes and their production strategies. In *Biomass, biofuels, biochemicals* (pp. 31–48). Elsevier.

Sheldon, R. A., & Woodley, J. M. (2017). Role of biocatalysis in sustainable chemistry. *Chemical Reviews*, *118*(2), 801–838.

Siedhoff, N. E., Schwaneberg, U., & Davari, M. D. (2020). Machine learning-assisted enzyme engineering. *Methods in Enzymology*, *643*, 281–315.

Singh, J., & Chiphang, C. (2025). Biocatalysts in green chemistry: Enzymes as sustainable reaction tools. *International Journal of Advanced Research and Interdisciplinary Scientific Endeavours*, 2(6), 770–774.

Siregar, N. A., Choudhary, M. I., Wahab, A. T., Salar, U., & Munir, E. (2024, Feb). Urease from *Vigna umbellata* seeds: Isolation, partial purification, characterization and antifungal activity. In *IOP conference series: Earth and environmental science* (Vol. 1302, No. 1, p. 012091). IOP Publishing.

Snajdrova, R., Buller, R., Lutz, S., Moore, J., Kazlauskas, R., & Bornscheuer, U. (2023). Surfing the biocatalysis wave to new applications changed to: From nature to industry: Harnessing enzymes for biocatalysis, as per request of Editor. *Science*, *382*(6673).

Solomon, W., Janda, T., & Molnár, Z. (2024). Unveiling the significance of rhizosphere: Implications for plant growth, stress response, and sustainable agriculture. *Plant Physiology and Biochemistry*, *206*, 108290.

Sudheer, S., Bai, R. G., Usmani, Z., & Sharma, M. (2020). Insights on engineered microbes in sustainable agriculture: Biotechnological developments and future prospects. *Current Genomics*, *21*(5), 321–333.

Tawfik, D. S. (2014). Accuracy-rate tradeoffs: How do enzymes meet demands of selectivity and catalytic efficiency. *Current Opinion in Chemical Biology*, *21*, 73–80.

Tiwari, P. (Ed.). (2022). *Endophytes: Types, potential uses and mechanism (s) of action*. Nova Publishers.

Tiwari, P., Muhammad, A., & Bae, H. (2023). Endophyte-mediated bioremediation-an efficient biological strategy in ecological subsistence and agriculture. In Erwin, D. J. (Ed.), *Endophytic and arbuscular mycorrhizal fungi and their role in sustainable agriculture*,. Nova Publishers.

Toogood, H. S., Hollingsworth, E. J., Brown, R. C., Taylor, I. N., Taylor, S. J., McCague, R., & Littlechild, J. A. (2002). A thermostable L-aminoacylase from *Thermococcus litoralis*:

Cloning, overexpression, characterization, and applications in biotransformations. *Extremophiles*, *6*(2), 111–122.

United States Environmental Protection Agency (EPA). (2025). Amano Enzyme. https://www.amano-enzyme.com/news/green-chemistry-the-role-of-enzymes-in-sustainable-solutions/

Viswanath, B., Rajesh, B., Janardhan, A., Kumar, A. P., & Narasimha, G. (2014). Fungal laccases and their applications in bioremediation. *Enzyme Research*, *2014*(1), 163242.

Wang, B., Sun, D., Yuan, T. Q., Song, G., & Sun, R. C. (2021). Recent advances in lignin modification and its application in wastewater treatment. *Lignin Utilization Strategies: From Processing to Applications*, 143–173.

Wang, Q., Cui, Y., & Dong, Y. (2002). Phytoremediation of polluted waters potentials and prospects of wetland plants. *Acta Biotechnologica*, *22*(1–2), 199–208.

Wu, Z., Kan, S. J., Lewis, R. D., Wittmann, B. J., & Arnold, F. H. (2019). Machine learning assisted directed protein evolution with combinatorial libraries. *Proceedings of the National Academy of Sciences*, *116*(18), 8852–8858.

Yang, J., Li, F. Z., & Arnold, F. H. (2024). Opportunities and challenges for machine learning assisted enzyme engineering. *ACS Central Science*, *10*(2), 226–241.

Yang, K. K., Wu, Z., & Arnold, F. H. (2019). Machine-learning-guided directed evolution for protein engineering. *Nature Methods*, *16*(8), 687–694.

Yang, Y., & Arnold, F. H. (2021). Navigating the unnatural reaction space: Directed evolution of heme proteins for selective carbene and nitrene transfer. *Accounts of Chemical Research*, *54*(5), 1209–1225.

Yu, T., Boob, A. G., Singh, N., Su, Y., & Zhao, H. (2023). In vitro continuous protein evolution empowered by machine learning and automation. *Cell Systems*, *14*(8), 633–644.

Zetzsche, L. E., Chakrabarty, S., & Narayan, A. R. (2022). The transformative power of biocatalysis in convergent synthesis. *Journal of the American Chemical Society*, *144*(12), 5214–5225.

Zhou, Q., Chin, M., Fu, Y., Liu, P., & Yang, Y. (2021). Stereodivergent atom-transfer radical cyclization by engineered cytochromes P450. *Science*, *374*(6575), 1612–1616.

Index

For Product Safety Concerns and Information please contact our EU
representative GPSR@taylorandfrancis.com Taylor & Francis Verlag GmbH,
Kaufingerstraße 24, 80331 München, Germany

Batch number: 10731490

Printed by Printforce, the Netherlands